Combinatorics, Complexity, and Logic

Springer
Berlin
Heidelberg
New York
Barcelona
Budapest
Hong Kong
London
Milan
Paris
Santa Clara
Singapore
Tokyo

D.S. BRIDGES
S. REEVES
C.S. CALUDE
I.H. WITTEN
J. GIBBONS
(EDITORS)

Combinatorics, Complexity, & Logic

Proceedings of DMTCS'96

Springer

Douglas S. Bridges
Departement of Mathematics
Univesity of Waikato
Hamilton
New Zealand

Cristian S. Calude
Department of Computer Science
University of Auckland
Auckland
New Zealand

Jeremy Gibbons
School of Computing and
Mathematical Sciences
Oxford Brookes University
Oxford
England

Steve Reeves
Department of Computer Science
University of Waikato
Hamilton
New Zealand

Ian H. Witten
Department of Computer Science
University of Waikato
Hamilton
New Zealand

Library of Congress Cataloging-in-Publication Data

DMTCS '96 (1996 : Auckland, N.Z.)
Combinatorics, complexity, and logic : proceedings of DMTCS '96 /
D.S. Bridges ... [et al.].
p. cm.
Papers from the First International Conference on Discrete
Mathematics and Theoretical Computer Science held in Auckland, N.Z.
in Dec. 1996.
Includes bibliographical references.
ISBN 981308314X (singapore)
1. Combinatorial analysis--Congresses. 2. Computational
complexity--Congresses. 3. Logic, Symbolic and mathematical-
-Congresses. I. Bridges, D. S. (Douglas S.), 1945- . II. Title.
QA164.D62 1997
511'.6--dc21

ISBN 981-3083-14-X

Printed in Singapore

Typesetting: Camera-ready by authors
SPIN 10560222 5 4 3 2 1 0

Preface

The Centre for Discrete Mathematics and Theoretical Computer Science, a joint venture involving the Computer Science and Mathematics Departments of the Universities of Auckland and Waikato, was founded in 1995 to support basic research on the interface between mathematics and computing.

DMTCS'96, the first of a planned series of conferences arranged by the Centre, was organised at the University of Auckland (city campus) from 9 to 13 December 1996. The city campus of the University of Auckland is situated on a rise overlooking the Waitemata Harbour, only five minutes walk from the restaurant, shopping and entertainment facilities of the central business district. Auckland is antipodal to Malaga in Spain and is known as the "City of Sails". It has a population of close to one million and a climate moderated by the adjacent Tasman Sea and the Pacific Ocean. December corresponds to the start of the summer holiday period in New Zealand, with daytime temperatures in the range of 16–23 degrees Celsius.

The **Programme Committee**, consisting of the following people:

L. Babai, D. S. Bridges (co-chair), C. S. Calude (co-chair), G. Chaitin, M. Conder, J. Dinitz, J. Gibbons, P. Gibbons, R. L. Graham, S. Hayashi, K. Heinrich, S. Reeves, G. Rozenberg, A. Salomaa, H. Siegelmann, K. Weihrauch,

had a difficult task selecting 28 papers out of 45 submissions covering a fairly large spectrum of subjects in discrete mathematics and theoretical computer science. The help of the following distinguished referees is gratefully acknowledged:

J. An, H. Barendregt, P. Bonnington, K. Broughan, V. E. Căzănescu, J. N. Crossley, R. W. Doran, R. Downey, P. Fenwick, R. Goldblatt, H. Güsgen, P. Hafner, M. D. Hendy, P. Hertling, H. J. Hoogeboom, M. Ito, H. Jürgensen, B. Khoussainov, E. V. Krishnamurthy, M. Lennon, N. Leslie, C. Little, R. Nicolescu, P. Odifreddi, J. Paris, K. Salomaa, R. Shore, J. Siran, L. Staiger, J. Staples, K. Svozil, I. Tomescu, W. Vogel, I. H. Witten, S. Yu.

The opinions expressed by the authors are not necessarily shared by the editors.

The invited speakers at the conference were: Gregory Chaitin, Fan Chung, Jeffrey Dinitz, Ron Graham, Susumu Hayashi, Piergiorgio Odifreddi, Grzegorz Rozenberg, Hava Siegelmann, Klaus Weihrauch.

We extend our thanks to all members of the **Conference Committee**, P. Bonnington, D. S. Bridges (co-chair), C. S. Calude (co-chair), M. Conder, R. W. Doran, P. Gibbons, H. Güsgen, S. Reeves (secretary), for their tremendous dedication in performing a multitude of tasks.

It goes without saying that we could not think of organising this conference without the sponsorship of the University of Auckland (the Science Faculty through the Dean of Science, Professor R. Cooney, the Departments of Computer Science and Mathematics), the School of Computing and Mathematical

Science of the University of Waikato (through its director Professor I. Graham), New Zealand Mathematical Society. We extend our gratitude to all.

We express our warm thanks to P. Barry, A. Lai and N. Raniga, for their invaluable organisational work, and to R. Burrowes and C. West, for the computer assistance.

Least, but not last, the cooperation of Springer-Verlag, Singapore team, especially I. Shelley and G. Chee, was essential for producing this volume in time for the conference. Thank you Ian, thank you Gillian!

D. S. Bridges
C. S. Calude
J. Gibbons
S. Reeves
I. H. Witten

Auckland, New Zealand
October 1996

Table of contents

Invited papers

Contributed papers

An Invitation to Algorithmic Information Theory*

G. J. Chaitin[1]

IBM Research, P.O. Box 218, Yorktown Heights, NY 10598, USA.
chaitin@watson.ibm.com, http://www.research.ibm.com/people/c/chaitin.

Abstract. I'll outline the latest version of my "limits of math" course. The purpose of this course is to illustrate the proofs of the key information-theoretic incompleteness theorems of algorithmic information theory by means of algorithms written in a specially designed version of LISP. The course is now written in HTML with Java applets, and is available at http://www.research.ibm.com/people/c/chaitin/lm. The LISP now used is much friendlier than before, and because its interpreter is a Java applet it will run in the Netscape browser as you browse my "limits of math" Web site.

1 Introduction

Hi everybody! It's a great pleasure being back in this beautiful, beautiful state. You guys were nice enough to invite me up a bunch of times in the past. And I've usually tried to explain general ideas. I thought this time I'd try to give a different kind of a talk.

I've been working for several years on a project, on a course I call "The Limits of Mathematics." And this course in the past three years has gone through a number of different versions. It has a lot of software attached to it. And I've been changing this software. I've been trying to explain this software to people as the course. In fact a few weeks from now I'll be giving this course in Rovaniemi, Finland, on the arctic circle, for two weeks.

The way this course is now is that it's on this Web site that I gave you guys the URL for in the abstract. It's an HTML document and it has a lot of LISP code hanging off of it, and just to make life worse this is not normal LISP. It's pretty close to pure LISP, you know, the heart of LISP that everybody learns. But I've had to add a few features to LISP. So this stuff is all available and you're welcome to play with it. The course is there.

What's also there by the way on my Web site are the transcripts of the previous talks I've given here at the University of New Mexico. One of them is "The Berry Paradox," it's in HTML. Another one is called "Randomness in Arithmetic and the Decline and Fall of Reductionism in Pure Mathematics."

* Lecture given Wednesday 24 April 1996 at a Computer Science Colloquium at the University of New Mexico. The lecture was videotaped; this is an edited transcript prepared for DMTCS'96.

That's another talk that I gave here. And there's a talk that I should have given here but it didn't go well. The version I gave at the Santa Fe Institute is there. It's called "How to Run Algorithmic Information Theory on a Computer." So these are three very understandable papers and I give them out when I give this course.

Now there is a small question: Once a paper is published in a journal are you allowed to have it anymore on your Web site? At my lab at this moment the view is that you should erase it. Los Alamos has a wonderful physics preprint server and they don't erase anything. And all over the academic community everybody has all their key papers on their Web site even after they're published, right? So I don't know. At this moment all this stuff is on my Web site. Some of it may disappear temporarily while the matter is discussed.

2 Recursive function theory revisited

My work basically is doing recursive function theory at the kind of level it was done in the early days in the 1930's before it got technical. At the kind of level that Gödel and Turing were doing it where the concern was what are the limits of mathematical reasoning. And basically I've added two new things. This is sort of revisited sixty years later, this stuff. I've added two new things.

One new thing I add to the stew is that an idea that was missing which is significant is the idea of program-size complexity, or how big a program is, how many bits of information there are. Not the time! In that sense it's just like the old recursive function theory and Turing machines. I don't care about time. But I do care about the size of a program in bits. Okay? So that's one new idea.

The other new idea I've added just in the past three years is that I want to really be able to run the programs on interesting examples on a computer. In other words, it's not enough to wave your hands and talk about an algorithm. I not only want a universal Turing machine that I can prove theorems with, but one that is fun to actually program and that you can run interesting examples on with a computer.

John McCarthy invented LISP not just for use in AI [artificial intelligence], not just for use as a practical language. If you look at his 1960 paper on LISP in the *Communications of the ACM,* he said, "This is a better universal Turing machine. Let's do recursive function theory that way!"[2] And the funny thing is that nobody except me has really I think taken that seriously. And the reason of course is that theoreticians didn't really much care about programming, about really playing with the computer. So the thing I've added is I think nowadays you have no excuse! If a theory has to do with the size of computer programs, you want to damn well see the programs, you want to be able to run them, it ought to be a programming language that is easy to use.

So I've done that using LISP because LISP is simple enough, LISP is in the intersection between theoretical and practical programming. Lambda calculus

[2] Of course these aren't McCarthy's words, it's my paraphrase.

is even simpler and more elegant than LISP, but it's unusable. Pure lambda calculus with combinators A and K, it's beautifully elegant, but you can't really run programs that way, they're too slow. So I would say that I believe that LISP is a powerful language that can actually be used on a computer to run stuff in a finite amount of time, but on the other hand it's simple enough and elegant enough that you can prove theorems about it. So that's the kind of game that I've been playing. Okay, so you'll find this on my Web site.

And as general background then I suggest these three papers: "The Berry Paradox," which was published in the first issue of *Complexity* actually. This other lecture on "The Decline and Fall of Reductionism. . . " which was published, among other places, by Oxford University Press in *Nature's Imagination,* that's the name of the volume. Oh there's also a paper called "A New Version of Algorithmic Information Theory," which is more technical, which was published in the latest issue, the current issue, the fourth issue of *Complexity* magazine, the one which just came out. And the paper "How to Run Algorithmic Information Theory on a Computer" which kicks around these ideas is going to come out in a future issue of *Complexity.* But for now they're all on the Web site, until they get thrown out!

3 My new LISP & complexity measure

Okay, so let me leap in and give you the general idea. What is this big new thing you add to LISP, this change you make to LISP to be able to talk about program-size complexity? Well, you could just take LISP expressions and measure how big are they in characters, right? That's not a bad complexity measure. But it's not the best complexity measure.

The best complexity measure says, you have a LISP S-expression, which is a computer program, and then you have binary data on the side:

```
lisp        binary
S-exp       data
```

The problem with measuring the size of a LISP program is that due to LISP syntax the LISP S-expression is redundant. So you really want to add to the LISP S-expression, which is a powerful way to express an algorithm, some raw binary data. That's like having data on a tape in addition to your program. So in the data the bits can be 0 and 1 independently, and there's some way that the program can get access to the data, and then you look at the size of the whole thing.

So basically the universal Turing machine I'm using as my yardstick to talk about program size, to use to measure program size, has programs which are big binary programs:

```
BINARY PROGRAM
lisp        binary
S-exp       data
```

That sounds bad, right? We have this awful machine language in binary. You're not going to want to write in binary. Well, it's not so bad. The beginning of a program is a LISP S-expression converted to binary, 8 bits per character. And you know where the LISP S-expression ends, because by convention there is always a special ASCII character at the end.

```
  BINARY PROGRAM
  lisp        binary
  S-exp       data
 8 bits/char
```

So the universal Turing machine starts reading the binary tape bit by bit. It doesn't sort of gobble it all in and start running. It reads each bit of the program as it needs to—that's very important. And it starts off by reading a LISP S-expression, and then every time the LISP S-expression says "Read another bit," it gets it from what's left on the tape, that is, in the binary program. And if the S-expression attempts to read beyond the end of the binary data, the program is a failure and is aborted. So that's how it goes. And then you just add the total number of bits, which is 8 times the number of characters in the S-expression plus the number of bits in the data. That gives you the total number of bits in the program, okay?

So that's done because just using the size in characters of a LISP S-expression (or multiplying it by some factor to have it in bits) is no good. LISP syntax is pretty simple, but even so there's redundancy there. So you've got to add this extra binary data on the side.

4 Programming my UTM in LISP

So let me show you my universal Turing machine programmed out in my new LISP. When I came here a year ago, I had a LISP, it was pretty awful. It was a LISP where atoms and variable names were only one character long. And it was a LISP where there was no arithmetic. The way you had to do integers was as lists of 0's and 1's and you had to define, you had to program out arithmetic. I thought this was cute, this is a fun game to play once in your life, but the problem is it's an obstacle if I want to explain this to anybody else! So now I have a LISP like all other LISP's with long names for atoms, and my universal Turing machine is now programmed like this.

Define U of p, U is the function, p is the program, as follows:

```
define (U p)
cadr try no-time-limit
         'eval read-exp
         p
```

Now this is it, this is my universal Turing machine that I described in words. `Try` is a primitive function in this LISP sort of like `eval` in normal LISP. `Try` has three arguments, here they are:

```
no-time-limit
'eval read-exp
p
```

So this, argument 2,

```
'eval read-exp
```

is a LISP expression that we're going to evaluate with a time limit. In fact there's no time limit here:

```
no-time-limit
```

You also have associated binary data. `Try` is the way that you give a LISP expression binary data. So this

```
p
```

is the binary data, it's the program for the universal Turing machine. It's a list of 0's and 1's. This is `try`'s third argument.

Oh by the way, in normal LISP you have to put lots of parentheses. In my LISP I leave out a lot of them because in my LISP every built-in function has a fixed number of arguments, so I don't bother to write all the parentheses.

Back to the LISP code for my UTM.

```
(read-exp)
```

is a primitive function with no arguments. So we're trying to evaluate this

```
('(eval(read-exp)))
```

normally with a time limit, but in this case with no time limit. And `read-exp` reads a LISP expression from the beginning of the binary data `p`. And then we evaluate it, in other words, we run it. And while you're running it, if within the LISP expression that you read you ask for more binary data, you get it from what's left of the binary data.

It works! `Try` does exactly what it should, I added `try` to normal LISP just for this. And `try` is the real difference between normal LISP and my LISP. Okay, so this

```
(define (U p)
(cadr (try no-time-limit
           ('(eval(read-exp)))
           p)))
```

is my universal Turing machine, which I think is not a very big, complicated program.

And the nice thing about this is that you see lots of examples in my course. You have there a Java applet for the LISP interpreter. Which means you can write LISP and see what it does. But most of the course is "canned" LISP. What

I do is I define some big LISP functions then I run them on examples. And there are lots of comments saying why I'm doing it and what it proves.

Actually there are two versions of each of these canned LISP runs, one of them with no comments and running it just on the case that counts, and another one with all the comments in the world and with lots of test cases for all the auxiliary functions. I do this because without comments a program is incomprehensible, but when you put all the comments in it becomes incomprehensible for another reason—it's so big that you can't find anything. So I present it both ways.

Okay, so one of the things you can do is you can actually get a program and run it on this machine. So how do you do it? Well, you take a LISP expression, and then you want to convert it to binary, so I have a primitive function for doing that, and then you concatenate it to the binary data that it needs, and then you feed this into the function U of p that we defined. And in my Web site I have runs of the universal Turing machine that show how it works.

5 A simple program for U that proves a theorem

Let me show you a particularly interesting example of this, and I'm also going to prove a theorem, a very simple theorem about my version of program-size complexity. This theorem says that the program-size complexity of a pair of objects is bounded by the sum of the individual program-size complexities of the two objects plus a constant:

$$H(x, y) \leq H(x) + H(y) + c.$$

What this means is this. $H(x)$ is the size in bits of the smallest program for the universal machine U of p that calculates x; x is an S-expression. And $H(y)$ is the size of the one that calculates y. And putting them together and adding a fixed number of bits you get a program that calculates the pair (x, y), which won't have a comma by the way, since we're doing LISP, which is without commas.

So how does this work? Well, let me show you what the program looks like. What you do is you take the minimum size program for x, the minimum size program for y, and I'll call that x^* and y^*, and you concatenate in front a prefix ϕ_c, a magic prefix that's going to be c bits long, and I'll show you this prefix. This gives you

$$\phi_c \, x^* y^*.$$

Then when you give it to the universal machine U of p it's going to produce the pair (x, y).

Now what is this prefix ϕ_c that you're concatenating in front of these two bit strings x^* and y^*? Actually you can put any program for x and any program for y here in $\phi_c \, x^* y^*$. I don't know the minimum ones. You'll see that this prefix works because if you give it any program x^* to calculate x and any program y^* to calculate y, this prefix ϕ_c when you feed it to my universal machine U of p is going to end up being a program that calculates the pair (x, y).

So what is this prefix ϕ_c? Well, the prefix is this. Start with

```
(cons (eval (read-exp))
(cons (eval (read-exp))
      nil))
```

Here I'm putting in all the parentheses. Then convert this to a bit string ϕ_c. So you write this expression and use a primitive function to convert it to a bit string and then you use `append` to concatenate these lists of bits and get this long bit string:

$$\phi_c\, x^* y^* .$$

And then you feed it into U and it produces (x, y).

How does this

```
(cons (eval (read-exp))
(cons (eval (read-exp))
      nil))
```

work? It says, "Read from the binary data." The binary data when ϕ_c is running is going to be x^*y^*, the rest of the program. So it's going to read and run the program x^* to calculate x, and that's going to give it x. And then it's going to read off of the rest of the binary data and run the program y^* to calculate y, and that's going to give it y. And then it's going to `cons` these two things x and y up into a pair. Okay?

So it's the number of bits in this S-expression ϕ_c

```
(cons (eval (read-exp))
(cons (eval (read-exp))
      nil))
```

which is the constant c in

$$H(x, y) \leq H(x) + H(y) + c.$$

So count the number of characters here in ϕ_c, multiply by 8, and that's the constant c in this theorem. Okay? So this is pretty straight-forward.

6 "Computing" the halting probability Ω

Now, what do I do next in my course on my Web site? Okay, so I've defined this universal Turing machine U of p. The next major step in my theory is to define the halting probability Ω. This is a number that a lot of people get excited about [laughter], the halting probability. Well, I'm going to write down a program to calculate the halting probability now. Let me actually write it down for you! Here it goes. Let's write this program out. So I'm going to write a LISP program as we go.

Okay, so we're going to define a function that counts how many programs halt, let's call it `count-halt`, which has two arguments, one is a `prefix`, and the other is `bits-left`.

```
define (count-halt prefix bits-left)
```

What's the idea? I'm going to count how many program for my universal Turing machine halt... Oh, I should put in time... Let me put in `time`, there's another argument here:

```
define (count-halt time prefix bits-left)
```

So I'm going to count how many programs halt on my universal Turing machine that have this `prefix` with `bits-left` bits added on afterwards, within this amount of time. So this `count-halt` is the auxiliary function that I use. Okay?

So the first thing to show you is how you get from this the halting probability. So before I define the auxiliary function, let me show you how to define a function `omega`, the main function, that uses this auxiliary function. It's only one main function with one auxiliary function. Let's call it `omega` of n:

```
define (omega n)
```

This is the nth approximation, the nth lower bound on the halting probability. As n goes to infinity, this will give you the halting probability in the limit from below. The problem is it converges very, very, very, very slowly! [Laughter] Noncomputably slowly! Thanks for the laughs.

Let's define this function. Well, what I want to do is I want to `cons` up a rational number. And the way I do it is, I count how many programs halt within time n that have as prefix `nil`, the empty bit string, and that add n bits to it.

```
define (omega n)
cons (count-halt n nil n)
```

And I `cons` that with division.

```
define (omega n)
cons (count-halt n nil n)
cons /
```

And `cons` that with, I use `^` for power, 2 raised to the power n, and `nil`.

```
(define (omega n)
(cons (count-halt n nil n)
(cons /
(cons (^ 2 n)
      nil))))
```

This depends on the representation you pick for rational numbers; I just write out a fraction. And so `(omega n)` is a triple, I'm creating a triple. So it's (the number of programs n bits in size that halt within time n) divided by 2^n. Okay?

So that's the main function. Now let me define the auxiliary function. The auxiliary function goes like this. You start off with the empty prefix `nil` and recursively you start adding bits until you get a full n-bit program and then you run it for time n and you see whether it halts. Now how do you program that out in LISP? Well, recursively it's very easy, you go like this.

Let's first see if `bits-left` is not equal to 0. If `bits-left` is greater than 0, then add a 0 bit to the prefix and subtract one from the number of bits left, see how many of these programs halt within time n, and add that to the same thing when you `append` a 1 to the prefix instead of a 0.

So we've got this:

```
define (count-halt time prefix bits-left)
if > bits-left 0
+ (count-halt time (append prefix '(0)) (- bits-left 1))
  (count-halt time (append prefix '(1)) (- bits-left 1))
```

I'll put parentheses in occasionally and other times not. [Laughter] Well, actually in my LISP you don't put them in for primitive functions and the interpreter adds them, and writes it out for you just so that you can check that it's what you meant.

So this is the recursion. This is going to make `count-halt` look at all n-bit programs. And what happens when it finally has no bits left to add? It started off with n bits to append to the prefix, and with the empty prefix. So now you've finally got an n-bit string, you get all possible n-bit strings. So then you take the n-bit string, and you `try` for time n `eval read-exp` applied to the n-bit data string which is the prefix. If this `try` is a `success` then `count-halt` is a 1 otherwise it's a 0.

So now we've got this:

```
define (count-halt time prefix bits-left)
if > bits-left 0
+ (count-halt time (append prefix '(0)) (- bits-left 1))
  (count-halt time (append prefix '(1)) (- bits-left 1))
if = success car (try time
                      'eval read-exp
                      prefix)
   1
   0
```

This recursion is going out and seeing whether all possible n-bit programs halt or not. When you've got the whole n-bit program, how do you see whether it halts or not? You `try` running it for the time that you were given, which is going to be n for `omega` of n. And this part

```
(try time
     'eval read-exp
     prefix)
```

does that. It's taking the prefix as the S-expression and the binary data of a program for my universal Turing machine. It's reading an S-expression off of the beginning of the prefix and it's running the S-expression with the rest of the prefix as binary data. Okay?

I don't really have the time to explain this properly. If you were giving this to students in a class you'd probably take a whole class to explain this. And then the students would go and they would play with it and they would use it. And if you look at my Web site you'll find that I'm running `count-halt` on examples to show that it works. I also have a different definition of `omega` of n which is more traditional, which builds up the list of all n-bit strings, and then sees which halt, and then counts how many halt. But I think that what I have here is more elegant, it's really all you need.

Let me say it in words again.

What you see here in this version of LISP is the nth lower bound on Ω. With increasing positive integers n, `omega` of n gives you better and better lower bounds on the halting probability. The problem is that you never know how far out to go to get the halting probability with a given degree of accuracy. To get n bits of the halting probability right, you have to go out Busy Beaver function of n, that is basically the idea.

You look at all n-bit programs, as prefix you start off with the empty list of bits, and the bits left to add is n initially, and you're going to count how many of them halt within time n. That's what

```
(count-halt n nil n)
```

does.

7 Discussion of the program for Ω

Let's look at this conceptually and let's look at it pedagogically. Pedagogically, I certainly learned a lot writing this program, right? I had to invent the programming language, I had to write the interpreter! So I learned a lot from this. Is this a good way to transmit an idea to a student? I don't know! Reading somebody else's code is not great, right?

You really understand an algorithm when you've programmed it and tried it on examples and debugged it yourself. So I think that to use this in a class effectively you need exercises. You need to ask the students to do variations on what you've presented, so they make it their own. I don't do this here. This is a very concise version of the course.

If someone really wanted to use this as a course, they could start with this, which it is at least feasible that someone could understand. Before it wasn't. I could understand it. A few very committed people could sort of try to think they understood it, but I think there's more of a chance with this, it looks more like normal LISP. So that's from the pedagogic point of view.

Now let's look at this philosophically. The point is this. This program is Ω's definition. This is really it. So if this number is going to be very mind-boggling, you'd like to pin down very concretely how it's defined. And we have.

You'd like to pin down very concretely how Ω's defined before you start being astonished by it. Here Ω is really very concrete and you can actually run `(omega 0)`, `(omega 1)`, `(omega 2)`, ... Of course the time grows exponentially as n goes up, but maybe you can get to one interesting example. You can try variations of this program to convince yourself that you understand it. And you can try debugging `count-halt` separately by running examples to show that it works. That's how to convince yourself that this works: run lots of examples. I haven't tried giving a formal proof that this program is correct. I wonder if that would be an interesting project?

Okay, now I once tried explaining this to an astrophysicist who amazingly enough was briefly interested [laughter], a very bright guy. He was so bright that he could understand the previous version of the program for Ω, the one in which the names of built-in functions are a single character, and where you had to program out arithmetic, you didn't have positive integers like we do here. He took the definition of `(omega n)` and `count-halt`, printed them on a small sheet of paper, folded it up and put it in his wallet. He said that it was like a mantra, "I have Ω in my pocket!" But it was Ω for him only because he was very bright and he took the trouble to go through this course with me. I think that this new version is much easier to understand than the one that he put in his pocket.

8 Proof that Ω is algorithmically irreducible

Okay, so what is the next thing that one does in this course on the limits of mathematics? I've tried to throw out all the inessential stuff and just give the main ideas. And now we'll start to see why this number Ω is significant.

Well, the next thing I do in the course is I point out that `(omega 0)`, `(omega 1)`, `(omega 2)`, ... is a monotone increasing sequence of rational numbers. So, this isn't constructive, but we all know that there is a real number which is the least upper bound of this sequence. So `(omega n)` defines in a nonconstructive way a real number Ω. It's almost constructive. It's pretty close to being constructive. Really constructive would be if you could calculate each digit of Ω, right? I'm getting better and better lower bounds, the only thing I don't have is I don't have a way to calculate how far out to go to get a given degree of accuracy.

Okay, so this defines a real number that I call Ω. Now let's imagine this number being written in base-two binary, so there's going to be a 0, a binary point, and then there're going to be some bits, right?

$$\Omega = 0.011101\ldots$$

And let's think of this bit string as a LISP S-expression, it's going to be a list of 0's and 1's separated by blanks. And let's define Ω_N to be the first N bits of this, the LISP S-expression for the first N bits of Ω. It's a list with N elements.

The theorem you want to show is that $H(\Omega_N)$ is pretty complicated, that there's a lot of information in those first N bits of Ω. In fact it's algorithmically irreducible. You cannot compress the first N bits of Ω into a program substantially less than N bits in size. The exact result is that

$$H(\Omega_N) > N - 8000.$$

This turns out to be the case.

Why is there a lot of information in the first N bits of Ω? It's not difficult to see why. If you knew the first N bits of Ω, it solves the halting problem. It would enable you, slowly, but it would enable you to solve the halting problem for all programs up to N bits in size. So once you do that, you run all the programs up to N bits in size that halt, you see what they calculate, and then you just put what they calculate together into a list. That gives you something which cannot be done by a program less than or equal to N bits in size.

And it takes 8000 bits to explain how to do this. There's a prefix ϕ_{8000} which is 8000 bits long that if you put it in front of a program that calculates the first N bits of Ω, this gives you a program to calculate something whose complexity is greater than N. This shows rather concretely that this inequality

$$H(\Omega_N) > N - 8000$$

has to follow. Okay?

So I'm giving my universal Turing machine U a program which starts with an 8000-bit prefix ϕ_{8000}. That sounds like a lot, but in fact it's only a thousand characters of LISP. And then next to it I concatenate any program Ω_N^* for calculating the first N bits of Ω, if somehow I had one. I don't know where to get one! So this is what we've got:

$$U(\phi_{8000}\,\Omega_N^*).$$

And then you run this and what the machine U does because of ϕ_{8000} is this. First it reads in and runs Ω_N^* to calculate the first N bits of Ω. Then it uses `(omega n)` to get better and better lower bounds on Ω until Ω_N is correct, until it gets the first N bits of Ω right. God, or an oracle, is giving us the program for the first N bits of Ω!

Once U has calculated `(omega n)` for n high enough that the first N bits of Ω are correct, at that point it's seen all N-bit programs that are ever going to halt. In fact, if they ever halt, they halt within time n.

And then U sees what these programs produce, and it forms the list of the output of all programs up to N bits in size that halt. And to finish up everything, U outputs this list and halts. In other words, this list is the value of

$$U(\phi_{8000}\,\Omega_N^*).$$

It's precisely the list of all LISP S-expressions x with program-size complexity $H(x) \leq N$.

This list cannot itself have program-size complexity $\leq N$, because it can't be contained within itself. So it can't be the output of a program less than or equal to N bits in size, it's got to have program-size complexity greater than N. Therefore this program $\phi_{8000}\,\Omega_N^*$ for producing the list

$$U(\phi_{8000}\,\Omega_N^*)$$

must be greater than N bits in size. In other words, (the size of any program that calculates the first N bits of Ω) plus 8000 has got to be greater than N. So you get this inequality

$$H(\Omega_N) > N - 8000.$$

Okay?

And in my Web site I actually show you this program. I program out in LISP the algorithm that I just described in words. You program it out, it's not a big deal, ϕ_{8000} is about one page of LISP, of my new LISP. But if you put in a lot of comments and run lots of examples then it's more than one page. Okay?

Martin Gardner had an explanation of this algorithm in an article in *Scientific American* on Ω that you'll find in one of the Martin Gardner collections. In my Web site this algorithm is actually written out in LISP and you can run it on some examples. You can certainly run the auxiliary functions and other pieces of the algorithm and convince yourself that it works. But when you put it all together, you're not usually going to be able to run it on the real data Ω_N, because the whole point of this is to show that the bits of Ω are hard to know, they're algorithmically irreducible.

By the way, it's easy to see that

$$H(\Omega_N) > N - 8000$$

implies that Ω is violently non-computable. Let's compare the two real numbers Ω and π. π has the property that the string of its first N bits has very small program-size complexity. Given N, you can calculate the first N bits of π. So the first N bits of π only have about $\log_2 N$ bits of complexity. But the first N bits of Ω are irreducible, can be reduced at most 7999 bits, maybe down to $N - 7999$, we've proved that.

By the way, program-size irreducibility implies statistical randomness. So Ω's irreducibility implies that Ω is a normal real number. In any base, its digits all have the same limiting relative frequency.

9 Proof that you can't even deduce what the bits of Ω are

Okay, so the next thing you do with this, and this is the next (and last) program in this course, is you get an incompleteness result for Ω from this. From this inequality

$$H(\Omega_N) > N - 8000$$

which states that Ω is algorithmically irreducible. By the way, I don't show in the course that this implies statistical randomness. That's a detour. I want to go straight to the incompleteness result.

(But I think you can sort of wave your hands and try to convince people that irreducibility implies statistical randomness. Or you can actually go and develop algorithmic information theory in excruciating detail! But in my course my goal is to get to the major incompleteness results as quickly as possible, illustrating them with interesting LISP programs, not to develop the entire theory. Now you could also take the proof of every theorem in my Cambridge University Press book *Algorithmic Information Theory* and write down a LISP program for the algorithm in it. I once started doing that but I quickly gave it up. It was too much work, and you don't really want to see every algorithm in the theory in such detail. I'm concentrating on programming in LISP what I think are the fun algorithms in the theory, especially the ones connected with the basic incompleteness results, the ones I want to use to present what I think are the fundamental concepts in my theory.)

So the last program in this course is the one that yields an incompleteness result for Ω.

Let's start with this question: How do I represent a formal axiomatic system in LISP? Well, this is the way I do it. Think of it as a black box which every now and then throws out a theorem. So it's a program that you start running, and it may or may not halt, normally it doesn't halt, and every now and then it prints out a theorem.

So why is this a formal axiomatic system? Well, the idea is that I don't care about the details of what's going on inside, what the axioms are, or what the logic used is like. What I care about is the criterion, which I think was enunciated rather clearly by Hilbert, that states that the essence of a formal axiomatic system is that there should be a proof-checking algorithm. So if that's the case, you can run through all possible proofs in size order, see which ones are correct, and print out all the theorems. Of course this is impractical. The time it takes to do it grows exponentially, and I wouldn't really want to run such a program.

But you can cheat and you can just have a LISP S-expression which every now and then outputs a theorem...It won't be the final value...Every now and then it uses a primitive function, I call it `display`, but you could also say "output intermediate result." So I think of a formal axiomatic system as a LISP S-expression that every now and then calls a particular new LISP primitive function which outputs an intermediate result. And that way it puts out a lot of LISP S-expressions, which are the theorems.

And you can cheat and instead of putting a real formal axiomatic system in there with a proof-checking algorithm and running through all possible proofs, which is terribly slow (and never halts), you can cheat and put in a little example of something that just `display`'s a few sample theorems and then halts. That way you can run a little example and debug algorithms which work with formal axiomatic systems.

How can algorithms work with these toy formal axiomatic systems? Let's say that you have a formal axiomatic system. We've agreed that it's a LISP program that is going to put out intermediate results using a new primitive function called `display`. Actually all normal LISP's have a function for outputting intermediate results; you use it for debugging, right? But it becomes more important in this framework. It plays an official role in my computerized version of algorithmic information theory, because `try` captures all the intermediate results. That's very important.

Try is a time-bounded execution/evaluation of a LISP expression. You use `try` to run a program with a time limit, giving the program raw binary data on the side, and *capturing all the intermediate output* as well as the final value (if any). That's very important. You see, the new primitive function `try` that I added to LISP is the way that you can run a formal axiomatic system for a while and see what are the theorems that it produces. That's put in as a primitive function in this LISP. The value returned by `try` is a 3-element list $(\alpha\,\beta\,\gamma)$. α is `success` or `failure`. β is the value of the expression being tried if the `try` is a success, and is `out-of-time` or `out-of-data` if the `try` is a failure. And γ is the list of captured `display`'s, the list of theorems. The theorems don't get displayed, they end up in this list instead. Okay?

Let me say by the way that this LISP started off as three-hundred lines of Mathematica. I invented this LISP using Mathematica as my programming tool; I wrote the LISP interpreter in Mathematica. That way I could play with my LISP and try it out as the design evolved. Mathematica is the most powerful programming language that I know. But it's slow. So the next thing I did was I rewrote the interpreter for this LISP in C. And it's a thousand lines of C, it's a hundred times faster than in Mathematica, but the program is incomprehensible of course.

So the next thing was, I rewrote the interpreter in Java, because by this time HTML, the World Wide Web, and Java had appeared! And in Java it's 750 lines of code. But I cheat, I didn't take the trouble to program out `bignum`'s, arbitrarily large integers. I did that in C, Mathematica comes with it built in. And in Java you get 18-digit decimal integers built in, so I said, "That's enough!" Probably somebody someday is going to add arbitrarily large integers to the Java class libraries, it's object oriented, a lot of stuff is in those libraries. I don't want to do that work.

So it's 750 lines of Java, and I think that the Java code is pretty nice. The C code is incomprehensible! Like all good C programs, it's too clever. You can deal with it while you're writing it, and immediately afterwards it's already incomprehensible, even for the programmer! The Mathematica code is much easier to understand, but you have to know Mathematica, which is a pretty big language, and you have to know the subset of it that I'm using, which may not be the subset that you like to play with normally.

The Java code I have to say I really like. It's 750 lines of Java. I think it's pretty understandable, I think it's pretty clean, and I think that if I were presenting this as a course I would go through the Java interpreter with the students. I think it's important that they should see the code. It's only 750 lines of Java, Java's a fairly reasonable language, and they could make changes to the interpreter as a way of showing that they understand it. So it's there, the Java source code is there on my Web site.

Okay, so what's the next (and the last) program? The next program works with a formal axiomatic system, which is just a mechanism for producing an infinite set of theorems. It's given to us as a binary program for U, FAS*, which is a LISP expression plus binary data, and we're going to measure its complexity in bits the same way that we did before. I'll prove that if the formal axiomatic system has program-size complexity N, then it can enable you to determine, to prove what is the value of at most $N + 15328$ bits of the halting probability Ω.

> If a FAS has program-size complexity N, then it can enable you to determine at most $N + 15328$ bits of Ω.[3]

So how do I show this? Well, I have a Berry paradox kind of proof. The idea is that if you could prove a lot of bits of Ω, then Ω wouldn't be this irreducible

$$H(\Omega_N) > N - 8000.$$

That would give you a way to compress Ω into the axioms of the FAS. It would give you too concise a way to calculate Ω. If you could prove what the bits of Ω are, you'd do it systematically by searching through all possible proofs, and that would give you a way to calculate the bits of Ω. That's all we're saying, that it would contradict this

$$H(\Omega_N) > N - 8000.$$

So deduction and computation are very close, as Turing already noticed in his famous 1936 paper "On Computable Numbers..." where he proves an incompleteness result using the unsolvability of the halting problem. My argument is at the same level, it's analogous.

The new business here is roughly like the paradox of the first uninteresting positive integer. You know, that's sort of an interesting number. So if you could prove that a number's uninteresting, that would be an interesting fact about it, and "uninteresting" is the notion of algorithmically incompressible. So it turns out that you can't prove that an N-bit string is algorithmically incompressible if N is larger than the complexity of your axioms. You can't prove that an N-bit string is algorithmically irreducible if it has more bits than the axioms you're using for the proof. And similarly it turns out that with N bits of axioms you get at most $N + 15328$ bits of Ω.

That's the general idea. Here are the details.

I write out a 7328-bit program ϕ_{7328}, it's about one page of LISP code. Why 7328 bits? Because you have to add that to the constant in the inequality for $H(\Omega_N)$ to get the constant in our incompleteness result. So the difference between these two constants 8000 and 15328 is the size of the next program in this course, the last program. Divided by 8, you get the size of a LISP expression in characters.

And what this LISP program ϕ_{7328} does is this. Using `try` it starts running the formal axiomatic system that you're assumed to be given that's N bits of code. So we're looking at this:

$$U(\phi_{7328}\,\text{FAS}^* \ldots).$$

[3] Perhaps one should refer to this as the *logical (or deductive) irreducibility* of Ω, to distinguish it from the *computational (or algorithmic) irreducibility* of Ω, namely the fact that $H(\Omega_N) > N - 8000$.

The prefix ϕ_{7328}, the program that outputs the theorems of the FAS, and some extra stuff that I'll explain later are concatenated and fed to U.

ϕ_{7328} starts running the formal axiomatic system using larger and larger time bounds, capturing the intermediate output, which are the theorems. And it looks at the theorems to see how many bits of Ω it got. And I allow partial determinations where you get some of the bits but you leave holes with unknown bits between some of the bits that you do know.

And what ϕ_{7328} does is it looks for a small set of axioms that enable you to prove substantially more bits of Ω than there are in those axioms—at least 15329 bits more. And then it just fills in the holes, the missing bits, which costs just one bit per bit, one bit for each missing bit. So ϕ_{7328}'s final output and the value of

$$U(\phi_{7328}\,\text{FAS}^*\ \text{missing bits of}\ \Omega)$$

will be one of the Ω_N's. For some N, it'll be the list of the first N bits of Ω. By the way, ϕ_{7328} may not need all the bits of FAS*, because it only keeps a finite part of the potentially infinite computation for the formal axiomatic system.

So if you could use N bits of axioms to get essentially more than N bits of Ω—more than $N + 15328$ bits, in fact—then you could fill in the missing bits at a cost of one bit each, and you get into trouble with this inequality

$$H(\Omega_N) > N - 8000.$$

That's the point.

So that's how it goes, the basic idea is straight-forward. And I actually have this 7328-bit LISP program ϕ_{7328} and you can run it. Well, you can certainly test the auxiliary functions. You can certainly run all the auxiliary functions on examples to convince yourself that they work. But to test the whole algorithm, its main function, you have to cheat, you don't give it a real formal axiomatic system, because if you really try to run through all possible proofs in size order it would take too long. So you cheat, you give the algorithm a LISP expression that just throws out a few test theorems that will put the algorithm ϕ_{7328} that's running the formal axiomatic system through it's paces.

That's how you convince yourself that this all works. And you may have to "tweak" things a little bit, you may have to slightly change the LISP code so that it works with your simple test cases. So you can convince yourself by running little examples that this would all work if you really gave it a genuine formal axiomatic system, say Zermelo-Fraenkel set theory.

I would like to have somebody program out Zermelo-Fraenkel set theory in my version of LISP, which is pretty close to a normal LISP as far as this task is concerned, just to see how many bits of complexity mathematicians normally assume. You see, the whole effort here has been to get these constants 8000 and 15328. And if you programmed ZF, you'd get a really sharp incompleteness result. It wouldn't say that you can get at most $H(\text{ZF})+15328$ bits of Ω, it would say, perhaps, at most 96000 bits! We'd have a much more definite incompleteness theorem. I hope that somebody will program ZF in LISP. I stop at this point [laughter], you know, programming fatigue! And you want to leave something for the students to do, right? So this is a little job for them. And if they're really clever maybe it'll put out the theorems in Zermelo-Fraenkel set theory reasonably quickly, not by running through all possible proofs in size order, which would take too long, but by doing some kind of tree search so that you can actually have something interesting happen in a reasonable amount of time.

10 Discussion

Okay, so why is all this interesting? Well, some of you may know already, but let me wave my hands frantically in the last few minutes and say why.

This looks like a kind of programming madness, right? Up to know, it looks like what I've been telling you is, "Oh how much fun it was to program something that nobody's ever programmed before!" Or maybe just that this is neater than it was in the previous versions of my course. Yes, programming can be an obsession, it can ruin your life. You can sit at your terminal watching your life fall apart as you stay there hacking away until four in the morning! But is there any other justification for this except as a kind of drug? Well, I think so! I think this has some philosophical significance.

What's the point about this incompleteness theorem, about this technical result?

> If a FAS has program-size complexity N, then it can enable you to determine at most $N + 15328$ bits of Ω.

Well, what it's really telling you is that you might get some bits of Ω, the first ones, say. You might be able to calculate the first few bits of Ω. In fact, in a different version of this theory, with a different UTM, (this was with my one-character LISP, my one character per atom LISP), I did get the first 7 bits of Ω, and they were all 1 bits. You see, once your lower bound on Ω is 127/128ths, you know that these bits can't change. The first 7 bits were 1's. And now I have a much friendlier LISP, but I lost this, I can no longer determine the first 7 bits of Ω this way.

But anyway, you might be able to get some bits of Ω without contradicting Ω's logical and computational irreducibility, without contradicting this incompleteness result and this inequality:

- You can determine at most $H(\text{FAS}) + 15328$ bits of Ω.
- $H(\Omega_N) > N - 8000$.

But in spite of this, the basic point is that Ω really shows that some areas of mathematics have no structure, have no pattern at all.

Let me put it this way. Normally you think that if something is true, it's true for a reason, right? In mathematics, the reason is called a proof, and the job of a mathematician is to find the reason that something is true, to find a proof. But the bits of Ω are mathematical facts that are true for no reason, they're accidental!

Now this is a very specific Ω, I've programmed it out. Imagine it being written in binary. And you ask, an individual bit, say the 33rd bit, is it a 0 or a 1? Let's say you're trying to prove which it is. And the answer is, you can't! The reason you can't is because, whether that particular bit is a 0 or a 1 is true for no reason, it's true by accident. It's so delicately balanced whether it's going to be a 0 or a 1, that we will never know!

It's like independent tosses of a fair coin. Independent tosses of a fair coin has got to come out heads or tails in each case, but there's no reason that it comes out one or the other, right? So it's exactly the same story with these mathematical facts, with the bits of Ω. There is no pattern or structure in the sequence of bits of Ω.

I don't know why anybody would want to try to determine bits of Ω. Although people have played with the Busy Beaver function, which in a way is like trying to calculate the bits of Ω. I don't know why you'd try to determine the bits of Ω. But if you were to try to do this, what this incompleteness result shows you is that you're in big, big trouble! Because essentially the only way to prove what a bit of Ω is, is to add the theorem that you want to prove as a new axiom. It's irreducible mathematical information. Now you can prove *anything* by adding it as a new axiom. The point here is that for Ω that's essentially *the only way* to do it. No compression is possible.

So there is no structure, there is no pattern, Ω has maximum entropy, it mirrors independent tosses of a fair coin. Now to a physicist what I'm saying sounds pretty reasonable, right? Ω has maximum entropy, the bits of Ω are completely uncorrelated. But to a mathematician this all sounds weird!

My new Ω is a particular well-defined real number. It's a real number with a rather simple definition, I can even write the LISP program that defines Ω on one computer screen. So you believe, thinking in Platonic terms, that each bit is either a 0 or a 1, even if I'm never going to know which. But it's black or white, right? And what I'm saying is that I think that it's really better to think that it's *grey*. It's really better to think that each bit of Ω has probability one-half of being 0 or of being 1—even though it's a particular well-determined bit, because I've written out the program that defines this number. Defines it, not by enabling you to calculate it bit by bit—that would contradict Ω's unknowability. But the program for Ω does enable you to calculate better and better lower bounds on Ω, so Ω is *almost* a computable real number in the same sense that π is. Almost, but not quite!

So the game is like this. I'm trying very hard to be constructive, as constructive as possible. Writing out programs is a sure sign that you're a constructivist, right? [Laughter] I want to settle all the programming details. But I'm trying to be as constructive as possible about non-constructivity! I want to exhibit something that escapes the power of constructivity, of mathematical reasoning, that you can't calculate, but that's just over the border between the constructible and the non-constructible. And I think that Ω is pretty damn good, it's just on the border.

Now this doesn't mean that all of mathematics falls down in a heap! But the normal notion of mathematics was that there were a small, finite set of axioms and rules of inference that we could all agree on, from which all the infinite mathematical truth would follow. This is the tradition that goes back to Euclid, to Leibniz, to Peano, to Frege, to Russell and Whitehead, to Hilbert. And Gödel showed that there was a problem. And Turing showed that there was a problem, using a different method involving computers. And I think that Ω follows in that tradition and shows that the problem is even bigger. However I don't think this means that you should stop doing mathematics! So what is the significance of Ω and of incompleteness? Should it affect how we actually do mathematics? I'll give you my opinion.

Of course the nature of mathematics has been discussed for a long time! Every generation of mathematicians have their own answer. But let me share with you my own feelings about this, my tentative conclusions.

There's a word that a philosopher coined that's very good. He says that there's an emerging new school, and emerging new *quasi-empirical* view of the foundations of mathematics. One talks about the formalist school, the logicist school, and the intuitionist school. Well, Thomas Tymoczko has a book, *New Directions in the Philosophy of Mathematics,* which has a whole bunch of articles, including two of mine, and he thinks that all these articles tend to support a new quasi-empirical view of the foundations of mathematics.

Now what does quasi-empirical mean? I'll tell you what it means to me. Quasi-empirical means that *pure math ain't that different from physics!* The normal notion of pure math is that mathematicians have some kind of direct pipeline to God's thoughts, to absolute truth! But poor physicists! You know, they tried Newtonian mechanics, it looks good for a while, then Einstein shows it's all wrong. Then—surprise!—quantum mechanics shows that Einstein was wrong! And now there's superstring theory. And is it right? Is it wrong? And mathematicians laugh and say, "Oh those poor physicists! It's such a messy subject! They always have to backpedal, they don't know what they're doing! It's all so tentative!"

Well, I think that mathematics and physics are not really that different!—Physicists love it when I say this! [Laughter]

Let me try explaining this another way. Euclid said that mathematics is based on self-evident truths. But my impression is that maybe axioms are not self-evident truths. I don't believe in self-evident truths. Maybe it's more like in physics. Maybe mathematics should be done more like physics, where you're willing to add new axioms because they're useful, not because they're self-evident. And then of course you have to be prepared to say "I goofed!" and remove an axiom, which mathematicians don't like to have to do.

This may sound completely crazy, but in fact it's not just my opinion. Gödel makes very similar remarks in Volume II of his *Collected Works,* in his essay on "Russell's Mathematical Logic." This essay was originally published in the Paul Arthur Schilpp volume *The Philosophy of Bertrand Russell.*

I talked about new axioms to a mathematician once, and he replied, "Okay, I'm willing to add the Riemann hypothesis as a new axiom if you can prove to me that it doesn't follow from the usual axioms." Well, that's hard to do, because if the Riemann hypothesis were false, then there would be a numerical counter-example that one could easily verify that shows that it's false. So if you could show that the Riemann hypothesis is beyond the power of the usual axioms, that would imply that the Riemann hypothesis is true!

So these ideas are very controversial.

But the whole point of algorithmic information theory, the whole point of my information-theoretic approach to incompleteness, is that sometimes to get more information out of a set of axioms, you've just got to put more in. So let's put more axioms in! Physicists have always done that.

I had these ideas a long time ago. I proved my first information-theoretic incompleteness theorem in 1970. Although it's only in the past two or three years that I discovered how to actually program the algorithm in my original proof and run it on examples. And I was recently surprised to discover or rediscover that there are highly relevant quotes from Einstein and from Gödel. Let me throw them into this discussion, let me end with that.

Einstein has a very nice remark that I angered some mathematicians with. But what do they care, he's only a physicist, right? [Laughter] In his essay "Remarks on Bertrand Russell's Theory of Knowledge" in the Paul Arthur Schilpp volume *The Philosophy of Bertrand Russell,* Einstein says that "the series of integers is obviously an invention of the human mind, a self-created tool which simplifies the ordering of certain sensory experiences." So you can see that Einstein's attitude is very empirical.

I think that Einstein's position is that the positive integers are not *a priori,* they're not God given, we invent them like we invent all of physics. But the positive integers look more *a priori* than other concepts because they've been around longer, we invented them a long time ago. After all, when an idea has been around for a few thousand years, it's not surprising that people think that it's obvious, that it's just common sense, that it's "a necessary tool of thought." The other extreme is the latest field theory in physics. That looks a lot more tentative. It hasn't been here long, and it'll probably be shot down next week, right? And there are probably thirteen different versions! But in Einstein's view there is no fundamental difference. The positive integers have been around for a long time, but they're still just an invention.

So I like this quote from Einstein, but it doesn't convince mathematicians.

Then there are some very interesting remarks made by Gödel. You can find them in Gödel's *Collected Works.* Gödel's philosophical position is the exact opposite of Einstein's. Gödel believed in the Platonic universe of mathematical ideas, he believed that the positive integers are just as real as tables and chairs. There are an infinity of positive integers, and they're out there somewhere. They're in the Platonic universe of mathematical ideas, that's where mathematical objects are!

I don't know! I used to laugh at all of this when I was a kid. But think about it seriously. You're young, you're trying to learn mathematics, you're doing elementary number theory, and you don't really start to worry about the fact that elementary number theory presupposes arbitrarily large positive integers, and how do they fit in the universe? Imagine a positive integer that is

$$10^{10^{10^{10}}}$$

digits long. Does it exist? In what sense does it exist? You don't care—right?—you prove theorems about it, you know that it would be commutative, right? $a + b$ is going to be equal to $b + a$ even if neither number fits in the universe! [Laughter] But then later on in life you start to worry about this! [Laughter]

So I'm not sure if the positive integers exist anymore. But Gödel thinks that they do, and that philosophical position is associated with the name of Plato. But it's really just the classical mathematical position that the positive integers really exist, that an infinity of them are really out there. And starting from that, Gödel comes to a very surprising conclusion. Since the positive integers are just as real as tables and chairs, you can do experiments with them by doing calculations, and if you see a pattern, you can just go ahead like a scientist would in dealing with electrons. Since integers are just as real as electrons, why can't we use the same kinds of methods that scientists use?

Here are Gödel's exact words, taken from a previously unpublished manuscript **1951* that is in Volume III of his *Collected Works:*

> "If mathematics describes an objective world just like physics, there is no reason why inductive methods should not be applied in mathematics just the same as in physics."

And in his essay "What is Cantor's Continuum Problem?" in Volume II of his *Collected Works,* Gödel says that maybe we'll come up with new versions of set theory, maybe we'll come up with new ideas about sets the same way that physicists come up with new ideas about physical objects. The justification for these new principles would be that they're useful, that they help us to organize our mathematical experience, just as the ultimate justification for physical principles is that they help us to organize our physical experience.

I think it's very funny! Here's Einstein, who's a diehard empiricist, and here's Gödel, who's a diehard Platonist, and they sort of come to the same conclusion! But they were buddies at the Institute for Advanced Study, so it's not surprising that they influenced each other. And the information-theoretic viewpoint that I've explained to you today leads me in the very same direction.

Another funny thing is that I don't believe that all this work on the foundations of mathematics has changed at all the way mathematicians actually work. Gödel's incompleteness theorem was initially very shocking. But then mathematicians noticed that the kind of assertion that Gödel constructed that's true but unprovable is not the kind of assertion that you deal with in your everyday work as a mathematician. And I think it's fair to say that about Ω too. The longer I live with Ω, the more natural it looks to me. But a skeptic would say, "I don't care about the bits of Ω, so what if there's trouble there!"

But I do think that something else is making a difference in how mathematicians carry on their everyday work. It's the fact that the computer has so vastly expanded mathematical experience, that we just have to deal with it somehow. A lot of physicists do experimental work on the computer now. There's even a journal called *Experimental Mathematics*.

The computer has so vastly increased mathematical experience, that how do we keep up with all of it? Well, the answer is that sometimes you see that something seems to be the case, and it'd be nice if you could prove it, but for the moment you can't, so you sort of conjecture it. And if you're doing mathematical physics, then you're on the borderline between math and physics, and it's certainly okay to behave in this way. But if you're a mathematician, if you're just over the border, then it's not so clear anymore.

Well, I once had a conversation like this with a mathematician, and the Riemann hypothesis came up, and he said, "It works fine the way we do things now. You have a paper, and you say the paper is 'modulo the Riemann hypothesis.' So why do you need to call it a new axiom?" And he has a point. But I do think that if a principle is very helpful for a long time, and nobody shoots it down, why not call it a new axiom? But one shouldn't do this too quickly. One has to be careful, the way physicists are, although I sometimes wonder how careful physicists really are! [Laughter]

Okay, that's the idea. And if you go to my Web site you'll find there in HTML all my less technical papers. And you've also got this course there, and the Java source code and byte code for my LISP interpreter. So you're welcome to go to my Web site and play with it, and you're welcome to send me e-mail about it, and I'd be pleased as Punch if one of you tried giving my course on the limits of mathematics to real students in a normal university setting.

Anyway, you guys have been nice enough to invite me for many years to give talks, and it's been very stimulating to talk with you, and for the moment I think I've exhausted this train of thought. Of course, I've been thinking that since I was fifteen years old, but fortunately every now and then I get a new idea! [Laughter] But for now, that's all I have to say, and maybe someday I'll have more to say, and I'll be lucky enough to come out here again and have a chance to kick it around with you! Thank you very much! [Applause]

Discrete Isoperimetric Inequalities (abstract)

Fan Chung

University of Pennsylvania
Email: chung@math.upenn.edu

Most important properties of a graph are related to its eigenvalues. However, it is only recently that it has been possible to make this connection precise. In this talk, we will describe a variety of new techniques which allow us for the first time to control many graph invariants in terms of the eigenvalues and eigenfunctions. In particular, this involves a strong two-way interaction between concepts from spectral geometry such as heat kernels, Sobolev and Harnack inequalities and their emerging discrete counterparts.

N(n) and $\nu(n)$: Similarities and Differences

Charles J. Colbourn[1] and Jeffrey H. Dinitz[2]

[1] Department of Computer Science and Electrical Engineering, University of Vermont, Burlington, VT 05405
[2] Department of Mathematics, University of Vermont, Burlington, VT 05405

1 Introduction and Definitions

Much has been written about the construction of sets of mutually orthogonal latin squares (MOLS). In [8], a lengthy survey of these constructions is given. Existence of MOLS is tabulated in [1], historical information appears in [9, 21], and proofs of many of the existence results appear in [1, 4, 21]. Rather than repeat these surveys here, we instead explore how some of the available constructions can be applied to a similar problem that has not been studied quite as extensively. We begin with some definitions.

A *latin square* of *side* n is an $n \times n$ array in which each cell contains a single element from an n-set S, such that each element occurs exactly once in each row and exactly once in each column. A latin square L of side n is *symmetric* if $L(i,j) = L(j,i)$ for all $1 \leq i,j \leq n$. A latin square L of side n is *idempotent* if $L(i,i) = i$ for all $1 \leq i \leq n$. Two latin squares L and L' of the same order are *orthogonal* if $L(a,b) = L(c,d)$ and $L'(a,b) = L'(c,d)$, implies $a = c$ and $b = d$. An equivalent definition for orthogonality is: Two latin squares of side n, $L = (a_{i,j})$ (on symbol set S) and $L' = (b_{i,j})$ (on symbol set S'), are *orthogonal* if every element in $S \times S'$ occurs exactly once among the n^2 pairs $(a_{i,j}, b_{i,j})$, $1 \leq i,j \leq n$.

A set of latin squares $L_1, \ldots, L_m$ is *mutually orthogonal*, or a set of *MOLS*, if for every $1 \leq i < j \leq m$, L_i and L_j are orthogonal. One of the most fundamental questions in combinatorics is: What is the size of the maximum set of mutually orthogonal latin squares of order n? Let $N(n)$ denote this number. It is easily shown that for all n, $N(n) \leq n-1$. Good *lower* bounds for $N(n)$ are more difficult to establish. In Table 1 we give the current best lower bound for $N(n)$ for $n \leq 50$ when n is not a power of a prime. In the case that n is a prime power, that $N(n) = n-1$ is well known, as we see later. Lower bounds for $N(n)$ for $n \leq 10{,}000$ are tabulated in [1], along with explicit constructions for small values of n. The dates of the results in Table 1 are interesting, pointing again to the fact that much research has been done recently in this area.

We consider the analogue of orthogonal latin squares in the case where the latin squares are required to be symmetric. Suppose that L and M are idempotent, symmetric latin squares of the same order. Then L and M are *orthogonal symmetric* latin squares if, for any two elements x and y, there exists at most one ordered pair (i,j) with $i < j$ such that $L(i,j) = x$ and $M(i,j) = y$. Orthog-

Order	$N(n) \geq$	Year	Reference(s)
6	1	1900	[25]
10	2	1960	[5]
12	5	1961	[17]
14	3	1985	[27]
15	4	1978	[24]
18	3	1978	[30]
20	4	1989	[28]
21	5	1991	[20]
22	3	1978	[30]
24	5	1992	[2, 22]
26	4	1994	[7]
28	5	1995	[1]
30	4	1993	[3]

Order	$N(n) \geq$	Year	Reference(s)
33	5	1994	[1]
34	4	1994	[1]
35	5	1995	[32]
36	5	1994	[23]
38	4	1993	[3]
39	4	1977	[18]
40	7	1994	[2]
42	5	1995	[1]
44	5	1995	[1]
45	6	1995	[1]
46	4	1974	[31]
48	6	1996	[33]
50	6	1974	[31]

Table 1. Best Current Lower Bounds on $N(n)$

onal symmetric latin squares are not orthogonal latin squares, but they are as "orthogonal" as possible, given that they are symmetric. Necessarily, n is odd.

A set of symmetric latin squares $L_1, \ldots, L_m$ is *pairwise orthogonal symmetric*, or a set of *POSLS*, if for every $1 \leq i < j \leq m$, L_i and L_j are orthogonal. For odd n, let $\nu(n)$ denote the maximum size of a set of POSLS(n). Questions about $\nu(n)$, analogous to those asked about $N(n)$, are natural. One remark should be made immediately. While it is trivial to prove an attainable upper bound for $N(n)$, at this time a comparable upper bound for $\nu(n)$ has eluded researchers. In fact, there are two conflicting conjectures. It has been conjectured by Gross, Mullin, and Wallis [15] that $\nu(n) \leq (n-1)/2$, while Teirlinck [26] has conjectured that $\nu(n) \leq n-2$, with equality infinitely often. We return to the discussion of upper bounds for ν in Section 4.

In Table 2, we give the current best lower bound for $\nu(n)$ for small values of n. These bounds come essentially from only three sources. In [19] it was shown that if $q \equiv 3 \pmod 4$, then $\nu(q) \geq (q-1)/2$. In [11] it was shown that if $n \geq 11$ ($n \neq 15$), then $\nu(n) \geq 5$. The other results are all in the case when q is a prime power and $q \equiv 3 \pmod 4$; these can be found in [10]. One can contrast this small number of sources with the much larger number of sources for the small values of N(n) given in Table 1. An extensive survey for all results on $\nu(n)$ is [13].

2 Transversal Designs and Villa Designs

Many combinatorial objects are intimately related to sets of MOLS. In this section and the next we examine some parallels between these objects, and similar objects that are related to sets of mutually orthogonal symmetric latin squares.

The first and most fundamental of these structures is the transversal design. A *transversal design* of *order* or *groupsize* n, and *blocksize* k, denoted $TD(k, n)$,

n	$\nu(n) \geq$	n	$\nu(n) \geq$	n	$\nu(n) \geq$	n	$\nu(n) \geq$
1	$= \infty$	27	13	53	17	79	39
3	$= 1$	29	13	55	5	81	5
5	$= 1$	31	15	57	5	83	41
7	$= 3$	33	5	59	29	85	5
9	$= 4$	35	5	61	21	87	5
11	5	37	15	63	5	89	11
13	5	39	5	65	5	91	5
15	4	41	9	67	33	93	5
17	5	43	21	69	5	95	5
19	9	45	5	71	35	97	5
21	5	47	23	73	9	99	5
23	11	49	5	75	5	101	31
25	7	51	5	77	5	103	51

Table 2. Best Current Lower Bounds on $\nu(n)$

is a triple $(V, \mathcal{G}, \mathcal{B})$, where

1. V is a set of kn elements;
2. $\mathcal{G}$ is a partition of V into k classes (called *groups*), each of size n (in each group the elements are labeled $1, \ldots, n$);
3. $\mathcal{B}$ is a collection of k-subsets of V (called *blocks*);
4. every unordered pair of elements from V is either contained in exactly one group, or is contained in exactly one block, but not both.

A transversal design is *idempotent* if it contains each block in which every entry is i, for $1 \leq i \leq n$. The existence of a $TD(k, n)$ is equivalent to the existence of $k - 2$ mutually orthogonal latin squares of order n. To motivate our analogue in the case of symmetric latin squares, we review this equivalence.

Let $(V, \mathcal{G}, \mathcal{B})$ be a $TD(k, n)$. Now given any block $B \in \mathcal{B}$, say that B contains the element r from the first group and c from the second group. For $1 \leq t \leq n-2$ construct a latin square L_t by letting $L_t(r, c) = x$ if block B contains the element x in group number $t + 2$. The conditions on the TD imply that each of the L_t are indeed latin and are pairwise orthogonal. This construction can be reversed. Obviously, $N(n)$ is the maximum k such that there exists a $TD(k + 2, n)$.

We now define a new structure which is intended to be the analogue of transversal designs in the case where the latin squares in the construction of the preceding paragraph are orthogonal symmetric latin squares.

A *villa design* of *order* or *groupsize* n, and *blocksize* k, denoted $VD(k, n)$, is a triple $(V, \mathcal{G}, \mathcal{B})$, where

1. V is a set of kn elements;
2. $\mathcal{G} = \{G_1, \ldots, G_k\}$ is a partition of V into k classes (called *groups*), each of size n (denote $G_i = \{g_{i1}, \ldots, g_{in}\}$);

3. $\mathcal{B}$ is a collection of k-subsets of V (called *blocks*), each containing exactly one element from each group;
4. For $\{i,j\} \subset \{1,2,\ldots n\}$, $i \neq j$, there is either
   (a) exactly one block containing $\{g_{1i}, g_{2j}\}$ and no block containing $\{g_{2i}, g_{1j}\}$; or
   (b) exactly one block containing $\{g_{2i}, g_{1j}\}$ and no block containing $\{g_{1i}, g_{2j}\}$;
5. For $3 \leq h \leq k$ and $1 \leq i,j \leq n$, with $i \neq j$, element g_{hi} is contained in exactly one block intersecting $\{g_{1j}, g_{2j}\}$;
6. If $3 \leq h < \ell \leq k$ and $1 \leq i,j \leq n$, with $i \neq j$, the pair $\{g_{hi}, g_{\ell j}\}$ appears in at most one block;
7. If $1 \leq h < \ell \leq k$ and $1 \leq i \leq n$, the pair $\{g_{hi}, g_{\ell i}\}$ appears in no block.

There is a natural connection between villa designs and orthogonal symmetric latin squares.

Theorem 1. *There exists a villa design $VD(k,n)$ if and only if there exist $k-2$ pairwise orthogonal symmetric latin squares of order n.*

Proof. The proof is very similar to the proof that a $TD(k,n)$ is equivalent to $k-2$ MOLS. Let $(V, \mathcal{G}, \mathcal{B})$ be a $VD(k,n)$. Given any block $B \in \mathcal{B}$ say that B contains g_{1r} and g_{2c}. For $1 \leq t \leq k-2$ construct a latin square L_t by letting $L_t(r,c) = x$ *and* $L_t(c,r) = x$ if block B contains the element $g_{(t+2)x}$ in group G_{t+2}. Also let $L_t(i,i) = i$ for $1 \leq i \leq n$. Conditions 4 and 7 of the definition of VD imply that these squares are well defined and that they are symmetric and idempotent. Condition 5 implies that each of these squares is indeed a latin square, while condition 6 ensures that these squares are orthogonal symmetric latin squares. Again, this construction is easily reversed, proving the theorem. □

Obviously, $\nu(n)$ is the maximum k such that there exists a $VD(k+2,n)$. Much of the progress in finding new lower bounds for $N(n)$ has been facilitated by the use of recursive and direct constructions for transversal designs. Villa designs are introduced here in order to help researchers make renewed progress on questions concerning $\nu(n)$. Indeed we are able for the first time to give an analogue of what is arguably the most powerful TD construction, the Wilson construction, to the symmetric latin square case. This is done in Section 5.

3 Other Equivalences

We now examine several other analogies between MOLS and POSLS. Let S be a set of $n+1$ elements (*symbols*). A *Room square* of side n (on symbol set S) is an $n \times n$ array, F, which satisfies the following properties:

1. every cell of F either is empty or contains an unordered pair of symbols from S,
2. each symbol of S occurs once in each row and column of F,

3. every unordered pair of symbols occurs in precisely one cell of F.

A Room square of side n is *standardized* (with respect to the symbol ∞) if the cell (i,i) contains the pair $\{\infty, i\}$. The existence of a pair of orthogonal symmetric latin squares A and B on the symbol set $\{1,2,\ldots n\}$ is equivalent to the existence of a Room square R of side n. Merely put the pair $\{x,y\}$ in cell (r,c) of R if $A(x,y) = r$ and $B(x,y) = c$. Similarly, to construct a Room square from a $VD(4,n)$, put the pair $\{x,y\}$ in cell (r,c) if the block B of the VD contains $g_{1x} \in G_1$, $g_{2y} \in G_2$, $g_{3r} \in G_3$ and $g_{4c} \in G_4$. Both of these constructions are easily reversed.

The notion of Room square generalizes to higher dimensions. A *Room d-cube* of side n is a d-dimensional cube of side n with the property that every 2-dimensional projection is a Room square of side n. The constructions of the previous paragraph carry over to this higher dimensional case. For example, given a set of d pairwise orthogonal symmetric latin squares $A_1, A_2, \ldots, A_d$ all on the symbol set $\{1,2,\ldots n\}$ now put the pair $\{x,y\}$ in cell $(c_1, c_2, \ldots c_d)$ of R if $A_i(x,y) = c_i$ for all $1 \leq i \leq d$.

A *graeco-latin square* of side n is an $n \times n$ array where each cell contains an ordered pair of elements and in addition, if a new square L is made from the first element from each of the cells and a second square R is made from the second element of each of the cells, then L and R are orthogonal latin squares of side n. With this definition, the existence of a graeco-latin square of side n is equivalent to the existence of a pair of orthogonal latin squares of side n. (The term graeco-latin is derived from the the fact that the first elements are often taken from the Greek alphabet, while the second elements are from the Latin alphabet).

Room squares and graeco-latin squares again evoke the similarity between these two concepts. One of the main differences is that Room squares have unordered pairs in each cell, while the pairs in the graeco-latin squares are ordered.

A further set of analogous objects are best described in graph terminology. A *1-factor* of a graph G is a set of edges from G which contain each vertex exactly once. A *1-factorization* of a graph G is a set $\mathcal{F} = \{F_1, \ldots, F_k\}$ of edge-disjoint 1-factors of G whose edge-sets partition the edge-set of G. Two 1-factorizations $\mathcal{F} = \{F_1, \ldots, F_k\}$ and $\mathcal{G} = \{G_1, \ldots, G_\ell\}$ of a graph are *orthogonal* if each 1-factor of $\mathcal{F}$ has at most one edge in common with each 1-factor of $\mathcal{G}$.

The existence of k orthogonal 1-factorizations of the complete bipartite graph $K_{n,n}$ is equivalent to the existence of k orthogonal latin squares of side n, while the existence of k orthogonal 1-factorizations of the complete graph K_n is equivalent to the existence of k orthogonal symmetric latin squares of side n.

The proof is essentially the same in both cases. Given a 1-factorization $\mathcal{F} = \{F_1, \ldots, F_n\}$ of $K_{n,n}$ construct a square $L = L(i,j)$ by $L(i,j) = x$ if the edge (i,j) is in F_x. An $n \times n$ square so constructed is indeed latin square. Given a set of k orthogonal 1-factorizations, the k latin squares formed by this procedure can be shown to be orthogonal.

When the underlying graph is K_n, in order to have a 1-factor it is necessary that n be even. So a 1-factor contains $n/2$ edges and a 1-factorization contains

$n-1$ 1-factors. Given a 1-factorization $\mathcal{F} = \{F_1, \ldots, F_{n-1}\}$ of K_n construct a square $L = L(i,j)$ by $L(i,j) = x$ and $L(j,i) = x$ if the edge $\{i,j\}$ is in F_x. Also let $L(i,i) = i$ for $1 \leq i \leq n-1$. Here we have constructed a symmetric idempotent latin square of side $n-1$. Given a set of k orthogonal 1-factorizations of K_n the k symmetric latin squares formed by this procedure can be shown to be a set of symmetric orthogonal latin squares.

We summarize our discussions in the following two corollaries.

Corollary 2. *The existence of the following are equivalent:*

1. $N(n) \geq k$,
2. *k pairwise orthogonal latin squares of side n,*
3. *A transversal design $TD(k,n)$,*
4. *A set of k pairwise orthogonal 1-factorizations of the graph $K_{n,n}$.*

Corollary 3. *The existence of the following are equivalent:*

1. $\nu(n) \geq k$;
2. *k pairwise orthogonal symmetric latin squares of side n;*
3. *A villa design $VD(k,n)$;*
4. *A set of k pairwise orthogonal 1-factorizations of the graph*
5. *A Room k-cube of side n.*

There are even more equivalent combinatorial structures. See [1, 12, 13] for further examples.

4 Finite Fields and the Upper Bounds

In this short section we briefly discuss the direct construction of MOLS and POSLS from algebraic structures like groups and fields.

The most basic construction gives a pair of orthogonal latin squares for every *odd* order n. Merely take the addition and subtraction tables in the group Z_n to get two orthogonal latin squares of side n. This construction was known to Euler [14], the first mathematician who had an interest in latin squares. There is no known comparable direct method (in Z_n or any other group) to get a pair of orthogonal symmetric latin squares.

Now let q be a prime power. In this case there is a well-known construction that gives $q-1$ MOLS of side q. For each $\alpha \in \mathrm{GF}(q)\backslash\{0\}$, define the latin square $L_\alpha(i,j) = i + \alpha j$, where $i, j \in \mathrm{GF}(q)$ where all the algebra is performed in GF(q). The set of latin squares $\{L_\alpha | \alpha \in \mathrm{GF}(q)\backslash\{0\}\}$ is a set of $q-1$ MOLS of side q.

In the case of orthogonal symmetric latin squares there is also a construction in the Galois field. The interested reader can find this construction in [13]. The theorem states that if $q = 2^k t + 1$ where t is odd, then there are t POSLS of side q. When $k = 1$, then $t = (q-1)/2$ which is one of the conjectured upper bounds for sets of POSLS of side q.

A set of $n-1$ MOLS of side n is called a *complete set* of MOLS(n) since it meets the upper bound. In this case it is known that the existence of a complete set of MOLS is equivalent to the existence of a projective plane of order n and indeed a complete set of MOLS exists for all prime power orders. Interestingly, in the case where the upper bound for the number of MOLS is met, there exists an extremely nice geometric object (the projective plane) that is equivalent to this set of MOLS. This is not known to be the case for symmetric latin squares. Not only is there no upper bound known that is attained infinitely often, but no nice geometric object is known to be equivalent to a "complete" set (whatever that might mean) of orthogonal symmetric latin squares. This is a crucial difference between the status of $N(n)$ and $\nu(n)$.

5 Splitting a Transversal Design

In this section we set up a framework for dealing with questions about $\nu(n)$ by using known constructions for MOLS and TDs. Our goal is to exploit the substantial knowledge about transversal designs to generate new discoveries concerning lower (and hopefully also upper) bounds for $\nu(n)$. This framework revolves around a connection between villa designs and transversal designs.

Let $V_1 = (V, \mathcal{G}, \mathcal{B}_1)$ be a villa design $VD(k,n)$ and let $\mathcal{B}_3$ be the set of idempotent blocks B_i (i.e. for each $1 \leq i \leq n$, B_i is the block containing the element i from each group). If there exists another villa design $V_2 = (V, \mathcal{G}, \mathcal{B}_2)$ with the property that $T = (V, \mathcal{G}, \mathcal{B}_1 \cup \mathcal{B}_2 \cup \mathcal{B}_3)$ is a transversal design $TD(k,n)$, then V_1 is called an *embeddable* villa design (an EVD). Conversely, if a idempotent transversal design $TD(k,n)$ can be decomposed into two villa designs $VD(k,n)$ (plus the idempotent lines), then the transversal design is called *splittable* (STD).

Suppose that R is a standardized Room square of side r in which the entry $\{\infty, x\}$ has been replaced by $\{x, x\}$ for every x. Wallis [29] calls R an *embedded Room square* if it is possible to order every entry of R, replacing $\{x, y\}$ by either (x, y) or (y, x), and then to add further entries in such a way that the resulting array is a graeco-latin square of side r. We have the following proposition whose proof is left as an exercise for the interested reader.

Proposition 4. *The existence of an embedded Room square of side r is equivalent to the existence of an embeddable villa design of order r and blocksize 4, an $EVD(4, r)$.*

A villa design $VD(k,n)$ has $\binom{n}{2}$ blocks while a $TD(k,n)$ has n^2 blocks so since $\binom{n}{2} + \binom{n}{2} + n = n^2$, the existence of a $EVD(k,n)$ is at least numerically possible. Indeed the following theorem proves their existence in the case when $k = 3$.

Lemma 5. *All $VD(3,n)$ are embeddable in a $TD(3,n)$.*

Proof. Let V_1 be a $VD(3,n)$. We construct V_2, also a $VD(3,n)$, which is disjoint from V_1. Let $\{g_{1a}, g_{2b}, g_{3c}\}$ be any block in V_1. Then let $\{g_{1b}, g_{2a}, g_{3c}\}$ be a block in V_2. From the properties of a villa design the blocks of V_1 are disjoint from the blocks of V_2 and the union of all these blocks, plus the idempotent blocks, gives an idempotent $TD(3,n)$. □

In the next theorem we give an infinite class of EVDs for larger blocksize. We first need some definitions. Let G be an additive abelian group of odd order n. A *starter* in G is a set of unordered pairs $S = \{\{s_i, t_i\} : 1 \leq i \leq (n-1)/2\}$ which satisfies the following two properties:

1. $\{s_i : 1 \leq i \leq (n-1)/2\} \cup \{t_i : 1 \leq i \leq (n-1)/2\} = G \backslash \{0\}$
2. $\{\pm(s_i - t_i) : 1 \leq i \leq (n-1)/2\} = G \backslash \{0\}$

Let $S = \{\{s_i, t_i\} : 1 \leq i \leq (n-1)/2\}$ and $T = \{\{u_i, v_i\} : 1 \leq i \leq (n-1)/2\}$ be two starters in G. Without loss of generality, we may assume that $s_i - t_i = u_i - v_i$, for all i. Then S and T are said to be *orthogonal* starters if $u_i - s_i = u_j - s_j$ implies $i = j$, and if $u_i \neq s_i$ for all i.

Orthogonal starters were developed to construct sets of POSLS (see [13]). In our new terminology we have: The existence of k pairwise orthogonal starters in an abelian group G of order n implies the existence of a VD(k, n). It is instructive to see how this construction works. Let $S_1, S_2, \ldots S_k$ be a set of orthogonal starters in an abelian group of order n. From each starter $S = \{\{s_i, t_i\} : 1 \leq i \leq (n-1)/2\}$ define the translate $S + \alpha = \{\{s_i + \alpha, t_i + \alpha\} : 1 \leq i \leq (n-1)/2\}$. Then from the properties of starters, given any unordered pair $\{x, y\} \subset G$ there is precisely one $\alpha \in G$ such that $\{x, y\} \in S + \alpha$. Now to construct a VD(k, n), for each $\{x, y\} \in G$ define the block $B_{x,y}$ as the set of points $\{g_{1x}, g_{2y}, g_{3\alpha_1}, g_{4\alpha_2}, \cdots g_{k+2,\alpha_k}\}$ where $S_i + \alpha_i$ is the unique translate of the starter S_i that contains the pair $\{x, y\}$. Using the conditions on orthogonality of starters, it can be shown that the set of blocks $B_{x,y}$ where $\{x, y\} \in G$ defines a VD(k, n).

The most important class of orthogonal starters are the Mullin-Nemeth starters. Suppose q is a prime power with $q \equiv 3 \pmod 4$. Let R denote the set of residues in the multiplicative group of $GF(q)$ and let N denote the set of nonresidues. Now, for every $a \in N$, define $T_a = \{\{x, ax\} : x \in R\}$. The $(q-1)/2$ starters thus obtained are termed the *Mullin-Nemeth* starters. They have the property that they are all pairwise orthogonal. Thus there exists a $VD((q+3)/2, q)$ for all prime powers $q \equiv 3 \pmod 4$. The following theorem states that this VD is indeed embeddable.

Theorem 6. *Let $q \equiv 3$ (mod 4) be a prime power. Then there exists a $EVD((q+3)/2, q)$.*

Proof. Begin with $V_1 = (V, \mathcal{G}, \mathcal{B}_1)$, the $VD((q+3)/2, q)$ that is constructed from the Mullin-Nemeth starters. We assume that the groups are labeled G_1, G_2 and G_a for each $a \in N$, the nonresidues of GF(q). For a block $B \in V_1$, define $-B$ by $g_{i(-x)} \in -b$ if $g_{ix} \in B$ and let $\mathcal{B}_2 = \{-B | B \in \mathcal{B}_1\}$. We show that $(V, \mathcal{G}, \mathcal{B}_1 \cup \mathcal{B}_2 \cup \mathcal{B}_3)$ is an idempotent TD$((q+3)/2, q)$, where $\mathcal{B}_3$ is the set of idempotent lines.

Let $a \in N$. For every $x \in R$ there is exactly one block B_x of V_1 that comes directly from the Mullin-Nemeth starter T_a (and not a translate of it). The block B_x contains the points $g_{1x}, g_{2(ax)}$ and for each $c \in N$ it contains $g_{c\alpha_c}$ where $T_c + \alpha_c$ is the unique translate of the starter T_c that contains the pair

$\{x, ax\}$. Let $\mathcal{B}_{T_a}$ denote this set of blocks. Then $\alpha_a = 0$ for all blocks $b_x \in B_{T_a}$; in other words, all blocks in $\mathcal{B}_{T_a}$ contain the point g_{a0}.

Let $c \in N$. We must show, given any $u \neq v$, that exactly one block from either V_1 or V_2 contains $\{g_{au}, g_{cv}\}$. Now, all blocks of $\mathcal{B}_1$ are translates of the blocks in $\mathcal{B}_{T_a}$ and all blocks in $\mathcal{B}_2$ are translates of the blocks $\{-B|B \in \mathcal{B}_{T_a}\}$. So, it suffices to show that if a block $B_x \in \mathcal{B}_{T_a}$ contains the point $g_{c\omega}$, then there is no block in $\mathcal{B}_{T_a}$ that contains the point $g_{c(-\omega)}$.

The definition of α_c ensures that $\{x, ax\} = \alpha_c + \{z, cz\}$ for some $z \in R$; again, the block B_x contain the point $g_{c\alpha_c}$. Let B_y $(y \in R)$ be another block from the Mullin-Nemeth starter T_a. Since x and y are both quadratic residues, there exists another residue $r \in R$ where $xr = y$. So $\{y, ay\} = \{xr, axr\} = r\alpha_c + \{rz, crz\}$. Since $r\alpha_c$ is in the same residue class as α_c, the points in group c that are contained in blocks from $\mathcal{B}_{T_a}$ are precisely either the set of residues or the set of nonresidues. In either case, (since -1 is not a residue) if a block $B_x \in \mathcal{B}_{T_a}$ contains the point $g_{c\omega}$, then there is no block in $\mathcal{B}_{T_a}$ that contains the point $g_{c(-\omega)}$. The proof is complete when the conditions on groups 1 and 2 are checked. This is straightforward and is omitted here. □

It can be shown further that the idempotent TD$((q+3)/2, q)$ constructed in Theorem 6 can itself be extended to an idempotent TD(q, q) — an affine plane of order q. The $q-2$ idempotent MOLS that correspond to this plane can each be described by the equation $L_a(x, y) = ax + (1-a)y$ where the square L_a corresponds to the the ath group of the transversal design. The original villa design and its negative are embedded in the rows indexed by the nonresidues of $GF(q)$.

The basic motivation for the study of EVDs or STDs is to develop a general machinery which can, for example, apply standard recursive constructions for TDs to generate STDs, and hence to obtain new VDs and Room cubes. Indeed, the following theorem is a variant of Wilson's celebrated recursive construction for transversal designs. In this case, however, when certain ingredient TDs are splittable, so is the resulting design. In this way, we translate theorems on TDs into theorems on Room cubes.

Again we need some definitions in order to proceed. We must define the notion of holes in both TDs and VDs. For transversal designs it has been well studied; see [8]. A TD$(k, n+u; u)$ is a triple $(V, \mathcal{G}, \mathcal{B})$, where

1. V is a set of $k(n+u)$ elements;
2. $\mathcal{G}$ is a partition of V into k classes (called *groups*); each of size $n+u$ (in each group the elements are labeled $1, \ldots, n, \infty_1, \ldots \infty_u$);
3. $\mathcal{B}$ is a collection of k-subsets of V (called *blocks*);
   every unordered pair of non-infinity elements from V is either contained in exactly one group, or is contained in exactly one block, but not both;
5. every infinity element and every non-infinity element from different groups are in exactly one block;
6. no block contains two infinity elements.

A VD$(k, n+u; u)$ is defined similarly. Essentially it has the same properties as a VD, except that two infinity type elements occur together in no blocks. The meaning of STD$(k, n+u; u)$ and EVA$(k, n+u; u)$ is also clear.

Theorem 7.

1. *If there exists a* $STD(k+1, n)$, *a* $STD(k, m+u_i; u_i)$ *for* $1 \leq i \leq n$, *and a* $STD(k, u)$ *where* $u = \sum u_i$, *then there exists a* $STD(k, mn+u)$.
2. *If there exists a* $EVD(k+1, n)$, *a* $EVD(k, m+u_i; u_i)$ *for* $1 \leq i \leq n$, *and a* $EVD(k, u)$ *where* $u = \sum u_i$, *then there exists a* $EVD(k, mn+u)$.

Proof. Obviously these two statements are equivalent. We prove (1). Without the requirement that the TD be splittable, this statement is the simplest variant of Wilson's theorem beyond simple multiplication. A proof of Wilson's theorem is given in [6]; the statement here employs the $STD(k+1, n)$ as the master design, the $STD(k, m+u_i; u_i)$ for $1 \leq i \leq n$ as ingredient designs in the inflation, and the $STD(k, u)$ to fill the hole which results.

Now the split of the $STD(k+1, n)$ partitions the transversal design into $\mathcal{B}_1$, $\mathcal{B}_2$, and $\mathcal{B}_3$ where $\mathcal{B}_1$ and $\mathcal{B}_2$ are $VD(k+1, n)$s and $\mathcal{B}_3$ is the set of idempotent blocks. To split the final design, we find the $EVD(k, mn+u)$ which it contains. First, for each block in $\mathcal{B}_2$, erase all blocks which arise from the inflation of the block. Second, for each block in $\mathcal{B}_3$, employ the fact that the $STD(k, m+u_i; u_i)$ is splittable to retain only an $EVD(k, m+u_i; u_i)$ from the blocks obtained in the inflation. Finally, employ the fact that the $STD(k, u)$ is splittable to retain only an $EVD(k, u)$ from the blocks obtained in the inflation. All blocks obtained in the inflation of blocks from $\mathcal{B}_1$ are retained. It is now easy to verify that this is a $VD(k, mn+u)$, and that interchanging the roles of $\mathcal{B}_1$ and $\mathcal{B}_2$, and retaining the alternate $EVD(k, m+u_i; u_i)$ and $EVD(k, u)$, we obtain a disjoint $VD(k, mn+u)$ which supplies the embedding. □

In the previous theorem, it is not necessary that the VDs be embeddable. Of course, if the ingredients are not embeddable then neither is the resulting design. The following is a Wilson-type construction for VDs.

Theorem 8. *If there exists a* $VD(k+1, n)$, *a* $VD(k, m+u_i; u_i)$ *for* $1 \leq i \leq n$, *a* $TD(k, m+u_i; u_i)$ *for* $1 \leq i \leq n$, *and a* $VD(k, u)$ *where* $u = \sum u_i$, *then there exists a* $VD(k, mn+u)$.

Naturally, the previous two theorems extend to more sophisticated applications of Wilson's theorem. However, we content ourselves here with this basic one.

6 Conclusion

MOLS and transversal designs have been extensively explored, and difficult and elegant construction techniques have been found. POSLS and Room cubes have not been attacked with the same success, or with the same diligence. We believe that one reason for this is the difficulty of exploiting results on transversal designs in the symmetric case. For this reason, we have introduced villa designs (designs with many "rooms"), and shown that one of the more powerful recursive techniques for MOLS, Wilson's theorem, can now be employed for a restricted class of sets of POSLS. The eventual value of this "translation", by embedding villa designs into transversal designs, remains very much to be seen. Nevertheless, since examples of embeddable villa designs, and a powerful recursive construction, have both been found here, we expect that a study of Room cubes via this embedding approach will prove fruitful.

Acknowledgments

Thanks to Alan Ling and Dan Archdeacon for valuable observations on this research.

References

1. R.J.R. Abel, A.E. Brouwer, C.J. Colbourn, and J.H. Dinitz, Mutually orthogonal latin squares, in: *CRC Handbook of Combinatorial Designs* (C.J. Colbourn, J.H. Dinitz; eds.), CRC Press, Boca Raton FL, 1996, pp. 111-142.
2. R.J.R. Abel and Y.W. Cheng, Some new MOLS of order $2^n p$ for p a prime power, *Austral. J. Combin.* 10 (1994), 175-186.
3. R.J.R. Abel and D.T. Todorov, Four MOLS of order 20, 30, 38 and 44, *J. Combinat. Theory (A)* 64 (1993), 144-148.
4. T. Beth, D. Jungnickel and H. Lenz, *Design Theory*, Cambridge University Press, Cambridge, 1986.
5. R.C. Bose, S.S. Shrikhande and E.T. Parker, Further results on the construction of mutually orthogonal latin squares and the falsity of Euler's conjecture, *Canad. J. Math.* 12 (1960), 189-203.
6. A.E. Brouwer and G.H.J. van Rees, More mutually orthogonal latin squares, *Discrete Math.* 39 (1982), 263-281.
7. C.J. Colbourn, Four MOLS of order 26, *J. Comb. Math. Comb. Comput.* 17 (1995), 147-148.
8. C.J. Colbourn and J.H. Dinitz, Making the MOLS table, in: *Computational and Constructive Design Theory*, (W.D. Wallis, ed), Kluwer Academic Publishers, Boston, 1996, 67-134.
9. J. Dénes and A.D. Keedwell, *Latin Squares: New Developments in the Theory and Applications*, North-Holland, Amsterdam, 1991.
10. J.H. Dinitz, Room $n-$cubes of low order, *J. Austral. Math. Soc. (A)* 36 (1984), 237-252.
11. J.H. Dinitz, The existence of Room 5-cubes, *J. Combinat. Theory (A)* 45 (1987), 125-138.
12. J.H. Dinitz, Room Squares, in: *CRC Handbook of Combinatorial Designs* (C.J. Colbourn, J.H. Dinitz; eds.), CRC Press, Boca Raton FL, 1996, pp. 437-442.

13. J.H. Dinitz and D.R. Stinson, Room squares and related designs, in: *Contemporary Design Theory: A Collection of Surveys* (J.H. Dinitz and D.R. Stinson, eds), John Wiley & Sons, New York, 1992, 137-204.
14. L. Euler, Recherches sur une nouvelle espèce de quarrés magiques, *Verh. Zeeuw. Gen. Weten. Vlissengen* 9 (1782), 85-239.
15. K.B. Gross, R.C. Mullin and W.D. Wallis, The number of pairwise orthogonal symmetric latin squares, *Utilitas Math.* 4 (1973), 239-251.
16. T.P. Kirkman, On the perfect r-partitions of $r^2 - r + 1$, *Transactions of the Historic Society of Lancashire and Cheshire* (1850), 127-142.
17. D.M. Johnson, A.L. Dulmage and N.S. Mendelsohn, Orthomorphisms of groups of orthogonal latin squares, *Canad. J. Math.* 13 (1961), 356-372.
18. W.H. Mills, Some mutually orthogonal latin squares, *Proc. Eighth S.E. Conf. Combin. Graph Theory Computing*, 1977, pp. 473-487.
19. R. C. Mullin and E. Nemeth, An existence theorem for Room squares, *Canad. Math. Bull* 12 (1969), 493-497.
20. A.V. Nazarok, Five pairwise orthogonal latin squares of order 21, *Issled. oper. i ASU*, 1991, pp. 54-56.
21. D. Raghavarao, *Constructions and Combinatorial Problems in Design of Experiments*, John Wiley & Sons, New York, 1971.
22. C.E. Roberts Jr., Sets of mutually orthogonal latin squares with 'like subsquares', *J. Combinat. Theory (A)* 61 (1992) 50-63.
23. C.E. Roberts Jr., Sets of mutually orthogonal latin squares with 'like subsquares' II, preprint.
24. P.J. Schellenberg, G.H.J. van Rees and S.A. Vanstone, Four pairwise orthogonal latin squares of order 15, *Ars Combinat.* 6 (1978), 141-150.
25. G. Tarry, Le problème de 36 officiers, *Ass. Franc. Av. Sci.* 29 (1900), 170-203.
26. L. Teirlinck, Orthogonal one-factorizations, ordered designs and related structures, in: *Coding Theory, Design Theory, Group Theory*, (D. Jungnickel, S.A. Vanstone, et al, eds.), John Wiley & Sons, New York, 1993, 77-106.
27. D.T. Todorov, Three mutually orthogonal latin squares of order 14, *Ars Combinat.* 20 (1985), 45-48.
28. D.T. Todorov, Four mutually orthogonal latin squares of order 20, *Ars Combinat.* 27 (1989), 63-65.
29. W.D. Wallis, A.P. Street and J.S. Wallis, *Combinatorics: Room Squares, Sum-Free Sets, Hadamard Matrices*, Springer-Verlag, Berlin, 1972.
30. S.P. Wang, *On self-orthogonal latin squares and partial transversals of latin squares*, Ph.D. thesis, Ohio State University, 1978.
31. R.M. Wilson, A few more squares, *Proc. Fifth Southeastern Conf. Combin. Graph Theory Computing* (1974), pp. 675-680.
32. M. Wojtas, Five mutually orthogonal latin squares of order 35, *J. Combin. Designs* 4 (1996), 153-154.
33. M. Wojtas, Six Mutually Orthogonal Latin Squares of Order 48, preprint, 1996.

Forbidding and Enforcing
(abstract)

A. Ehrenfeucht[1] and G. Rozenberg[2]

[1] Department of Computer Science, University of Colorado at Boulder, Campus 430, Boulder, CO 80309, U.S.A.
[2] Department of Computer Science, Leiden University, P.O. Box 9512, 2300 RA Leiden, The Netherlands

Formal language theory is perhaps the most mature area of theoretical computer science. Defining languages by (generative) grammars and (accepting) automata is very well understood by now.

The prevalent Old Axiom of formal language theory is:

everything that is not (explicitly) allowed is forbidden.

Thus, for example, if a language L is generated by a grammar G, then unless a word is listed (allowed) by G it is not in L (it is forbidden). The same philosophy holds for automata.

In some applications (such as in the now very fashionable area of DNA computing) this axiom does not work very well. We propose a New Axiom:

everything that is not (explicitly) forbidden is allowed.

In our talk we discuss a new approach to formal language theory that obeys the New Axiom. In particular we propose two mechanisms for defining languages: *forbidding* and *enforcing*. Some applications are also discussed.

Constructive Programming: A Personal View

Susumu Hayashi*

Department of Computer and Systems Engineering,
Faculty of Engineering,
Kobe University,
1-1 Rokko-dai, Nada, Kobe, Japan

Abstract. Constructive programming is a kind of formal method for functional programming; its aim is to develop verified programs through the creation of formalized proofs that use constructive mathematics. This paper is a personal perspective of the research area. It consists of four parts: a brief account of formal methods, an account of constructive programming, correspondences between these two areas, and a conclusion. An appendix contains some recent thoughts that were added after the body of the paper was finished.

1 What are formal methods?

Constructive programming is a kind of formal method for functional programming whose aim is to develop verified programs through the creation of formalized proofs that use constructive mathematics. Formal methods comprise that part of software engineering that employs formal logic to certify that programs have the properties expected of them.

Let me give an example. Suppose you are developing a library function `div`(a, b) for a commercial system. The expected output is a pair comprising the quotient q and the remainder r of the division of the integer a by a non-zero integer b. The function `div` may be used for some important purposes. One day, a client shows up and asks, "I am going to use your function for bank accounting. If the calculation is incorrect, my boss will fire me! Can you guarantee that your `div` function is correct?" Can you?

The conventional way to certify programs correct is by *testing* them with sample inputs. But testing is finite. The function `div` is supposed to support arbitrary precision numbers. Thus the space of possible inputs is infinite. Even if the inputs were finite, testing may not be good enough, for there are many finite problems that cannot be solved—even by supercomputers—because of their massive size.

How can we certify that such programs are correct? The only known way is through the use of *formal methods*. Formal methods employ formal logic or some related proof discipline to *prove* the correctness of programs, just as we prove

* Supported by Nod868367, Monbusyo, Kaken-hi, (the aid of Scientific Research, The Ministry of Education).

the correctness of theorems in mathematics. Thus they can handle infinite input spaces.

Many logics have been developed to prove programs correct. The first, and oldest, is Hoare logic. Early attempts to prove correctness of programs by such logics are called *program verification*. Program verification is known to have *failed* to achieve its goal. Proving the *complete* correctness of programs is a much more difficult task than was expected. Though possible in theory, it is impossible in practice.

Nevertheless, since the late 80's there have been many applications of "program verification," applications which fall within the scope of formal methods. Often they do not involve proving the correctness of existing programs completely: rather, they use program logic to *increase* the correctness of programs as they are developed. A typical example is the development of a new version of the CICS transaction processing system at IBM Hursley Laboratory, Britain in which some parts were specified in the Z language (see [5] for more details). Z is a system of notation for expressing typed formulas of Zermelo set theory [14]. Although specifications were written, no proof was constructed. Nonetheless, bug reports from users on the parts of the system that had been specified in Z decreased by 60%. This is reminiscent of the experience of mathematicians in areas like proof theory or abstract algebra that once a theorem is formulated correctly, the task is often half finished.

Other applications of formal methods can be found in a recent survey [5]. [19] gives a precise account of formal methods from a general software engineering point of view. Many textbooks on Z are now available, e.g. [14].

2 What is constructive programming?

Hoare logic and other logics for program verification take the approach of "first write the program; then verify it." They use logic to prove the correctness of existing programs. In the case of the `div` function, `div` is written first and then we prove that the result, $\mathtt{div}(a, b)$, is a pair comprising the quotient and remainder. Although different languages used in formal methods express this in various ways, it is essentially contained in the following logic formula:

$$\forall a : N.\forall b : N^+.\exists q : N.\exists r : N.(\langle q, r\rangle = \mathtt{div}(a, b) \wedge a = b * q + r \wedge 0 \leq r < |b|) \quad (1)$$

N is the set of natural numbers (including zero), and N^+ is the set of non-zero natural numbers. The task is (i) writing the program `div` and then (ii) proving the formula (1).

Constructive programming takes another tack. First, we formulate the goal as follows:

$$\forall a : N.\forall b : N^+.\exists q : N.\exists r : N.(a = b * q + r \wedge 0 \leq r < |b|) \quad (2)$$

Then we prove this using a proof checker that works within a formal system of constructive mathematics. The proof is expressed as a formal derivation. Next we apply a program extractor to it, and this automatically produces a function. If we name it `div`, then it satisfies the property expressed in (1). The program is correct so long as the proof checker and extractor are correctly implemented.[2] We do not include the actual development here: some example developments for the `div` function can be found in [8].

2.1 Advantages and disadvantages

Developing a constructive proof corresponds to writing a program and verifying it *at the same time*. This contrasts sharply with the "first program, then verify" paradigm. The difference implies advantages and disadvantages of the constructive programming paradigm. For brevity, we abbreviate the constructive programming paradigm to CP, and the "first program, then verify" paradigm to PtV.

Advantages. The advantages of CP over PtV are as follows.

1: It operates in a single integrated pass.
2: It is more abstract.
3: It uses only ordinary logic.

In the PtV paradigm, the task consists of two passes. A program and its correctness proof are often "isomorphic." When writing a program, every programmer does some kind of "reasoning" about why it works. In PtV, the programmer—or someone else—verifies it afterwards by reconstructing the reasoning in a more precise way: thus the reasoning is repeated. In CP, such repetition is unnecessary. This is the first advantage.

Another feature of constructive programming is its abstract nature. A proof represents not a concrete program but the idea behind it. There are many ways of extracting programs from a proof. It is the extractor that makes concrete the proof's abstract computational content. We can make extractors that "compile" proofs into various programming languages, and even extract different programs from them. This is the second advantage.

Program logics like Hoare logic are specialized for program verification. But normally such a logic includes within it an ordinary logic like predicate logic. For example, to implement a proof checker in Hoare logic, two levels are involved: (i) Hoare's axioms and rules to derive the properties of program constructs, and (ii) predicate logic to prove the verification conditions. In CP, these are integrated into a single logic: constructive logic. This is the third advantage.

Disadvantages: The disadvantages of CP are as follows:

1: It is indirect.

[2] Critics doubt whether we can implement proof checkers and extractors correctly. Our experience indicates that these tools become very reliable after some years of use.

2: It is very abstract.
3: It is difficult because one must prove first, whereas most people prefer programming to proving.

In CP, you write a proof and then the extractor produces a program. This indirection step means that the program might not be the one you expected. This is the first disadvantage.

Real programming is concrete. Programmers take account of such things as hardware architecture to write high-performance programs. Abstract programs do not reflect this mindset.

For ordinary people, constructing proofs is more difficult than constructing programs. Even programmers who know nothing about verification can write a program and leave the task of verification to experts in formal methods. But CP binds programming and verification together so tightly that programmers must make proofs.

2.2 Developments in constructive programming

I began research in this area in 1981. Soon I started to develop a system called PX, for "Program eXtractor"; this line of work was completed in 1987 with the publication of the monograph [8]. PX was the first large-scale system on which real programs could be built using constructive proof development. Details of the PX system, and the history of the project up to 1987, can be found in [8] and will not be repeated here. (It should be noted that in [8] the area was called "program extraction"—the term "constructive programming" was coined at a later date by Masahiko Sato, a pioneer in the area.) The development of CP after the PX project will be reviewed below.

Optimization. The main feature of PX was its "optimization" method for codes. This consisted of two techniques:

1. specifying the structure of recursion by inductive definitions,
2. eliminating irrelevant "computational" aspects of proofs from programs.

The first technique, called *CIG-induction*, made it possible to extract efficient recursive programs, demonstrating that inductive definition is a central device for CP. It was later generalized by Tatsuta and Kobayashi [11].

The second technique was called PX-realizability with *rank 0 formula*. Its origin is the concept of Harrop formulae. This idea was taken by C. Paulin [13] in the Coq project [4] and extended in an elegant way to higher-order type theory.

My and Paulin's methods make it possible to identify redundant parts of a specification. For example, (2) dictates that `div` should return both quotient and remainder. But if the remainder alone is required, this can be expressed in PX as follows:

$$\forall a : N.\forall b : N^{+}.\exists r : N.\Diamond\exists q : N.(a = b * q + r \land 0 \leq r < |b|) \qquad (3)$$

The $\diamondsuit$ operator specifies that the computational content of the rest of the formula is unnecessary. $\diamondsuit$ is essentially a double negation, and so (3) is directly provable from (2). But with this derivation the program extracted will be $proj_2(\mathtt{div}(a,b))$, which computes both quotient and remainder and then discards the former. However, the remainder is computable without computing the quotient. To change the program to accomplish this, it is necessary to *reprove* formula (3) directly.

Takayama [20] showed that the reproving step is unnecessary. He introduced the new idea of an "extended projection method" (EPM) which automatically extracts the program computing only the remainder from the proof of (2). It was later shown that EPM can be thought as translating a proof of (2) into a direct proof of (3).

S. Berardi [2] further developed this idea, showing that simple types are sufficient for such optimization. EPM analyzes proofs to determine computationally unnecessary parts. Then, a program is extracted based on this analysis. Berardi showed that simple types are sufficient to accomplish the same optimization. He showed that simply-typed functional programs have the optimally simplified form with respect to an optimization order, and this can be computable by an algorithm similar to Takayama's.

Thus one may extract a *typed* program from a proof of the specification (2) and forget about the proof completely. When we wish to compute only the remainder, we optimize the program $proj_2(\mathtt{div}(a,b))$ by Berardi's method. The result is essentially the same as the program extracted by Takayama's method.

My student K. Shii and I are developing a new version of the PX system to realize Berardi's idea. Since Berardi's method needs types and PX was an untyped system based on LISP, we are changing the PX system so that its terms are typed in the same way as Z. Then all the terms are "typed LISP programs," and the new PX system can develop typed LISP programs via proofs. The realizability of the original PX system was a kind of Kleene's q-realizability, and it is changed to a kind of Kreisel's modified realizability (à la q-realizability). Whereas the original extractor does many optimizations, the new version does not do any optimization, but extracted programs are optimized by Berardi's method. Although the new version is not yet complete, we expect it to do essentially the same optimization as the original version. Since the original version does some *ad hoc* optimizations not considered by Berardi, they must be handled separately.

Elimination of tuples. Programs extracted are pure functional, and pure functional programs tend to carry around tuples corresponding to "state" in imperative programming. Since tuples were implemented as lists in the extracted programs of the PX system, this caused allocation of temporary cells and thus garbage collection. But this is not an essential disadvantage of CP. The mvPX system, described in [10], shows that such records can be replaced by multiple values in Common Lisp. An experiment in [10] shows that temporary cell allocation never happens if programs extracted by mvPX are executed by a standard implementation of Common Lisp, since the multiple values are allocated as "segments" on the stack.

Control. The theoretical basis of CP is Curry-Howard isomorphism or a related principle like realizability interpretation. Since these principles assume pure functional programs, the extracted programs were limited to such programs.

But Griffin [7] and others extended the Curry-Howard isomorphism to correspondences between classical logic and a control mechanism called "continuation." It does not seem that there has yet been any attempt to apply the new correspondence to extract *real* programs yet, but it may be possible to do so. It will be very interesting to see whether one can really write natural programs with continuations using classical logic.

On the other hand, H. Nakano [12] showed that propositional LK, with a limited implication introduction rule such that the right hand side of a sequent has at most one formula, corresponds to a simply-typed call-by-value lambda calculus with a catch/throw mechanism. His correspondence is very natural, and one can easily write programs using the catch/throw mechanism via proofs. An extension of the PX system using a variant of Nakano's logic has been implemented, and it has been shown that programs with catch/throw can be written by natural proofs with exceptions [10].

Other work. The work described above is only that in which I myself was somehow involved. There is much other interesting work in CP.

S. Kobayashi is developing a theory integrating CP and Moggi's theory of monad and Pitts' evaluation logic. It will implant various computational notions, e.g. assignment, concurrency, etc, into CP in a standard way.

Sato's group have done interesting work in CP. Sato introduced a series of systems for CP: Qute[15], Quty[16], SST[17], SPT [18]. Unlike PX, these systems were not intended to extract conventional programs (PX extracts LISP programs). The target languages were specially designed experimental languages. These systems, including the experimental languages, were implemented by Sato's students and coworkers.

As mentioned earlier, Tatsuta extensively studied the induction principle for CP [21]. He also showed that coinduction in CP produces programs with streams. Kameyama further extended the induction principle so that the generation of type systems can be formalized.

CP has been mainly worked on in Japan, but there is some related work in Europe and the USA. These include projects building type-theoretic proof checkers based on Curry-Howard isomorphism. Notable examples are Nuprl [3], Coq [4] and Alf [1]. In early articles on these systems, program extraction is included as one of their aims and some work was done by Sasaki and Paulin, etc. But these systems are now considered to be proof checkers rather than an environment for CP.[3]

[3] Footnote added on Oct. 2 1996: After finishing this paper, I learned this is not quite true for the Coq project; see the Appendix.

2.3 Constructive programming is possible

CP was believed to be impossible. It was believed that programs cannot be expressed naturally by proofs, because to do so is so indirect. Extracted programs were believed to be very inefficient. I remember the first time I heard about CP. Sato told the audience that he had extracted the factorial function and that *fact*(3) or something took some minutes to compute! Now we can extract a LISP program for the factorial function which is just as efficient as we expect. Now we can write programs with exceptions. Assignment is still missing, but it can be simulated by functional programs as shown in [8]. There are also studies by Sato and Kobayashi suggesting that it may be included into the framework of CP.

After the developments in CP mentioned above, I am quite sure that CP can be made industrially strong. It could be used to produce verified and safe software, usable in the real world. The cost of developing verified programs will be the same as for other formal methods.

Nevertheless I do *not* think that we should start to apply CP in the industry. I will explain why in the rest of the paper.

3 Correspondences

In this section, I will point out that developments of CP are parallel to developments in other formal methods, especially the works by those working on the Z notation. The correspondence between CP and other methods is the main reason why I think CP is not indispensable. A precise explanation of this correspondence is beyond the scope of this short survey and will be published somewhere else. Here, I will give only the general idea.

3.1 Refinement

The first advantage claimed above for CP was the integration of programming and verification. This is also true for "refinement," which is now popular in the PtV paradigm. The idea originates in Dijkstra's work on the weakest precondition [6] and is inherited in the Z notation and other formal methods.

Let us compare developments in Z and in CP (via PX system). In Z, a specification is given by a structured logical formula called a *schema*, which specifies the relation between the "input and pre-state" and the "output and post-state." A schema *Div* for `div` function is as follows:

Div

$a?, b?, q!, r! : N$

$b? \neq 0$

$a = b * q + r \wedge 0 \leq r < |b|$

Since `div` is a pure function, it does not have any state. Thus *Div* specifies only the relation between the input $a?$ and $b?$, and the output $q!$ and $r!$. The schema is a structured presentation of the following formula:

$$a?, b?, q!, r! : N \wedge b? \neq 0 \wedge (a = b * q + r \wedge 0 \leq r < |b|) \tag{4}$$

By definition, the domain of the inputs of *Div*, "pre *Div*" in Z notation, is as follows:

$$\exists q!, r!. \{a?, b?, q!, r! : N \wedge b? \neq 0 \wedge (a = b * q + r \wedge 0 \leq r < |b|)\} \tag{5}$$

The specifier (and/or verifier) is obliged to show that pre *Div* is the *intended* domain, that is, $a? : N$, $b? : N^+$.

To develop a `div` program (function) by refinements, a series of schemas $Div = S_1, S_2, \ldots, S_n$ refining *Div* will be defined. A schema C refines another schema A iff C is more concrete than A. C is more concrete, if it is closer to the program (implementation). (See [14] for the exact definition.) The specification S_1 is gradually refined (rewritten) to a program S_n. Each intermediate S_i is *either* a more concrete specification *or* a mixture of specification and program.[4]

Refinement is a constructive proof. Now, let us explain a development of the division function in 4.3.2.1 of [8]. Two developments are given there. The second, from the end of p.78, is the most typical example of the development methodology that I call "CIG-methodology." It enabled PX to extract more efficient programs than the other systems for CP. In the development, a "domain" $D(b)$ of a of the recursion

$$foo(a, b) = \textbf{if } a < b \textbf{ then } \cdots \textbf{ else } \cdots foo(a - b, b) \cdots \tag{6}$$

is inductively defined: (i) if $a < b$ then $a : D(b)$, and (ii) if $a \not< b$ and $a - b : D(b)$ then $a : D(b)$. Then, by the induction on $D(b)$, the following new specification is proved for $a : D(b)$, $b : N^+$:

$$\exists q!. \exists r! : Z.(a = b * q + r \wedge 0 \leq r < |b|) \tag{7}$$

The new specification corresponds to a refinement by giving an inductive definition of the input domain. Proving it by the induction rule corresponds to an application of the refinement rule of repetition (see [14]). More precisely, it corresponds to a refinement rule of repetition based on Theorem 4 of [6].

A schema C refining A must satisfy "pre $A \Rightarrow$ pre C" by the definition of refinement. In our case, it corresponds to showing that $a : N \wedge b : N^+ \Rightarrow a : D(b) \wedge b : N^+$, which was mentioned as a "termination condition" in [8]. Note that the termination condition is "necessary" to derive the original specification (2) from the refined specification (7), and this corresponds to the condition "pre $A \Rightarrow$ pre C."

[4] The official Z only allows formulas (specifications, as we call them) as schemas. Abrial's B-method and Morgan's refinement calculus defines formal languages that intertwine logical formulae and programs.

To prove (7) by induction, we use some lemmas. These correspond to components of further refinements. As the repetition refinement rule corresponds to the induction rule, there is a good correspondence between some refinement rules and proof rules of predicate logic (see the table in the proof of Theorem 1, Chapter 4 of [8]).

A refinement can be considered as a series of "lemmas." A series of lemmas is almost a proof. Thus a refinement corresponds to the proof structure. In CP it is indeed presented as a constructive proof.

Refinement is abstract. The second advantage of CP identified above is its abstractness. Refinements can, however, be very abstract. Refinement rules may be divided into two kinds. One is for further refinement of specifications (as in the step from (7) from (4)), and the other is for introducing (abstract) program constructs such as repetition (as the proof of (7) by induction rule). The latter corresponds to computationally meaningful inference rules, and the extraction algorithm for them.

In CP, we tend to prove specification directly by computationally meaningful inference rules. This means that the latter kind of refinement rules are applied at early stages of refinement. It is discouraged in refinement methodologies, since concrete design decisions at early stages often affect later developments. Refinement methodology tries to refine specifications so that their computational meaning becomes clear. Then they are transformed to programs. Such an approach is taken less often in CP, although it is possible to do so.

Furthermore, in CP with PX and other systems, data reification (refinement of implementation of data structures) is not considered. PX and other systems of CP are so tightly related to particular programming languages that they are formal theories of particular programming languages. Thus the idea of data reification does not fit this kind of systems.

We can thus conclude that refinement methodology is even more abstract than CP.

Refinement is done by ordinary logic. The third advantage of CP above was that it can be done by ordinary logic alone. Refinements can also be developed by ordinary logic alone. Z books encourage one to refine schemas as far as possible so that the translation of schema to programs become trivial. If such an approach is taken, the main tasks are accomplished in ordinary logic. Recall that Z is more or less ordinary set theory, and can be used as practical specification logic.

3.2 Animation and testing proofs

A technique called "animation" is known in Z-like formal methods. It is to "execute" specifications for testing and validation. In PX system, a program can be extracted before a proof is completed and execute it. If the extracted program has a bug, it means that the unfinished proof is going wrong because of some bad definitions, specifications or lemmas. In this way, an unfinished proof can be tested. This is quite an efficient way to find errors, when you are going into a wrong way. Indeed the large proof developments with PX system have been done in this way. Many bugs were found in this way, in the course of the development of a propositional theorem prover in the Appendix B in [8]. After fixing these bugs, only one bug was found by verification of hypotheses by a proof checker.

3.3 Z and ATTT

We may say that "refinement is doing logical inference and its program extraction explicitly at the same time." The realizability interpretation is the technical device for program extraction (of PX and other Japanese systems), and type theory is known to be a logical device to carry out realizability interpretation explicitly in a structured way. Thus development of programs in type theory can be similar to refinement approach.

In fact, my type theory ATTT [9], which was designed for CP after the experience with PX system, is remarkably similar to Z. A type of ATTT specifies a type in the ordinary sense and its subset. These correspond to the declaration part and the predicate part of Z-schema. Usages of singleton to define sets is common. The type of ATTT is even called "refinement type." And I did not know about Z, when I designed it. The name "refinement type" was borrowed from a work on ML-language and not from refinement of formal methods. The difference is that Z is for conventional imperative language and ATTT is for polymorphic second order functional language.

In ATTT, just as in PX, programs were restricted to executable programs. Nakano pointed out that it should be extended to non-executable programs as well. If we do so, ATTT will become something like Abrial's B-language. It will be noteworthy that Sato's Quty [16] had such non-executable programs, although it is not clear Sato intended to use them as specifications.

4 Conclusion

In the preceding section, we considered correspondences between CP and other formal methods. Now we draw some conclusions.

CP has been well developed, even though only a few researchers took the subject seriously. We can be proud of it. On the other hand, it is our own fault that for a long time we were ignorant of work outside CP. Now we know that there is a good correspondence between CP and other formal methods. Generally speaking, the capability of CP is not very different from that of other methods. Although there are some features which are found in other methods but not in CP, they can be easily added to CP. However, the refinement approach seems more flexible than CP. It is quite well developed and application has already started. There is no reason to repeat in CP what has been achieved in other methods. Thus, I conclude that

CP is possible but not indispensable.

Nevertheless, there are some areas for which CP is better suited than other methods.

1. CP fits higher-order functional languages.
2. CP might be useful as a development tool for formal proofs.

The higher-order type theories such as Quty and ATTT developed in CP will be useful formal development methods for higher-order functional languages. It will be particularly worthwhile to establish a method of refinement in these type theories, or in their extension following B-method.

As pointed out in Section 3.2, CP can animate incomplete proofs. This means that a formal proof with computational significance can be *tested.* If we find a bug in the test, the proof has a bug.[5] This gives an efficient way to test incomplete formal proofs. At the final stage of the development of a formal proof, we often find an error in a definition, or in the formulation of an axiom or theorem, at the very first stage, and the proof must be reworked from scratch. "Proof animation" will help in finding such mistakes at relatively early stages, as my experiences with PX demonstrate. Note that "proof animation" is restricted to constructive proofs: it is also possible to animate classical proofs. One way is to use an interactive proof animator that asks when a non-computational decision is necessary. Another is to use the recent studies of execution of classical proofs. It will be great fun to trace the execution of classical proofs.

Note added on Oct. 2 1996: After sending this paper to the editor, I discussed the subject with members of the Coq project. These discussions and their recent work led me to reconsider my conclusion above. Since I have not yet reached a new conclusion, I will include in the Appendix some issues arising out of the discussions. When I reach a new conclusion, I will publish it in some way.

References

1. Alf project: *A user's guide to ALF*, Department of Computer Science, University of Göteborg/Chalmers, June, 1994, see http://www.cs.chalmers.se/ComputingScience/Research/Logic/alf/guide.html
2. Berardi, S.: Pruning simply typed lambda terms, to appear in *Journal of Logic and Computation*
3. Constable, R.L. and others: *Implementing Mathematics with the Nuprl Proof Development System*, Prentice-Hall, 1986
4. Coq project: *The Coq Proof Assistant Reference Manual*, Version 5.10, Technical report, No. 177, INRIA, 1995, see http://pauillac.inria.fr/coq/coq-eng.html
5. Craigen, D., Gerhart, S., Ralston, T.: *An International Survey of Industrial Applications of Formal Methods*, U.S. National Institute of Standards and Technology, March 1993, Technical Report NIST GCR 93/626 (Volumes 1 and 2). Also published by the U.S. Naval Research Laboratory (Formal Report 5546-93-9582, September 1993) and the Canadian Atomic Energy Control Board, reports

[5] Note that this is the reverse of the idea of CP, which is "if a program is extracted from a (complete) proof, then it will be correct."

INFO-0474-1 (vol. 1) and INFO-0474-2 (vol. 2), January 1995. Also available at http://www.ora.on.ca/

6. Dijkstra, E.W.: Guarded Commands, Nondeterminacy and Formal Derivation of Programs, *Communications of the ACM*, 453-457, 1975
7. Griffin, T. G.: *A formulae-as-types notion of control*, Conf. Rec. ACM Symp. on Principles of Programming Languages, 47-58, 1990
8. Hayashi, S. and Nakano, H.: *PX: A Computational Logic*, The MIT Press, 1988
9. Hayashi, S.: Singleton, Union and Intersection Types for Program Extraction, *Information and Computation*, vol. 109, 174-210, 1994
10. Hayashi, S., Ishikawa, M., Kobayashi, S., Nakano, H., Nakazaki, S.: Two extensions of PX system (extended abstract), Linear Logic 96 Tokyo Meeting, Keio University, Tokyo, 1996, to appear in *Electronic Notes of Theoretical Computer Science*, Vol. 3, http://www.elsevier.nl/mcs/tcs/pc/Menu.html, Elsevier-Science Publishing
11. Kobayashi, S., Tatsuta, M.: Realizability Interpretation of Generalized Inductive Definitions, *Theoretical Computer Science* 131, 121-138, 1994.
12. Nakano, H.: *A Constructive Formalization of the Catch and Throw Mechanism*, in Proc. 7th Annual IEEE Symp. on Logic in Computer Science, 82-89, 1992
13. Paulin-Mohring, C.: *Extracting F_ω's Programs from Proofs in the Calculus of Constructions*, Proceedings of 16th Annual ACM Symposium on Principles of Programming Languages, 1989
14. Potter, B., Sinclair J., Till, D.: *An Introduction to Formal Specification and Z*, Prentice-Hall, 1991
15. Sato, M. and Sakurai, T.: *Qute: A Functional Language Based on Unification*, in Proceedings of International Conference on Fifth Generation Computer System '84, 157-165, 1984
16. Sato, M.: *Quty: A Concurrent Language Based on Logic and Function*, in J-L Lassez ed., Logic Programming, Proceedings of the Fourth International Conference, MIT Press, 1034-1056, 1987
17. Sato, M. and Kameyama, Y. *Constructive Programming in SST*, in Proceedings of the Japanese-Czechoslovak Seminar on Theoretical Foundations of Knowledge Information Processing, INORGA, 23-30, 1990
18. Sato, M.: *Adding Proof Objects and Inductive Definition Mechanisms to Frege Structures*, in T. Ito, A.R. Meyer eds., Theoretical Aspects of Computer Science, International Conference TACS'91 Proceedings, Lecture Notes in Computer Science 526, 53-87, 1991
19. Sommerville, I.: *Software Engineering*, fourth edition, Addison-Wesley, 1992
20. Takayama, Y.: *Extended Projection: a new technique to extract efficient programs from constructive proofs*, Proceedings of 1989 Conference on Functional Programming Languages and Computer Architecture, 1989
21. Tatsuta, M.: *Realizability of Inductive Definitions for Constructive Programming*, Ph.D. thesis, Tokyo University, 1993

Appendix

Having finished the paper, I had a chance to give a talk to the members of Coq project at INRIA Rocquencourt. It was based on the materials of the paper. They gave me some inspiring comments, and I learned recent interesting progresses on CP in Coq project. These had me reconsider my conclusion. I have not reached a new conclusion yet. Thus I do not change the materials of the paper. But I would like to include here two issues from discussions with Coq group which made me reconsider my conclusion. I am grateful to B. Barras, G. Dowek, G. Huet and B. Werner for their comments and discussions.

Program extracted as byproduct

Coq project has built a proof checker based on a calculus of inductive definitions. The name of the proof checker is Coq. Paulin's work mentioned in the paper was achieved in the project. Although many formal proofs have been developed in Coq, I had not known any significant programs extracted from the proofs.

The situation was changed by Barras' Master Thesis (Bruno Barras: *Coq en Coq*, Oct. 1995, INRIA-Rocquencourt). He extracted a type checker of CoC from a proof. It is considered as a type checker of Coq without inductive definitions. Proof checking of Coq is done by such a type checker with inductive definitions. The extracted program was an ML-program. Barras claims that the extracted program is more efficient than the one used in Coq. As he notes, it was quicker than Coq's as it does not include inductive definitions. Nevertheless it is fair to say that it is efficient enough.

It is a part of a project to bootstrap proof checkers. Proof checkers may have bugs. If they have bugs, certification by formal proofs can be meaningless. Thus it is hoped that proof checkers themselves are verified by bootstrapping-verification. Barras's work is a good step in this direction.

A proof checker is designed and programmed, assuming some metamathematical properties of the base logical system such as the type decidability (existence of type checking algorithm) and Church-Rosser property. If these are incorrect, the proof checker is incorrect. It is thus desirable that they be verified. In such an area, an existence proof of algorithm is often presented as a proof of decidability rather than giving an explicit algorithm. If applicable programs are extracted from such proofs as *byproducts*, it will be very useful. Barras' work suggests such a possibility.

There are some attempts of mechanized formal mathematics, e.g., MIZAR project and QED project. We may find good applications of program extraction as byproducts of such projects.

Program extraction as tool

I was reminded the importance of the approach by C. Parent of Lyon group of Coq project (Catherine Parent, Ph.D. Thesis, L'Ecole Normale Supérieure de Lyon, Jan. 1995). Her approach takes the two steps:

program $\to$ proof $\to$ verified program

Note that CP (in my sense) intends only the second step "proof $\to$ verified program." Parent has implemented a tactic which generates a proof and unproved lemmas (verification conditions) from a (unverified) program and its specification. Then a new program is extracted from the proof. The new program may be different from the program inputted, but it does the task specified as far as the verification conditions are correct. Barras used Parent's tactic in his work mentioned above.

The program inputted is used as a proof guideline as in Boyer-Moore's NQTHM theorem prover. More importantly, the program inputted can be considered as a prototype of the extracted program, as it is usable for validation by testing. It may be quite useful to debug incorrect prototype programs by combinations of testing and formal proof checking in this method. Such a validation method will need a good environment in which we can execute a program and, when a bug is found, can relate it to the location of the proof where the bug comes from. Such an environment will be very similar to an environment for proof animation. I think that development of such an environment will be very important and interesting.

Gilles Dowek pointed out to me that, in the last Logic Programming Conference, W. Biebel claimed that "Logic programming was a mistake and program extraction must be done instead." Of course, this is Gilles' version of Biebel's statement. It means that logic programs are not real programs but they are executable specifications or prototypes and thus they should be compiled to more deterministic programs. Note that this is again close to Parent's idea explained above. It might be more useful if the prototype program in Parent's method can be replaced by a logic program. Then the method will become closer to refinement and animation techniques of Z-like methods.

In the ideas illustrated above, proof extraction technology will be a pivotal *tool* for compilation or animation. It is not, however, a programming method. I may thus say that program extraction will be a useful tool, although constructive programming in my strict sense, which intends that developments of certified programs are done via proof developments, is not indispensable.

Indiscreet Applications of Discrete Mathematics

Piergiorgio Odifreddi

University of Torino, Italy
Email: piergior@di.unito.it

Mathematics, as it has developed in the modern era, is based on two fictions: infinity, and classical logic. The two were originally introduced, with a lot of misgivings, by the early Greeks: infinity was involved in the Pythagorean proof of irrationality of $\sqrt{2}$, as well as in Zeno's paradoxes, and aroused well-known scandals; classical logic was Parmenides' reaction to Heraclitus' dialectics, and its novelty was witnessed by the fact that Aristotle still needed to argue in its favor in book Γ of the *Metaphysics*, centuries after its first proposal. In due time, however, infinity and classical logic became the main tools of mathematics, as the adjective attached to the latter implies.

The success of mathematics in the applied sciences has tended to conceal the fact that both infinity and classical logic are fruitful *simplifications*, allowing a satisfactory treatment of suitably modified versions of problems otherwise unapproachable in their original form. The advent of computers has vastly widened the scope of the finitistic problems that can be solved by a direct computations, and has thus provoked the need for a Copernican revolution in mathematics: a return (as the original Copernican revolution was too, to the eliocentric system of Aristarchus of Samos) to genuinely finitistic mathematics and logics.

Two main areas involved in the preparation of the revolution are *combinatorics* on the one hand, and *intuitionistic logic* on the other. As for the latter, Gödel [11] has proved that it is indeed not a fragment, but an *extension* of classical logic: more precisely, classical logic is the fragment of intuitionistic logic dealing only with negation, conjunction and universal quantifiers, all other connectives and quantifiers being classically defined in terms of these (the other intuitionistic connectives and quantifiers are instead genuinely new introductions, not reducible to the remaining ones).

The change of perspective needed for a passage from modern to 'post-modern' mathematics is obviously still in the becoming (appropriately, since classical logic was introduced by Parmenides as a logic of Being). For example, the tension between the new concerns and the old methods are witnessed by the hybrid features of Computational Complexity (see Odifreddi [20]), in which finitistic problems of feasibility of computations are still treated by infinitistic means (in the form of asymptotic behavior, typically involved in the definitions of lower and upper bounds, as well as of complexity classes).

The switch from infinitistic to finitistic mathematics has to override an inertia expressed in the (conscious or unconscious) feeling that discrete mathematics is ugly, or it requires getting one's hands dirty. We hope to show here that this is far from being necessarily the case, by presenting a number of celebrated applications of simple finitistic methods in the applied sciences in which, reversing current general trends, highly interesting statements are proved by embarrassingly elementary proofs.

1 Biology

The most obvious concerns of biology are life and reproduction, and the most epochal achievement in the field was probably the one obtained by Crick and Watson [10], with the well-known discovery of the double-helix structure of DNA that earned them the 1962 Nobel prize for medicine. Less well-known is the influence on them of the works of two other leading scientists of the century: Schrödinger [23], and von Neumann [27].

1.1 Self-reproduction

The first piece of discrete mathematics we are going to deal with is from the latter work, and concerns the mechanism of self-reproduction: von Neumann approached the problem by analogy, using the ideas of the Fixed-Point Theorem of Recursion Theory (Kleene [17], see Odifreddi [19]). He tried to build a *self-reproducing mechanical automaton*, as follows.

First of all, he postulated the existence of a universal constructor A with the property that, when it is fed the description d_X of a machine X, it searches in the surrounding environment for the needed mechanical parts, and builds X. In symbols:

$$(A, d_X) \longrightarrow X.$$

Of course universality refers to a fixed class of machines, all particular specimen from a common species, that can be described in a uniform way. Thus A has the ability to understand all plans of a given type, and realize them. A alone is not yet self-reproducing because, when it is given its own description, it does build a copy of itself, but the result is lacking the description that was fed at the beginning:

$$(A, d_A) \longrightarrow A.$$

Von Neumann thus introduced a second machine B, with the simple task of reproducing any description given to it:

$$(B, d_X) \longrightarrow d_X.$$

To coordinate the two machines A and B, a third one C was introduced, and the compound machine $A + B + C$ will now work in the following way. Given a description d_X, A builds a copy of X, while B reproduces a copy of d_X; then the copy of X is fed the copy of d_X:

$$(A + B + C, d_X) \longrightarrow (X, d_X).$$

Then, if we call D the resulting machine $A + B + C$, we get

$$(D, d_D) \longrightarrow (D, d_D).$$

Thus $S = (D, d_D)$ is self-reproducing.

The same ideas allow not only for self-reproduction, but even for production of more complicated machines (a sort of evolutionary process). Indeed, it is enough to insert in D a description d_{D+F} of a machine composed of D and some other machine F. Then S will produce itself, together with F.

What von Neumann guessed, and Crick and Watson confirmed, is that this is an abstract, simplified representation of *genetical reproduction*: living cells contain universal constructors, basically the same for plants and animals, and only the genetical material (the program) is different. More precisely, the biological reinterpretation of the construction of S is the following: d_X works like a gene (a segment of DNA) that codifies the reproduction information; B (a special enzyme—RNA polymerase) has the function of duplicating the genetic material into a segment of RNA; A (a set of ribosomes) builds proteins by following (a segment of RNA containing) the reproduction information; S is a self-reproducing cell.

One of the simplifications occurs in the fact that the gene has only a partial codification of the reproduction information, and this allows for a partial modification of the reproduced object. The additional parts possibly produced are the analogues of enzymes that the gene produces, or whose production it stimulates. The effect of alterations on S are reminiscent of *mutations*: they may be lethal or sterilizing (killing the organism or inhibiting reproduction), produce modified (and possibly sterile) successors, or produce fertile successors with changed hereditary strain (generating different by-products).

The troublesome part in the discussion above lies in the assumption of the existence of a *universal constructor*. One might reason by analogy, and think that an argument like the one used for universal Turing machines might suffice. However, for Turing machines the actions needed for universality are of quite limited complexity (writing, reading, moving), and at the same level as those of the simulated machines. For actual constructors much more is certainly needed (recognizing, moving, manipulating and assembling components). Thus the possibility arises of an infinite regress, in the sense of needing an additional level of complexity, to build machines at a certain level.

In a later work, von Neumann [28] was able to solve the problem by abstracting some of the properties of the mechanical model seen above. He noted that constructions can be seen as series of events taking place in a space, and he thus considered—by a sort of cartesian representation—a space of cells that can be in a certain number of states (intended to represent presence or absence of certain parts of the mechanical model, so that motion of parts is represented by change of state in cells). Von Neumann thus used particular *cellular automata*, which are simply potentially infinite, directed graphs (spaces), whose nodes (cells) are finite state machines: the global behavior consists of the simultaneous and coordinated behavior (change of state) of the single cells. Von Neumann's particular automaton consisted of a planar space with 29-state cells of a single type, each connected to the four orthogonally adjacent neighbors: what he did was to find a finite quiescent configuration (of around 200,000 cells) that, given any other finite quiescent configuration, reproduces it in a different part of the space, without erasing itself: he thus found a universal constructor for the class of quiescent configurations. Later models improved on von Neumann's original presentation, in particular Codd [8] and Arbib [1] (with opposite emphasis: the former on local simplicity of cells, the latter on global simplicity of construction).

The simplest automaton admitting self-reproducing configurations we know of is Conway's *Life automaton* (see Berlekamp, Conway and Guy [6], or Poundstone [21]), so called because modelled on life-like behavior. It consists of a planar space, with each cell connected to the eight adjacent ones. The cells have just two states: 0 (death) and 1 (life), and the Life rules are the following: a dead cell gets born when exactly 3 neighbors are alive; a live cell survives if 2 or 3 neighbors are alive, and it dies otherwise (by overpopulation or starvation). Conway has shown that *there are self-reproducing Life configurations*, and that *the Life automaton is universal*, in the sense that any cellular automaton can be constructed by taking sufficiently large squares of Life cells as its basic cells. Life is about as simple as it can be (it is known that 2-state cells with a von Neumann neighborhood do not admit non-trivial self-reproduction), and it shows that self-reproduction does not need a complicated universe (since it is logically possible from simple physical models).

1.2 The genetic code

Before turning to other applications of discrete mathematics to different fields, it is perhaps dutiful to discuss an unsuccessful attempt to use it to crack the *genetic code*, reported by Crick in his autobiography [9].

The code is written in a language whose alphabet consists of four letters (A, G, T, and C for DNA, and the same with U in place of T for RNA) standing for the nucleic acids, and whose words consist of triplets of letters (codons) coding for amino acids. While there are 64 ($4 \times 4 \times 4$) codons, there are only 20 amino acids: today we know that the code is redundant, and most amino acids are coded by more than one codon.

Crick originally conjectured that the reason for the number 20 was that the code was comma-free: that is, the division in codons could be reconstructed automatically, even from a fragment. This eliminated the four codons with the same letter, since otherwise the fragment XXXXXX could be divided in different ways. Of the remaining 60 codons, if one was significant then the two permutations obtained from it could not be: if XYZ was significant then neither YZX nor ZXY could be, otherwise the fragment XYZXYZ could be divided in different ways. Thus only $\frac{60}{3} = 20$ codons could be significant.

Unfortunately, this nice piece of discrete mathematics was put out of use by the experimental discovery that even the codons with three repeated letters can code for an amino acid, and eventually the code was cracked by experimental, not theoretical methods.

2 Political sciences

The mathematics of politics is a subject that is rapidly becoming popular, as an off-spring of the economical theory of social choice on the one hand, and of the mathematical game theory on the other. Here we will prove in elementary ways some of the crucial impossibility theorems, often considered (for example by Paul Samuelson, Nobel prize for economics in 1970) as analogues of Gödel's results in their respective fields.

2.1 Apportionment

Our first concern is the problem of apportionment, examples of which are the assignments of seats to districts on the basis on their population (before an election), or to parties on the basis of the votes they receive (after an election).

Since the number of seats is usually smaller than the number of voters, in general the ratio will not be an integer. To be fair, the apportionment should satisfy a *proportionality principle*: the number of seats assigned to a district or a party should be one of the two integral approximations (by excess or by defect) of the ratio. For example, if there are 10 seats to be assigned and one party gets one third of the votes, it should be assigned either 3 or 4 seats.

A second obvious condition that apportionment should satisfy is *monotonicity*: if a party gets more votes than another one, it should not get less seats. And this should hold both synchronically, in a single election (relatively to the percentage of votes), and diachronically, in different elections (relatively to the ratio of the percentages of votes).

Theorem 1 (Balinski and Young [3]). *There is no apportionment method satisfying both the principles of proportionality and monotonicity.*

Proof. Suppose there are at least 4 parties (or districts) and 7 seats, and that the ratios of the number of seats to the number of votes for each parties is the following, where $\varepsilon > 0$:

parties	first election	second election
A	$5+\varepsilon$	$4-\varepsilon$
B	$\frac{2}{3}$	2
C	$\frac{2}{3}$	$\frac{1}{2}$
D	$\frac{2}{3}-\varepsilon$	$\frac{1}{2}+\varepsilon$
total	7	7

In the first election party A must get at least 5 seats: since at most 2 are left, they cannot go to party D by monotonicity, since D has a smaller percentage than both B and C.

In the second election party A can get at most 4 seats, and party B at most 2: since at least 1 is left, it must go to party D by monotonicity, since D has a greater percentage than C.

In the step from the first to the second election, party A has thus lost at least one seat, and party D has won at least one. By monotonicity, A should have lost in percentage relatively to party D, i.e.

$$\frac{4-\varepsilon}{\frac{1}{2}+\varepsilon} < \frac{5+\varepsilon}{\frac{2}{3}-\varepsilon},$$

but this is possible only if $\varepsilon > \frac{1}{61}$. By choosing $\varepsilon \leq \frac{1}{61}$ one thus gets a contradiction. □

The following is a numerical example of seats distribution that would look fair at first sight, but according to which A loses one seat while D gains one, although A has increased its votes relatively to D from 7.5 times to almost 8:

parties	%	seats	%	seats
A	$5,01$	5	$3,99$	4
B	$0,67$	1	$2,00$	2
C	$0,67$	1	$0,50$	0
D	$0,65$	0	$0,51$	1

2.2 Rights

A simple and efficient way of formalizing the problem of social choice is to consider a number of possible alternatives, ranked according to preference or indifference by each individual i in an *individual order* $\leq_i$: on the basis of these individual orders, society will produce its own *social order* $\leq_s$.

We write $A <_i B$ to mean that individual i prefers B to A, and $A =_i B$ to mean that the two alternatives are indifferent to him: thus $A \leq_i B$ means that either i prefers B to A, or they are indifferent to him. Similarly for $<_s$, $=_s$ and $\leq_s$.

The various orders should somehow mirror at least some of the intuitive properties of preferences, and in particular they should be:

- *Total*: given two alternatives A and B, either one is preferred to the other, or they are indifferent. Formally:

  $$A <_i B \text{ or } B <_i A \text{ or } A =_i B$$

  or, equivalently,

  $$A \leq_i B \text{ or } B \leq_i A,$$

  and similarly for $<_s$ or $\leq_s$.
- *Transitive*: given three alternatives A, B and C, if A is not preferred to B and B is not preferred to C, then A is not preferred to C. Formally:

  $$A \leq_i B \text{ and } B \leq_i C \implies A \leq_i C,$$

  and similarly for $\leq_s$.

Moreover, if the social order has somehow to mirror the individual orders, at least the following *unanimity principle* (introduced by Wilfried Pareto) should hold: if all individuals prefer one of A and B to the other, society does not prefer the other. Formally:

$$A <_i B \text{ for every } i \implies A \leq_s B \qquad \text{and} \qquad B <_i A \text{ for every } i \implies B \leq_s A.$$

Given the individual and social orders, one can formalize the notion of *right* by saying that an individual i has a right over two alternatives if, whenever he prefers one to the other, so must do society. Formally:

$$X <_i Y \implies X <_s Y.$$

Finally, an individual has *freedom of choice* if he can order the alternatives according to any order he likes.

Theorem 2 (Sen [24]). *If both unanimity and freedom of choice hold, at most one individual can have rights.*

Proof. Suppose individual 1 has a right over A and B, and individual 2 has a right over C and D. We want to build individual orders $\leq_1$ and $\leq_2$ so that

$$A <_s B \leq_s C <_s D \leq_s A,$$

contradicting transitivity. Note that:

- to have $A <_s B$ it is enough that $A <_1 B$, since 1 has a right over A and B;
- to have $B \leq_s C$ it is enough that $B <_1 C$ and $B <_2 C$, by unanimity;
- to have $C <_s D$ it is enough that $C <_2 D$, since 2 has a right over C and D;
- to have $D \leq_s A$ it is enough that $D <_1 A$ and $D <_2 A$, by unanimity

It is thus enough to define the individual orders of 1 and 2 as follows, which is possible by freedom of choice:

$$D <_1 A <_1 B <_1 C \qquad \text{and} \qquad B <_2 C <_2 D <_2 A.$$

□

2.3 Elections

Usually democracy is identified with the synthesis of the social order from the individual ones, by the process of elections (often taking the form of a choice between two alternatives).

In this case, one implicitly uses a *vote-dependence principle* according to which society's choice between two alternatives must depend only on the individual preferences about them (as expressed by the votes). In other words, $\leq_s$ on A and B must depend only on the various $\leq_i$ on A and B.

Theorem 3 (Arrow [2]). *If unanimity, freedom of choice and vote-dependence hold, then there must be a dictator.*

Proof. Let's consider two alternatives A and B that are not socially indifferent, i.e. such that $A <_s B$ or $B <_s A$. If nobody had a right over A and B, for every individual i it would be either $A <_i B$ but $A \not<_s B$ (i.e. $B <_s A$), or $B <_i A$ but $B \not<_s A$ (i.e. $A <_s B$). By vote-dependence, the social order on A and B is determined by the individual orders on them: let's consider the situation in which every individual has on A and B the preference that produces an opposite social order. Everybody must have the same preference, since it is impossible to have simultaneously both $A <_s B$ and $B <_s A$: thus for example $A <_i B$ for every individual but $B <_s A$, against unanimity.

This proves that, for every pair of alternatives not socially indifferent, there must an individual having a right on them. By the previous result, however, at most an individual can have rights at all: thus, for every pair of alternatives it is always the *same* individual that has rights on them. Then he must be a dictator, since he is the only one to have rights, and has them on every pair of non-socially-indifferent alternatives (the name 'dictator' is justified by the fact that society does everything he wishes, disregarding what everybody else wishes). □

As announced in the title, a little discrete mathematics has indeed been sufficient to prove a very indiscreet result: social decisions can reflect only the wishes of at most one individual, and always of the same one! Hardly a situation that justifies the self-christened name of 'democratic' that many societies of the 'free world' boast when describing themselves.

3 (Meta)physics

Bell's theorem has been described as "the deepest discovery of science" (by Stapp [25]), and "the most important recent development of physics" (by Brian Josephson, Nobel prize for physics in 1973). It is not necessary to share such an enthusiasm to realize that, in any case, the result is really exceptional, since it allows an *empirical* refutation of *metaphysical* assumptions!

After its original formulations by Bell himself, collected in [5], the theorem has been variously refined and simplified, most recently by Greenberger, Horne and Zeilinger [13], and Hardy [14]. The increased simplicity of the theoretical framework is however counterbalanced by the increased complication of the experimental setting: we thus choose to prove optimally balanced versions, one of which new.

3.1 The quadrangular inequality

The setting consists of two physical systems: on one two types of measurements are made (A and B), and on the other two other types (C and D). Each measurement can give a positive (+) or negative (−) outcome. Measurements are made repeatedly, to obtain statistically relevant results, and are said to exhibit *correlation* when they give the same outcome (i.e. either both + or both −).

We let $p(xy|ij)$ be the probability that, when making measurements i on the first system and j on the second, the outcome will respectively be x and y (obviously, i will be either A or B, j will be either C or D, and x and y will take values + or $-$). Probabilities $p(x|i)$, $p(x|ij)$, $p(x|ijy)$, as well as those with y in place of x, are defined similarly.

The main hypothesis of Bell's theorem, to be discussed later on, is the following.

Axiom 4 (Factorizability). *The outcomes of the experiments on the two systems are independent:*

$$p(xy|ij) = p(x|i) \cdot p(y|j).$$

Theorem 5 (Quadrangular Inequality; Bell [4], Stapp [26]). *If factorizability holds, then lack of correlation in one of the two types of measures cannot exceed the lack of correlation in the remaining three types. For example,*

$$\begin{aligned} p(+-|AC) + p(-+|AC) \ &\leq \\ p(+-|AD) + p(-+|AD) \ &+ \\ p(+-|BD) + p(-+|BD) \ &+ \\ p(+-|BC) + p(-+|BC)&. \end{aligned}$$

Proof. By factorizability, the quadrangular inequality becomes

$$\begin{aligned} p(+|A) \cdot p(-|C) + p(-|A) \cdot p(+|C) \ &\leq \\ p(+|A) \cdot p(-|D) + p(-|A) \cdot p(+|D) \ &+ \\ p(+|B) \cdot p(-|D) + p(-|B) \cdot p(+|D) \ &+ \\ p(+|B) \cdot p(-|C) + p(-|B) \cdot p(+|C)&. \end{aligned}$$

For simplicity we write a^+ for $p(+|A)$, a^- for $p(-|A)$, and similarly in the other cases. We thus get

$$a^+c^- + a^-c^+ \leq a^+d^- + a^-d^+ + b^+d^- + b^-d^+ + b^+c^- + b^-c^+.$$

Since measure A always gives positive or negative result, $a^+ + a^- = 1$: we can thus further write a for a^+, $1 - a$ for a^-, and similarly in the other cases. We thus get

$$a(1-c) + (1-a)c \leq a(1-d) + (1-a)d + b(1-d) + (1-b)d + b(1-c) + (1-b)c,$$

i.e.

$$ad + bd + bc \leq ac + b + d.$$

To prove the latter inequality, it is enough to remember that the numbers a, b, c and d are between 0 and 1, and to consider three cases:

1. if $\boldsymbol{b \leq a}$, we prove
$$ad + bd + c(b - a) \leq b + d.$$
Since $b \leq a$ and $c \geq 0$, $c(b - a) \leq 0$: thus
$$ad + bd + c(b - a) \leq ad + bd.$$
Moreover, $ad \leq d$ because $a \leq 1$, and $bd \leq b$ because $d \leq 1$.

2. if $\boldsymbol{d} \leq \boldsymbol{c}$, similarly one proves

$$a(d-c)+bd+bc \leq b+d.$$

3. if $\boldsymbol{b} \geq \boldsymbol{a}$ and $\boldsymbol{d} \geq \boldsymbol{c}$, we consider two further subcases:
   (a) if $\boldsymbol{b} \leq \boldsymbol{d}$, we prove

   $$d(a+b)+c(b-a) \leq b+d.$$

   Since $b \geq a$ and $c \leq 1$, $c(b-a) \leq b-a$; and $d(a+b) \leq a+b$ because $d \leq 1$: thus

   $$d(a+b)+c(b-a) \leq (a+b)+(b-a)=2b.$$

   From $b \leq d$ we have $2b \leq b+d$.
   (b) if $\boldsymbol{b} \geq \boldsymbol{d}$, similarly one proves

   $$a(d-c)+b(d+c) \leq b+d.$$

□

3.2 Analysis of the hypothesis

Following Jarrett [15], one can *decompose* the factorizability hypothesis in the conjunction of two distinct hypotheses.

Axiom 6 (Separability). *The statistical outcomes of an experiment performed on one system are independent of the outcomes of an experiment performed on the other system.*

$$p(x|ijy)=p(x|ij) \qquad \textit{and} \qquad p(y|ijx)=p(y|ij).$$

Separability holds if the spatial-temporal separation of the two systems is sufficient to make them distinct, and thus to make the global state of the two reducible to their individual states ("the whole is equal to the sum of the parts").

Axiom 7 (Locality). *The statistical outcomes of an experiment performed on one system are independent of the types of experiments performed on the other system:*

$$p(x|ij)=p(x|i) \qquad \textit{and} \qquad p(y|ij)=p(y|j).$$

Locality holds if the spatial-temporal separation of the two systems is sufficient to forbid their instantaneous interaction, and thus to make the individual state of a system independent of the state of the other ("no action-at-distance").

Proposition 8. *Factorizability follows from separability and locality.*

Proof. It is enough to note that

$$\begin{aligned} p(xy|ij) &= p(x|ijy)\cdot p(y|ij) && \text{by definition of probability} \\ &= p(x|ij)\cdot p(y|ij) && \text{by separability} \\ &= p(x|i)\cdot p(y|j) && \text{by locality} \end{aligned}$$

□

In a different direction, it is also possible to *replace* factorizability by a classical assumption of probability theory.

Axiom 9 (Additivity). *The probabilities of compound events are the sums of the probabilities of their independent atomic events.*

Proposition 10. *The quadrangular inequality follows from additivity.*

Proof. The four types of measure and the two types of outcomes generate 16 types of atomic events, that can for example be recorded by a quadruple of +'s and −'s (correspondent to the measures A, B, C and D, in order). For example, $p(+--+)$ will then be the probability that the outcomes of measures A and D are positive, and those of measures B and C are negative.

From additivity follows, for example, that

$$p(+-|AC) = p(++-+) + p(++--) + p(+--+) + p(+---),$$

and similarly in the other cases. The quadrangular inequality follows from the fact that the 24 cases on the right cover the 8 cases on the left, as one can easily check. □

3.3 The quadrangular implication

The last proposition is a probability result apparently obtained by brute force. Actually, it follows from an independent logical principle.

Axiom 11 (Distributivity). *Logical conjunction is distributive w.r.t. conjunction:*

$$\alpha \wedge (\beta \vee \gamma) \;\Rightarrow\; (\alpha \wedge \beta) \vee (\alpha \wedge \gamma).$$

By considering probabilities as possible truth-values of propositions, we can write A and $\neg A$ for $p(+|A)$ and $p(-|A)$, and similarly in the other cases. The algebraic operations on truth-values can then be interpreted as logical operation in the usual way, i.e. by writing $\vee$, $\wedge$ and $\Rightarrow$ for $+$, $\cdot$ and $\leq$. We thus get a logical form of the quadrangular inequality.

Theorem 12 (Quadrangular Implication). *If distributivity holds, then*

$$\begin{array}{l}(A \wedge \neg C) \vee (\neg A \wedge C) \;\Rightarrow \\ (A \wedge \neg D) \vee (\neg A \wedge D) \;\vee \\ (D \wedge \neg B) \vee (\neg D \wedge B) \;\vee \\ (B \wedge \neg C) \vee (\neg B \wedge C).\end{array}$$

Proof. By distributivity,

$$\alpha \wedge (\neg\beta \vee \beta) \;\Rightarrow\; (\alpha \wedge \neg\beta) \vee (\alpha \wedge \beta).$$

By letting $\alpha = X \wedge \neg Y$ and $\beta = Z$, one gets

$$(X \wedge \neg Y) \wedge (\neg Z \vee Z) \;\Rightarrow\; (X \wedge \neg Y \wedge \neg Z) \vee (X \wedge \neg Y \wedge Z),$$

and so (by using the law of the excluded middle on the left, and weakening the conclusion on the right)

$$(X \wedge \neg Y) \Rightarrow (X \wedge \neg Z) \vee (Z \wedge \neg Y).$$

By a double application (first to A, C and D, then to C, D and B) of the principle just proved, one gets

$$(A \wedge \neg C) \Rightarrow (A \wedge \neg D) \vee (D \wedge \neg B) \vee (B \wedge \neg C),$$

and, similarly,

$$(\neg A \wedge C) \Rightarrow (\neg A \wedge D) \vee (\neg D \wedge B) \vee (\neg B \wedge C).$$

Together, the two conclusions prove the statement. □

3.4 Prolegomena to every future physics

It is easy to concoct experimental situations in which the quadrangular inequality and implication are not satisfied. For example, one can excite a calcium atom by means of two laser beams, and observe the light emitted (in the form of photons) when the atom goes back to the normal, non-excited state: by appropriately directing the beams, one obtains two correlated light rays that can be directed through distant polarizing filters. Quantum mechanics predicts, and experience confirms, that the fraction of photons of two correlated light rays going through two polarizing filters with polarization directions forming an angle α is $\cos^2(\alpha)$.

One can measure on the first system the polarization at angles $0°$ and $45°$, and on the second system the polarizations at angles $22,5°$ and $67,5°$. In three cases the angle between the polarization directions of the two filters will be $22,5°$, and they will thus let go $\frac{2+\sqrt{2}}{4} \approx 85\%$ photons: the lack of correlation will thus be 15%. In the remaining case the angle will instead be $67,5°$, thus letting go 15% photons: the lack of correlation will thus be 85%. This produces an experimental situation that has actually been performed, and in which one has indeed obtained the following failure of the quadrangular inequality:

$$\begin{aligned} p(+-|AC) + p(-+|AC) &= 85\% \\ p(+-|AD) + p(-+|AD) &= 15\% \\ p(+-|BD) + p(-+|BD) &= 15\% \\ p(+-|BC) + p(-+|BC) &= 15\%. \end{aligned}$$

Since the quadrangular inequality and implication turn out to be experimentally falsifiable, no theory satisfying their hypotheses can be physically correct. In particular, it is necessary to drop at least one of the principles of *locality* and *separability*: since dropping the first would require a revision of relativity theory, the solution adopted by quantum mechanics is to drop the second, thus turning to a holistic view of the world.

From the failure of the principle of factorizability follows the necessity of dropping the *causality principle*, at least in the following form due to Reichenbach [22]: whenever two events are correlated (not independent), they must have a common cause (i.e. there must exist an event in their common past relatively to which they are independent). Indeed, the experimental conditions falsifying the quadrangular inequality provide a counterexample: the events cannot be independent relatively to any common cause, otherwise factorizability would hold relatively to it, and the quadrangular inequality would follow. This provides an example of what Jung and Pauli [16] called *synchronicity*: events that are correlated in a way that is neither casual nor causal.

From the failure of the principle of additivity follows the impossibility of using *classical probability*, as developed by Kolmogorov [18], in quantum mechanics: no assignment of probabilities to atomic events can generate the probability of compound events in a way accounting for experience. Moreover, it also follows the necessity of dropping the *truth-functionality* of classical logic: no assignment of truth-values to atomic propositions can generate the assignment of truth-values to compound propositions in a way accounting for experience.

Failure of *distributivity*, first proved for quantum logic by Birkoff and von Neumann [7] in a different way, is here extended to any logic that purports to be adequate for the treatment of the physical world, thus excluding classical logic as viable candidate.

With the proofs of the inadequacy of classical logic we have thus come full circle, and have hopefully justified the claim made at the beginning: that the study of the physical and social worlds benefits from (and actually requires) a switch of emphasis from infinity and classical logic to discrete mathematics and non-classical logics.

References

1. Arbib, M.A. *Theories of Abstract Automata*, Prentice Hall, 1969.
2. Arrow, K. *Social Choice and Individual Values*, Yale University Press, 1951.
3. Balinski, M., and Young, P. *Fair Representation*, Yale University Press, 1982.
4. Bell, J. On the Einstein-Podolsky-Rosen paradox, *Physics*, 1 (1964) 195–200, also in [5], pp. 14–21.
5. Bell, J. *Speakable and Unspeakable in Quantum Mechanics*, Cambridge University Press, 1987.
6. Berlekamp, E., Conway, J., and Guy, R. *Winning Ways*, Academic Press, 1982.
7. Birkoff, G., and von Neumann, J. The logic of quantum mechanics, *Annals of Mathematics*, 37 (1936) 823–843.
8. Codd, E.F. *Cellular Automata*, Academic Press, 1968.
9. Crick, F. *What Mad Pursuit*, Basic Books, 1988.
10. Crick, F., and Watson, J. Molecular structure of nucleic acids, *Nature*, 171 (1953) 737–738.
11. Gödel, K. Zur intuitionistischen Arithmetik und Zahlentheorie, *Ergebnisse eines mathematischen Kolloquiums*, 4 (1933) 34–38, translated in [12], pp. 287–295.
12. Gödel, K. *Collected works*, Oxford University Press, 1986.
13. Greenberger, D.M., Horne, M., and Zeilinger, A. Going beyond Bell's theorem, in *Bell's theorem, quantum theory, and conceptions of the universe*, Kafatos ed., Kluwer Academic, 1989, pp. 69–72.

14. Hardy, L. Non-locality for two particles without inequalities for almost all entangled states, *Physical Review Letters*, 71 (1993) 1665–1668.
15. Jarrett, J. On the physical significance of the locality condition in the Bell argument, *Noûs*, 18 (1984) 569–589.
16. Jung, C.G., and Pauli, W. *The interpretation of nature and psyche*, 1952.
17. Kleene, S.K. On notations for ordinal numbers, *Journal of Symbolic Logic*, 3 (1938) 150–155.
18. Kolmogorov, A. *Foundations of Probability Theory*, 1933.
19. Odifreddi, P. *Classical Recursion Theory*, North Holland, 1989.
20. Odifreddi, P. *Classical Recursion Theory*, volume II, North Holland, 1997.
21. Poundstone, W. *The Recursive Universe*, Contemporary Books, 1986.
22. Reichenbach, H. *The Direction of Time*, University of California Press, 1956.
23. Schrödinger, E. *What Is Life?*, Cambridge University Press, 1944.
24. Sen, A. *Collective Choice and Social Welfare*, Holden-Day, 1970.
25. Stapp, H. Are superluminar connections necessary?, *Nuovo Cimento*, 40 (1977) 191–204.
26. Stapp, H. Whiteheadian approach to quantum theory and the generalized Bell's theorem, *Foundations of Physics*, 9 (1979) 1–25.
27. Von Neumann, J. *The Computer and the Brain*, Yale University Press, 1958.
28. Von Neumann, J. *The Theory of Self-Reproducing Automata*, University of Illinois Press, 1966.

A Foundation for Computable Analysis

Klaus Weihrauch

FernUniversität Hagen, D–58084 Hagen
e–mail: Klaus.Weihrauch@Fernuni-Hagen.de

Abstract. While for countable sets there is a single computability theory (ordinary recursion theory), for computability in Analysis several mutually non–equivalent theories have been proposed, none of which, however, has been accepted by the majority of mathematicians or computer scientists.
One of these theories, TTE (Type 2 Theory of Effectivity), is presented in this contribution. TTE intends to characterize and study exactly those functions, operators etc. known from Analysis, which can be realized correctly by digital computers. The paper gives a short introduction to basic concepts of TTE and shows its general applicability by some selected examples. First, Turing computability is generalized from finite to infinite sequences of symbols. Assuming that digital computers can handle (without loss of generality) only sequences of symbols, infinite sequences of symbols are used as names for "infinite objects" such as real numbers, open sets, compact sets or continuous functions. Naming systems are called representations. An easy but general principle for defining effective representations for T_0–spaces with countable bases is introduced. The concepts are applied to real numbers, open sets, compact sets and continuous functions. The problem of determining zeroes is considered. Finally, computational complexity is discussed and computability in measure theory is sketched.

1 Introduction

Classical computability theory considers the natural numbers ω or the words Σ^* over some finite alphabet Σ as basic sets. Many definitions of computability have been proposed and justified by intuition, most of which turned out to be equivalent to computability by Turing machines. This lead to a claim generally referred to as Church's Thesis, which can be formulated as follows: A partial function $f :\subseteq \Sigma^* \longrightarrow \Sigma^*$ is computable in the intuitive sense or by a physical device such as a digital computer, if and only if it can be computed by a Turing machine. Although Church's Thesis is not provable it is generally accepted, and today "computable" is the usual abbreviation of "Turing computable".

Church's Thesis does not apply to functions on the real numbers. Several theories for studying aspects of effectivity in Analysis have been developed in the past (for example, intuitionistic logic [TD88], Bishop [Bis67], Grzegorczyk [Grz55], Ceitin [Cei59], Traub et al. [TWW88], Pour–El/Richards [PR89], Ko [Ko91], Blum/Shub/Smale [BSS89], [WS81], [KW85], and so on), and several generalizations of Church's Thesis have been suggested. Although each of these generalizations is based on easily comprehensible intuition, they are partly inconsistent. Moreover, the relations between the theories and computability definitions are not yet fully explored. None of the theories has been accepted by the majority of mathematicians or computer scientists. This somewhat confusing situation may explain but does not excuse the fact that computer scientists have neglected Computable Analysis in research and almost disregarded it in teaching.

In this article we present one of the existing approaches to Computable Analysis, henceforth called "Type 2 Theory of Effectivity" (TTE for short). TTE is based on a definition of computable real functions given by the Polish logician A. Grzegorczyk in 1955 [Grz55], where computable operators transform fast converging sequences of rational numbers representing real numbers. TTE extends this original approach in several directions:

1. Also continuous operators are considered where "continuity" can be interpreted as a very fundamental kind of "constructivity".
2. By using representations effectivity can be studied on many other spaces used in Analysis.
3. A general concept of "admissible" (effective) representations is derived from basic principles.
4. Computational complexity ("bit-complexity") is included as a part of TTE.

Since intuition has not lead to a single concept of computability for real functions, TTE intends to characterize and study those real functions which can be computed by physical devices. The main difficulty stems from the fact that the set of real numbers is not countable and that therefore real numbers cannot be coded by natural numbers or finite words but must be considered as infinite objects (for example, infinite decimal fractions). By laws of physics, however, any device can store only finitely many bits of information at any moment, and any input or output channel can tranfer only finitely many bits in a finite amount of time. TTE solves this by considering finite objects as approximations of infinite ones and by representing infinite objects by infinite sequences of finite objects.

In recursion theory computability is defined explicitly (without loss of generality) on Σ^*. Computability is transferred to other countable sets M by notations $\nu :\subseteq \Sigma^* \longrightarrow M$, where words serve as names. TTE considers additionally the set Σ^ω of infinite sequences as standard infinite objects, defines computability (and other effectivity concepts) on Σ^ω explicitly and transfers them to sets M with maximally continuum cardinality by representations $\delta :\subseteq \Sigma^\omega \longrightarrow M$, where infinite sequences of symbols serve as names (example: infinite decimal fractions as names for real numbers).

In Section 2 we introduce computable functions from tuples of finite or infinite sequences to finite or infinite sequences $f :\subseteq Y_1 \times \ldots \times Y_k \longrightarrow Y_0$ $(Y_0 \ldots Y_k \in \{\Sigma^*, \Sigma^\omega\})$ by means of Type 2 machines, which are Turing machines with one-way finite or infinite input and output tapes. We observe that for such a computable function, every finite portion of the result depends only on a finite portion of the input. This finiteness property can be expressed as continuity with respect to the Cantor topology on Σ^ω. We give examples of computable functions and prove that multiplication of infinite decimal fractions is not computable.

Section 3 introduces the vocabulary for transferring effectivity concepts from Σ^ω to represented sets. Although every representation $\delta :\subseteq \Sigma^\omega \longrightarrow M$ of a set M induces some type of effectivity on M, most of the countless representations are not interesting, only "effective" ones are important. Effectivity is usually based on some structure on M. A type of structure we consider is a topological T_0-space with a notation of a subbase [Eng89]. For each such "information structure" we introduce a standard representation with very interesting natural topological and computational properties, which justify to call them "effective" or "admissible". Only admissible representations will be used in the following. They provide a very natural and sufficiently general concept to introduce computability in almost all branches of Analysis.

In Section 5 we investigate computability on the real line, based on the Cauchy representation which is admissible for this topological space. We show that the representation by infinite decimal fractions is not admissible, we give examples for computable functions, and we show that no (non–trivial) set of real numbers is recursive.

In Section 6 we apply the general concept to the set of compact subsets of $\mathbb{R}$. Several admissible representations, which express different types of knowledge, are introduced and compared. Compactness is an instructive example for a classical concept which branches into several parts in TTE.

In Section 7 we introduce effective representations of spaces of continuous functions, and as a further instructive example in Section 8 the effectivity of determining zeroes under different restrictions is studied.

Computational Complexity is considered in Section 9. For functions on Σ^ω, complexity and input lookahead (input information) are measured as functions of the output precision. For transferring complexity and input lookahead, distinguished admissible representations are needed. For the real line we introduce the "modified binary representation" (which uses the digits 0, 1 and -1). The resulting complexity concept ("bit complexity") has been used by Ko [Ko91] and others.

Computability in measure theory is sketched very briefly in Section 10.

The first presentation of the concepts of TTE is [KW84]. More complete versions are [Wei85] and [KW85] as well as Part 3 of [Wei87]. A formally simpler access to TTE is "A simple introduction to computable analysis" [Wei95], much more details can be found in [Wei94]. Some further publications are added without reference from the text.

We shall denote the set $\{0, 1, 2, \ldots\}$ of natural numbers by ω. A partial function f from A to B is denoted by $f :\subseteq A \longrightarrow B$. $x\Sigma^\omega$ abbreviates the set $\{p \in \Sigma^\omega \mid x \text{ is a prefix of } p\}$. Σ will usually be a sufficiently large finite alphabet.

2 Computability and continuity on infinite sequences of symbols

For defining computability on finite and infinite sequences of symbols we introduce *Type 2 machines.*

Definition 2.1 *A Type 2 machine is a multi-tape Turing machine M with k ($k \in \omega$) one-way input tapes and a single one-way output tape together with a type specification $(Y_1, \ldots, Y_k, Y_0)$, where $Y_0, \ldots, Y_k \in \{\Sigma^*, \Sigma^\omega\}$ and $Y_i = \Sigma^*(\Sigma^\omega)$ means that Tape i is provided for finite (infinite) inscriptions. The function $f_M :\subseteq Y_1 \times \ldots \times Y_k \longrightarrow Y_0$ computed by a Type 2 machine M is defined as follows.*

Case $Y_0 = \Sigma^*$ (finite output): $f_M(y_1, \ldots, y_k) = w :\Longleftrightarrow$ M with input $(y_1, \ldots, y_k)$ halts with result $w \in \Sigma^*$ on the output tape.

Case $Y_0 = \Sigma^\omega$ (infinite output): $f_M(y_1, \ldots, y_k) = p :\Longleftrightarrow$ M with input $(y_1, \ldots, y_k)$ computes forever writing the sequence $p \in \Sigma^\omega$ on the output tape.

The above semantics considers infinite input and output tapes which do not exist physically and infinite computations which cannot be completed in reality. Notice, however, that for a computation of a Type 2 machine any finite prefix of the output can be obtained in finitely many computation steps from finite prefixes of the inputs. Finite initial parts of computations concerning only finite prefixes of inputs and outputs, however, can be realized by digital computers. There are several other computability concepts which are equivalent (via appropriate encodings) to Type 2 computability (enumeration operators, partial recursive operators, partial recursive functionals, ..., see [Rog67]). This supports the following generalization of Church's Thesis:

> A function $f_M :\subseteq Y_1 \times \ldots \times Y_k \longrightarrow Y_0$ $(Y_0, \ldots, Y_k \in \{\Sigma^*, \Sigma^\omega\})$ is computable informally or by a physical device, if and only if it can be computed by a Type 2 machine.

In the following we shall say "computable" instead of "computable by a Type 2 machine". We illustrate the concepts by some examples.

Example 2.2

1. *Let $A \subseteq \Sigma^*$ be recursively enumerable. Define $f :\subseteq \Sigma^\omega \longrightarrow \Sigma^*$ by $f(p) :=$ (0 if $p \in x\Sigma^\omega$ for some $x \in A$, div otherwise). Then f is computable (easy proof).*
2. *There is a computable function $f :\subseteq \Sigma^\omega \longrightarrow \Sigma^\omega$ which divides infinite decimal fractions by 3. (Use school method beginning from the left.)*
3. *The function $f :\subseteq \Sigma^\omega \longrightarrow \Sigma^*$ defined by $f(p) =$ (0 if $p = 0^\omega$, 1 otherwise) is not computable. For a proof, assume that some Type 2 machine M computes f. Consider the input $p = 0^\omega$. After some number t of computation steps M has written 0 and halts. During this computation M has read at most the first t symbols of the input 0^ω. Now, consider the input $q := 0^t10^\omega$. Also for this input M must write 0 and halt after t computation steps. Therefore $f_M(q) = 0$. But $f(q) = 1$, hence $f \neq f_M$.*

4. No computable function multiplies infinite decimal fractions by 3. For a proof assume that some Type 2 machine M multiplies infinite decimal franctions by 3. Consider the input $p = 0.3^\omega$. Then $f_M(p) = 0.9^\omega$ or $f_M(p) = 1.0^\omega$. Consider the case $f_M(p) = 0.9^\omega$. After some number t of computation steps M has written "0." on the output tape. During this computation M has read at most the first t symbols from the input tape. Now, consider the input $q := 0.3^t40^\omega$. Also for this input M must write "0." during the first t steps. But for correct multiplication the output must begin with "1.". (The case $f_M(p) = 1.0^\omega$ is treated accordingly.)

The above two proofs of non–computability are very typical for TTE. They use only the following basic finiteness property for computable functions on Σ^ω or Σ^*: If $f(z) = x$, then every finite portion of x depends only on (is already completely defined by) a finite portion of z. This finiteness property can be expressed by continuity with respect to two topologies [Eng89] which we consider as fixed in the following:

- the discrete topology on $\Sigma^* : \tau = \{A \mid A \subseteq \Sigma^*\}$
- the Cantor topology on $\Sigma^\omega : \tau_C := \{A\Sigma^\omega \mid A \subseteq \Sigma^*\}$. (The set $\{w\Sigma^\omega \mid w \in \Sigma^*\}$ is a basis of τ_C.)

Theorem 2.3 *Every function $f :\subseteq Y_1 \times \ldots \times Y_k \longrightarrow Y_0$ computable by a Type 2 machine is continuous.*

Already the conventions for Type 2 machines that inputs must be read symbol by symbol and the outputs must be written one–way symbol by symbol suffice to prove the theorem. Since the finiteness property, or more generally speaking continuity, can be interpreted as a very fundamental kind of effectivity or constructivity, it is of separate interest in TTE. We extend well known definitions from recursion theory.

Definition 2.4 *Consider $X \subseteq W \subseteq Y_1 \times \ldots \times Y_k$ $(Y_1, \ldots, Y_k \in \{\Sigma^*, \Sigma^\omega\})$. X is called recursively enumerable (r.e.) in W, iff $X = W \cap dom(f)$ for some computable function $f :\subseteq Y_1 \times \ldots \times Y_k \longrightarrow \Sigma^*$. X is called recursive in W, iff there is some computable function $f :\subseteq Y_1 \times \ldots \times Y_k \longrightarrow \Sigma^*$ such that $f(x) = 0$ if $x \in X$ and $f(x) = 1$ if $x \in W \setminus X$. (For $W = Y_1 \times \ldots \times Y_k$ we omit "in W".)*

Every r.e. set is open. For every set $X \subseteq \Sigma^\omega$, X is r.e. $\iff$ $X = A\Sigma^\omega$ for some r.e. set $A \subseteq \Sigma^*$. The standard numbering $\varphi : \omega \longrightarrow P^{(1)}$ of the partial recursive number functions is an important object in recursion theory [Rog67]. It is determined uniquely except for equivalence by the smn– and the utm–theorem. Correspondingly, for any $a, b \in \{*, \omega\}$ there is a notation $\xi^{ab} : \Sigma^* \longrightarrow P^{ab}$ of the set of computable functions $f :\subseteq \Sigma^a \longrightarrow \Sigma^b$ with similar properties. Moreover, for continuous functions from Σ^ω to Σ^* or Σ^ω there are *admissible* representations [Wei85, Wei87, Wei94]. Let $F^{\omega *}$ be the set of all continuous functions $f :\subseteq \Sigma^\omega \rightarrow \Sigma^*$ with open domain, and let $F^{\omega\omega}$ be the set of all continuous functions $f :\subseteq \Sigma^\omega \rightarrow \Sigma^\omega$ with G_δ-domain. Consider $a \in \{*, \omega\}$.

- Every continuous function $f :\subseteq \Sigma^\omega \to \Sigma^a$ has an extension in $F^{\omega a}$;

There is a total function $\eta^{\omega a}$ from Σ^ω onto $F^{\omega a}$ with the following properties:

- The function $u :\subseteq \Sigma^\omega \times \Sigma^\omega \to \Sigma^a$ defined by $u(p,q) := \eta^{\omega a}(p)(q)$ is computable (the utm-theorem);
- For any computable function $g :\subseteq \Sigma^\omega \times \Sigma^\omega \to \Sigma^a$ there is a computable function $s : \Sigma^\omega \to \Sigma^\omega$ such that $g(p,q) = \eta^{\omega a}(s(p))(q)$ (the computable smn-theorem);
- For any continuous function $g :\subseteq \Sigma^\omega \times \Sigma^\omega \to \Sigma^a$ there is a continuous function $s : \Sigma^\omega \to \Sigma^\omega$ such that $g(p,q) = \eta^{\omega a}(s(p))(q)$ (the continuous smn-theorem).

3 Notations and representations, induced effectivity

In TTE it is assumed that computers can only read and write finite or infinite sequences of symbols. Computability is transferred to other sets by notations and representations where finite or infinite sequences are used as names. We introduce the vocabulary now.

Definition 3.1 *A notation of a set M is a surjective function $\nu :\subseteq \Sigma^* \longrightarrow M$, a representation of a set M is a surjective function $\delta :\subseteq \Sigma^\omega \longrightarrow M$. A naming system is a notation or a representation.*

Examples for notations are the binary notation $\nu_{bin} :\subseteq \Sigma^* \longrightarrow \omega$ of the natural numbers and the representation $\nu_Q :\subseteq \Sigma^* \longrightarrow \mathbb{Q}$ of the rational numbers by signed binary fractions (for example, $\nu_{bin}(101) = 5$, $\nu_Q(-10/-111) = 2/7$). In the following we shall abbreviate $\nu_Q(w)$ by $\bar{w}$. Well known is the decimal representation $\delta_{dec} :\subseteq \Sigma^\omega \longrightarrow \mathbb{R}$ of the real numbers by infinite decimal fractions.

Naming systems can be compared by reducibility, where we consider continuous and computable translations.

Definition 3.2 *For functions $\gamma :\subseteq Y \longrightarrow M$ and $\gamma' :\subseteq Y' \longrightarrow M'$ with $Y, Y' \in \{\Sigma^*, \Sigma^\omega\}$ we call γ reducible to γ', $\gamma \leq \gamma'$, iff $(\forall y \in dom(\gamma))\ \gamma(y) = \gamma' f(y)$ for some computable function $f :\subseteq Y \longrightarrow Y'$. We call γ and γ' equivalent, $\gamma \equiv \gamma'$, iff $\gamma \leq \gamma'$ and $\gamma' \leq \gamma$.*

Topological reducibility $\leq_t$ and topological equivalence $\equiv_t$ are defined accordingly by substituting "continuous" for "computable".

We have, for example, $\nu_{bin} \leq \nu_Q \leq \delta_{dec}$, but $\delta_{dec}|^{\mathbb{Q}} \not\leq_t \nu_Q$. Next, we define how naming systems transfer effectivity concepts from Σ^* and Σ^ω to the named sets.

Definition 3.3 *Let $\gamma_i :\subseteq Y_i \longrightarrow M_i$ $(i = 1, \ldots, k)$ be naming systems.*

1. *$x \in M_1$ is γ_1-computable, iff there is a computable element $y \in Y_1$ with $\gamma_1(y) = x$.*

2. *$X \subseteq M_1 \times \ldots \times M_k$ is $(\gamma_1, \ldots, \gamma_k)$–open (–r.e., –recursive), iff*

$$\{(y_1, \ldots, y_k) \in Y_1 \times \ldots \times Y_k \mid (\gamma_1(y_1), \ldots, \gamma_k(y_k)) \in X\}$$

is open (r.e., recursive) in $dom(\gamma_1) \times \ldots \times dom(\gamma_k)$.

For any naming system $\gamma :\subseteq Y \longrightarrow M$, the set $\tau_\gamma := \{X \subseteq M \mid X$ is γ–open$\}$ is called the final topology of γ.

For a γ–open set X, a property $\gamma(p) \in X$ is guaranteed already by a finite prefix of p, for a γ–r.e. set there is additionally a "proof–system" for showing this. Examples will be given below.

Definition 3.4 *For $i = 0, \ldots, k$ let $\gamma_i :\subseteq Y_i \longrightarrow M_i$ be naming systems.*

1. *A relation $Q \subseteq M_1 \times \ldots \times M_k \times M_0$ is $(\gamma_1, \ldots, \gamma_k, \gamma_0)$–computable (–continuous) iff there is some computable (continuous) function $f :\subseteq Y_1 \times \ldots \times Y_k \longrightarrow Y_0$ with*

$$(\gamma_1(y_1), \ldots, \gamma_k(y_k), \gamma_0 f(y_1, \ldots, y_k)) \in Q$$

if $\exists x.(\gamma_1(y_1), \ldots, \gamma_k(y_k), x) \in Q$.

2. *A function $F :\subseteq M_1 \times \ldots \times M_k \longrightarrow M_0$ is $(\gamma_1, \ldots, \gamma_k, \gamma_0)$–computable (–continuous) iff there is some computable (continuous) function $f :\subseteq Y_1 \times \ldots \times Y_k \longrightarrow Y_0$ with*

$$F(\gamma_1(y_1), \ldots, \gamma_k(y_k)) = \gamma_0 f(y_1, \ldots, y_k)$$

whenever $F(\gamma_1(y_1), \ldots, \gamma_k(y_k))$ exists.

As an example, the real function $x \mapsto x/3$ is $(\delta_{dec}, \delta_{dec})$–computable, but $x \mapsto x \cdot 3$ is not even $(\delta_{dec}, \delta_{dec})$–continuous (see Section 2). Computability of the relation Q is an effective version of the mere existence statement $(\forall x_1, \ldots, x_k)(\exists x_0)$ $Q(x_1, \ldots, x_k, x_0)$. A function $f :\subseteq M_1 \times \ldots \times M_k \longrightarrow M_0$ is called a *choice function* for Q, iff $Q(x_1, \ldots, x_k, f(x_1, \ldots, x_k))$ for all $(x_1, \ldots, x_k) \in dom(Q)$. A relation may be computable without having a computable choice function.

For instructive examples consider the enumeration representation *En* and the characteristic function representation *Cf* of $2^\omega = \{A \mid A \subseteq \omega\}$ defined by

$$En(p) := \{n \mid 10^{n+1}1 \text{ is a subword of } p\},$$
$$Cf(p) := \{n \mid p(n) = 0\}.$$

The following statements can be proved easily. A set $A \subseteq \omega$ is r.e., iff it is *En*–computable, and it is recursive, iff it is *Cf*–computable; $Cf \leq En$ but $En \not\leq_t Cf$ (an enumeration cannot be converted to a characteristic function); the set $El := \{(A, n) \in 2^\omega \times \omega \mid n \in A\}$ is (En, ν_{bin})–r.e., it is (En, ν_{bin})–computable, has no (En, ν_{bin})–continuous choice function but a (Cf, ν_{bin})–computable choice function. (Informally, *El* corresponds to the existence statement "every non–empty subset of ω has an element".)

4 Effective representations

Simple considerations show that two naming systems of a set M are equivalent (topologically equivalent), iff they induce the same computational (topological) properties on M.

Which among the numerous representations of the set $\mathbb{R}$ of real numbers induce the "right" kind of computability? This question has no answer for the bare set $\mathbb{R}$ but a definite answer for any *information structure* on $\mathbb{R}$.

Definition 4.1

1. *An information structure on a set M is a triple (M, σ, ν), where $\nu :\subseteq \Sigma^* \longrightarrow \sigma$ is a notation of a set $\sigma \subseteq 2^M$ of subsets of M, which identifies points (that is, $\bigcup\sigma = M$ and $\{Q \in \sigma \mid x \in Q\} = \{Q \in \sigma \mid y \in Q\} \Rightarrow x = y$ for all $x, y \in M$).*
2. *The standard representation $\delta_\nu :\subseteq \Sigma^\omega \longrightarrow M$ for an information structure (M, σ, ν) is defined by:*

$$\delta_\nu(p) = x :\iff \{w \mid x \in \nu(w)\} = \{w \mid \sharp w \sharp\sharp \text{ is a subword of } p\}$$

We assume tacitly that $dom(\nu) \cap \Sigma^*\{¢, \sharp\}\Sigma^* = \emptyset$. The elements Q of σ are called *atomic properties*. A standard name $p \in \Sigma^\omega$ of $x \in M$ is a list of all those atomic properties $Q \in \sigma$ which hold for x. Since σ identifies points, x is uniquely defined by p. There are many important examples for information structures:

Example 4.2 *The following triples (M, σ, ν) are information structures.*

(1) $M = \Sigma^\omega$, $p \in \nu(w) \iff w$ *is a prefix of* p;
(2) $M = 2^\omega$, $A \in \nu(w) \iff \nu_{bin}(w) \in A$;
(3) $M = \mathbb{R}$, $x \in \nu(w) \iff \bar{w} < x$ (*remember* $\bar{w} = \nu_Q(w)$);
(4) $M = \mathbb{R}$, $x \in \nu(w) \iff x < \bar{w}$;
(5) $M = \mathbb{R}$, $x \in \nu(w) \iff (w = u¢v, \bar{u} < x < \bar{v})$;
(6) $M = \mathbb{P}$, $f \in \nu(w) \iff (w = 0^i¢0^j, f(i) = j)$;
(7) $M = \tau_{\mathbb{R}}$, $U \in \nu(w) \iff (w = u¢v, [\bar{u}; \bar{v}] \subseteq U)$;
(8) $M = C(\mathbb{R})$, $f \in \nu(w) \iff (w = u¢v¢y¢z, f[\bar{u}; \bar{v}] \subseteq (\bar{y}; \bar{z}))$

(*where* $\mathbb{P} = \{f \mid f :\subseteq \omega \longrightarrow \omega\}$ *and* $C(\mathbb{R}) = \{f : \mathbb{R} \longrightarrow \mathbb{R} \mid f$ *continuous*$\}$).

Any information structure defines a T_0–topology τ_σ on M, which has σ as a subbase [Eng89]. On the other hand, for any T_0–space (M, τ) with notation $\nu :\subseteq \Sigma^* \longrightarrow \sigma$ of a subbase, (M, σ, ν) is an information structure.

The user, who wants to operate effectively on a set M, has to specify the set σ of atomic properties and the notation ν of σ. By σ a concept of "approximation" is introduced on M (mathematically by the topology τ_σ), and the notation ν expresses how atomic properties can be handled concretely, that is, ν fixes computability on σ and M. The case of real numbers in the above example shows that for a set M different information structures may be of interest. Of course, only the most relevant ones can be investigated in more detail. The choice of the notation ν of σ should be justified, but usually there is some "standard notation" which is generally accepted as "effective".

In Example 4.2 the topologies τ_σ are (1) the Cantor topology, (2) the Scott cpo topology, (3) the left topology $\{(x;\infty) \mid x \in \mathbb{R}_\infty\} \cup \{\mathbb{R}\}$, (4) the right topology, (5) the standard real line topology, (8) the compact–open topology. The standard representation δ_ν is continuous and open, and, as an outstanding property it is t–complete in the class of continuous representations:

Theorem 4.3 *For any representation δ of M: $\delta \leq_t \delta_\nu \iff \delta$ is continuous.*

Since δ is continuous iff $\{(u,p) \mid \delta(p) \in \nu(u)\}$ is open in $dom(\nu) \times dom(\delta_\nu)$, this theorem has a striking similarity to the following characterization of the "admissible Gödel numberings" $\varphi : \omega \longrightarrow P^{(1)}$: For any $\psi : \omega \longrightarrow P^{(1)}$ we have $\psi \leq \varphi \iff \{(i,j,k) \mid (j,k) \in graph(\psi_i)\}$ is r.e. (Here "$\Leftarrow$" is the smn–theorem and "$\Rightarrow$" is the utm–theorem, see [Rog67].)

The representations which are t–equivalent to the standard representation δ_ν are called *admissible*. The admissible representations for an information structure are defined already by their final topology. They are the "topologically effective" representations. Most important is the following theorem:

Theorem 4.4 *For admissible representations $\delta_i :\subseteq \Sigma^\omega \longrightarrow M_i$ with final topologies τ_i $(i = 0, \dots, k)$, a function $f :\subseteq M_1 \times \dots \times M_k \longrightarrow M_0$ is continuous, iff it is $(\delta_1, \dots, \delta_k, \delta_0)$–continuous.*

Consequently, for admissible representations $(\delta_1, \dots, \delta_k, \delta_0)$–computable functions are continuous. For separable metric spaces, the *Cauchy representations* are admissible.

Theorem 4.5 *Let (M,d) be a metric space with topology τ and let $\alpha :\subseteq \Sigma^* \longrightarrow A$ be a notation of a dense subset. Then the Cauchy representation $\delta_\alpha :\subseteq \Sigma^\omega \longrightarrow M$ defined as follows is admissible with final topology τ.*

$$\begin{aligned}\delta_\alpha(p) = x :\iff\ & p = u_0 \sharp u_1 \sharp \dots \ (u_i \in dom(\alpha)) \text{ with} \\ & (\forall k)(\forall i > k) d(\alpha(u_i), \alpha(u_k)) < 2^{-k} \text{ and } x = \lim_{i\to\infty} \alpha(u_i).\end{aligned}$$

Notice, that only "fast" converging sequences of elements of the dense set A are used as names.

5 Computability on the real line

The main subject in Analysis is the real line, that is, the topological space $(\mathbb{R}, \tau_\mathbb{R})$, where $\tau_\mathbb{R}$ is the set of open subsets of $\mathbb{R}$. We consider the information structure $(\mathbb{R}, \sigma, \nu)$ from Example 4.2(5) with $x \in \nu(w) \iff (w = u¢v$ and $\bar{u} < x < \bar{v})$ as basis of our effectivity theory on the real line. The standard representation δ_ν (Def. 4.1) is equivalent to the Cauchy representation, where fast converging Cauchy sequences of rational numbers serve as names. We define $\rho_C := \delta_{\nu_Q}$ explicitly:

Definition 5.1 *The Cauchy representation* $\rho_C :\subseteq \Sigma^\omega \longrightarrow \mathbb{R}$ *is defined by* $\rho_C(p) = x$ *iff* $p = u_0\sharp u_1\sharp\ldots$ *with* $u_0, u_1, \ldots \in dom(\nu_Q)$ *and* $|\bar{u}_i - \bar{u}_k| < 2^{-k}$ *whenever* $i > k$ *and* $x = \lim \bar{u}_k$.

Like δ_ν, the Cauchy representation is admissible with final topology $\tau_\mathbb{R}$. The requirement of admissibility with final topology $\tau_\mathbb{R}$ excludes many other representations. Since $x \mapsto 3x$ is not $(\delta_{dec}, \delta_{dec})$–continuous (see Example 2.2.(4)), δ_{dec} cannot be admissible with final topology $\tau_\mathbb{R}$ by Theorem 4.4. Since $\tau_\mathbb{R}$ is the final topology of δ_{dec}, it cannot be admissible at all.

Furthermore, there is no injective and no total admissible representation of $\mathbb{R}$ with final topology $\tau_\mathbb{R}$. The "naive" Cauchy representation $\delta_{naive} :\subseteq \Sigma^\omega \longrightarrow \mathbb{R}$, where $\delta_{naive}(p) = x :\Longleftrightarrow p = u_0\sharp u_1\sharp\ldots$ and $(\bar{u}_0, \bar{u}_1, \ldots)$ is a Cauchy sequence with limit x, has $\{\emptyset, \mathbb{R}\}$ as its final topology and is not admissible. If $\delta_{naive}(p) = x$, then no finite prefix of p contains any information about x.

We shall now study effectivity induced on $\mathbb{R}$ by the Cauchy representation ρ_C. We shall also consider the notations ν_{bin} of ω and ν_Q of $\mathbb{Q}$ and the representation δ_C of $\mathbb{R}$ as standard and say, by convention, "computable" instead of "ρ_C–computable", "r.e." instead of "$(\nu_{bin}, \rho_C, \nu_Q)$–r.e." etc. (see the definitions in Chapter 3).

By Definition 3.3(1), a real number x is computable, iff $x = \rho_C(p)$ for some computable sequence $p \in \Sigma^\omega$. Every rational number is computable. $\sqrt{2}, e, \pi$ and many other important real numbers are computable. Every "fast" converging (ν_{bin}, ρ_C)–computable Cauchy sequence has a computable limit. For any set $A \subseteq \omega$, $x_A := \Sigma\{2^{-i} \mid i \in A\}$ is computable, iff A is recursive.

Let A be r.e., but not recursive. There is some injective computable function $f : \omega \longrightarrow \omega$ with $A = range(f)$. Define $a_n := \Sigma\{2^{-f(i)} \mid i \leq n\}$. Then $(a_n)_{n\in\omega}$ is an increasing (ν_{bin}, ν_Q)–computable sequence converging to the non–computable number x_A. Since the function f must enumerate many numbers $k \in A$ very late the sequence $(a_n)_{n\in\omega}$ converges very slowly. However, the number x_A is $\rho_<$–computable (see below).

Next, we study which properties are recursive or r.e. (with respect to to ρ_C).

Theorem 5.2 *For any* $X \subseteq \mathbb{R}$,

1. X *is* ρ_C*–open* $\Longleftrightarrow$ X *is open (that is,* $X \in \tau_\mathbb{R}$*).*
2. X *is* ρ_C*–r.e.* $\Longleftrightarrow$ $X = \bigcup\{(\bar{u}; \bar{v}) \mid (u, v) \in Y\}$ *for some r.e.* $Y \subseteq \Sigma^* \times \Sigma^*$.
3. X *is* ρ_C*–recursive* $\Longleftrightarrow$ $X = \emptyset$ *or* $X = \mathbb{R}$.

By (3), no non–trivial property on $\mathbb{R}$ can be decided, if only ρ_C–names are available. For a proof, let X be ρ_C–recursive. Then X and $\mathbb{R}\setminus X$ must be ρ_C–open, hence open. But $\emptyset$ and $\mathbb{R}$ are the only subsets of X with this property. Theorem 5 can be generalized easily from $\mathbb{R}$ to $\mathbb{R}^n$. $\{(x, y) \mid x < y\}$ and $\{x \mid x < b\}$ (b computable) are examples of r.e. sets. For example, there is a Type 2 machine M which for input (p, q), $p = u_0\sharp u_1\sharp\ldots \in dom(\rho_C)$, $q = v_1\sharp v_2\sharp\ldots \in dom(\rho_C)$, halts as soon as it has found some k with $\bar{u}_k + 2^{-k} < \bar{v}_k - 2^{-k}$).

The above sets are not recursive with respect to our representation ρ_C. Are there representations δ which make them recursive? For the set $\{x \in \mathbb{R} \mid x < 0\}$

the solution is easy: the set is δ_{dec}–recursive ($\delta_{dec}(p) < 0 \iff$ "–" is the first symbol of p and some digit different from 0 appears in p). However, there is no representation δ of $\mathbb{R}$ such that $\{(x,y) \mid x < y\}$ is (δ,δ)–recursive [Wei95]. The relation $x < y$ is "easily definable" but "absolutely" non–recursive. (Notice that some computability models on $\mathbb{R}$ have decidability of $x < y$ as a basic assumption [BSS89]).

Occasionally, the left representation $\rho_<$ derived from the information structure from Example 4.2(3) and the corresponding right representation $\rho_>$ (Example 4.2(4)) are useful. The following can be proved easily: ($\delta \le \rho_<$ and $\delta \le \rho_>$) $\iff \delta \le \rho_C$; $\rho_< \not\le_t \rho_C$; $\rho_> \not\le_t \rho_C$; a real number x is ρ_C–computable $\iff$ x is $\rho_<$–computable and $\rho_>$–computable; $\Sigma\{2^{-i} \mid i \in A\}$ is $\rho_<$–computable if A is r.e. (where $A \subseteq \omega$).

By our definitions, a function $f :\subseteq \mathbb{R} \longrightarrow \mathbb{R}$ is (ρ_C, ρ_C)–computable, iff some Type 2 machine transforms any Cauchy sequence of rational numbers converging fast to $x \in dom(f)$ to some Cauchy sequence of rational numbers converging fast to $f(x)$. Since ρ_C and δ_ν from Example 4.2(4) are equivalent, f is computable, iff some Type 2 machine transforms any enumeration of all basic neighbourhoods of x to an enumeration of all basic neighbourhoods of $f(x)$. Notice that by Theorem 4.4 every real function computable with respect to ρ_C is continuous. Many of the commonly used functions from "classical" analysis are computable.

Theorem 5.3 *1. The real functions $(x,y) \mapsto x + y$, $(x,y) \mapsto x \cdot y$, $(x,y) \mapsto \max(x,y)$, $x \mapsto -x$, $x \mapsto 1/x$ are computable.*

2. Let $(a_i)_{i\in\omega}$ be a (ν_{bin}, ρ_C)–computable sequence and let $R_0 > 0$ be the radius of the power series $\Sigma a_i x^i$. Then for any $R < R_0$ the real function $f_R(x) :=$ ($\Sigma a_i x^i$ if $|x| < R$, div otherwise) is computable.

A Type 2 machine M for computing $x \mapsto 1/x$ may work for input $u_0\sharp u_1\sharp\ldots \in dom(\rho_C)$ as follows. First M determines some $N \in \omega$ with $|\bar{u}_N| > 2 \cdot 2^{-N}$. Then M writes $v_0\sharp v_1\sharp\ldots$ where $\bar{v}_k = 1/\bar{u}_{2N+k}$. If $x = \rho_C(u_0\sharp u_1\sharp\ldots)$ then $1/x = \rho_C(v_0\sharp v_1\sharp\ldots)$. The other cases of (1) are proved accordingly. A Type 2 machine M for computing the power series (2) may dermine for input $u_0\sharp u_1\sharp\ldots \in dom(\rho_C)$ a sequence $v_0\sharp v_1\sharp\ldots$ such that $\bar{v}_n = \sum_{i=0}^{N} c_i b^i$, where N is sufficiently large and $b, c_0, \ldots, c_N$ are rational numbers with $|b - x|$ and $|a_i - c_i|$ sufficiently small such that $|f(x) - \bar{v}_n| < 2^{-n-1}$ (where $x = \rho_C(u_0\sharp u_1\sharp\ldots)$. For details see [Wei95]. By 5.3.(2), exp, sin, cos and many other functions are computable. Notice also, that the computable functions are closed under composition. Theorem 5.3(2) cannot be generalized to the case $R = R_0$! By identifying the complex plane $\mathbb{C}$ with $\mathbb{R}^2$, computability of complex functions can be reduced to computability of real functions. The complex functions addition, multiplication, inversion, exp, sin, log and so on are computable.

6 Open, closed and compact subsets of the reals

The power set $2^{\mathbb{R}}$ of $\mathbb{R}$ has no representation since its cardinality is too large. Therefore, it is impossible within TTE to express computability of functions such as sup $:\subseteq 2^{\mathbb{R}} \longrightarrow \mathbb{R}$. The open, the closed and the compact subsets of $\mathbb{R}$, however, have representations since they have continuum cardinality. We introduce and discuss some useful admissible representations of these sets.

A subset $U \in \mathbb{R}$ is open (that is, $U \in \tau_{\mathbb{R}}$), iff it is the union of a set of open intervals with rational boundaries. The following definition suggests itself.

Definition 6.1 *Define* $\delta_{op} : \Sigma^\omega \longrightarrow \tau_{\mathbb{R}}$ *by*

$$\delta_{op}(p) := \bigcup\{(\bar{u};\bar{v}) \mid u, v \in dom(\nu_Q), \sharp u¢v\sharp \text{ is a subword of } p\}$$

Thus, $\delta_{op}(p)$ is the open set "enumerated" by p. If the notation ν_Q is replaced by some standard notation of, for example, the finite binary fractions, a representation equivalent to δ_{op} is obtained. Thus, the definition is invariant under change of unimportant parameters. Let δ_7 be the standard representation for the information structure from Example 4.2(7). Then $\delta_7 \equiv \delta_{op}$, hence δ_{op} is admissible, where the final topology on $\tau_{\mathbb{R}}$ is generated by the subbase consisting of all sets $\{U \in \tau_{\mathbb{R}} \mid [a;b] \subseteq U\}$ (where a, b are rational numbers).

Operations on $\tau_{\mathbb{R}}$ such as union and intersection are $(\delta_{op}, \delta_{op}, \delta_{op})$-computable. The function $S :\subseteq \tau_{\mathbb{R}} \longrightarrow \mathbb{R}$, defined by $S(U) = x :\iff x = \sup U$, is $(\delta_{op}, \rho_<)$-computable ("least upper bounds of open sets can be approximated from below") but not (δ_{op}, ρ_C)-continuous. An interval $(a;b)$ with $a, b \in \mathbb{R}$ is δ_{op}-computable, iff a is $\rho_>$-computable and b is $\rho_<$-computable. For any computable function $f : \mathbb{R} \longrightarrow \mathbb{R}$, the function $F : \tau_{\mathbb{R}} \longrightarrow \tau_{\mathbb{R}}$, defined by $F(U) := f^{-1}(U)$, is $(\delta_{op}, \delta_{op})$-computable. This is a computable version of the fact that $f^{-1}(U)$ is open, if f is continuous and U is open.

A natural representation of the closed sets is obtained from δ_{op} by complementation: $\delta_{cl}(p) := \mathbb{R} \setminus \delta_{op}(p)$. Compactness is a good example for a classical concept which branches into many parts in TTE. A subset $X \subseteq \mathbb{R}$ is compact, iff it is closed and bounded or, equivalently by the Heine/Borel theorem, iff every open covering has a finite subcovering. Several representations of the set $K(\mathbb{R})$ of the compact subsets of $\mathbb{R}$ are of interest. We characterize some of them by information structures (Def. 4.1).

Definition 6.2 *Define information structures* $(K(\mathbb{R}), \sigma_{..}, \nu_{..})$ *and the corresponding standard representations:* $\delta_{..} :\subseteq \Sigma^\omega \longrightarrow K(\mathbb{R})$ *as follows:*

$$\begin{aligned}
\delta_o : X \in \nu_o(w) : &\iff w = u¢v \text{ and } X \cap [\bar{u};\bar{v}] = \emptyset \\
\delta_n : X \in \nu_n(w) : &\iff w = u¢v \text{ and } X \cap (\bar{u};\bar{v}) \neq \emptyset \\
\delta_{ob} : X \in \nu_{ob}(w) : &\iff w = y¢0^n \text{ and } X \in \nu_o(y) \text{ and } X \subseteq (-n;n) \\
\delta_{obn} : X \in \nu_{obn}(w) : &\iff w = x¢z \text{ and } X \in \nu_{ob}(x) \cap \nu_n(z) \\
\kappa_w : X \in \nu_w(w) : &\iff w = u_1¢v_1¢\ldots¢u_k¢v_k \\
&\qquad \text{and } X \subseteq (\bar{u}_1;\bar{v}_1) \cup \ldots \cup (\bar{u}_k;\bar{v}_k) \\
\kappa_s : X \in \nu_s(w) : &\iff w = u_1¢v_1¢\ldots¢u_k¢v_k \\
&\qquad \text{and } X \subseteq (\bar{u}_1;\bar{v}_1) \cup \ldots \cup (\bar{u}_k;\bar{v}_k) \\
&\qquad \text{and } X \cap (\bar{u}_i;\bar{v}_i) \neq \emptyset \text{ for } i = 1, \ldots, k
\end{aligned}$$

Thus: $\delta_o(p) = X$, iff p enumerates the "outside" of X (that is, δ_o is equivalent to a restriction of δ_{cl}); $\delta_n(p) = X$, iff p enumerates the open "neighbours" of

X; $\delta_{ob}(p) = X$, iff p enumerates the outside and additionally bounds of X; $\delta_{obn}(p) = X$, iff p enumerates the outside, the bounds and the neighbours of X; $\kappa_w(p) = X$, iff p enumerates all finite coverings (with rational intervals) of X; $\kappa_s(p) = X$, iff p enumerates all minimal finite coverings of X. The above namings $\delta_{..}$ represent different types of knowledge about the named objects. Reducibilities express dependences between these types. For the first four representations we have $\delta_{obn} \leq \delta_{ob} \leq \delta_o$ and $\delta_{obn} \leq \delta_n$, while $\not\leq_t$ holds for all other combinations. The following equivalences can be interpreted as two computable versions of the Heine/Borel theorem [KW87]:

Theorem 6.3 $\delta_{ob} \equiv \kappa_w$; $\delta_{obn} \equiv \kappa_s$.

Every non-empty compact set $X \in K(\mathbb{R})$ has a maximum. Which type of information suffices to determine it?

Theorem 6.4 *Let* $Max :\subseteq K(\mathbb{R}) \longrightarrow \mathbb{R}$ *be the maximum function on the compact subset of* $\mathbb{R}$, *then*

- *Max is* $(\delta_n, \rho_<)$*-computable and* $(\delta_{ob}, \rho_>)$*-computable,*
- *Max is* **not** $(\delta_n, \rho_>)$*-continuous,* $(\delta_0, \rho_>)$*-continuous or* $(\delta_{ob}, \rho_<)$*-continuous.*

Notice in particular, that without knowing a concrete upper bound not even a $\rho_>$-name of $Max(X)$ can be determined from a δ_0-name of X.

The above representations behave differently with respect to union and intersection. While union is computable with respect to all of them, intersection is for example $(\kappa_w, \kappa_w, \kappa_w)$-computable but not even $(\kappa_s, \kappa_s, \delta_n)$-continuous.

We have already mentioned that no non-trivial subset of $\mathbb{R}$ is ρ_C-recursive. The κ_s-computable compact sets may be used as substitutes for ρ_C-recursive sets. This view is affirmed by the following fact: X is κ_s-computable, iff the distance function $d_X : \mathbb{R} \longrightarrow \mathbb{R}$, $d_X(y) := \min\{|x - y| \mid x \in X\}$, is (ρ_C, ρ_C)-computable. The function d_X can be considered as a generalized characteristic function. (Notice the similarity to locatedness in [Bis67].)

As a basic fact from topology, $f(X)$ is compact, if X is compact and $f : \mathbb{R} \longrightarrow \mathbb{R}$ is continuous. We have the following computable version: For computable functions $f : \mathbb{R} \longrightarrow \mathbb{R}$ the function $F : K(\mathbb{R}) \longrightarrow K(\mathbb{R})$, $F(X) := f(X)$, is (κ_s, κ_s)-computable.

All the concepts can be generalized easily from $\mathbb{R}$ to $\mathbb{R}^n$ $(n > 1)$. Some more details can be found in [KW87, Wei94, Wei95].

7 Spaces of continuous functions

For $X \subseteq \mathbb{R}$, $C(X)$ denotes the class of continuous functions $f :\subseteq \mathbb{R} \longrightarrow \mathbb{R}$ with $X = dom(f)$. We consider the classes $C(\mathbb{R})$ and $C[0;1]$. First we define a representation of $C(\mathbb{R})$ by giving an information structure.

Definition 7.1 *Let $\delta_{\mathbb{R}}$ be the standard representation for the information structure $(C(\mathbb{R}), \sigma, \nu)$ where*

$$f \in \nu(w) :\iff w = u¢v¢x¢y \text{ and } f[\bar{u};\bar{v}] \subseteq (\bar{x};\bar{y}) \text{ (see Example 4.2.(8))}$$

Thus $\delta_{\mathbb{R}}(p) = f$, iff p enumerates all $(a, b, c, d) \in \mathbb{Q}^n$ with $f[a; b] \subseteq (c; d)$. Any atomic information about f is a rectangle bounding the graph of f. $\delta_{\mathbb{R}}$ is admissible, where the final topology is the well-known compact-open topology [Eng89]. The following theorem emphasizes the effectivity of $\delta_{\mathbb{R}}$ and additionally is a useful tool.

Theorem 7.2 *Define apply* : $C(\mathbb{R}) \times \mathbb{R} \longrightarrow \mathbb{R}$ *by apply*$(f, x) := f(X)$. *Then for any representation δ of $C(\mathbb{R})$:*

- *apply is (δ, ρ_C, ρ_C)-continuous* $\iff \delta \leq_t \delta_{\mathbb{R}}$,
- *apply is (δ, ρ_C, ρ_C)-computable* $\iff \delta \leq \delta_{\mathbb{R}}$.

Notice the formal similarity with the smn– and the utm–theorem from recursion theory. The following theorem summarizes some effectivity properties [KW87]:

Theorem 7.3

1. $f : \mathbb{R} \longrightarrow \mathbb{R}$ *is (ρ_C, ρ_C)-computable* $\iff$ *f is $\delta_{\mathbb{R}}$-computable.*
2. *The composition $(f, g) \mapsto fg$ on $C(\mathbb{R})$ is $(\delta_{\mathbb{R}}, \delta_{\mathbb{R}}, \delta_{\mathbb{R}})$-computable.*
3. *The function $G : C(\mathbb{R}) \times \tau_{\mathbb{R}} \longrightarrow \tau_{\mathbb{R}}$, $G(f, X) := f^{-1}(X)$, is $(\delta_{\mathbb{R}}, \delta_{op}, \delta_{op})$-computable.*
4. *The function $H : C(\mathbb{R}) \times K(\mathbb{R}) \longrightarrow K(\mathbb{R})$, $H(f, X) := f(X)$, is $(\delta_{\mathbb{R}}, \kappa_s, \kappa_s)$-computable.*

On $C[0; 1]$ we introduce a Cauchy representation based on the "maximum metric" $d(f, g) := \max\{|f(x) - g(x)| \mid 0 \leq x \leq 1\}$. Let $\alpha :\subseteq \Sigma^* \longrightarrow Pg$ be the following notation of all polygon functions from $C[0; 1]$ with rational vertices: $\alpha(v_0¢u_1¢v_1¢ \ldots ¢u_{n-1}¢v_{n-1}¢v_n) :=$ the polygon with vertices $(0, \bar{v}_0), (\bar{u}_1, \bar{v}_1), \ldots, (\bar{u}_{n-1}, \bar{v}_{n-1}), (1, \bar{v}_n)$ $(0 < \bar{u}_1 < \ldots < \bar{u}_{n-1} < 1)$. It is known that Pg is dense in the metric space $(C[0; 1], d)$. By Theorem 4.5 the Cauchy representation δ_α of $C[0; 1]$ is admissible. If $\delta_\alpha(u_0 \sharp u_1 \sharp \ldots) = f$, then $\alpha(u_0), \alpha(u_1), \ldots$ is a sequence of polygons moving (with speed 2^{-n}) closer and closer to f. The representation δ_α is equivalent to the "restriction" δ of $\delta_{\mathbb{R}} :\subseteq \Sigma^\omega \longrightarrow C(\mathbb{R})$ defined by $\delta(p)(x) := (\delta_{\mathbb{R}}(p)(x)$ if $0 \leq x \leq 1$, *div* otherwise). This affirms the naturality of δ_α. We state three properties.

Theorem 7.4

1. *Max* : $C[0; 1] \longrightarrow \mathbb{R}$ *with Max*$(f) := \max\{f(x) \mid 0 \leq x \leq 1\}$ *is (δ_α, ρ_C)-computable.*
2. *Int* : $C[0; 1] \times \mathbb{R} \longrightarrow \mathbb{R}$ *with Int*$(f, a) := \int_0^a f(x)dx$ *is $(\delta_\alpha, \rho_C, \rho_C)$-computable.*
3. *Diff* $:\subseteq C[0; 1] \longrightarrow C[0; 1]$, *defined by Diff*$(f) = g$ *iff $g \in C[0; 1]$ is the derivative of f, is not $(\delta_\alpha, \delta_\alpha)$-continuous.*

In accordance with experience from numerical mathematics, the maximum can be determined, integration is easy but differentiation cannot be performed effectively on δ_α–names.

8 Determination of zeroes

As a case–study we investigate the determination of zeroes for functions from $C[0;1]$. For details see [Wei87, Wei95]. A subset $X \subseteq [0;1]$ is closed, iff it is the set of zeroes of some $f \in C[0;1]$. For the Cauchy representation δ_α of $C[0;1]$ and the representation δ_o from Definition 6.2 a computable version can be proved:

Theorem 8.1 *Define $S := \{(f, X)) \mid f \in C[0;1],\ X$ closed, $X = f^{-1}\{0\}\}$. Then S is $(\delta_\alpha, \delta_o)$–computable and S^{-1} is $(\delta_o, \delta_\alpha)$–computable (see Definition 3.4.(1)). But S^{-1} has no $(\delta_o, \delta_\alpha)$–continuous choice function.*

We consider the problem of determining zeroes. As a positive result, the minimal zeroes can be approximated from below (and the maximal zeroes from above):

Theorem 8.2 *The function $Z_{\min} :\subseteq C[0;1] \longrightarrow \mathbb{R}$, where $Z_{\min}(f) = x :\iff x = \min f^{-1}\{0\}$, is $(\delta_\alpha, \rho_<)$–computable but not (δ_α, ρ_C)–continuous. (Correspondingly for $Z_{\max}$)*

Corollary 8.3 *The function $Z_1 :\subseteq C[0;1] \longrightarrow \mathbb{R}$, where $Z_1(f) = x :\iff x$ is the only zero of f, is (δ_α, ρ_C)–computable.*

Thus, a computable function determines the zeroes for all functions which have exactly one zero. By the intermediate value theorem, every function $f \in C[0;1]$, which changes its sign, has a zero; however:

Theorem 8.4 *The relation $R := \{(f, x) \mid f \in C[0;1], x \in \mathbb{R}, f(0) < 0, f(1) > 0, f(x) = 0\}$ is not (δ_α, ρ_C)–continuous.*

Thus, there is not even a continuous function, which for any δ_α–name of a function f (which changes its sign) determines some ρ_C–name of a zero of f. We obtain a weak positive result, if we consider only functions f, for which $f^{-1}\{0\}$ contains no proper interval.

Theorem 8.5 *The relation $R := \{(f, x) \mid f \in C[0;1], x \in \mathbb{R}, f(0) < 0, f(1) > 0, f(x) = 0, f^{-1}\{0\}$ contains no proper interval$\}$ is (δ_α, ρ_C)–computable but has no (δ_α, ρ_C)–continuous choice function (see Chapter 3).*

A special case is the problem to determine a zero of $f_a : \mathbb{R} \longrightarrow \mathbb{R}$ from $a \in [-1;1]$ where $f_a(x) := x^3 - x + a$. The relation $\{(a, x) \mid f_a(x) = 0\}$ is (ρ_C, ρ_C)–computable but has no (ρ_C, ρ_C)–continuous choice function. Every Type 2 machine, which for any ρ_C–name of some $a \in [-1;1]$ determines a ρ_C–name of some zero of f_a, is not "(ρ_C, ρ_C)–extensional", that is, there are different names of some number a which are mapped to names of different zeroes.

9 Computational complexity

We introduce computational complexity for Type 2 machines and transfer it by representations to other sets. For an (ordinary) Turing machine M, $Time_M(x)$ is the number of steps which M with input x works until it halts. Such a definition is useless for Type 2 machines with infinite output, since valid computations never reach a HALT statement. We introduce as a further parameter a number $k \in \omega$ and measure the time until M has produced the first k output symbols. Another important information is the input lookahead, that is, the number of input symbols which M requires for producing the first k output symbols. The parameter k can be interpreted as precision of the intermediate result.

Definition 9.1 *Let M be a Type 2 machine with $f_M :\subseteq (\Sigma^\omega)^m \longrightarrow \Sigma^\omega$. For all $y \in (\Sigma^\omega)^m$ and all $k \in \omega$ define time and input lookahead by:*

$Time_M(y)(k) :=$ *the number of steps which M with input y needs for writing the first $k+1$ output symbols,*

$Ila_M(y)(k) :=$ *the maximal j such that M with input y reads the jth symbol from some input tape during the first $Time_M(y)(k)$ computation steps.*

For every $y \in dom(f_M)$, $Time_M(y) : \omega \longrightarrow \omega$ and $Ila_M(y) : \omega \longrightarrow \omega$ are functions which may vary considerably as functions of the input y. Bounds for time or input lookahead, which are uniform on some $X \subseteq (\Sigma^\omega)^m$ are of particular interest. Call a set $X \subseteq (\Sigma^\omega)^m$ co–r.e., iff $(\Sigma^\omega)^m \setminus X$ is r.e.. (The co–r.e. sets are of the same type as the κ_w–computable compact subsets of $\mathbb{R}$.) The co–r.e. subsets of $dom(f_M)$ are those sets for which uniform computable time bounds exists.

Theorem 9.2 *Let M be a Type 2 machines with $f_M :\subseteq (\Sigma^\omega)^m \longrightarrow \Sigma^\omega$.*

1. *If $X \subseteq dom(f_M)$ is co–r.e., then $(\forall y \in X)(\forall k)\, Time_M(y)(k) \leq t(k)$ for some computable function $t : \omega \longrightarrow \omega$.*
2. *If $t : \omega \longrightarrow \omega$ is computable, then $X := \{y \mid (\forall k)\, Time_M(y)(k) \leq t(k)\}$ is a co–r.e. subset of $dom(f_M)$.*

We would like to call a real function computable in time $t : \omega \longrightarrow \omega$, iff it is determined by a function on ρ_C–names which is computable in time t. Unfortunately, this definition is unreasonable since precision on $\mathbb{R}$ and precision on Σ^ω are not recursively related. Every ρ_C–name p can be padded arbitrarily with redundant symbols. Therefore, every computable real function can be computed in linear time (pad the output!). Also the definition "$g : \mathbb{R} \longrightarrow \mathbb{R}$ is computable in time $t : \omega \longrightarrow \omega$, iff some Type 2 machine computes g on ρ_C–names such that $g(x)$ is determined with error 2^{-k} in at most $t(k)$ steps" is not reasonable. In this case not even the identity is computable in time t (for some t) since on the input tape arbitrarily redundant (padded) names are admitted.

We solve the problem by using an admissible representation with uniformly bounded redundance, the "modified binary representation", which uses the binary digits 0, 1 and -1 (which we denote by $\bar{1}$).

Definition 9.3 *Define* $\rho :\subseteq \Sigma^\omega \longrightarrow \mathbb{R}$ *by:* $dom(\rho) := \{a_n \ldots a_0 \cdot a_{-1}a_{-2} \ldots \mid n \geq -1, a_i \in \{0,1,\bar{1}\}$ *for* $i \leq n$, $a_n \neq 0$ *if* $n \geq 0$, $a_n a_{n-1} \notin \{1\bar{1}, \bar{1}1\}$ *if* $n \geq 1\}$ *and* $\rho(a_n \ldots a_0 \cdot a_{-1}a_{-2} \ldots) = \sum\{a_i \cdot 2^i \mid i \leq n\}$.

Any ρ–name $p = a_n \ldots a_0 \cdot a_{-1} \ldots$ of some x defines a nested sequence of closed intervals converging to x. The prefix $a_n \ldots a_0 \cdot a_1 \ldots a_k$ determines an interval of length $2 \cdot 2^{-k}$. The digit a_{k+1} selects the left half if $a_{k+1} = \bar{1}$, the right half if $a_{k+1} = 1$ and the middle half if $a_{k+1} = 0$.

Theorem 9.4

1. $\rho \equiv \rho_C$.
2. $\rho^{-1}X$ *is compact if* X *is compact.*
3. $\rho^{-1}X$ *is co–r.e., if* X *is* κ_w*–computable.*

By (1), ρ induces the same computability theory on $\mathbb{R}$ as the Cauchy representation. As a consequence of (2), for any compact set $X \subseteq \mathbb{R}$ the redundance of ρ–names is uniformly bounded. We define computational complexity on $\mathbb{R}$ relative to the modified binary representation.

Definition 9.5 *A Type 2 machine* M *computes* $f :\subseteq \mathbb{R} \longrightarrow \mathbb{R}$ *on* $X \subseteq dom(f)$ *in time* $t : \omega \longrightarrow \omega$, *with input lookahead* $s : \omega \longrightarrow \omega$, *iff (1)* $f\rho(p) = \rho f_M(p)$, *(2)* $Time_M(p)(n) \leq t(n)$, *(3)* $Ila_M(p)(n) \leq s(n)$ *for all* $n \in \omega$, *whenever* $\rho(p) \in X$. *(Accordingly for* $f :\subseteq \mathbb{R}^k \longrightarrow \mathbb{R}$*)*

This kind of computational complexity is sometimes called "bit–complexity". It is essentially equivalent to definitions used by Brent [Bre76], Schönhage [Sch90], Ko [Ko91] and others. We state some simple results. Let $M : \omega \longrightarrow \omega$ be some "regular" upper time bound for binary integer multiplication (for example $M(n) = n \cdot \log n \cdot \log\log n$).

Theorem 9.6 *For every compact subset* X *of its domain, (1) addition can be computed in time* $c_1 \cdot n + c_1$ *with input lookahead* $n + d_1$, *(2) multiplication can be computed in time* $c_2 \cdot M(n) + c_2$ *with input lookahead* $2n + d_2$ *and inversion can be computed in time* $c_3 \cdot M(n) + c_3$ *with input lookahead* $2n + d_3$.

The constants d_1, d_2, d_3 depend only on X, while c_1, c_2 and c_3 may depend also on details of the Type 2 machine model. For proofs see [Wei95].

Theorem 9.7 *If* $f :\subseteq \mathbb{R} \longrightarrow \mathbb{R}$ *is analytic and computable in polynomial time, then its derivative is also computable in polynomial time.*

Proofs can be found in [Ko91] and [Mue87]. In Section 6 we have introduced the κ_s–computable compact subsets of $\mathbb{R}^n$, which are those compact sets X, for which the distance function $d_X : \mathbb{R}^n \longrightarrow \mathbb{R}$, $d_X(y) := d(y, X)$, is computable.The computational complexity of a compact set X can be defined by the computational complexity of the distance function d_X on some ball $[-k, k]^n$ which includes X. For subsets X of the plane this is the complexity of "plotting" the "picture" X pointwise [Wei95]. There may be a very strong trade–off between time complexity and input lookahead [Wei91]. There is a function $f \in C[0;1]$ computable in linear time with input lookahead $2n$, such that every machine which computes f with minimal lookahead requires more than polynomial time for many inputs iff $P \neq NP$. Therefore, if input information is cheap, f should be computed by a linear time machine. If, however, the input bits are expensive (for example, as results of physical measurements or an expensive preceeding computation), then an algorithm with minimal input lookahead may be advisible.

Admissible representations δ such that $\delta^{-1}K$ is compact, if K is compact, exist for all separable metric spaces, but not for more general topological spaces [Sch95]. This limits the possibility of studying computational complexity in Analysis.

10 Measure and integration

The concepts introduced so far can be used to define natural computability concepts in measure theory. In this section we sketch some ideas briefly. For details see [Wei96]. We only use some basic concepts from measure theory which can be found in any textbook on this topic, such as [Bau74].

For simplicity, we consider only the real interval $[0;1]$ as basic space. Let $Int := \{(a,b), [0;a), (b,1], [0;1] \mid 0 < a < b < 1, a, b \in \mathbb{Q}\}$ be the set of open subintervals of $[0;1]$ with rational boundaries. Let $\mathbf{B}$ be the set of Borel subsets of $[0;1]$, that is, the smallest σ–Algebra containing *Int*. Let $\mathbf{M}$ be the set of probability measures on the space $([0;1], \mathbf{B})$. Every measure $\mu \in \mathbf{M}$ is uniquely determined by its restriction to *Int*.

We define explicitly an information structure $(\mathbf{M}, \sigma, \nu)$ (Section 4) which induces canonically a topology and a computability theory on $\mathbf{M}$. Let $I :\subseteq \Sigma^* \longrightarrow Int$ be some standard notation of the set *Int* of intervals (with $dom(I) \cap \Sigma^*\{¢, \sharp\}\Sigma^* = \emptyset$). (Remember $\bar{u} := \nu_Q(u)$, see Section 3.)

Definition 10.1 *Define the information structure* $(\mathbf{M}, \sigma, \nu)$ *by*

$$\mu \in \nu(u¢v) :\Longleftrightarrow \bar{u} < \mu(I_v)$$

for all $u \in dom(\nu_Q)$, $v \in dom(I)$ *and* $\mu \in \mathbf{M}$.

Let τ_m *be the topology on* $\mathbf{M}$ *with subbase* σ *and let* δ_m *be the standard representation for this information structure (Definition 4.1).*

Notice, that σ identifies points in $\mathbf{M}$: $\mu = \mu'$ iff $\bar{u} < \mu(I_v) \iff \bar{u} < \mu'(I_v)$ for all $u \in dom(\nu_Q)$ and $v \in dom(I)$. Notice also, that only (rational) lower bounds of $\mu(J)$ ($J \in Int$) are available from δ_m–names. The effectivity theory induced on $\mathbf{M}$ by this information structure does not depend too much on the particular notation I of intervals. Let ρ_C, $\rho_<$ and $\rho_>$ be the representations of $\mathbb{R}$ introduced in Section 5.

Theorem 10.2 (stability). *Let $I' :\subseteq \Sigma^* \longrightarrow Int'$ be a notation of subintervals of $[0;1]$ with recursive domain such that*

1. *for some dense subset $D \subseteq [0;1]$,*

$$Int' = \{(a;b), [0;a), (a;1], [0;1] \mid 0 < a < b < 1; a, b \in D\},$$

and

2. sup *and* inf *on Int' are (I', ρ_C)-computable (Definition 5.1).*

Define $\delta_{m'}$ and $\tau_{m'}$ by substituting I' for I in Definition 10.1. Then $\tau_{m'} = \tau_m$ and $\delta_{m'} \equiv \delta_m$.

Remember: $\delta_{m'} \equiv \delta_m$ iff $\delta_{m'}$ and δ_m induce some effectivily theory on $\mathbf{M}$ (Section 3). The notation $I :\subseteq \Sigma^* \longrightarrow Int$ itself satisfies the conditions in Theorem 10.2. An important other example for I' is obtained by choosing for D the set of finite binary fractions $m/2^n$ $(m, n \in \omega, 1 \leq m < 2^n)$ instead of $\mathbb{Q}$.

The representation δ_m of the set of measure $\mathbf{M}$ has several interesting effectivity properties.

Theorem 10.3 *The function $(a, \mu, \mu') \mapsto a\mu + (1-a)\mu'$ is $(\rho_C, \delta_m, \delta_m, \delta_m)$-computable for $0 \leq a \leq 1$.*

Let δ_α be the standard representation of $C[0;1]$ from Section 7.

Theorem 10.4 *The function $(f, \mu) \mapsto T_f(\mu)$ for continuous $f : [0;1] \longrightarrow [0;1]$ and $\mu \in \mathbf{M}$ is $(\delta_\alpha, \delta_m, \delta_m)$-computable (where $T_f(\mu)(A) := \mu f^{-1}A$ for all $A \in \mathbf{B}$).*

As an application consider any iterated function system $([0;1], f_1, \ldots, f_N, p_1, \ldots, p_n)$ with computable $f_1, \ldots, f_N : [0;1] \longrightarrow [0;1]$ and with computable $p_1, \ldots, p_N \in \mathbb{R}_+$ $(p_1 + \ldots + p_N = 1)$. Then $G : \mathbf{M} \longrightarrow \mathbf{M}$, defined by

$$G(\mu)(A) := \sum_{i=1}^{N} p_i \mu f^{-1}(A)$$

is (δ_m, δ_m)-computable.

Also integration of continuous functions is computable. We formulate a fully effective version.

Theorem 10.5 *The function $(f, \mu) \mapsto \int f d\mu$ for $f \in C[0;1]$ and $\mu \in \mathbf{M}$ is $(\delta_\alpha, \delta_m, \rho_C)$-computable.*

Notice, that ρ_C-names can be determined and not only $\rho_<$-names. Measures of open sets are uniformly $\rho_<$-computable. Let δ_{op} be the representation of $\tau_\mathbb{R}$ from Definition 6.1, and define the representation $\delta_\tau :\subseteq \Sigma^\omega \longrightarrow \tau$ of the set τ of open subsets of $[0;1]$ by $\delta_\tau(p) := \delta_{op}(p) \cap [0;1]$.

Theorem 10.6 *The function $(\mu, U) \mapsto \mu(U)$ for $\mu \in \mathbf{M}$ and $U \in \tau$ is $(\delta_m, \delta_\tau, \rho_<)$-computable.*

This means, for given $\mu \in \mathbf{M}$ and $U \in \tau$, $\mu(U)$ can be approximated from below by rational numbers.

As a consequence (or by definition of δ_m), $(\mu, J) \mapsto \mu(J)$ is $(\delta_m, I, \rho_<)$-computable on $\mathbf{M} \times Int$. It seems to be desirable to approximate $\mu(J)$ also from above, that is, to obtain ρ_C-names.

An explicit way is to choose a "richer" representation. Replace the definition of ν in Definition 10.1 by

$$\mu \in \nu(u¢v¢w) : \Longleftrightarrow \bar{u} < \mu(I_v) < \bar{w}$$

For the resulting standard representation δ', the function $(\mu, J) \mapsto \mu(J)$ is (δ', I, ρ_C)-computable. Unfortunately, the stability property (Theorem 10.2) does not hold for δ'. Also Theorem 10.4 is false for δ' instead of δ_m. The difficulties come from measures with $\mu\{x\} \neq 0$ for some points $x \in [0;1]$. We shall show that for "reasonable" representations δ of $\mathbf{M}$, $(\mu, J) \mapsto \mu(J)$ cannot be $(\delta, I, \rho_>)$-computable. First we formulate a further effectivity property of δ_m: For any $x \in [0;1]$ define $\mu_x \in \mathbf{M}$ by $\mu_x(A) :=$ (1 if $x \in A$, 0 otherwise) for all $A \in \mathbf{B}$.

Theorem 10.7 *The function $x \mapsto \mu_x$ on $[0;1]$ is (ρ_C, δ_m)-computable.*

It seems to be natural that every reasonable representation δ of $\mathbf{M}$ satisfies Theorem 10.7. We show that this excludes the possibility of approximating measures of rational intervals from above. Let $\mathbf{M}' := \{\mu_x \mid x \in [0,1]\}$.

Theorem 10.8 *There is no representation δ of $\mathbf{M}'$ (or of $\mathbf{M}$) such that (1) and (2) hold.*

1. *$x \mapsto \mu_x$ on $[0;1]$ is (ρ_C, δ)-continuous.*
2. *$\mu \mapsto \mu(\frac{1}{2};1]$ on $\mathbf{M}'$ is $(\delta, \rho_>)$-continuous.*

Since (2) holds for the representation δ' defined above, (1) must fail for δ'. Since (1) holds for δ_m, (2) must fail for δ_m, that is, not even $\mu \mapsto \mu(\frac{1}{2};1]$ for $\mu \in \mathbf{M}'$ is $(\delta_m, \rho_>)$-computable (or continuous).

Let $f : [0;1] \longrightarrow \mathbb{R}$ be the characteristic function of $(1/2;1]$. Then $\mu(1/2;1] = \int f d\mu$ for $\mu \in \mathbf{M}$. Notice that f is not continuous. Restricted to continuous functions f, the integration operator is computable according to Theorem 10.5.

We introduce a further representation of the space $\mathbf{M}$ which might be useful for concrete computations.

For $n \in \omega, 0 \leq m \leq 2^n$ define the triangle function $t_{n,m} \in C[0;1]$ by

$$t_{n,m}(x) := \begin{cases} x - (m-1)2^{-n} & \text{if } (m-1)2^{-n} \leq x \leq m \cdot 2^{-n} \\ (m+1) \cdot 2^{-n} - x & \text{if } m \cdot 2^{-n} \leq x \leq (m+1) \cdot 2^{-n} \\ 0 & \text{otherwise} \end{cases}$$

for all $x \in [0;1]$. Let $Tr := \{t_{n,m} \mid n \in \omega, 0 \leq m \leq 2^n\}$ be the set of these triangle functions. Consider the information structure $(\mathbf{M}, \sigma', \nu')$ defined by

$$\mu \in \nu'(0^n ¢ 0^m ¢ u ¢ v) :\iff \bar{u} < \int t_{n,m} d\mu < \bar{v}.$$

(Indeed, $\sigma' = range(\nu')$ identifies points.) Let $\delta'_m :\subseteq \Sigma^\omega \longrightarrow \mathbf{M}$ be the associated standard representation of $\mathbf{M}$.

Theorem 10.9

$$\delta_m \equiv \delta'_m$$

For proving $\delta_m \leq \delta'_m$ use Theorem 10.5(2). For proving $\delta'_m \leq \delta_m$ notice that for any open interval $J \in Int$, the characteristic function cf_J can be approximated from below with arbitrary precision by finite sums of triangle functions from Tr.

11 Conclusion

The approach presented here (TTE) connects abstract analysis and Turing computability. It provides a simple language for expressing a variety of types of effectivity ranging form continuity (interpreted as "constructivity" or "finite dependence") via computability to computational complexity of points, sets, relations and functions. It is claimed that "Type 2 computability" models "computability by machines" adequately.

Computability aspects in Analysis have been discussed already from the beginning of Computability Theory. In his famous paper [Tur36], A. Turing defines computable real numbers by computable infinite binary fractions. In a correction [Tur37], he proposes to use computable sequences of nested intervals with rational boundaries instead, and he proves that no computable function translates programs for such sequences to equivalent binary fractions. As we already mentioned, TTE extends a definition of computable real functions by A. Grzegorczyk [Grz55]. Two years later, he introduced several definitions equivalent to the first one [Grz57]. Grzegorczyk's characterization "$f : [0;1] \to \mathbb{R}$ is computable, iff f maps every computable sequence of computable numbers to a computable sequence of computable numbers and f has a computable modulus of uniform continuity" has been generalized to computable Banach spaces by Pour–El and Richards [PR89]. TTE includes this generalization. Another characterization by Grzegorczyk (by means of monotone (with respect to inclusion) computable functions mapping intervals with rational boundaries to intervals with rational boundaries) is the computational background of Interval Analysis [Moo79, Abe88]. Domain approaches consider computability on spaces the elements of which are real numbers as well as approximations of real numbers (for example, intervals with rational boundaries) [WS81, Bla95, Hot95]. Domain computability restricted to the real numbers coincides with TTE–computability.

The "Russian" approach to Computable Analysis handles programs of computable real numbers. Let ψ be some standard notation (by "programs") of all computable elements of Σ^ω. Then $\rho_C\psi$ is a partial notation of the set of all computable real numbers $\mathbb{R}_C$. A function $f :\subseteq \mathbb{R}_C \to \mathbb{R}_C$ is Ceitin computable, iff it is $(\rho_C\psi, \rho_C\psi)$–computable. Every TTE–computable function $f :\subseteq \mathbb{R} \to \mathbb{R}$ with $dom(f) \subseteq \mathbb{R}_C$ is Ceitin computable. The converse holds only under certain conditions on $dom(f)$ [Cei59]. A necessary and sufficient condition has been given in [Her96]. Computational complexity based on the bit–model [Bre76, Sch90, Ko91] can be considered as a part of TTE. While in the Russian approach only the computable real numbers are considered as entireties, the "real RAM" model [BM75, BSS89, PS90] considers all real numbers as entireties. A real RAM is a random access machine operating on real numbers with assignments "$x := y+z$", "$x := y \cdot z$", ... and branchings "$x < 0$". Time complexity is measured by the number of operations as a function of the input dimension. The real RAM model is unrealistic since almost all real RAM–computable functions are not continuous, hence not realizable by physical machines, and almost all TTE–computable (that is, realizable) functions (such as exp, sin) are not real RAM–computable.

The computational model used in "Information Based Complexity" (IBC) [TWW88] extends the real RAM model [Nov95]. Also this model cannot be realized mathematically correctly by digital computers. Moreover, the output of one machine (in general) cannot be used as the input for a second machine. A slightly modified real RAM, the "feasible real RAM", however, can be realized physically [HB95]. Here, the precise tests "$x < 0$" are replaced by finite precision tests.

A different way to study effectivity in analysis is the use of constructive logic; see for example [TD88]. Bishop's "Constructive analysis" [Bis67, BB85] is the most famous example. Its close relation to TTE still has to be investigated.

References

[Abe88] Aberth, Oliver: Precise numerical analysis, Brown Publishers, Dubuque, Iowa, 1988.

[Bau74] H. Bauer: Wahrscheinlichkeitstheorie und Grundzüge der Maßtheorie, de Gruyter, Berlin, New York, 1974.

[Bis67] Bishop, Errett: Foundations of constructive analysis, McGraw-Hill, New York, 1967.

[BB85] Bishop, Errett; Bridges, Douglas: Constructive analysis, Springer-Verlag, Heidelberg, 1985.

[Bla95] Blanck, Jens: Domain representibility of metric spaces, in: Informatik-Berichte 190, Fernuniversität Hagen, 1995.

[BM75] Borodin, A.; Munro, I.: The computational complexity of algebraic and numeric problems, Elsevier, New York, 1975.

[Bre76] Brent, R.P.: Fast multiple precision evaluation of elementary functions, J. ACM 23, 242–251 (1976).

[BSS89] Blum, L.; Shub, M.; Smale, S.: On a theory of computation and complexity over the real numbers, Bull. Amer. Math. Soc. 21, 1–46, 1989.

[Cei59] Ceitin, G.S.: Algorithmic operators in constructive complete separable metric spaces (in Russian), Doklady Akad. Nauk 128, 49–52 (1959).

[Eng89] Engelking, R.: General Topology, Heldermann, Berlin, 1989.

[Grz55] Grzegorczyk, A.: Computable functionals, Fund. Math. 42, 186–202 (1955).

[Grz57] Grzegorczyk, A.: On the definition of computable real continuous functions, Fund. Math. 44, 61–71 (1957).

[HB95] Hertling, Peter; Brattka, Vasco: Feasible real random access machines, in: Informatik-Berichte 190, Fernuniversität Hagen, 1995.

[Her96] Hertling, Peter: Effectivity and effective continuity of functions between computable metric spaces. in: Proceedings of DMTCS, Springer-Verlag, 1996.

[Hot95] Hötzel Escardo, Martin: PCF extended with real numbers, in: Informatik-Berichte 190, Fernuniversität Hagen, 1995.

[Ko91] Ko, Ker-I: Complexity Theory of Real Functions, Birkhäuser, Boston, 1991.

[KW84] Kreitz,Ch.; Weihrauch,K.: A unified approach to constructive and recursive analysis. In: Computation and proof theory, (M.M. Richter et al., eds.), Springer-Verlag, Berlin, Heidelberg, 1984.

[KW85] Kreitz, Ch.; Weihrauch, K.: Theory of representations, Theoretical Computer Science 38, 35–53, (1985).

[KW87] Kreitz, Ch.; Weihrauch, K.: Compactness in constructive analysis revisited, Informatik Berichte, Fernuniversität Hagen (1984) and Annals of Pure and Applied Logic 36, 29–38 (1987).

[Moo79] Moore, Ramon E.: Methods and applications of interval analysis, SIAM, Philadelphia, 1979.

[Mue87] Müller, N.Th.: Uniform computational complexity of Taylor series. In: Lecture notes in Computer Science 267, Springer-Verlag, Berlin, Heidelberg, 435–444 (1987).

[Nov95] Novak, Erich: The real number model in numerical analysis, Journal of Complexity 11, 57–73, 1995.

[PR89] Pour–El, Marian B.; Richards, Jonathan I.: Computability in Analysis and Physics, Springer–Verlag, Berlin, Heidelberg, 1989.

[PS90] Preparata, F.P.; Shamos, I.S.: Computational Geometry, Springer-Verlag, New York, Berlin, Heidelberg, 1990.

[Rog67] Rogers, Hartley: Theory of recursive functions and effective computability, McGraw-Hill, New York, 1967.

[Sch90] Schönhage, A.: Numerik analytischer Funktionen und Komplexität, Jahresberichte der Deutschen Math. Verein. 92, 1–20 (1990).

[Sch95] Schröder, Matthias: Topological spaces allowing Type 2 complexity theory, in: Informatik-Berichte 190, Fernuniversität Hagen, 1995.

[TD88] Troelstra, A.S.; van Dalen, D.: Constructivism in Mathematics, Volume 1, North-Holland, Amsterdam, 1988.

[Tur36] Turing, Alan M.: On computable real numbers, with an application to the Entscheidungsproblem, in: Proceedings of the London Mathematical Society (2) 42, 230–265, (1936).

[Tur37] Turing, Alan M.: On computable real numbers, with an application to the Entscheidungsproblem. A correction, in: Proceedings of the London Mathematical Society (2) 43, 544–546, (1937).

[TWW88] Traub, J.F.; Wasilkowski, G.W.: Wozniakowski, H.: Information–based Complexity, Academic press, New York, 1988.

[Wei85] Weihrauch, Klaus: Type 2 recursion theory, Theoretical Computer Science 38, 17–33 (1985).

[Wei87] Weihrauch, Klaus: Computability, Springer-Verlag, Berlin, Heidelberg, 1987.

[Wei91] Weihrauch, Klaus: On the complexity of online computations of real functions, Journal of Complexity 7, 380–394, 1991.

[Wei93] Weihrauch, Klaus: Computabilitiy on computable metric spaces, Theoretical Computer Science 113, 191–210, 1993.

[Wei94] Weihrauch, K.: Effektive Analysis, correspondence course, FernUniversität, Hagen, 1994.

[Wei95] Weihrauch, K.: A simple introduction to computable analysis, Informatik-Berichte Nr. 171, Fernuniversität Hagen, 2nd ed. July 1995.

[Wei96] Weihrauch, K.: Computability on the probability measures on the Borel subsets of the unit interval, in: Proceedings of the Second Workshop on Computability and Complexity in Analysis (to appear).

[WS81] Weihrauch, Klaus; Schreiber, Ulrich: Embedding metric spaces into cpo's, Theoretical Computer Science 16, 5–24 (1981).

Quantum Electronic Devices Based on Metal-Dielectric Transition in Low-Dimensional Quantum Structures

I. Antoniou[1], B. Pavlov[2] and A. Yafyasov[3]

[1] Solvay Institute, Brussels, Belgium
[2] Institute for Physics, St. Petersburg University, Russia; Current address: Department of Mathematics, the University of Auckland, New Zealand
[3] Institute for Physics, St. Petersburg University, Russia

1 Introduction

Recently P. Shor [1] discovered an efficient quantum "polynomial-time algorithm" (length $(\lg_2 N)^k$ for some finite k) for factoring large numbers N in the form of a product of prime factors. This event makes the implementation of more efficient quantum gates (see [2]) an actual problem of quantum engineering.

The quality of the implementation of the gates is characterized by the energy $E = 1/T_0$ of interaction between pair of qubits and the damping rate $\Gamma = 1/\tau$ for the coherence of each qubit. The ratio of the single qubit coherence decay time τ ("coherence time") and the time T_0, required to perform the coherent change of state of qubits due to the interaction ("switching time") should be large for reversible coherent computations:

$$\frac{\tau}{T_0} \gg 1.$$

The quantum coherence has been observed for tunnel-coupled systems of several quantum dots. The dominant mechanism destroying coherence at low temperatures is the electron-electron scattering, which is sensitive to the geometry of devices [3, 4, 5]. The above ratio for a single quantum dot is approximately 10^3 at Helium temperatures and can be improved if smaller quantum dots could be constructed.

The superconducting quantum interference devices, constructed as a system of superconducting rings with Josephson contacts between them are dual to the quantum dots, trapping magnetic flux instead of electrons, however the loss of coherence in those devices is not properly understood yet (see [6]).

Cirac and Zoller [7] have proposed recently a laser-cooled array of Ions in the linear Pauli trap as a device for quantum computations. The coherence time for this device could be several seconds or even longer, and the switching time is defined by the optical transitions frequency. Thus the characteristic ratio of "coherence time" to "switching time" should be large at Helium temperatures. For a modern discussion of other optical gates see [8].

It was shown (see for instance [9]), that only two types of operations—"gates"—are necessary to perform any quantum computation: the single bit

"Rotate Operator" and two-bits "Controlled-Not Operator". In this paper we shall discuss *the implementation of quantum "Controlled-Not" gates in the form of a quasi-molecular chain of atoms or clusters, controlled by some sort of optical excitation.* Our analysis shows that, for proper choice of materials, the working temperatures of the proposed device might be essentially higher than the working temperatures of devices mentioned above.

In our device we shall use two qubits. The role of the first qubit is played by the states of inner atomic electrons which are transferred simultaneously by optical excitation in all atoms of the chain from the ground state into the excited state, but still remain inside atoms; the role of the second qubit is played by the asymptotic states (scattering states) of the conductivity electrons in quantum wires attached to the chain. Depending on the occupation of inner orbitals the transmission coefficient might be (close to) zero or to the transmission coefficient through semiinfinite periodic lattice. These cases correspond to the case of zero conductance of the device or, respectively, to the good conductance in accordance with Landauer formula [10, 11]. The corresponding ratio τ/T_0 is formed by the life-time of the excited state and the inverse optical frequency, and could vary in a wide range dependening on the choice of materials.

Strictly speaking, the scattering states of conductivity electrons are not representing qubits in proper sense, since the transmission coefficient through the chain is usually different from 0 or modulo 1. Nevertheless we can show, that the proper choice of materials and parameters of the device makes them reasonably close to 0 or to the transmission coefficient T_∞ for a semiinfinite lattice depending on the occupation of atomic orbitals. Due to the unitarity of the total scattering matrix, the difference $1 - |T_\infty|^2$ is equal to the square modulo of the reflection coefficient from the semiinfinite lattice. Formally it plays the role of the relative measure of necessary "garbage" which should be stored and used when performing invertible computations. But in our case the reflection coefficient is an analytic function of energy, hence it can be restored from the orthogonality condition, in unique way, up to some unimodular Blashke-factor, which is trivial, if the bound states with negative energies are absent. Thus the nonzero transmission coefficient contains the complete information of the incoming signal, and *nothing* is lost in form of a reflected wave, hence the computation act is reversible, though some part of the signal disappears forever.

The *irreversible* loss of information appeares when measuring the result of scattering by some classical device or even by some quantum device on a stochastic background. Then the part of information contained in the transmitted signal is destroyed and the computation *becomes irreversible.* More detailed discussion of comparison of quantum computations and classical computations on stochastic (Markov) background is contained in the last Section 4.

We are discussing here one-electron approximation, assuming, that life-time of the inner cluster electrons is large enough, hence the occupation of cluster orbitals is changed by resonance optical exitation only. But our model can be used beyond these limits as well, if we consider the corresponding two-electrons problem, as in [12]. It will be done elsewhere.

2 One-electron scattering by single quantum dot: Soluble model and physical reality

In this section we describe a soluble model of the simplest quantum "Controlled-not" quantum gate represented in the form a quasi-molecular cluster attached to some quantum wire by tunnel contacts. This model can be applied for theoretical analysis of constructions of nanoelectronic devices of narrow-band semiconductors.

We represent the one-electron Hamiltonian of quantum dot attached to the quantum wire as a result of perturbation of the orthogonal sum of "internal" and "external" Hamiltonians. The "internal" Hamiltonian of the electron in a cluster is determined by the occupation of the molecular orbitals

$$A_g \sim \alpha_e < \alpha_1 < \alpha_2 < \ldots; \quad A_g \phi_s^g = \alpha_s \phi_s^g$$

$$A_e \sim \alpha_g < \alpha_1 < \alpha_2 < \ldots; \quad A_e \phi_s^e = \alpha_s \phi_s^e.$$

Before the interaction is switched on, we represent the internal Hamiltonian by proper finite-dimensional operators (diagonal matrices) in each ground and excited states. The "external" Hamiltonian—the free Hamiltonian of an electron in a wire ("external channel")—is defined by the Schrödinger-type differential expression:

$$lU = -\frac{\hbar^2}{2m}\frac{\mathrm{d}^2 U_0}{\mathrm{d}^2 x},$$

where the effective electron mass m is determined both by the wire structure and the Fermi level λ_F energy near the conductivity band edge.

The quasi-molecular cluster attached to the wire could be represented of course by the equivalent potential well, separated from the outside of the wire by the potential barriers, simulating the Coulomb blockade, see [5]. But the analysis of solutions of the corresponding Schrödinger equation is generally possible either in the quasi-classical limit, which is rather far from realistic physical conditions, or for properly choosen soluble models. This is a reason why we use here a soluble model based on Operators Extensions Theory.

We use such a model here in form of so called *zero-range potentials with the inner structure*, proposed as an universal model for certain class of Schrödinger-type spectral problems in [13]. These models supply us with a wide range of spectrally interesting Hamiltonians and give a realistic physical picture of the spectrum and dynamics of perturbed systems, when the perturbations satisfy the following condition:

the de Broghlie wave-length in the nonperturbed system is greater than typical dimensions δ of the perturbation considered

$$\delta \ll \lambda \approx \frac{2\pi\hbar}{\sqrt{2mkT}},$$

where m is the effective mass of electrons at the Fermi level.

In a certain sense this condition is similar to the condition of the slow phase change, formulated for mesoscopic processes by Altshuler (see [3]).

Our analysis shows that the above condition holds for constructions of *narrow-gap and zero-gap semiconductors* , since for narrow-gap semiconductors the de Broglie wave-length is inversely proportional to $\sqrt{m}$ which is essentially smaller than the conventional mass of electron.

For the semiconductors presented in the following Table, the electron wave-length is some hundreds of Angstroms at the room temperatures and thousands of ones at the liquid nitrogen temperature. Creating relatively "small" clusters, which fulfill the above condition and can be considered as perturbations for quantum wires of these materials, is already possible using modern technical and technological tools: electronic lithography, ionic implantation, and selected chemical etching [14].

This allows our device to work at essentially higher temperatures as known world samples of quantum gates. Our calculations (see Section 3, [15]) are illustrated by the following Table.

Semiconductor	E_g (300 K) eV	$\frac{m_e^*}{m_0}$	$\frac{m_p^*}{m_0}$	ε_{sc}	λ (300 K) Å	λ (77 K) Å
$GaAs$	1.430	0.070	0.40	12.5	290	580
$InAs$	0.360	0.022	0.40	15.0	515	1020
$Cd_x Hg_{1-x} Te$						
$x = 0.20$	0.150	0.0130	0.45	17.0	670	1320
$x = 0.25$	0.220	0.0165	0.45	16.4	590	1160
$x = 0.27$	0.250	0.0180	0.45	16.2	570	1125
$x = 0.28$	0.260	0.0190	0.45	16.1	550	1085
$x = 0.29$	0.275	0.0200	0.45	16.0	540	1065
$x = 0.30$	0.290	0.0210	0.45	16.0	525	1035
$x = 0.31$	0.300	0.0220	0.45	15.9	515	1015
$x = 0.32$	0.315	0.0230	0.45	15.8	500	985
$x = 0.44$	0.510	0.0550	0.45	15.2	325	640
$HgTe$	-0.117	0.012	0.50	21.0	700	1380
$Zn_{0.15} Hg_{0.85} Te$[1]	0.190	0.015	0.45	17.0	620	1220

Thus the properties of the mentioned semiconductive materials allow to realize the low-dimensional quantum system for temperatures *higher than the nitrogen boiling temperature.* The characteristic dimensions of such structures are available for the theoretical analysis of the construction of devices using the developed zero-range potential techniques and their fabrication by modern technology methods.

[1] In series of triple compounds A_2B_6, this material is more prospective in technology, as its properties are considerably more stable in comparison with $Cd_x Hg_{1-x} Te$.

Currently the basic problem requiring the experimental solution remains the problem of the influence of charge heterogeneities near the low-dimensional system. The possibility to achieve low densities of the charge heterogeneity was shown in our experiments for the Ge surface at the room temperature [18].

The proposed model can give us realistic spectral properties of objects under consideration, but of course *not the local differential properties of the solutions* of the corresponding Schrödinger equation, hence it is essentially a "spectral modelling". On the other hand, our models help producing a lot of explicit formulas for spectral characteristics such as scattering matrices and resonances, both in asymptotically homogeneous and periodic cases.

According to the general scheme of zero-range potentials with internal structure, developed in [13] the connection of the model cluster determined by matrix A and the wire is described by the boundary conditions, which connect the value of the external component of the wave function and its derivative jump

$$\left[\frac{\mathrm{d}U_0}{\mathrm{d}x}\right](0) = \frac{\mathrm{d}U_0}{\mathrm{d}x}(+0) - \frac{\mathrm{d}U_0}{\mathrm{d}x}(-0)$$

at the point 0 of connection of both channels (the cluster and the wire):

$$\left\{-U_0(0), \left[\frac{\mathrm{d}U_0}{\mathrm{d}x}\right](0)\right\}$$

to the *boundary values* $\xi_\pm$ of the internal component U_1 determined by the choice of the corresponding *deficiency vector* h in the *internal* (cluster) "channel":

$$U_1 = U_{10} + \frac{A}{A - iI}\xi_+ \cdot h + \frac{1}{A - iI}\xi_- \cdot h$$

$$(A - iI)U_{10} \perp h.$$

This boundary condition is given by some Hermitian matrix $\{\gamma_{ik}\}$, which is a parameter of the model as well:

$$\begin{pmatrix} -U_0(0) \\ \xi_- \end{pmatrix} = \begin{pmatrix} \gamma_{00} & \gamma_{01} \\ \gamma_{10} & \gamma_{11} \end{pmatrix} \begin{pmatrix} \left[\frac{\mathrm{d}U_0}{\mathrm{d}x}\right](0) \\ \xi_+ \end{pmatrix}.$$

The above boundary condition describes the electron's transition from the internal (cluster) channel to the external one $L_2(R)$ (the wire) and backwards. The operator l_A which is defined on the linear set of elements (U_0, U_1), which satisfy the above boundary condition

$$l_A \begin{pmatrix} U_0 \\ U_1 \end{pmatrix} = \begin{pmatrix} -\frac{\hbar^2}{2m}\frac{\mathrm{d}^2U_0}{\mathrm{d}x^2} \\ AU_{10} - \frac{1}{A-iI}\xi_+ h + \frac{A}{A-iI}\xi_- h \end{pmatrix}$$

proves to be self-adjoint. It will serve as a Hamiltonian of a single electron in a wire with a cluster attached to it. This Hamiltonian can be used to describe the electron dynamics defined by the solutions of the corresponding nonstationary Schrödinger equation:

$$\frac{1}{i}\frac{\mathrm{d}}{\mathrm{d}t}\begin{pmatrix} U_0 \\ U_1 \end{pmatrix} = l_A \begin{pmatrix} U_0 \\ U_1 \end{pmatrix}.$$

The constructed model is obviously soluble: the eigenfunctions of the Hamiltonian can be given in an explicit form

$$\Psi_\lambda = \begin{pmatrix} \Psi_\lambda^0 \\ \Psi_\lambda^1 \end{pmatrix}.$$

In particular, for the scattered waves we have:

$$\Psi_\lambda^0 = \begin{cases} \mathrm{e}^{ikx} + R\mathrm{e}^{-ikx}, x > 0, \\ T\mathrm{e}^{ikx}, x < 0;\ \lambda = k^2 \end{cases}$$

$$R = T - 1,$$

$$T(\lambda) = \frac{\frac{ik\hbar^2}{m}[\gamma_{00} - |\gamma_{01}|^2(\gamma_{11} - Q(\lambda))^{-1}]}{\frac{ik\hbar^2}{m}[\gamma_{00} - |\gamma_{01}|^2(\gamma_{11} - Q(\lambda))^{-1}] - 1},$$

$$\Psi_\lambda^1 = \frac{A + iI}{A - \lambda I} h (Q(\lambda) - \gamma_{11})^{-1} \frac{\gamma_{10}}{m} \hbar^2 (1 - T).$$

Here $Q(\lambda) \equiv \langle \frac{1+\lambda A}{A-\lambda I} h, h \rangle$ is the rational function which has a positive imaginary part in the upper half-plane (and the negative one in the lower one). If the parameters of the model are connected by the condition

$$\gamma_{11} + \langle Ah, h \rangle = 0,$$

then the transmission coefficient T has the physically correct high-energy asymptotics which is typical for the Schrödinger equation with smooth potentials:

$$T(k) = 1 + o(1/k).$$

The conductivity of the wire with the model cluster attached to it is given by the Landauer formula [10, 11]:

$$\sigma(\lambda) = \frac{e^2}{\hbar}\frac{|T|^2}{1 - |T|^2}.$$

Thus the conductivity vanishes when $T(\lambda) = 0$. In particular, neglecting the analogs of "chemical shifts" γ_{00}, γ_{11} caused by the joining of the cluster to the quantum wire we get:

$$T(\lambda) = \frac{\frac{ik\hbar^2}{m}|\gamma_{01}|^2}{\frac{ik\hbar^2}{m}|\gamma_{01}|^2 + Q(\lambda)}, \quad \lambda = k^2.$$

We see that the conductivity of the device considered vanishes for electrons having the energies equal to ones of the non-occupied orbitals corresponding energy levels α_s of the attached cluster. In particular, if the cluster is in the ground state, $A = A_g$, and the non-occupied energy levels are α_e, α_1, $\alpha_2, \ldots$, then the transmission coefficient vanishes for electrons with the corresponding energies. If it is in the excited state $A = A_e$, then the nonoccupied levels are equal to α_g, α_1, $\alpha_2, \ldots$, hence the transmission coefficient vanishes for electrons having these energies.

If the Fermi level λ_F in the wire coincides with α_e, then in the first case the device conductivity vanishes at cryogen temperature, and it is positive in the second case:

$$\sigma_g = 0,$$

$$\sigma_e = \frac{e^2}{\hbar^2} \frac{|T_e(\lambda_F)|^2}{|T_e(\lambda_F)|^2 - 1},$$

where T_e is the transmission coefficient corresponding to the excited cluster.

The switching on and off of this device can be managed by any type of optical excitations. When excited by the resonance light, the cluster electron is transferred from the level α_g to the level α_e and is supposed to remain there until the next exposition with resonant light induces the inverse transition.

At a higher temperature the considerable part of conductivity electrons has the energy λ that differs from λ_F, so the real conductivity should be calculated as a mean value of conductivity $\sigma(\lambda)$ with Fermi distribution $\rho(\lambda, t)$ as follows:

$$\int \sigma(\lambda) \mathrm{d}\rho(\lambda, t) = \bar{\sigma}(t).$$

Obviously the operating of the described gate is not efficient for higher temperatures, since the value of the mean conductivity could be far from zero in the ground state. But being joined in a chain, these gates seem to be essentially more efficient even for higher temperatures.

3 Soluble model for optically simulated metal-dielectric transition in an infinite quasi-onedimensional lattice

In this section we consider the infinite chain of model molecular clusters, similar to one discussed above. We assume that the clusters are chosen such that in the ground state all of them have the same Hamiltonians

$$A^s = A^s_g = A_g$$

represented by the finite-dimensional operators (diagonal matrices) in the spaces $E^s = E$. Further we assume that the optical excitation with the frequency ν is in resonance with odd clusters only, hence only clusters with the odd numbers $s = 2l + 1$ get excited,

$$\alpha_e^{2l+1} - \alpha_g^{2l+1} = 2\pi\hbar\,\nu,$$

and all excited clusters are equivalent, $A_e^{2l+1} = A_e^1$. Thus, if in the ground state the chain of clusters has the period L, then in the excited state it has the period

$2L$. At last we assume that the deficiency elements $h = h_s$ and the self-adjoint boundary conditions describing the connection of neighboring clusters are the same for all clusters of the chain:

$$\xi_-^s = \gamma\xi_+^{s-1} + \bar{\gamma}\xi_+^{s+1}.$$

We consider a selfadjoint operator $\mathcal{A}_\gamma$ defined in the space $\mathcal{E} = l_2(E)$ of all infinite sequences U^s of E-vectors:

$$U = \{U^s\}_{s=-\infty}^{s=\infty} \subset \mathcal{E} = \sum \oplus E_s.$$

Writing down each component in a form parametrised by the boundary values, as it was done in the previous section

$$U^s = U_0^S + \frac{A^s}{A^s - iI}\,\xi_+^s h + \frac{1}{A^s - iI}\,\xi_-^s h$$

we define the operator by the formula

$$(\mathcal{A}_\gamma U)^{(s)} = A^s U_0^s - \frac{1}{A^s - iI}\,\xi_+^s h + \frac{A^s}{A^s - iI}\,\xi_-^s h$$

and the proper translation invariant self-adjoint boundary conditions imposed onto the boundary values $\xi_\pm^s$:

$$\gamma\xi_+^{s-1} + \bar{\gamma}\xi_+^{s+1} = \xi_-^s.$$

Then the operator $\mathcal{A}_\gamma^g$ is the one-particle lattice model with the period L, and the operator $\mathcal{A}_\gamma^e$ is the lattice model with the period $2L$. In the generic case the spectrum of the operator $\mathcal{A}_\gamma^g$ consists of $\dim A_g$ bands which are determined by the solutions of the algebraic equation:

$$Q_g(\lambda) = \langle\frac{I + \lambda A_g}{A_g - \lambda I}h, h\rangle = \gamma\theta^{-1} - \bar{\gamma}\theta,$$

with the uni-modular unknown function $\theta = e^{i\kappa}$, κ playing the role of quasimomentum and $\theta = e^{i\kappa}$ playing the role of quasimomentum exponential describing the translation-symmetry of the corresponding wave functions :

$$U = \{U^s\} = \{U^0\theta^s\}$$

which generally are represented by Bloch-type solutions of the corresponding homogeneous equation, that is the eigenfunctions of $\mathcal{A}_\gamma^g$ on the corresponding spectral bands. The condition of existence of nontrivial bounded solutions of the last algebraic equation takes the form

$$|Q_g| \equiv \left|\langle\frac{I + \lambda A_g}{A_g - \lambda I}h, h\rangle\right| \leq 2|\gamma|.$$

The spectrum of the operator $\mathcal{A}^g_\gamma$ consists of spectral bands, constituted by the (real) values of the energy λ, which satisfy the last condition. Two Bloch eigenfunctions correspond to each value λ from the band:

$$\Psi^s_\pm = \frac{A^g + iI}{A^g - \lambda I} L\theta^{\pm s}, \quad s = 0, t, \ldots .$$

According to the conventional quantum-mechanical interpretation the electrons in the Bloch states, corresponding to spectral bands of nonexcited chain are not "localized" but are "moving" in the infinite one-dimensional conductor formed by the periodic chain of clusters in the ground state. Under the values λ, lying outside of the spectrum $|\theta(\lambda)| < 1$ or $|\theta(\lambda)| > 1$, the corresponding Bloch solutions $\Psi_\pm$ are exponentially decreasing in one direction and exponentially increasing in the opposite one so they can't be eigenfunctions of the continuous spectrum. In real lattices the electrons with the corresponding energies are either absent or localized on impurities in the proper bound states,which are combined of the decreasing branches of the corresponding Bloch solutions.

If each odd cluster is excited, and each even one is in the ground state

$$A^{2l} = A_g,$$
$$A^{2l+1} = A_e,$$

then the corresponding Hamiltonian $\mathcal{A}^e_\gamma$ is also self-adjoint one, and its spectrum also has a band structure. However, the condition of belonging λ to the spectrum of the $2L$-periodic lattice is described technically in other form. Now the role of the period of the unperturbed lattice operator is played by the orthogonal sum $A_g \oplus A_e$ in $E \oplus E$, and the Bloch-vector of the perturbed lattice with the boundary conditions switched on is represented as a sequence of two-component E-vectors

$$\Psi = \{\theta^n \binom{v_0}{v_1}\}.$$

Denoting by $A^*_{e,g}$ the formal expression for the operator's action on elements of E, represented as in the first section,

$$U_1 = U_{10} + \frac{A}{A - iI}\xi_+ \cdot h + \frac{1}{A - iI}\xi_- \cdot h,$$
$$(A - iI)U_{10} \perp h,$$

we write down the corresponding homogeneous equation for the boundary values $\xi_+, \xi_- = Q(\lambda)\xi_+$ of Bloch vectors of the exited chain

$$\Psi_\pm = \left\{ \begin{matrix} \frac{A_g+iI}{A_g-\lambda I}\xi^g_+ \\ \frac{A_e+iI}{A_e-\lambda I}\xi^e_+ \end{matrix} \theta^{\pm l} \right\}_{l=-\infty}^{l=+\infty} .$$

in the following form

$$\begin{bmatrix} Q_g(\lambda) & -\bar\gamma - \gamma\bar\theta \\ -\gamma - \bar\gamma\theta & Q_e(\lambda) \end{bmatrix} \begin{pmatrix} \xi_g \\ \xi_e \end{pmatrix}_+ = 0,$$

$$\begin{bmatrix} Q_g(\lambda) & -\bar{\gamma} - \gamma\theta \\ -\gamma - \bar{\gamma}\bar{\theta} & Q_e(\lambda) \end{bmatrix} \begin{pmatrix} \xi_g \\ \xi_e \end{pmatrix}_- = 0.$$

In particular, from the second equation we have the following condition for the corresponding quasi-momentum exponential $\theta = e^{i\kappa}$

$$Q_g(\lambda)Q_e(\lambda) = 4|\gamma|^2 cos^2(\varphi + \frac{\kappa}{2}).$$

Here $\varphi = \arg\gamma$. Thus $cos^2(\varphi + \frac{\kappa}{2}) = \pm\frac{\sqrt{Q_g Q_e}}{2|\gamma|}$, and

$$\xi_g^- = \frac{1}{\sqrt{|Q_g|}}, \; \xi_e^- = \frac{e^{i\kappa}}{\sqrt{|Q_e|}}.$$

It follows from here, that the spectral bands of the excited lattice coincide with the intervals of real axis, where $0 \leq Q_g(\lambda)Q_e(\lambda) \leq 4|\gamma|^2$. Another important fact concerning the boundary values $\xi_{g,e}^e$ of Bloch vectors will be used in the following section for estimates of the transmission coefficient:

> *The linear combination of the boundary values with pure imaginary coefficient k:*
>
> $$\xi_g^- + ik\gamma\theta\xi_e^-$$
>
> *does not have roots on the real axis* λ.

Assuming that the spectral properties of clusters in the ground and excited states are the same as in the previous section we see that the functions Q_g, Q_e are essentially distinguished by the position of one pole only:

$$\frac{1+\alpha_e^2}{\alpha_e - \lambda}|\langle h, \phi^e\rangle|^2 \longleftrightarrow \frac{1+\alpha_g^2}{\alpha_g - \lambda}|\langle h, \phi^g\rangle|^2.$$

This leads to the partition of the lower spectral band of the L-periodic chain of clusters A_g into two subbands corresponding to the $2L$-periodic chain of A_g, A_e-clusters. The arising lacuna (spectral gap) δ is found from the condition:

$$Q_e(\lambda)Q_g(\lambda) < 0,$$

and shifts $\lambda_e^\pm \to \lambda^\pm$ of the upper and lower bounds are determined from the equation:

$$Q_e(\lambda)Q_g(\lambda) = 4|\gamma|^2.$$

If the residues $(1+\alpha^2)|\langle h, \phi^e\rangle|^2$ at the poles α_e α_g, which correspond to the ground and excited states of the clusters, are approximately equal, then the dispersion curves $\lambda = \lambda(\kappa)$ corresponding to the excited and the nonexcited chain are approximately parallel and hence the lacuna (spectral gap) width $|\delta|$ proves to be approximately equal to the distance between the ground and exited energy levels of odd clusters:

$$\delta \sim \alpha_e - \alpha_g.$$

The described chain exhibits a remarkable type of behavior under resonance optical excitations, which can be interpreted as a "*Simulated Mott-Pejerls transition*". If the Fermi level λ_F is situated at the center of the gap δ (thus *inside* the conductivity band of the chain in ground state), then the chain in the ground state is equivalent to one-dimensional metallic conductor. On the other hand the Fermi level is situated inside the gap δ of the exited chain, hence the *exited chain is a dielectric one for electrons on Fermi level* λ_f. This is exactly Mott-Pejerls-type behavior, see [19]. It is important, that the chain remains dielectric within some interval of energies near λ_f, defined by the width of the gap $|\delta|$, hence the described chain *can play the role of the gate in some interval of temperatures* $0 \leq T \leq \frac{|\delta|}{2k}$. Under conditions above this interval is approximately determined by the distance between the levels $\alpha_g - \alpha_e$. In Section 2 the advantages of narrow-band semiconductors were discussed for creating quantum wires with large values of de Broghlie wave length. Here we underline another essential advantage of the narrow-gap semiconductors with small values m: they manifest a high resolution of the levels in quantum well which can exceed kT already in the range of room temperatures. It means, that the distance between energy-levels of cluster orbitals is large, and thus, according to observation above, the critical temperature $|\delta|/2k$ may be significantly higher than the helium temperature or even higher than nitrogen one.

4 Scattering by finite chain of clusters

The quantum gate can't be technologically implemented in the form of infinite cluster chain. We consider now the finite chain of clusters length $N\ L = 2n\ L$ inserted into a quantum wire. Assuming that the clusters of the chain possess properties described in the previous section, we solve the scattering problem for exited chain and for the chain in ground state. The corresponding transmission coefficient proves to be exponentially small for large N outside the spectral bands of the exited infinite chain, but is close to the transmission coefficient "through the semi-infinite chain" in the ground state. Practically it means that the finite chain inserted in the quantum wire can serve as a quantum gate under certain conditions on the parameters of the clusters and the chain.

The model Hamiltonian of a single electron on the quantum wire with the inserted finite lattice is constructed by the same method of as above (Sections 2, 3) of two components: the free Hamiltonian on the wire and lattice Hamiltonian on the inserted periodic chain with proper boundary conditions at the points of contacts to provide the selfadjointness of the total Hamiltonian :

$$\begin{pmatrix} 0 & \bar\gamma & 0 & 0 & \dots & \dots & \dots & \dots & 0 & 0 & 0 \\ \gamma & 0 & \bar\gamma & 0 & 0 & \dots & \dots & \dots & & 0 & 0 \\ 0 & \gamma & 0 & \bar\gamma & 0 \dots & \dots & \dots & \dots & & 0 & 0 \\ \dots & \dots & \dots & \dots & \dots & \dots & \dots & \dots & \dots & \dots & \dots \\ \dots & \dots & \dots & \dots & \dots & \dots & \dots & \dots & \dots & \dots & \dots \\ \dots & \dots & \dots & \dots & \dots & \dots & \dots & \dots & \dots & \dots & \dots \\ \dots & \dots & \dots & \dots & \dots & \dots & \dots & \dots & \dots & \dots & \dots \\ \dots & \dots & \dots & \dots & \dots & \gamma & 0 & \bar\gamma & 0 & 0 & \\ \dots & \dots & \dots & \dots & \dots & \dots & \gamma & 0 & \bar\gamma & 0 & \\ \dots & \dots & \dots & \dots & \dots & \dots & \dots & \gamma & 0 & \bar\gamma & \\ \dots & \dots & \dots & \dots & \dots & \dots & \dots & \dots & \gamma & 0 & \end{pmatrix} \begin{pmatrix} u'(0) \\ \xi_g^0 \\ \xi_e^0 \\ \xi_g^1 \\ .. \\ .. \\ \xi_e^{n-1} \\ \xi_g^n \\ \xi_e^n \\ -u'(NL) \end{pmatrix} = \begin{pmatrix} u(0) \\ Q_g\xi_g^0 \\ Q_e\xi_e^0 \\ Q_g\xi_g^1 \\ .. \\ .. \\ Q_e\xi_e^{n-1} \\ Q_g\xi_g^n \\ Q_e\xi_e^n \\ u(NL) \end{pmatrix}.$$

The scattered waves are combinations of solutions of the stationary Schrödinger equation on the quantum wire

$$\begin{array}{rl} Te^{ikx}, & -\infty < x \le 0, \\ e^{ikx} + Re^{-ikx}, & NL \le x < \infty, \end{array}$$

and Bloch solutions on the inserted lattice $0 \le l \le N$:

$$\Psi = \alpha\Psi_+ + \beta\Psi_-.$$

A straightforward calculation gives the connection between α and β:

$$\frac{\beta}{\alpha} = -\frac{\theta^{-1}\xi_e^+ - ik\gamma\xi_g^+}{\theta\xi + g^- - ik\gamma\xi_g^-},$$

and the explicit expression for the transmission coefficient. Let us denote

$$\Delta_+, \equiv \frac{\theta\xi_g^+ + ik\gamma\xi_e^+}{\theta^{-1}\xi_e^+ - ik\gamma\xi_g^+}, \quad \Delta_- \equiv \frac{\theta^{-1}\xi_g^- + ik\gamma\xi_e^-}{\theta\xi_e^- - ik\gamma\xi_g^-}.$$

Then

$$T = \frac{-2ik\gamma e^{iLNk}}{\theta\xi_e^- - ik\gamma e_g^-}\,\frac{\xi_g^+\xi_e^-\theta - \xi_g^-\xi_e^+\theta^{-1}}{\theta^n\Delta_+ - \theta^{-n}\Delta_-}.$$

Due to the statement from the previous section the coefficient in front of θ_n in the denominator does not have real zeroes, hence the transmission coefficient has the following asymptotics on the sequence of spectral gaps of the exited chain for large N:

$$T \approx 2ik\frac{\bar\gamma}{\gamma}\frac{e^{ikLN}\theta^n}{\theta^{-1}\xi_g^- + ik\gamma\xi_g^-},$$

so it is exponentially small inside the spectral gaps.

On the other hand, the similar calculation of the reflection coefficient R^N of the chain in the ground state gives

$$R_g^N = e^{2ikLN}\frac{\theta^N\Delta_1 - \theta^{-N}\Delta_2}{\theta^N\Delta_3 - \theta^{-N}\Delta_4},$$

where

$$\Delta_1 \equiv \frac{Q+ik|\gamma|^2-\theta^{-1}\gamma}{Q-ik|\gamma|^2-\theta\bar{\gamma}},$$
$$\Delta_2 \equiv \frac{Q+ik|\gamma|^2-\theta\gamma}{Q-ik|\gamma|^2-\theta^{-1}\bar{\gamma}},$$
$$\Delta_3 \equiv \frac{Q-ik|\gamma|^2-\theta^{-1}\gamma}{Q-ik|\gamma|^2-\theta\bar{\gamma}},$$
$$\Delta_4 \equiv \frac{Q-ik|\gamma|^2-\theta\gamma}{Q-ik|\gamma|^2-\theta^{-1}\bar{\gamma}}.$$

Notice that $|\Delta_{1,2}| = 1$, hence the reflection coefficient corresponding to the inserted into quantum wire finite chain of N equivalent non-excited clusters has exactly N roots on each spectral band of the corresponding infinite lattice. Due to the unitarity of the complete scattering matrix it means that the corresponding transition coefficient T_g is modulo one at N spectral points on each spectral band of the corresponding infinite lattice.

The Reflection Coefficient $R_g^N(k)$ is a contracting analytic function on the lower halfplane of momentum k. It can be shown, that on every compact domain of it it is converging uniformly to the reflection coefficient R_g^∞, corresponding to the semi-infinite lattice.

From the observation above follows, that this convergence does not take place on real axis of momentum,in particular on spectral bands in the complex plane λ of energy. Nevertheless for any problem for the corresponding nonstationary Schrödinger equation with smooth initial data given the in form of incoming waves in quantum wire one can use the fact that both reflection coefficients from the semiinfinite lattic and from the inserted finite chain of N clusters are close for large N:

$$R_g^N(k) \approx R_g^\infty.$$

Thus the measure of "formal garbage" produced in any scattering act by finite chain is limited by the corresponding reflection coefficient of the semi-infinite lattice. Similarly to the the situation described in Section 2 we see that the information about the reflected waves is contained actually in the transmitted waves, so it is never lost, thus it is "potentially observable".

The real garbage in a single act of computation is caused by the measurement of the scattering state, that is, by projection of the state onto the states of measuring device, or by averaging them over stochastic states of the random background. This garbage is *really* lost. Hence the corresponding information is irreversibly lost which makes the computation irreversible. In [18] efficient probabilistic algorithms for classical computations are discussed. Precizeli as the discussed quantum computations, these algorithms are "polynomial-time algorithms". Professor C. Calude formulated the question if there any "typical features" of quantum computations which make them different, and more efficient, than classical probabilistic computations? Our model permits to compare the working regime of the described quantum gates with the corresponding regime of classical gates as an element of classical circuit acting in frames of a "random" algorithm. Really, the single act of quantum computation consists in opening or closing the quantum gate. Assuming that the life-time of inner cluster electrons on the excited level is much longer than typical time-intervals of the process under consideration, we see that the the quantum evolution of the conductivity electron is unitary, and the computation is invertible, unless we do not project the result onto the states of the measuring device. *The inversibility of the quantum computation is produced by the act of measurement only.*

If the constructed self-adjoint operator is positive, we can consider it as a generator of evolution of the transition probabilities of some Markov process. Then the relevant parabolic equation is interpreted as a Fokker-Plank equation for transition probabilities. The corresponding evolution is contracting, hence the conservation law is absent, and *the garbage, produced by the process, is killed each moment in the course of evolution thus resulting in global irreversibility of the process.* So the important difference between the quantum computations and similar probabilistic algorithms is the moment when the unrecoverable garbage is produced, which corresponds to the *irreversible* loss of information. For classical computations it is produced permanettly *in course of evolution*, but for quantum ones it is produced just in the last moment, when projecting the results of computation (evolution) onto the states of device. The "potentially observable garbage", caused by outgoing waves in quantum process does not destroy the reversibility of the process of quantum computations, though these waves remove some part of wave-packet to infinity. Possibly this difference can help defining the proper domain of applications of both computational techniques.

The implementation of the construction of quantum gates described in Section 4 and other constructions based on simulated Mott-Pejerls transitions (see [20]) require solving numerous difficult physical and technological problems such as the choice of proper quasi-molecular clusters and their location inside the isolating medium, the information output/input from the macroscopic level, the choice of the proper optical window for each type of gates etc. Possibly, the most prospective constructions of such kind may be created as polymers or biomolecules. The general advantage of the Mott-Pejerls gates is higher stability of their characteristics and a relatively weak dependence of their working regimes on temperature.

Acknowledgment. The authors are grateful to Professor P. Zoller for the materials supplied and to Professor K. Svozil for the fruitful discussion.

The authors are grateful to Professor C. Calude who read and commented the text and formulated the interesting question.

This paper was written as a part of the Collaboration Project managed by Solvay Institute, under support of the system of Grants ESPRIT, awarded by the Commission in Research and Development of European Community.

References

1. P. Shor, *"Algorithms for Quantum Computation: Discrete Logarithms and Factoring"*. In Proceedings 35 Annual Symposium of Computer Science, edited by S. Goldwasser (IEEE Computer Society Press, Los Alamitos, CA, 1994), p.124.
2. R. P. Feynman, *Quantum Mechanical Computers.* Optics News, **11**(1985), p.11.

3. B. L. Altshuler, P. A. Lee and R. A. Webb, eds., *Mesoscopic Phenomena in Solids*, North Holland, New-York (1991).
4. T. Ando, A. B. Fowler, and F. Stern, Rew. Mod. Phys. **54** (1982),p.437.
5. H. Grabert, M. H. Devoret, *Single Charge Tunneling—Coulomb Blockade Phenomena in Nanostructures.* Plenum Press, New York (1992).
6. M. Tinkham, *Introduction to Superconductivity.* McGraw Hill, New York (1995).
7. J. I. Cirac, P. Zoller, *Quantum computations with cold trapped ions.* Phys. Rew. Letters, **24** (1995) p.4091.
8. P. Berman, ed. *Advances in Atomic,Molecular and Optical Physics* Suppl.2. Academic Press, New York (1991).
9. S. Ishihara, *Optical Computers. New Science Age.* Ivanami Shoten Publishers. 1989.
10. R. Landauer, *IBM Journal for Research and Development* 1 (1957), 1. C.223.
11. M. Buttiker, *Nanostructured systems* , ed. M. A. Reed, NY Academ, 1990.
12. Yu. Melnikov, B. Pavlov, *Two-body scattering on a graph and application to simple nanoelectronic device.* J. Math. Phys. 1995. Vol.36. (6), P.2813-2825.
13. B. S. Pavlov, *The theory of extensions and explicitly-soluble models*, Russian Math. Surveys 42:6 (1987),127-168.
14. A. M. Yafyasov, V. B. Bogevolnov, T. V. Rudakova, *Quantum interferentional electronic transistor (QIET). Physical principles and technological limitations.* Draft. St. Petersburg University (periodical report,1995).
15. A. M. Yafyasov, V. B. Bogevolnov, T. V. Rudakova, *Quantum interferentional electronic transistor (QIET). Theoretical analysis of electronic properties for low-dimensional systems* (Progress report, 1995) Preprint IPRT N 99-95.
16. A. M. Yafyasov, V. B. Bogevolnov, T. V. Rudakova, *Low-dimensional effects in the Ge semiconductor surface at room temperature.* Thesis of ICSOS-5. July 8-12, 1996. Aix en Provence, France.
17. P. P. Konorov, A. M. Yafyasov, V. B. Bogevolnov, *Low-dimensional effects on the GaSb Semiconductor Surface.* Phys.Low-Dim.Structures. 2/3. 1995. P.133-138.
18. C. Calude, *Information and Randomness.* Springer-Verlag, 1994.
19. R. E. Pejerls, *Quantum Theory of Solids.* London, 1955.
20. B. S. Pavlov, G. P. Miroshnichenko, Patent Application 5032981/25 (0103431) (Russia) from 12.03.1992.

Recognizing Diameter Critical Graphs

R.P. Anstee[1]* and L. Caccetta[2] **

[1] Department of Mathematics, University of British Columbia,
#121-1984 Mathematics Road, Vancouver, B.C., Canada, V6T 1Z2.
[2] School of Mathematics and Statistics, Curtin University of Technology,
G.P.O. Box 1987, Perth, Australia.

Abstract. We say a graph $G = (V, E)$ is d-critical if G has diameter d and for each edge $e \in E$, the graph $G - e$ has diameter greater than d. We give an algorithm for recognizing d-critical graphs that runs in $O(n^2m)$ where $n = |V|$ and $m = |E|$. For $d = 2$, $O(\text{MULT}(n))$ suffices where $\text{MULT}(n)$ is the complexity of multiplying two matrices of order n (which is known to be $O(n^{2.377})$). For $d = 3$, $O(nm + \text{MULT}(n))$ suffices.

Key words: diameter critical, matrix multiplication

1 Introduction

Graph Theory has proven useful (see [1, 3, 4]) in many network design problems such as: the design of minimal cost reliable networks; the design of integrated circuits; the design of data communication networks capable of supporting large scale on-line applications; determining optimal routes in a given network; and many more. Usually, the problem that arises is that of constructing a graph which satisfies certain properties (representing the requirements of the system) and which is optimal according to some criterion such as cost, output or performance. Graph Theory is particularly useful when the requirements of the network can be expressed in terms of bounds on certain graph parameters. The solution of such a problem involves the development of good characterizations of the graph parameters involved and the utilization of these characterizations in the construction of speedy algorithms. Characterizing graph parameters is the subject of extremal graph theory.

A fundamental problem in extremal graph theory is finding the relationships between a given set of graph parameters. This is usually accomplished by fixing some of the parameters and investigating how the others vary. In the characterization problem it is often useful to study a restricted class of graphs, the so called 'critical graphs'. The term 'critical' is usually used with respect to a specified graph parameter P and applies when the graph G under consideration has property P but alteration of G (such as edge deletion, vertex deletion, edge addition) results in a graph not having property P. The resultant class of critical

* support provided by NSERC. Work done while on sabbatical at [2].
** support provided by Australian Research Council Grant no. A49532661

graphs often has a structure which can be utilized to yield a considerable amount of useful information. Often there is no loss in generality and quite a lot to gain by considering critical graphs.

Critical graphs have been studied by many researchers; a good survey is given in [2]. In this paper we consider graphs that are diameter critical with respect to the operation of edge deletion. The diameter is an important parameter in networks in which the time delay or signal degradation is approximately proportional to the length of the path being used.

For our purposes, $G = (V, E)$ is a simple connected graph. The *distance* $\mathrm{d}_G(x, y)$ between two vertices x, y in G is the length of the shortest x-y-path in G. The *diameter* $\mathrm{diam}(G)$ of G is defined as:

$$\mathrm{diam}(G) = \max\{\mathrm{d}_G(x, y) \,:\, x, y \in V\}. \tag{1}$$

G is *d-critical* if $\mathrm{diam}(G) = d$ and $\mathrm{diam}(G - e) > d$ for every $e \in E$.

The elementary approach to recognizing diameter critical graphs is to compute $\mathrm{diam}(G)$ and $\mathrm{diam}(G - e)$ for each $e \in E$. We can compute $\mathrm{diam}(G)$ using breadth first search (BFS) once for each $v \in V$ and hence in $O(nm)$ time. Alternatively it is well known folklore that $\log n$ matrix multiplications suffice using binary search on the formula

$$\mathrm{diam}(G) = \min\{d : (A + I)^d > 0\}, \tag{2}$$

where A is the usual $n \times n$ adjacency matrix of G and we write $B > 0$ if each entry of a matrix B is greater than zero. We denote the complexity of multiplying two matrices of order n as $\mathrm{MULT}(n)$ and note that the result of Coppersmith and Winograd [5] shows that $\mathrm{MULT}(n)$ is $O(n^{2.377})$.

Theorem 1. *We can compute diam(G) in* $O(\min\{nm, \log n MULT(n)\})$. □

Using the definition of d-critical we obtain a recognition algorithm.

Proposition 2. *We can recognize if G is diameter d-critical with respect to edges in* $O(\min\{nm^2, m \log n MULT(n)\})$. □

It is reasonable to hope that you might get a more efficient algorithm by combining some of the work in the m diameter computations. In what follows, we find it convenient to solve the directed case of this problem. We consider directed graphs G where $\mathrm{d}_G(x, y)$ is the length of the shortest directed x-y-walk and $\mathrm{diam}(G)$ is defined as the maximum over all ordered pairs x, y of vertices of $\mathrm{d}_G(x, y)$. We refer to directed walks as walks unless there is some confusion. We can associate with every undirected graph $G = (V(G), E(G))$ a directed graph $G' = (V(G), E(G'))$ where if $(i, j) \in E(G)$ then $(i, j), (j, i) \in E(G')$. We continue to use the adjacency matrix A (it is now an arbitrary (0,1)-matrix with 0's on the main diagonal) and the case where A is symmetric is equivalent to the undirected case. We say an arc (i, j) is *critical* if $\mathrm{diam}(G - (i, j)) > d$ i.e. there exist two vertices k, l with $k \neq l$ such that every k-l-walk uses the arc (i, j). To show a directed graph is d-critical, we must show every arc is critical. The following lemma shows that in the undirected graph case, an edge (i, j) is critical if either the arc (i, j) or the arc (j, i) is critical. We can solve the undirected graph case by showing that for each edge (i, j) that both arcs $(i, j), (j, i)$ are critical.

Lemma 3. *In an undirected graph G, if every undirected k-l-walk of at most t edges uses the edge (i, j), then in the associated directed graph, either every directed k-l-walk of at most t arcs uses the arc (i, j) or every directed k-l-walk of at most t arcs uses the arc (j, i)*

Proof. Among all (directed) k-l-walks of at most t arcs using the arc (i, j) but not using the arc (j, i), let W be a shortest such walk consisting of an initial k-i-walk P of a arcs not using arc (i, j) followed by the arc (i, j) and then a final j-l-walk Q of b additional arcs. We deduce that $a + 1 + b \leq t$ and Q doesn't use the arc (i, j) since Q is a shortest j-l-walk. Among all k-l-walks of at most t arcs using the arc (j, i) but not using the arc (i, j), let W' be a shortest such walk consisting of an initial k-j-walk P' of a' arcs not using arc (j, i) followed by the arc (j, i) and then a final i-l-walk Q' of b' additional arcs. As above, Q' does not use the arc (j, i) and $a' + 1 + b' \leq t$. But now PQ' and $P'Q$ are k-l-walks not using the arcs $(i, j), (j, i)$ and so $a + b' > t$, $a' + b > t$, yielding a contradiction. □

Now in the case when G has been obtained from an undirected graph, if every k-l-walk uses the arc (i, j) then any l-k-walk (which is necessarily a reversal of a k-l-walk) uses the arc (j, i). So to test if an undirected graph is critical, we need only check for the criticality of the associated directed graph.

Section 2 considers the problem, given vertices k, l, i and integer t, of whether every k-l-walk of at most t arcs uses vertex i. In doing so we obtain an algorithm for recognizing diameter d graphs critical with respect to vertex deletion. Section 3 solves the problem of recognizing d-critical graphs. Section 4 gives faster algorithms adapted for $d = 2$ and $d = 3$.

2 Critical with respect to vertices

We need the dn^2 entries of the matrices $A, A^2, A^3, \ldots, A^d$. We can get these entries in $O(d\text{MULT}(n))$ or just $O(dn^3)$. As well we can compute $\text{d}_G(k, l)$ for all pairs of vertices k, l in $O(nm)$ using BFS from each vertex and we can compute $\text{d}_{G-i}(k, l)$ for all triples of vertices k, l, i in $O(n^2 m)$. Our algorithm tests whether an arc (i, j) is critical by checking all pairs of vertices k, l. There are $O(n^2 m)$ cases to consider and so our final algorithm is $\Omega(n^2 m)$. Thus it is reasonable to assume the above computations have been done. Nonetheless, we carefully analyze pieces of our algorithm since they become useful in Theorems 5, 9 and 10. Also note that d can be $\Theta(n)$ so a dependence on d should be computed.

As a first step to seeing if every k-l-walk of at most t arcs ($t \leq d$) uses the arc (i, j), we can test if every k-l-walk of at most t arcs uses a vertex i.

Let $A - i$ denote the order $n - 1$ matrix obtained from A by deleting row and column i. We can compute the matrix $((A - i) + I)^t$ and then test if the kl-entry is zero or not. This takes $O(n \log d\mathrm{MULT}(n))$ in total over the n choices for i and all possible pairs k, l.

We can do better, at least for small d. To test if every k-j-walk of at most t arcs uses vertex i, use the following formulas. Let $n^t_{kl}(i)$ be the number of k-l-walks of at most t arcs not using vertex i (except as an origin ($k = i$) or terminus ($l = i$)). We may compute this for $k = i$ by having the last position for vertex i as the $s+1$st vertex in an i-l-walk (where we don't require $(i,l) \in E$) and using the formula (recursive in t)

$$n^t_{il}(i) = (A^t)_{il} - \sum_{s=1}^{t-1} (A^s)_{ii} n^{t-s}_{il}(i). \tag{3}$$

We have used the notation $(B)_{kl}$ to denote the k, l entry of matrix B. Over all pairs i, l and for all d choices for t, $O(d^2n^2)$ work suffices. Similarly we have the formula

$$n^t_{kl}(i) = (A^t)_{kl} - \sum_{s=1}^{t} (A^s)_{ki} n^{t-s}_{il}(i), \tag{4}$$

which is not recursive in t given the computations in (3). We have obtained the following useful result.

Lemma 4. *After $O(n^2d^2)$ preprocessing work then for a given triple of vertices i, k, l and given $t (\leq d)$ we can compute if every k-l-walk of at most t arcs uses vertex i in $O(d)$. Alternatively $O(n^2m)$ preprocessing work enables us to check in $O(1)$ for a given triple.*

Proof. The second estimate comes from testing if $\mathrm{d}_{G-i}(k,l) > t$ in which case every k-l-walk of at most t arcs uses vertex i. □

We can now tackle the problem of determining if G is diameter d-critical with respect to vertex deletion.

Theorem 5. *We can recognize if G is diameter d-critical with respect to vertices, namely diam$(G - v) > d$ for all $v \in V$, in $O(\min\{dn^3, n \log d MULT(n), n^2m\})$.*

Proof. Use Lemma 4 and only consider the case $t = d$. There are n^3 triples of vertices to consider. The other two estimates are applications of Theorem 1. □

Lemma 6. *We may compute*

$$f(k,l,i) = \min\{d_{G-i}(k,l), d+1\}, \tag{5}$$

for a given triple k, l, i in $O(d \log d)$ (given $O(d^2n^2)$ preprocessing) and hence for all triples in $O(\min\{n^3 d \log d, n^2m\})$.

Proof. We have

$$f(k,l,i) = 1 + \max_{t \le d}\{t : n^1_{kl}(i) = n^2_{kl}(i) = \cdots = n^t_{kl}(i) = 0\}. \tag{6}$$

Noting that if $n^t_{kl}(i) > 0$ then $n^{t+2}_{kl}(i) > 0$, we can use binary search (and Lemma 4) on the odd and even values of t separately to compute $f(k,l,i)$. The second estimate is also from Lemma 4. □

Note that we would require $O(nd\text{MULT}(n))$ using the matrix approach since we would need to compute $(A-i)^2, (A-i)^3, \ldots, (A-i)^d$ for each vertex i.

3 Critical with respect to arcs

Lemma 7. *We may compute the two functions*

$$h(i,j) = \min\{d_{G-(i,j)}(i,j), d+1\}, \tag{7}$$

$$g(i,j) = \min\{d_{G-(i,j)}(j,i), d+1\}, \tag{8}$$

for all $(i,j) \in E$ in $O(\min\{d^2m, m^2\})$.

Proof. Let $n^t_{kl}(i,j)$ denote the number of k-l-walks of exactly t arcs (where $t \le d$) not using the arc (i,j). In the following equation, the sum counts the number of i-i-walks with the arc (i,j) appearing for the first time as the $s+1$st arc. We may compute recursively

$$n^t_{ii}(i,j) = (A^t)_{ii} - \sum_{s=0}^{t-1} n^s_{ii}(i,j)(A)_{ij}(A^{t-s-1})_{ji}, \tag{9}$$

where we have included the term $(A)_{ij}$, even though it is 1, to improve readability. The above computation requires $O(d^2m)$, namely $O(d^2)$ for each arc (i,j). Similarly

$$n^t_{ij}(i,j) = (A^t)_{ij} - \sum_{s=0}^{t-1} n^s_{ii}(i,j)(A)_{ij}(A^{t-s-1})_{jj}, \tag{10}$$

$$n^t_{ji}(i,j) = (A^t)_{ji} - \sum_{s=0}^{t-1} n^s_{ji}(i,j)(A)_{ij}(A^{t-s-1})_{ji}. \tag{11}$$

Thus we can compute

$$h(i,j) = 1 + \max_{t \le d}\{t : n^1_{ij}(i,j) = n^2_{ij}(i,j) = \cdots = n^t_{ij}(i,j) = 0\}, \tag{12}$$

$$g(i,j) = 1 + \max_{t \le d}\{t : n^1_{ji}(i,j) = n^2_{ji}(i,j) = \cdots = n^t_{ji}(i,j) = 0\}, \tag{13}$$

in $O(d^2m)$.

An alternative approach is to compute $\mathrm{d}_{G-(i,j)}(i,j)$ and $\mathrm{d}_{G-(i,j)}(j,i)$ using two runs of BFS for each $(i,j) \in E$ which takes $O(m^2)$. □

This allows us to test for criticality in all cases:

Case 1. $k = i$ and $l = j$

We have that (i, j) is a critical arc if $h(i, j) = d + 1$.

Case 2. $k \neq i, l = j$.

Assume every k-j-walk of at most d arcs uses vertex i, which can be tested using Lemma 4 in $O(mnd)$. We assert that arc (i, j) is in every k-j-walk of d arcs if

$$\mathrm{d}_G(k,i) + \mathrm{d}_{G-(i,j)}(i,j) > d. \tag{14}$$

We can obtain d_G in $O(nm)$ and $\mathrm{d}_{G-(i,j)}$ as needed in $O(d^2m)$ by Lemma 7. For each of the mn pairs of vertices and arcs $k, (i, j)$ we can test for criticality in $O(1)$ and so $O(nmd)$ overall. A similar analysis handles the case $k = i$, $l \neq j$.

Case 3. $k \neq i, l \neq j$.

Now to test if every k-l-walk of at most d arcs uses the arc (i, j), we first verify that every k-l-walk of at most d arcs uses vertices i, j by table lookups in f at a cost of $O(\min\{dn^3, n^2m\})$. Then every k-j-walk of at most d arcs uses the arc (i, j) if

$$\mathrm{d}_G(k,i) + \mathrm{d}_{G-(i,j)}(i,j) + \mathrm{d}_G(j,l) > d, \tag{15}$$

$$\mathrm{d}_{G-i}(k,j) + \mathrm{d}_{G-(i,j)}(j,i) + \mathrm{d}_G(i,l) > d. \tag{16}$$

The idea is that if you consider a shortest k-l-walk of at most d arcs that doesn't use (i, j) and visits vertex i before j, then (15) will be violated. If j comes before i, then (16) will be violated. Thus a $O(1)$ test suffices for each $k, l, (i, j)$ using h, g and d_G (using $O(n^2m)$ preprocessing) and so $O(n^2m)$ overall.

We have provided an algorithm to prove the following result since we have tested for the criticality of (i, j) by all possible pairs k, l. Our faster approaches for small d get overwhelmed by the $O(n^2m)$ term from Case 3.

Theorem 8. *We can test if G is d-critical in $O(n^2m)$.* □

Note that we don't need fast matrix multiplication to achieve this complexity nor is the complexity a function of d.

4 Faster algorithms for the cases $d = 2, 3$

It is interesting that the cases $d = 2, 3$ can be made to run faster.

Theorem 9. *We can test if G is 2-critical in $O(MULT(n))$.*

Proof. As before, we compute A^2 in $O(\mathrm{MULT}(n))$. For each arc (i,j) we seek a pair k,l with $k \neq l$ so that every k-l-walk of at most 2 arcs uses (i,j). We can check the case $k = i$ and $l = j$ in $O(m)$ using Lemma 7. We need not consider the case $k \neq i$ and $l \neq j$ since then no k-l-walk of 1 or 2 arcs could use (i,j). For the case $k = i$, $l \neq j$ we proceed as follows: Construct a matrix $B = (b_{ij})$ of order n using

$$b_{ij} = \begin{cases} 1, & \text{if } (A + A^2)_{ij} = 1; \\ 0, & \text{otherwise,} \end{cases} \tag{17}$$

in $O(n^2)$. Now assume $(BA^T)_{ij} > 0$ (where A^T denotes the transpose of A). Then there is an l with

$$(B)_{il} = 1, (A^T)_{lj} = 1. \tag{18}$$

Since $(A)_{ij} = 1$, we have an i-l-walk ijl and so $(A^2)_{ij} \geq 1$. Then $(B)_{il} = 1$ implies $(A)_{il} = 0$ (hence $i \neq l$) and there is only one i-l-walk of at most 2 arcs, namely ijl, and so (i,j) is critical. We can handle the case $k \neq i$, $l = j$ using the matrix product $A^T B$.

Hence for arcs (i,j) not already shown to be critical, we can test if they are critical by computing $A^T B$ and BA^T in $O(\mathrm{MULT}(n))$ and then searching the $O(n^2)$ entries. □

Theorem 10. *We can test if G is 3-critical in $O(nm + MULT(n))$.*

Proof. We compute A^2, A^3 in $O(\mathrm{MULT}(n))$. As above, for each arc (i,j) we seek a pair k,l with $k \neq l$ such that every k-l-walk of at most 3 arcs uses (i,j). We can check the cases $k = i$, $l = j$ in $O(m)$ by Lemma 7. We can check the cases $k = i$, $l \neq j$ or $k \neq i$, $l = j$ in $O(nm)$ by Lemma 4 and the ideas of Case 2 in Section 3. We now consider the case $k \neq i$ and $l \neq j$. We note that a k-l-walk of at most 3 arcs using (i,j) must be the walk $kijl$. Construct a matrix $C = (c_{ij})$ of order n using

$$c_{ij} = \begin{cases} 1, & \text{if } i \neq j,\ (A^3)_{ij} = 1,\ (A + A^2)_{ij} = 0; \\ 0, & \text{otherwise,} \end{cases} \tag{19}$$

in $O(n^2)$. Now if $(A^T BA^T)_{ij} \geq 1$ then we have a pair k,l with

$$(A^T)_{ik} = 1, (B)_{kl} = 1, (A^T)_{lj} = 1, \tag{20}$$

hence $k \neq i$ and $l \neq j$. With $(A)_{ij} = 1$ we have the k-l-walk $kijl$. Then $(B)_{kl} = 1$ implies $k \neq l$ and $kijl$ is the only k-l-walk of at most 3 arcs so (i,j) is critical.

Hence for arcs (i,j) not already shown to be critical, we can test if they are critical by computing $A^T BA^T$ in $O(\mathrm{MULT}(n))$ and then searching the $O(n^2)$ entries. □

References

1. L. Caccetta, Graph Theory in Network Designs and Analysis, in *Recent Studies in Graph Theory*, (V. Kulli, editor), Vishwa International, 1989, 29-63.
2. L. Caccetta, On graphs that are critical with respect to the parameters: diameter, connectivity and edge-connectivity, *Le Mathematiche*, **XLVII**(1992), 213-229.
3. L. Caccetta, Graph Theory Modelling and Applications, in *Proceedings of International Congress on Modelling and Simulation*, (M. McAleer, A. Jakeman, editors), MSSA, **3**(1993), 1265-1270.
4. F.R.K. Chung, Diameters of communication networks, *Mathematics of Information Processing*, Amer. Math. Soc. Short Course Lecture Notes (1984), 1-18.
5. D. Coppersmith and S. Winograd, Matrix multiplication via arithmetic progressions, *Proceedings of the Nineteenth Annual ACM Symposium on Theory of Computing*, 1987, 1-6.

The Hadwiger Number for the Product of Two Cycles

Dan Archdeacon[1], C. Paul Bonnington[2], John Pearson[2], and Jozef Širáň[3]

[1] Department of Mathematics and Statistics, University of Vermont, Burlington VT 05401, USA, dan.archdeacon@uvm.edu
[2] Department of Mathematics, University of Auckland, Auckland, New Zealand, p.bonnington@auckland.ac.nz, pearson@scitec.auckland.ac.nz
[3] Department of Mathematics, Faculty of Civil Engineering, Slovak Technical University, 813 68 Bratislava, Slovakia, siran@lux.svf.stuba.sk

Abstract. The Hadwiger number of a graph G, is the order of the largest complete graph which is a minor of G. In this paper we calculate the Hadwiger number of those graphs that are the product of two cycles.

1 Introduction

In this paper all graphs are finite and do not contain loops or multiple edges.

Suppose that $e = uv$ is an edge in G. We write G/e for the graph obtained from G by contracting the edge e, that is, identifying the vertices u and v and deleting all resulting loops or multiple edges. A graph H is a *contraction* of G if it can be obtained from G by a finite sequence of edge contractions. A graph H is a *minor* of G if it can be obtained from G by a finite sequence of contractions and edge deletions.

The (*Cartesian*) *product* $G \times H$ of two graphs G and H is the graph with vertex set $V(G) \times V(H)$ and edge set

$$\{(u_1, u_2)(v_1, v_2) : (u_1 = v_1 \text{ and } u_2v_2 \in E(H)) \text{ or } (u_1v_1 \in E(G) \text{ and } u_2 = v_2)\}.$$

If $G_1, G_2, \ldots, G_n$ are graphs, then we write $G_1 \times G_2 \times \cdots \times G_n$ for the product

$$(G_1 \times G_2 \times \cdots \times G_{n-1}) \times G_n.$$

The *cycle of length* n, C_n $n \geq 3$, is the (unique) 2-regular connected graph with n edges (or equivalently n vertices). The *complete graph (on n vertices)*, K_n $n \geq 1$, is the unique graph on n vertices with an edge joining every pair of vertices. In this paper we are concerned primarily with products of two cycles, $C_m \times C_n$. For completeness, C_1 and C_2 sometimes denote K_1 and K_2 respectively.

The *Hadwiger number*, $h(G)$, of a connected graph G is the maximum of the cardinals h such that K_h is a minor of G. Hadwiger [2] conjectured that for any graph G, $h(G) \geq \chi(G)$, where $\chi(G)$ denotes the chromatic number of G. It is easy to see that this conjecture implies the four colour theorem. Furthermore, the conjecture has been settled by Robertson, Seymour and Thomas [4] for graphs with Hadwiger number at most five, however, this result requires the four colour theorem.

In this paper we compute the Hadwiger number for the product of two cycles. We begin in Section 2 by motivating the investigation of products of cycles. Specifically, we define the Hadwiger number of a group, and present a conjecture, due to Seifter [5], concerning the Hadwiger number for finite Abelian groups. In the remaining sections, we first establish a bound for the Hadwiger number of products of two cycles by exploiting the fact that K_7 can be embedded on the surface of genus one. This bound is met by all but a few special cases. We complete most of the remaining cases by counting arguments. Finally, we outline a computer algorithm that was employed to finish four isolated cases.

2 The Hadwiger number for finite Abelian groups

Products of cycles are an interesting class of graphs for the Hadwiger number. To see this, consider a finitely generated group Γ with generating set Λ not containing the identity. The *Cayley graph* of Γ (with respect to Λ) is the graph with vertex set Γ and edge set

$$\{gh : g \in \Gamma, h = gh' \text{ where } h' \in \Lambda \cup \Lambda^{-1}\}.$$

The *Hadwiger number* $h(\Gamma)$ of a finite group Γ is defined as the minimum of the Hadwiger numbers of its Cayley graphs.

A generating set Λ of a group Γ is said to be *irreducible* if no proper subset of Λ generates Γ. In computing the Hadwiger number of a group, we need consider only irreducible generating sets, since deleting edges cannot increase the Hadwiger number of a graph.

It is well known (see, for example, [3]) that any finite Abelian group Γ of rank r has a unique *canonical form*

$$\mathbb{Z}_{m_1} \times \cdots \times \mathbb{Z}_{m_r}$$

such that m_i divides m_{i+1} for $i = 1, \ldots, r-1$, and $m_1 > 1$. Each element of Γ is written as an r-tuple whose coordinates correspond to the canonical product form. The *canonical generating set* for Γ consists of the r distinct elements that have 1 as one coordinate and 0 for all others.

Conjecture 1 (Seifter [5]). *Let Γ be a finite Abelian group, and let G denote the Cayley graph of Γ with respect to the canonical generating set. Then $h(\Gamma) = h(G)$.*

It is not hard to see that the Cayley graphs of the form given in Conjecture 1 are products of cycles.

We conclude this section by noting that the Hadwiger number is unbounded over the class of graphs containing products of three cycles. To see this, first draw K_h in the plane. For the moment, replace each crossing with a new vertex. Now find a fine enough grid $C_n \times C_n$ which has this new graph as a minor. Using the second layer of this grid in $C_n \times C_n \times C_2$ we can easily remove the crossings.

3 An upper bound for products of two cycles

Let e be an edge in a graph G, and suppose G has been embedded on a surface S of minimum genus. Then G/e can be embedded on the homeomorphic surface S/e obtained by contracting S along e. This implies that the genus of G/e is at most the genus of G. We conclude that if G has Hadwiger number h, then the genus of G is at least that of K_h.

For example, every planar graph has no minor isomorphic to K_5 and hence has Hadwiger number at most four. Therefore $h(C_2 \times C_n) \leq 4$ for all $n \geq 3$. Furthermore, any cycle of length 3 or more has Hadwiger number 3, and thus $h(C_2 \times C_2) = 3$. Finally, by noticing that $C_2 \times C_3$ has a minor isomorphic to K_4 we get the following lemma.

Lemma 2. *For all integers $n \geq 3$, we have $h(C_2 \times C_n) = 4$. Furthermore, $h(C_2 \times C_2) = 3$.*

We now introduce a convenient method for representing the edge contractions in a product $C_m \times C_n$ required to produce the complete graph K_h on the torus. First colour the vertices of K_h with colours $1, 2, \ldots, h$. Now take an m by n grid where the squares correspond in the obvious way to the vertices of $C_m \times C_n$. (See, for example, Figure 1 where the grid represents $C_5 \times C_5$.) Two squares in the grid are regarded as *adjacent* if they share a boundary of positive measure, or if they occur at the top and bottom respectively (far left and far right respectively) of the same column (row). Thus two squares are adjacent if and only if the corresponding vertices in $C_m \times C_n$ are adjacent.

We use colours $1, 2, \ldots, h$ to represent the sequence of edge contractions. One or more squares in the grid are coloured with colour i if and only if their corresponding vertices become the single vertex i of K_h after the sequence of edge contractions. To verify that the coloured grid represents a complete graph it is necessary to check that all the squares of a given colour are adjacent and that each colour is adjacent to every other colour. Using this representation, Figure 1 shows that $C_5 \times C_5$ can be contracted onto K_7.

Now, it is well-known that K_7 is the largest complete graph with genus 1. Therefore, any graph with genus at most 1 has Hadwiger number at most 7. In particular, the Hadwiger number for a product of two cycles is at most 7, since all such products have a 2-cell embedding on the sphere or torus. (The reader interested in the Hadwiger number of an arbitrary toroidal graph should examine [1].)

Now, since $C_m \times C_n$ is a contraction of $C_{m+1} \times C_n$ and of $C_m \times C_{n+1}$ then $h(C_m \times C_n) \leq h(C_{m+1} \times C_n)$ and $h(C_m \times C_n) \leq h(C_m \times C_{n+1})$. Hence Figure 1 gives us the following lemma.

Lemma 3. *If n, m are integers such that $\min\{n, m\} \geq 5$, then*

$$h(C_m \times C_n) = 7.$$

1	1	7	7	3
5	6	6	7	5
5	5	6	7	7
4	5	6	6	4
4	2	2	3	3

Fig. 1.

4 The Hadwiger number for $C_3 \times C_n$ and $C_4 \times C_n$

In this section we establish the Hadwiger number of the two remaining infinite classes of graphs described in the heading. We begin by showing that $C_3 \times C_n$ cannot be contracted onto K_7.

The *face-width* (also called *representativity*) of an embedding of a graph G in a surface S is the minimum size of $|C \cap G|$ over all noncontractible cycles C in S. It is well-known (see, for example, [1]) that K_7 does not have an embedding with face-width 2 in the torus. Using this fact, we can derive the following proposition.

Proposition 4. *If n is a integer greater than 2, then*

$$h(C_3 \times C_n) \leq 6.$$

Proof. By way of contradiction, suppose $C_3 \times C_n$ can be contracted onto K_7. Embed $C_3 \times C_n$ on the torus according to Figure 2. (In this figure, the sides are identified in the usual way.)

If, in the sequence of edge contractions that produces K_7, a vertical edge in the figure were contracted, then K_7 would have a face-width two embedding on the torus, a contradiction. Hence, only horizontal edges from the figure are contracted. Since $|V(C_3 \times C_n)| = 3n$ then a total of $3n - 7$ edges must be contracted to produce a graph with 7 vertices. By the pigeonhole principle, some copy of C_n has at least $n - 2$ edges contracted. However, again this results in a face-width 2 embedding of K_7.

Figure 3(a) shows that the bound established in Proposition 4 is met by $C_3 \times C_6$. An exhaustive computer search revealed that $C_3 \times C_5$ cannot be contracted onto K_6. Figure 3(b) shows that $C_3 \times C_3$ can be contracted onto K_5 and hence so can $C_3 \times C_4$ and $C_3 \times C_5$. These results yield the following lemma.

Lemma 5. *For all n such that $3 \leq n \leq 5$, $h(C_3 \times C_n) = 5$. For all n such that $n \geq 6$, $h(C_3 \times C_n) = 6$.*

Now consider the graphs $C_4 \times C_n$, where $n \geq 4$. Figure 4 shows that $C_4 \times C_4$ and $C_4 \times C_6$ can be contracted onto K_6 and K_7 respectively. A computer search showed that $C_4 \times C_5$ cannot be contracted onto K_7. Hence we obtain the following lemma.

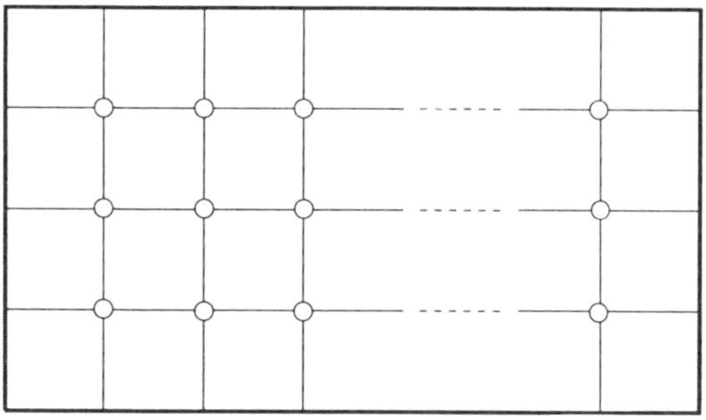

Fig. 2.

(a)

1	1	1	2	2	2
4	3	3	3	4	4
6	6	5	5	5	6

(b)

1	1	2
4	5	2
4	3	3

Fig. 3.

Lemma 6. *For $n = 4$ or $n = 5$, $h(C_4 \times C_n) = 6$. For all n such that $n \geq 6$, $h(C_4 \times C_n) = 7$.*

(a)

1	1	2	2
3	4	2	3
4	4	3	3
6	5	5	6

(b)

1	1	7	7	3	3
6	5	5	7	7	6
6	6	5	5	4	4
4	2	2	3	3	4

Fig. 4.

5 The algorithm

The authors used computer searches for several cases when calculating the Hadwiger number $h(C_m \times C_n)$. Constructively, a computer search found the contractions showing that $C_5 \times C_5$ and $C_4 \times C_6$ had Hadwiger number at least seven. These results are easily verified using Figure 1 and Figure 4(b) respectively. We also used computer searches to show that $C_3 \times C_5$ cannot be contracted onto K_6 and that $C_4 \times C_5$ cannot be contracted onto K_7. The validity of these results depends on the computer implementation of the algorithm used to obtain them, hence we provide the details of this algorithm here.

The backtracking type algorithm used is based on the procedure one could follow to colour the grids by trial and error. At first glance it would appear that this would not lead to an efficient algorithm; however it has meant the authors were able to implement a number of heuristics, which we describe at the end of this section.

We denote by $G(m,n) = G(n,m)$ the grid associated with the product $C_m \times C_n$ as defined above. A set of squares in $G(m,n)$ is said to be *connected* if the subgraph of $C_m \times C_n$ induced by the corresponding set of vertices is connected.

A *colouring* of $G(m,n)$ (*with h colours*) is a colouring of (not necessarily all) squares of $G(m,n)$ such that the set of squares coloured is connected, and the set of all squares coloured with a given colour is connected. A *complete colouring* of $G(m,n)$ (*with h colours*) is a colouring of $G(m,n)$ in which for every pair of colours $i, j \leq h$ there exist adjacent squares coloured i and j. A *total colouring* of $G(m,n)$ is a complete colouring in which every square is coloured. Thus the Hadwiger number for $C_m \times C_n$ is the maximum number of colours in a total colouring of $G(m,n)$. A *trivial colouring* of $G(m,n)$ is a colouring in which just one square is coloured — the colour used being 1.

Suppose we have a colouring C of $G(m,n)$ in which not every square is coloured. We can derive one or more colourings from C by colouring squares adjacent to those already coloured. Let C' be some colouring obtained from C by colouring, with colour i, an uncoloured square of $G(m,n)$. Furthermore, suppose the square coloured is adjacent to a square already coloured i, or, in the case no square was coloured i under C, the square is adjacent to a square already coloured. Then we say C' is a *derivative* of C with respect to colour i.

Now it is not hard to see that for every total colouring C of $G(m,n)$ there is a sequence of colourings

$$C_1, C_2, \ldots, C_{mn} = C$$

where C_1 is a trivial colouring, and C_{i+1} is a derivative of C_i for $1 \leq i < mn$. Furthermore, it is evident that the sequence may be chosen so that there exist distinct integers $z_0, z_1, z_2, z_3, \ldots, z_h$, where the number of squares coloured with colour k is $z_k - z_{k-1}$, which satisfy:

1. $0 = z_0 < z_1 < z_2 < z_3 < \cdots < z_h = mn$,
2. $z_1 \geq 2$
3. $z_j - z_{j-1} \leq z_{j+1} - z_j$ for all $1 < j < h$,
4. C_i is a derivative of C_{i-1} with respect to colour j for all $z_{j-1} \leq i < z_j$ where $j = 2, 3, \ldots, h$, and
5. $C_{z_j - 1}$ is a complete colouring for $j = 2, 3, \ldots, h$.

The following simple algorithm determines if such a sequence exists. It generates lists, L_i, of all acceptable colourings with i squares coloured, and outputs a positive result if the final list L_{mn} contains a total colouring.

procedure **colour**$(m, n, h, z_1, z_3 \ldots, z_h)$
    $c \leftarrow 1$
    $L_1 = \{C\}$ where C is a trivial colouring of $G(m,n)$

```
for all i such that 2 ≤ i ≤ mn do
        if i > z_c then
                c ← c + 1
                for each C ∈ L_{i-1} do
                        if C is not a complete colouring then
                                        L_{i-1} ← L_{i-1} − {C}
        for each C ∈ L_{i-1} do
                L_i ← L_i ∪ {D: D is a derivative of C with respect to colour c}
output "yes" if L_mn contains a total colouring; "no" otherwise
```

Now, to determine if K_h is a contraction of $C_m \times C_n$, one enumerates all sequences $z_1, z_2, \ldots, z_h$ satisfying the properties above, and then uses the colour procedure to test each such sequence. If the procedure outputs "no" to all sequences, then we conclude that $h(C_m \times C_n) < h$.

It is not difficult to enumerate the sequences required. For example, when determining if K_6 is a contraction of $C_3 \times C_5$ one needs to run the procedure on the following sequences only:

$$\begin{matrix} 2 & 4 & 6 & 8 & 10 & 15 \\ 2 & 4 & 6 & 8 & 11 & 15 \\ 2 & 4 & 6 & 9 & 12 & 15 \end{matrix}$$

To speed up the process of searching our implementation of this algorithm used the following heuristics;

1. Our implementation does not produce all colourings of a grid; only those that have the subgraph induced by each colour connected. This reduces the search space substantially. The ordering on the $z_i's$ reduces the number of isomorphic colourings produced, which again reduces the search space.
2. Since the graphs $C_m \times C_n$ are vertex transitive we only need to start off with one of the possible trivial colourings.
3. Our implementation was able to reuse the sets L_i of colourings it had generated. For example, in the three sequences given above used for showing that $C_3 \times C_5$ cannot be contracted onto K_6, we notice that the first three elements are the same. This means that in each of the three cases L_6 was the set of colourings of the grid $G(3,2)$, rooted at the same square, that had 2 squares coloured 1, 2 squares coloured 2, and 2 squares coloured 3. Hence this set needed to be generated only once.

6 Summary

We present the main result.

Theorem 7. *The Hadwiger number of the product $C_m \times C_n$ is given by the table:*

$h(C_m \times C_n)$ n / m	2	3	4	5	6	7	$\cdots$
2	3	4	4	4	4	4	4
3	4	5	5	5	6	6	6
4	4	5	6	6	7	7	7
5	4	5	6	7	7	7	7
6	4	6	7	7	7	7	7
7	4	6	7	7	7	7	7
$\vdots$	4	6	7	7	7	7	7

Acknowledgement

The second author wishes to thank Norbert Seifter of Montanuniversität Leoben for many interesting discussions on the Hadwiger number.

References

1. M. de Graaf and A. Schrijver, *Grid minors of graphs on the torus*, J. Combin. Theory Ser. B 61, no 1 (1994), 57–62.
2. Hadwiger H, *Über eine Klassifikation der Streckenkomplexe*, Vierteljahrsschrift der Naturf. Gesellschaft in Zürich (1943), 133–142.
3. S. Mac Lane and G. Birkhoff, *Algebra*, Macmillan, 1967.
4. N. Robertson, P. D. Seymour, and R. Thomas, *Hadwiger's conjecture for K_6-free graphs*, Combinatorica **13** (1993), 279–361.
5. N. Seifter, Personal communication.

Proper Bounded Edge-Colorings (Extended Abstract)

Claudia Bertram-Kretzberg[1], Hanno Lefmann[2], Vojtěch Rödl[3*] and Beata Wysocka[4]

[1] Universität Dortmund, Fachbereich Informatik, LS II, D-44221 Dortmund, Germany. bertram@ls2.informatik.uni-dortmund.de

[2] Universität Dortmund, Fachbereich Informatik, LS II, D-44221 Dortmund, Germany. lefmann@ls2.informatik.uni-dortmund.de

[3] Emory University, Department of Mathematics and Computer Science, Atlanta, Georgia 30322, USA. rodl@mathcs.emory.edu

[4] University of North Carolina at Greensboro, Department of Mathematical Sciences, Greensboro, NC 27412, USA. beata@maths.mu.oz.au

Abstract. For an n-element set X and a proper coloring $\Delta\colon [X]^k \longrightarrow \{0, 1, \ldots\}$ where each color class is a matching with cardinality bounded by u, we show that there exists a totally multicolored subset $Y \subseteq X$ with

$$|Y| \geq \max\left\{c_1 \cdot \left(n^k/u\right)^{\frac{1}{2k-1}}, \; c_2 \cdot \left(n^k/u\right)^{\frac{1}{2k-1}} \cdot \left(\ln\left(u/\sqrt{n}\right)\right)^{\frac{1}{2k-1}}\right\}$$

This bound is tight up to constant factors for $u = \omega(n^{1/2+\epsilon})$ for any $\epsilon > 0$. Moreover, for fixed k, we give a polynomial time algorithm for finding such a set Y of guaranteed size.

1 Introduction

On each of $\binom{3n}{3}/n$ school days, in a school attended by $3n$ students, the students are asked to line up in n rows, each containing three students. In 1851, Kirkman asked for the existence of such a schedule that would allow each triplet of students to occupy a row on exactly one of the school days [6]. This classical problem was answered completely by Ray-Chaudhuri and Wilson [10] who proved that such a schedule exists for each $n \equiv 1, 3 \bmod 6$. Here, we investigate a somewhat related combinatorial problem. Suppose that after such a schedule was prepared, the principle of the school wants (for unrevealed purposes) to select the largest group of, say, m students with the property that no two triplets of students occupy a row at the same day. Such an m must satisfy

$$c_1 \cdot n^{1/3} \cdot (\log n)^{1/3} \leq m \leq c_2 \cdot n^{2/3} \tag{1}$$

for any schedule. While the upper bound is straightforward, the lower bound follows from [2]. Here, we give a polynomial time algorithm which finds a group

* Research supported by NSF grant DMS 9401559. Part of this work was done during the author's visit of Humboldt-Universität, Berlin, with a Humboldt senior-fellowship.

of m students satisfying the lower bound in 1. Moreover, there are schedules which, up to a constant factor, are the best possible. We consider the general case in which one has n students which are asked to line up in at most u rows, each containing k people. We extend earlier results from [2] and [11] where the case $u = n/k$ respectively $k = 2$ was considered.

We formulate our problem in terms of edge-colored hypergraphs: vertices correspond to students, edges to rows, and the edges are colored by the day.

Definition 1. Let $\Delta: [X]^k \longrightarrow \omega$ where $\omega = \{0, 1, \ldots\}$ be a coloring of the k-element subsets of X. The coloring $\Delta: [X]^k \longrightarrow \omega$ with color classes $C_0, C_1, \ldots$, i.e., $\Delta^{-1}(i) = C_i$ for $i \in \omega$, is called *u-bounded* if $|C_i| \leq u$ for $i = 0, 1, \ldots$. The coloring $\Delta: [X]^k \longrightarrow \omega$ is called *proper* if each color class C_i, $i = 0, 1, \ldots$, is a matching, i.e., sets of the same color are pairwise disjoint, thus, $\Delta(U) = \Delta(V)$ implies $U \cap V = \emptyset$ for all distinct sets $U, V \in [X]^k$. A subset $Y \subseteq X$ is called *totally multicolored* if the restriction of the coloring Δ to the set $[Y]^k$ is a one-to-one coloring. For an n-element set X, define, minimizing over all proper u-bounded colorings $\Delta: [X]^k \longrightarrow \omega$, the following function

$$f_u(n,k) = min_{\Delta}\ max\{|Y|\ ;\ Y \subseteq X \text{ is totally multicolored}\}\ .$$

The first estimates on $f_u(n,k)$ were given by Babai [4], in connection with some Sidon-type problem. He showed for the case $u = n/2$ and $k = 2$ that $c_1 \cdot n^{1/3} \leq f_{n/2}(n,2) \leq c_2 \cdot (n \cdot \ln n)^{1/3}$. In [2], the lower bound was improved by the factor $O((\ln n)^{1/3})$. Here, we will show the following:

Theorem 2. *Let $k, u \geq 2$ be fixed integers. There exist positive constants c_1, c_2, c_3 such that for n large enough,*

$$\begin{aligned} &\max\left\{c_1 \cdot \left(n^k/u\right)^{1/(2k-1)},\ c_2 \cdot \left(n^k/u\right)^{1/(2k-1)} \cdot \left(\ln\left(u/\sqrt{n}\right)\right)^{1/(2k-1)}\right\} \\ &\leq f_u(n,k) \qquad (2) \\ &\leq c_3 \cdot \left(n^k/u\right)^{1/(2k-1)} \cdot (\ln n)^{1/(2k-1)}\ . \end{aligned}$$

Moreover, for every u-bounded proper coloring $\Delta: [X]^k \longrightarrow \omega$ with $|X| = n$, one can find in time $O(u \cdot n^{2k-1})$ a totally multicolored subset $Y \subseteq X$ with $|Y| \geq \max\left\{c_1 \cdot \left(n^k/u\right)^{1/(2k-1)},\ c_2 \cdot \left(n^k/u\right)^{1/(2k-1)} \cdot (\ln(u/\sqrt{n}))^{1/(2k-1)}\right\}$.

2 The Existence

Let $\mathcal{G} = (V, \mathcal{E})$ be a hypergraph with vertex set V and edge set $\mathcal{E}$. For a vertex $v \in V$, let $d(v)$ denote the *degree* of v in $\mathcal{G}$, that is, the number of edges $E \in \mathcal{E}$ containing v. Let $d = \sum_{v \in V} d(v)/|\mathcal{V}|$ denote the average degree of $\mathcal{G}$. The hypergraph $\mathcal{G}$ is called *k-uniform* if $|E| = k$ for each edge $E \in \mathcal{E}$. A *2-cycle* in $\mathcal{G}$ is a pair $E, E' \in \mathcal{E}$ of distinct edges which intersect in at least two vertices. The *independence number* $\alpha(\mathcal{G})$ is the largest size of a subset $I \subseteq V$ such that the induced hypergraph contains no edges.

Here, we will prove inequality (2) of Theorem 2. Some of our arguments are based on a result of Ajtai, Komlós, Pintz, Spencer and Szemerédi, [1]. We use a modified version proved in [7]; see [2, 12].

Theorem 3. *Let $k \geq 2$. Let $\mathcal{G}$ be a $(k+1)$-uniform hypergraph on n vertices. If (i) $\mathcal{G}$ contains no 2-cycles, and (ii) the average degree satisfies $d \leq t^k$ where $t \geq t_0(k)$, then for some positive constant $c = c(k)$,*

$$\alpha(\mathcal{G}) \geq c \cdot n/t \cdot (\ln\, t)^{1/k} . \tag{3}$$

Proof. We start by showing the lower bounds in (2). Let $\Delta\colon [X]^k \to \omega$ be a u-bounded proper coloring where $|X| = n$. We construct a $2k$-uniform hypergraph $\mathcal{H} = (X, \mathcal{E})$ on X where $U \in \mathcal{E} \subseteq [X]^{2k}$ if there exist two distinct sets $S, T \in [X]^k$, $S, T \subseteq U$ so that $\Delta(S) = \Delta(T)$. As Δ is u-bounded, we infer

$$|\mathcal{E}| = \sum_{i \in \omega} \binom{|\Delta^{-1}(i)|}{2} \leq \frac{\binom{n}{k}}{u} \cdot \binom{u}{2} . \tag{4}$$

If $I \subseteq X$ is an independent set of $\mathcal{H}$, then I is totally multicolored with respect to the coloring Δ. Hence, it is enough to show that $\mathcal{H}$ contains an independent set of size $c_1 \cdot (n^k/u)^{\frac{1}{2k-1}}$. This follows by an easy probabilistic argument: choose every vertex in X independently of the other vertices with probability

$$p = (n^{k-1} \cdot u)^{-1/(2k-1)} . \tag{5}$$

By Chernoff's and Markov's inequality, we know that there exists a subset $Y \subseteq X$ with $|Y| \sim (n^k/u)^{1/(2k-1)}$, and the subhypergraph induced on Y contains at most

$$2 \cdot p^{2k} \cdot |\mathcal{E}| \leq 2 \cdot p^{2k} \cdot \frac{\binom{n}{k}}{u} \cdot \binom{u}{2} \leq \frac{1}{2} \cdot \left(\frac{n^k}{u}\right)^{1/(2k-1)}$$

edges. We delete one vertex from each edge in $[Y]^{2k} \cap \mathcal{E}$, and we obtain a subset $Y' \subseteq Y$ with $|Y'| \geq |Y|/2 \geq pn/2$. Then, Y' is an independent set in $\mathcal{H}$, hence Y' is totally multicolored with respect to Δ.

If $u = \sqrt{n} \cdot \omega(n)$, where $\omega(n) \longrightarrow \infty$ with $n \longrightarrow \infty$, we can improve this lower bound by a logarithmic factor. Let $\Delta\colon [X]^k \to \omega$ be a u-bounded proper coloring. Consider the $2k$-uniform hypergraph $\mathcal{H} = (V, \mathcal{E})$ defined in the same way as above. Again, we want to find a large independent set in $\mathcal{H}$. The strategy is to find a random subset $Y \subseteq X$ such that the induced hypergraph has only a few 2-cycles. By deleting these 2-cycles, the desired result will follow from (3).

The number of edges of $\mathcal{H}$ satisfies inequality (4). With foresight we use a slightly larger value than in (5) for the probability p of picking vertices, namely,

$$p = \left(1/(n^{k-1} \cdot u)\right)^{1/(2k-1)} \cdot \left(u/\sqrt{n}\right)^{1/((k+1)(2k-1))} .$$

Let Y be a random subset of X obtained by choosing vertices $v \in X$ with probability p independently of the others. The expected size of Y is $E(|Y|) = p \cdot n$. Let $\nu_j(Y)$, for $j = 2, \ldots, 2k-1$, be random variables counting the number of $(2,j)$-cycles, that is, the number of pairs of edges in the subhypergraph of $\mathcal{H}$ induced on Y which intersect in exactly j vertices. The random variable $\mu_2(Y) = \sum_{j=2}^{2k-1} \nu_j(Y)$ counts the total number of 2-cycles of the subhypergraph induced on Y. We will give upper bounds on the expected values $E(\nu_j(Y))$. To do so, we estimate the total number ν_j of $(2,j)$-cycles in $\mathcal{H}$. Fix an edge $E \in \mathcal{E}$. The number of pairs of distinct sets $U, V \in [X]^k$ with $\Delta(U) = \Delta(V)$ and $|(U \cup V) \cap E| = j$ and $1 \leq |U \cap E|, |V \cap E| \leq j-1$ is at most

$$\sum_{i=\lceil j/2 \rceil}^{j-1} \binom{2k}{i} \cdot \binom{n-2k}{k-i} \cdot \binom{2k-i}{j-i} \leq c_1 \cdot n^{k-\lceil j/2 \rceil}\,, \tag{6}$$

as either $|U \cap E| \geq \lceil j/2 \rceil$ or $|V \cap E| \geq \lceil j/2 \rceil$, and every color class is a matching.

If $U \cap E = \emptyset$ or $V \cap E = \emptyset$, but $|(U \cup V) \cap E| = j$, then the number of such pairs U, V is bounded from above by

$$\binom{2k}{j} \cdot \binom{n-2k}{k-j} \cdot (u-1) \leq c_2 \cdot n^{k-j} \cdot u\,. \tag{7}$$

Now, (4), (6) and (7) imply that

$$\begin{aligned} \nu_j &\leq |\mathcal{E}| \cdot \left(c_1 \cdot n^{k-\lceil j/2 \rceil} + c_2 \cdot n^{k-j} \cdot u\right) \\ &\leq c_3 \cdot u \cdot \left(n^{2k-\lceil j/2 \rceil} + n^{2k-j} \cdot u\right)\,. \end{aligned} \tag{8}$$

As every color class is a matching, we have $u \leq n/k$, thus, $n^{2k-\lceil j/2 \rceil} \geq n^{2k-j} \cdot u$ for $j \geq 2$, and (8) becomes

$$\nu_j \leq c_4 \cdot u \cdot n^{2k-\lceil j/2 \rceil}\,. \tag{9}$$

We infer for $j = 2, \ldots, 2k-1$ that

$$\begin{aligned} E(\nu_j(Y)) &\leq p^{4k-j} \cdot c_4 \cdot u \cdot n^{2k-\lceil j/2 \rceil} \\ &= pn \cdot c_4 \cdot u^{\frac{j-2k+\frac{1}{k+1}(4k-j-1)}{2k-1}} \cdot n^{\frac{k(j+1-2\lceil j/2 \rceil)-\lfloor j/2 \rfloor-\frac{1}{2(k+1)}(4k-j-1)}{2k-1}}\,. \end{aligned}$$

As $u = \sqrt{n} \cdot \omega(n) \leq n/k$, we have $\omega(n) = O(\sqrt{n})$, and hence, for j odd,

$$\begin{aligned} E(\nu_j(Y)) &\leq pn \cdot c_4 \cdot u^{\frac{j-2k+\frac{1}{k+1}(4k-j-1)}{2k-1}} \cdot n^{\frac{-(j-1)/2-\frac{1}{2(k+1)}(4k-j-1)}{2k-1}} \\ &= pn \cdot c_4 \cdot \omega(n)^{\frac{j-2k+\frac{1}{k+1}(4k-j-1)}{2k-1}} \cdot n^{\frac{-k+1/2}{2k-1}} \\ &\leq pn \cdot c_4 \cdot \omega(n)^{\frac{k-1}{(k+1)(2k-1)}} \cdot n^{\frac{-k+1/2}{2k-1}} \qquad \text{as } j \leq 2k-1 \\ &= o(pn)\,. \end{aligned} \tag{10}$$

Similarly, for j even, we obtain

$$E(\nu_j(Y)) = o(pn)\,. \tag{11}$$

By (10) and (11), we infer $E(\mu_2(Y)) = \sum_{j=2}^{2k-1} E(\nu_j(Y)) = o(pn)$. Thus, there exists a subset $Y \subseteq X$ with $|Y| = c_5 pn$ such that the induced hypergraph contains at most $c_6 p^{2k}|\mathcal{E}|$ edges and has only $o(pn)$ 2-cycles. We omit one vertex from each 2-cycle in $\mathcal{H}_0$. The remaining subhypergraph $\mathcal{H}_1$ has $(c_5 - o(1)) \cdot pn$ vertices and by (4), the average degree d^{2k-1} satisfies $d \leq c_8 \cdot (u/\sqrt{n})^{\frac{1}{(k+1)(2k-1)}}$. As $u/\sqrt{n} \longrightarrow \infty$ with $n \longrightarrow \infty$, we obtain from (3) that

$$\alpha(\mathcal{H}) \geq \alpha(\mathcal{H}_1) \geq c \cdot \frac{(c_1 - o(1)) \cdot p \cdot n}{c_8 \cdot (u/\sqrt{n})^{\frac{1}{(k+1)(2k-1)}}} \cdot \left[\ln\left(c_8 \cdot \left(\frac{u}{\sqrt{n}}\right)^{\frac{1}{(k+1)(2k-1)}}\right)\right]^{\frac{1}{2k-1}}$$
$$\geq c' \cdot \left(n^k/u\right)^{1/(2k-1)} \cdot \left(\ln\left(u/\sqrt{n}\right)\right)^{1/(2k-1)} . \qquad \square$$

Next, we will show the upper bound in (2), extending some ideas from [4]. Let X be an n-element set where without loss of generality n is divisible by k. Set $m = \lceil c \cdot n^k/u \rceil$, where $c > 0$ is a constant. Let $M_1, \ldots, M_m$ be random matchings, chosen uniformly and independently from the set of all matchings of size u from $[X]^k$, and set $H_j = \bigcup_{i<j} M_i$. We define a coloring $\Delta: [X]^k \rightarrow \omega$ as follows: for $j = 1, \ldots, m$, color all sets in $M_j \setminus H_j$ by color j, and color all remaining elements in $[X]^k \setminus H_{m+1}$ in an arbitrary way, such that each color class is a matching. Let $Y \subseteq X$ be a fixed subset with $|Y| = x$ where $x = o(n/u^{1/k})$. We will prove that for $x \geq c_3 \cdot \left(n^k/u \cdot \ln n\right)^{1/(2k-1)}$ with probability approaching to 1 any such set Y is not totally multicolored, where $c_3 > 0$ is an appropriate constant. This will give the desired result. We split the proof into several claims.

Claim 4. *For $j = 1, \ldots, m$ and $t = 1, 2, \ldots$,*

$$\textit{Prob}\ \left[|M_j \cap [Y]^k| \geq t\right] \leq \left(u \cdot x^k/n^k\right)^t . \tag{12}$$

Proof. The left hand side of (12) does not depend on the particular choice of Y. Thus, assume that the matching M_j is fixed. The set Y can be chosen in $\binom{n}{x}$ ways. From M_j we can choose t edges in $\binom{u}{t}$ ways, and the remaining elements of Y can be chosen in at most $\binom{n-kt}{x-kt}$ ways, hence

$$\text{Prob}\ \left[|M_j \cap [Y]^k| \geq t\right] \leq \binom{u}{t} \cdot \binom{n-kt}{x-kt} / \binom{n}{x} \leq \left(u \cdot x^k/n^k\right)^t . \qquad \square$$

Claim 5. *For $t = 1, 2, \ldots$ and for large enough integers n,*

$$\textit{Prob}\ \left[|H_{m+1} \cap [Y]^k| \geq t\right] \leq \left(\frac{e \cdot (t+m) \cdot u \cdot x^k}{t \cdot n^k}\right)^t . \tag{13}$$

Proof. For $j = 1, \ldots, m$, consider the events $|M_j \cap [Y]^k| \geq t_j$. These events are independent. By Claim 4, we have Prob $[|\ M_j \cap [Y]^k\ | \geq t_j] \leq \left(u \cdot x^k/n^k\right)^{t_j}$. Since $|H_{m+1} \cap [Y]^k| \leq \sum_{j=1}^m |M_j \cap [Y]^k|$ we infer, using $\binom{n}{k} \leq (e \cdot n/k)^k$, that

$$\text{Prob}\ \left[|H_{m+1} \cap [Y]^k| \geq t\right]$$

$$
\begin{aligned}
&\leq \text{Prob}\left[\sum_{j=1}^{m} |M_j \cap [Y]^k| \geq t\right] \\
&\leq \sum_{(t_j)_{j=1}^m, t_j \geq 0, \sum_{j=1}^m t_j = t} \prod_{j=1}^{m} \text{Prob}\left[|M_j \cap [Y]^k| \geq t_j\right] \\
&\leq \sum_{(t_j)_{j=1}^m, t_j \geq 0, \sum_{j=1}^m t_j = t} \prod_{j=1}^{m} \left(u \cdot x^k / n^k\right)^{t_j} \\
&= \binom{t+m-1}{t} \cdot \left(u \cdot x^k / n^k\right)^t \\
&\leq \left(\frac{e \cdot (t+m) \cdot u \cdot x^k}{t \cdot n^k}\right)^t . \qquad \square
\end{aligned}
$$

Let E_i denote the event $|H_i \cap [Y]^k| \leq c_1 \cdot x^k$, where $c_1 > 0$ is a small constant.

Claim 6. *For large enough positive integers n,*

$$\textit{Prob}\ [E_{m+1}] \geq 1 - 2^{-c_1 \cdot x^k} .$$

Proof. For $t = c_1 \cdot x^k$ with $x = o\left(n/u^{1/k}\right)$, we have $t = o(n^k/u)$. If n is large, $m = \lceil c \cdot n^k/u \rceil$ and $ec/c_1 \leq 1/3$, then (13) is less than $(1/2)^t$, hence,

$$\text{Prob}\ [E_{m+1}] \geq 1 - \text{Prob}\ \left[|H_{m+1} \cap [Y]^k| \geq c_1 \cdot x^k\right] \geq 1 - 2^{c_1 \cdot x^k} . \qquad \square$$

We define another random variable $Y_j = |[M_j]^2 \cap [[Y]^k \setminus H_j]^2|$ for $j = 1, \ldots, m$.

Claim 7. *If n is a sufficiently large positive integer, then for $j = 1, \ldots, m$,*

$$E(Y_j | E_j) > c_5 \cdot u^2 \cdot x^{2k} / n^{2k} .$$

Proof. Clearly, we have Prob $[E_1] = 1$. As E_j holds, it is $|[Y]^k \setminus H_j| \geq \binom{x}{k} - c_1 \cdot x^k \geq c_2 \cdot x^k$. For each set $S \in [Y]^k$, there are less than $k \cdot \binom{x-1}{k-1}$ k-element subsets of Y which are not disjoint from S. Hence, for n large, the number of sets $\{S, T\} \in [[Y]^k \setminus H_j]^2$ with $S \cap T = \emptyset$ is at least

$$1/2 \cdot c_2 \cdot x^k \cdot \left(c_2 \cdot x^k - k \cdot \binom{x-1}{k-1}\right) > c_3 \cdot x^{2k} . \tag{14}$$

Two disjoint k-element sets S, T, are both in M_j with probability

$$\text{Prob}\ [S, T \in M_j] = \frac{u \cdot (u-1)}{\binom{n}{k} \cdot \binom{n-k}{k}} \geq c_4 \cdot \frac{u^2}{n^{2k}} . \tag{15}$$

By (14) and (15) for the conditional expected value $E(Y_j|E_j)$, we have $E(Y_j|E_j) \geq c_5 \cdot u^2 \cdot x^{2k}/n^{2k}$. $\square$

Claim 8. *For $j = 1, \ldots, m$, and large positive integers n, and $0 < \epsilon \ll c_5$,*

$$Prob\ [Y_j = 1 \mid E_j] \geq (c_5 - \epsilon) \cdot u^2 \cdot x^{2k}/n^{2k} .$$

Proof. For $t = 1, 2, \ldots$, we claim that

$$\text{Prob}\ [Y_j \geq t \mid E_j] \leq \left(u \cdot x^k/n^k\right)^{\lceil \sqrt{2t+1} \rceil} . \tag{16}$$

Namely, the statement $Y_j \geq t$ implies $|M_j \cap [Y]^k| \geq \lceil \sqrt{2t+1} \rceil$, hence,

$$\begin{aligned} &\text{Prob}\ [Y_j \geq t \mid E_j] \\ &\leq \text{Prob}\ \left[|M_j \cap [Y]^k| \geq \lceil \sqrt{2t+1} \rceil\right] \\ &\leq \left(u \cdot x^k/n^k\right)^{\lceil \sqrt{2t+1} \rceil} . \end{aligned}$$

For $i = 0, 1, \ldots$, set $p_i = \text{Prob}\ [Y_j = i \mid E_j]$. We infer from (16) that

$$\begin{aligned} E(Y_j \mid E_j) &= \sum_{i \geq 0} i \cdot p_i \\ &\leq p_1 + \sum_{i \geq 2} i \cdot \left(u \cdot x^k/n^k\right)^{\lceil \sqrt{2i+1} \rceil} \\ &= p_1 + O\left(\left(u \cdot x^k/n^k\right)^3\right) \\ &= p_1 + o\left(u^2 \cdot x^{2k}/n^{2k}\right) , \end{aligned}$$

as $x = o\left(n/u^{1/k}\right)$. By Claim 7, we obtain that $p_1 \geq (c_5 - \epsilon) \cdot u^2 \cdot x^{2k}/n^{2k}$ for some positive constant $\varepsilon < c_5$ and n large enough. □

Finally, let A_j denote the event ($Y_j = 0$ and E_{j+1}).

Claim 9.

$$Prob\ [A_1 \wedge \ldots \wedge A_m] \leq \exp\left(-c' \cdot u \cdot x^{2k}/n^k\right) .$$

Proof. By Claim 8, we have

$$\begin{aligned} \text{Prob}\ (A_1) &\leq \text{Prob}\ (Y_1 = 0\ |E_1) \\ &\leq \text{Prob}\ (Y_1 \neq 1\ |E_1) \\ &\leq 1 - (c_5 - \epsilon) \cdot u^2 \cdot x^{2k}/n^{2k} , \end{aligned} \tag{17}$$

while

$$\begin{aligned} &\text{Prob}\ [A_i\ |A_1 \wedge \ldots \wedge A_{i-1}] \\ &\leq \text{Prob}\ [Y_i = 0 \mid A_1 \wedge \ldots \wedge A_{i-1}] \\ &\leq \text{Prob}\ [Y_i \neq 1\ |A_1 \wedge \ldots \wedge A_{i-1}] \\ &\leq 1 - (c_5 - \epsilon) \cdot u^2 \cdot x^{2k}/n^{2k} . \end{aligned} \tag{18}$$

With $(1-x)^m \leq exp(-m \cdot x)$ and $m = \lceil c \cdot n^k/u \rceil$, inequalities (17), (18) imply

$$\begin{aligned}
&\text{Prob}\,[A_1 \wedge A_2 \wedge \ldots \wedge A_m] \\
&= \text{Prob}\,[A_1] \cdot \prod_{i=2}^{m} \text{Prob}\,[A_i \mid A_1 \wedge \ldots \wedge A_{i-1}] \\
&\leq \left(1 - (c_5 - \epsilon) \cdot u^2 \cdot x^{2k}/n^{2k}\right)^m \\
&\leq \exp\left(-(c_5 - \epsilon) \cdot m \cdot u^2 \cdot x^{2k}/n^{2k}\right) \\
&\leq \exp\left(-c \cdot (c_5 - \epsilon) \cdot u \cdot x^{2k}/n^k\right) . \qquad \square
\end{aligned}$$

Claim 10. *The probability that there exists a totally multicolored x-element subset is at most*

$$\binom{n}{x} \cdot \left(\exp\left(-c' \cdot u \cdot x^{2k}/n^k\right) + 2^{-c_1 \cdot x^k}\right) . \tag{19}$$

Proof. If Y is not totally multicolored, then either $A_1 \wedge \ldots \wedge A_m$ holds or E_{m+1} fails. As there are exactly $\binom{n}{x}$ x-element sets Y, by combining the estimates from Claim 3 and Claim 6, we obtain (19). □

For $x \geq c_6 \cdot \left(n^k/u\right)^{1/(2k-1)} \cdot (\ln n)^{1/(2k-1)}$, where $c_6 > 0$ is a big enough constant, expression (19) tends to 0 with $n \longrightarrow \infty$. For $n \leq n_0$, one can obtain asymptotically the same upper bound by taking an appropriately large constant.

3 The Algorithm

In this section, we sketch the idea how to find in time $O(u \cdot n^{2k-1})$ a totally multicolored subset as guaranteed by (2). The algorithm follows the probabilistic arguments given before. It is based on recent results from [5] and [8].

Let $k \geq 2$ be fixed and let $\Delta\colon [X]^k \to \omega$ with $|X| = n$ be a proper u-bounded coloring. First, we collect sets $S \in [X]^k$ of the same color. This can be done in time $O(n^k)$. We form a $2k$-uniform hypergraph $\mathcal{H} = (X, \mathcal{E})$, $\mathcal{E} \subseteq [X]^{2k}$, where $E \in \mathcal{E}$ if there exist distinct sets $S, T \in [X]^k$ with $S \cup T = E$ and $\Delta(S) = \Delta(T)$. Then, $|\mathcal{E}| = O(u \cdot n^k)$ by (4), hence, constructing $\mathcal{H}$ can be done in time $O(u \cdot n^k)$. First, assume that $u \leq C \cdot \sqrt{n}$ for some constant $C > 0$. We use the following algorithmic version of Turán's theorem for hypergraphs [5, 13].

Lemma 11. *Let $\mathcal{G} = (V, \mathcal{E})$ be a $(k+1)$-uniform hypergraph on n vertices with average degree d^k. Then, one can find in time $O(|V| + |\mathcal{E}|)$ an independent set $I \subseteq V$ with $|I| \geq c \cdot n/d$.*

Proof. We use the method of conditional probabilities [9, 3]. Let $V = \{v_1, \ldots, v_n\}$. Every vertex v_i will be assigned a probability p_i, $i = 1, \ldots, n$. Define a potential by

$$V(p_1, \ldots, p_n) = \sum_{i=1}^{n} p_i - \sum_{\{v_{i_1}, \ldots, v_{i_{k+1}}\} \in \mathcal{E}} \prod_{j=1}^{k+1} p_{i_j} .$$

In each step i, $i = 1, \ldots, n$, one after the other, we choose either $p_i = 0$ or $p_i = 1$ to maximize the value of $V(p_1, \ldots, p_n)$. As $V(p_1, \ldots, p_n)$ is linear in each p_i, for $i = 1$, for example, either $V(p_1, \ldots, p_n) < V(1, p_2, \ldots, p_n)$ or $V(p_1, \ldots, p_n) \leq V(0, p_2, \ldots, p_n)$. In the first case, we take vertex v_1, else we discard it. Choosing in the beginning, $p_1 = \ldots = p_n = p = 1/d$, the value of the potential is initially $V(p, \ldots, p) = p \cdot n - p^{k+1} \cdot n \cdot d^k/(k+1) = k/(k+1) \cdot n/d$. Finally, having chosen $p_1, \ldots, p_n \in \{0, 1\}$, the set $V' = \{v_i \in V \mid p_i = 1\}$ is an independent set in $\mathcal{G}$ of size as desired as can easily be seen. Each vertex and each edge in $\mathcal{G}$ is considered only a constant number of times, thus, the running time is $O(|V| + |\mathcal{E}|)$. □

By (4), the average degree d of $\mathcal{H}$ satisfies $d^{2k-1} \leq c' \cdot n^{k-1} \cdot u$. Thus, in $\mathcal{H} = (X, \mathcal{E})$, we can find in time $O(u \cdot n^k)$ an independent set of size at least $c \cdot \left(n^k/u\right)^{1/(2k-1)}$. This part of the algorithm can be done in time $O(u \cdot n^k)$.

Now, assume that $u = \sqrt{n} \cdot \omega(n)$ where $\omega(n) \longrightarrow \infty$ with $n \longrightarrow \infty$. First, we construct the sets $C_{2,j}$ of $(2, j)$-cycles in $\mathcal{H}$, $j = 2, \ldots, 2k-1$. Using (9), this can be done in time $O(u \cdot n^{2k-1})$. We use the following lemma [5].

Lemma 12. *Let $\mathcal{G} = (V, \mathcal{E})$ be a $(k+1)$-uniform hypergraph. Let $\mathcal{G}$ contain $\nu_j(\mathcal{G})$ many $(2, j)$-cycles for $j = 2, \ldots, k$. Then, for any real p, $0 \leq p \leq 1$, one can find in time $O(|V| + |\mathcal{E}| + \sum_{j=2}^{2k-1} \nu_j(\mathcal{G}))$ an induced subhypergraph $\mathcal{G}' = (V', \mathcal{E}')$ such that $|V'| \geq p/3 \cdot |V|$, $|\mathcal{E}'| \leq 3p^{k+1} \cdot |\mathcal{E}|$ and $\sum_{j=2}^{k} \nu_j(\mathcal{G}') \leq 3 \cdot \sum_{j=2}^{k} p^{2k+2-j} \cdot \nu_j(\mathcal{G})$ for $j = 2, \ldots, k$.*

Proof. Let $V = \{v_1, \ldots, v_n\}$. The proof is similar to that of Lemma 11. Step by step, one minimizes the potential $V(p_1, \ldots, p_n)$ given by

$$3^{p \cdot n/3} \cdot \prod_{i=1}^{n} (1 - 2p_i/3) + \frac{\sum_{E \in \mathcal{E}} \prod_{v_i \in E} p_i}{3 \cdot p^{k+1} \cdot |\mathcal{E}|} + \frac{\sum_{j=2}^{k} \sum_{c_j \in C_{2,j}} \prod_{v_i \in c_j} p_i}{3 \cdot \sum_{j=2}^{k} p^{2k+2-j} \cdot |C_{2,j}|},$$

which satisfies in the beginning $V(p, \ldots, p) < 1$. □

We apply Lemma 12 to the hypergraph $\mathcal{H}$ with $p = \left(1/(n^{k-1} \cdot u)\right)^{1/(2k-1)} \cdot (u/\sqrt{n})^{1/((k+1)(2k-1))}$, and we obtain in time $O(u \cdot n^{2k-1})$ an induced subhypergraph $\mathcal{H}' = (X', \mathcal{E}')$ with $|X'| \geq p/3 \cdot |X|$ and $|\mathcal{E}'| \leq 3p^{2k} \cdot |\mathcal{E}|$ and, by (10) and (11), we have $\sum_{j=2}^{2k-1} \nu_j(\mathcal{H}') \leq p/6 \cdot |X|$. In time $o\left(u \cdot n^{2k-1}\right)$, we can determine easily all 2-cycles in $\mathcal{H}'$ and delete from $\mathcal{H}'$ one vertex from each 2-cycle. The remaining hypergraph $\mathcal{H}''$ on at least $pn/6$ vertices contains at most $c \cdot p^{2k} \cdot n^k \cdot u$ edges, thus, has average degree $d^{2k-1} \leq c' \cdot p^{2k-1} \cdot n^{k-1} \cdot u$. Then, we apply the following result from [5] which gives an algorithmic version of the existence result from [7] and extends an algorithm of Fundia [8].

Theorem 13. *Let $k \geq 2$ be a fixed integer. Let $\mathcal{G} = (V, \mathcal{E})$ be a $(k+1)$-uniform ypergraph on n vertices with average degree d^k. If $\mathcal{G}$ does not contain any 2-les, then one can find for every fixed $\delta > 0$ in time $O(n \cdot d^k + n^3/d^{3-\delta})$ an dependent set of size at least $c(k, \delta) \cdot n/d \cdot (\ln d)^{1/k}$.*

Using Theorem 13, we can find an independent set in $\mathcal{H}''$, hence in $\mathcal{H}$, in time $O\left(p^{2k} \cdot n^k \cdot u + n^3/\left(p \cdot n^{(k-1)/(2k-1)} \cdot u^{1/(2k-1)}\right)^{3-\delta}\right) = o\left(u \cdot n^{2k-1}\right)$.

4 Final Remarks

For $u = \omega(\sqrt{n})$, the running time of the algorithm can be reduced to $O(u^{1-2/(k+1)} \cdot n^{2k-1-1/(k+1)})$ as follows. Similarly as in Lemma 12, for $p = (\sqrt{n}/u)^{2/((k+1)(2k-1))}$ one chooses first a subhypergraph $\mathcal{H}' = (X', \mathcal{E}')$ of $\mathcal{H} = (X, \mathcal{E})$, where we do not control the 2-cycles, but where $|X'| = p/3 \cdot |X|$ and $|\mathcal{E}'| \leq 3p^{2k} \cdot |\mathcal{E}|$. Hence, $\mathcal{H}'$ has $O(u \cdot (p \cdot n)^{2k-\lceil j/2 \rceil})$ many $(2, j)$-cycles. Then, we apply Lemma 12 with a different value p' and proceed as before. Thus, we save some time as we build the 2-cycles later. We omit the details here.

It might also be interesting to find the real growth rate of $f_u(n, k)$ and a corresponding fast algorithm or to give an explicit coloring which yields our upper bound for $f_u(n, k)$ or even a better upper bound for $u = O(\sqrt{n})$.

References

1. M. Ajtai, J. Komlós, J. Pintz, J. Spencer and E. Szemerédi, Extremal Uncrowded Hypergraphs, *Journal of Combinatorial Theory* Ser. A 32, 1982, 321-335.
2. N. Alon, H. Lefmann and V. Rödl, On an Anti-Ramsey Type Result, *Coll. Math. Soc. János Bolyai*, 60. Sets, Graphs and Numbers, 1991, 9-22.
3. N. Alon and J. Spencer, *The Probabilistic Method*, Wiley & Sons, New York, 1992.
4. L. Babai, An Anti-Ramsey Theorem, *Graphs and Combinatorics 1*, 1985, 23-28.
5. C. Bertram-Kretzberg and H. Lefmann, The Algorithmic Aspects of Uncrowded Hypergraphs, *8th ACM-SIAM Symp. on Discrete Algorithms*, to appear, 1997.
6. N. L. Biggs, T. P. Kirkman, *Bull. London Math. Soc.* 13, 1981, 97-120.
7. R. A. Duke, H. Lefmann and V. Rödl, On Uncrowded Hypergraphs, *Random Structures & Algorithms 6*, 1995, 209-212.
8. A. Fundia, Derandomizing Chebychev's Inequality to find Independent Sets in Uncrowded Hypergraphs, *Random Structures & Algorithms 8*, 1996, 131-147.
9. P. Raghavan, Probabilistic Construction of Deterministic Algorithms: Approximating Packing Integer Programs, *Journal of Computer and Systems Sciences.* 37, 1988, 130-143.
10. D. Ray-Chaudhuri and R. M. Wilson, Solution of Kirkman's Schoolgirl Problem, *Proc. Symp. Pure Math.* Vol. XIX, UCLA, AMS, Providence, 1968, 187-203.
11. H. Lefmann, V. Rödl, and B. Wysocka, Multicolored Subsets in Colored Hypergraphs, *Journal of Combinatorial Theory* Ser. A 74, 1996, 209-248.
12. V. Rödl and E. Šiňajová, Note on Independent Sets in Steiner Systems, *Random Structures & Algorithms 5*, 1994, 183-190.
13. J. Spencer, Turán's Theorem for k-Graphs, *Discrete Mathematics 2*, 1972, 183-186.

Sharply Bounded Alternation within P (Extended Abstract)

Stephen A. Bloch[1*], Jonathan F. Buss[2**], Judy Goldsmith[3*,***]

[1] Department of Mathematics and Computer Science,
Adelphi University, Garden City, NY 11530, USA.
[2] Department of Computer Science,
University of Waterloo, Waterloo, Ontario, Canada N2L 3G1.
[3] Computer Science Department,
University of Kentucky, Lexington, KY 40506–0027, USA.

Abstract. We define the *sharply bounded hierarchy*, $SBH(QL)$, a hierarchy of classes within P, using quasilinear-time computation and quantification over values of length $\log n$. It generalizes the limited nondeterminism hierarchy introduced by Buss and Goldsmith, while retaining the invariance properties. The new hierarchy has several alternative characterizations.
We define both $SBH(QL)$ and its corresponding hierarchy of function classes, *FSBH(QL)*, and present a variety of problems in these classes, including $\leq_m^{ql}$-complete problems for each class in $SBH(QL)$. We discuss the structure of the hierarchy, and show that certain simple structural conditions on it would imply $P \neq PSPACE$. We present characterizations of $SBH(QL)$ relations based on alternating Turing machines and on first-order definability. We also give recursion-theoretic characterizations of function classes corresponding to $SBH(QL)$.

1 Introduction

Problems not computable in polynomial time are intractable. Problems computable in polynomial time may or may not be feasibly computable in practice, principally because the exponent may be too large. We consider subclasses of polynomial-time-computable problems, with the ultimate goal of characterizing the set of feasible polynomial-time computations. These subclasses are quite natural, both in the sense of having interesting complete problems and the sense of having several equivalent characterizations.

Many classes within P have been studied, including those defined by circuit complexity, or those defined on Turing machines or RAM models by restricting

* Supported in part by Natural Sciences and Engineering Research Council (NSERC) of Canada while at the University of Manitoba.
** Supported in part by grants from NSERC Canada and the Information Technology Research Centre (an Ontario Centre of Excellence).
*** Supported in part by National Science Foundation (NSF) grant CCR-9315354.

the polynomial time bound; most of these classes are not robust. Quadratic time is not closed under composition, and linear time, while closed under composition, is highly sensitive to details of the computational model such as number of tapes. Furthermore, linear time does not seem to suffice for such basic problems as multiplication and sorting. We draw particular attention to sorting, since many simple, natural problems can sort as a by-product of their operations, and sorting itself is an extremely useful technique for problems that do not explicitly require it. (For complexity-theoretic examples, see [8, 13, 16, 26] and section 2.2 below.)

Several approaches have been considered for complexity classes slightly larger than linear time. Schnorr [26] and Gurevich and Shelah [16] have considered *quasilinear* and *nearly linear* time classes (QL and NLT), defined as $DTIME(\mathcal{O}(n \log^{\mathcal{O}(1)} n))$ on multi-tape Turing machines and on random-access machines, respectively. These classes are closed under composition, and are relatively machine-independent—QL is invariant under obliviousness assumptions [13] and the number of tapes of a multi-tape Turing machine, while NQL is equivalent under five different random-access models. It is an open question whether $QL = NLT$. In the current paper, we focus on the Turing-machine model (QL). We denote $\mathcal{O}(n \log^{\mathcal{O}(1)} n))$ by $Q(n)$.

What makes a polynomial-time algorithm inefficient? Given a polynomial-time decision algorithm, we may distinguish between two possible sources of complexity: is it performing an inherently time-consuming test, or is it merely testing a large number of possible candidates for problem solutions? For example, a naive algorithm to determine whether a graph contains a k-clique takes time $\mathcal{O}(n^k)$ (or slightly more, depending on the precise model). Most of the time cost comes from the loop over all possible k-vertex subgraphs, a cost we could eliminate with $k \log n$ bits of nondeterminism. This observation was developed by Buss and Goldsmith [8], who extended QL with *sharply bounded* existential quantifiers (of the form "$\exists y |y| \leq \log |x|$"). They developed a structure theory for these classes and showed that if the hierarchy contains all of P then $P \neq NP$.

In this paper we examine the implications of mixing sharply bounded universal ($\tilde{\forall}$) and existential ($\tilde{\exists}$) quantification over a quasilinear-time test. The resulting hierarchy of complexity classes, extending the Buss-Goldsmith hierarchy but still contained within P, we call the Sharply Bounded Hierarchy over Quasilinear Time and denote by $SBH(QL)$ or (when the time bound is clear from context) SBH.

The Sharply Bounded Hierarchy is partly inspired by Wrathall's *linear-time hierarchy* (LTH) [28]. Like her, we define a hierarchy based on bounded quantifiers over a small time-bounded class within P, and then give other characterizations of the hierarchy and the classes therein. There are several important differences, however. Quasilinear-time predicates give our classes more machine independence, and sharply bounded quantifiers keep our entire hierarchy within P. The sharp bounds apparently add to the intricacy of the hierarchy, because strings of the same quantifier do not obviously collapse (in fact, there are oracles relative to which they separate). Thus the $SBH(QL)$ does not mirror the polynomial-time hierarchy as neatly as does the LTH. (The "quasilinear-time hierarchy" of Naik, Regan and Sivakumar [22] differs from $SBH(QL)$ by allowing larger quantifiers and mimics LTH more closely.)

In the next section, we give definitions of the relevant complexity classes. We then present a variety of problems in the $SBH(QL)$, define a notion of completeness, and show that some of the problems considered yield quasilinear-complete sets for classes in the hierarchy. We also define function classes based on the sharply bounded hierarchies over linear and quasilinear time, and show that these function classes are appropriate generalizations of the relation classes. Later sections discuss the structure of the $SBH(QL)$, including upward collapse theorems, and show that if $SBH(QL)$ contains a non-trivial amount of P, then $P \neq PSPACE$. We present characterizations of both $SBH(QL)$ and $SBH(linear)$ in terms of a variant of alternating Turing machines. We also present machine-independent characterizations of both the relation and the function classes of both $SBH(QL)$ and $SBH(linear)$, using recursion-theoretic methods.

2 Definitions and example problems

2.1 The basic classes

Schnorr [26] introduced the classes $QL = \bigcup_k DTIME(n \log^k n)$ and $NQL = \bigcup_k NTIME(n \log^k n)$, with respect to multi-tape Turing machines. We use the notation of Buss and Goldsmith [8] and refer to these classes as P_1 and NP_1, respectively. Similarly, we define $P_r = \bigcup_k DTIME(n^r \log^k n)$.

Classes in the Sharply Bounded Hierarchy over Quasilinear Time, $SBH(QL)$, are defined by ***sharply bounded quantifiers*** over quasilinear-time predicates. A sharply bounded quantifier is one of the form $\exists y\ |y| \leq \log |x|$ or $\forall y\ |y| \leq \log |x|$, written as $\tilde{\exists}$ and $\tilde{\forall}$, respectively. If S is a string of sharply bounded quantifiers (such as $\tilde{\forall}\tilde{\exists}\tilde{\forall}\tilde{\forall}$), then $A \in SP_1$ iff there is a P_1 relation R so that $x \in A \Leftrightarrow S\mathbf{y}[R(x, \mathbf{y})]$. The notation $SBH(QL)$, or SBH for short, can mean either the collection of classes $\{SP_1 : S \in \{\tilde{\exists}, \tilde{\forall}\}^*\}$ or the single class which is the union of them; which is meant in a particular instance should be clear from context.

Note that the strings of quantifiers in the above may contain repetitions. Unlike the classes defined by bounded quantifiers, there is no evidence that iterations of a sharply bounded quantifier ($\tilde{\forall}\tilde{\forall}$, for example) can be collapsed to a single quantifier. Note also that, if S is a string of r sharply bounded quantifiers, then $SP_1 \subseteq P_{r+1}$. Hence $SBH \subseteq P \cap QLSPACE$.

We also consider function classes in Section 5. The functions btt-reducible in quasilinear time to SBH relations are denoted by $FQL^{SBH(QL)[1]}$. The analogous class with quasilinear time replaced by linear time, in both the relation class and the reduction, is denoted by $FL^{SBH(linear)[1]}$.

2.2 Languages in the classes

Buss and Goldsmith [8] have given various examples of problems in the classes $\tilde{\exists}^k P_1$. Many of the problems they considered are fixed-parameter variants of

NP-complete problems. For instance, they consider CSAT(k), a restriction of the circuit satisfiability problem. A Boolean circuit C with $k \log n$ inputs and n total gates is in CSAT(k) if there is an input string that causes it to output a 1. CSAT(0), the Circuit Value Problem, was shown to be in P_1 by Pippenger [24]. Buss and Goldsmith showed that CSAT(k) is complete for $\tilde{\exists}^k P_1$ (for all k); we observe that there are variants of this problem in all classes of *SBH*. For example, a Boolean circuit C is in $\tilde{\exists}\tilde{\forall}$CSAT if it has n gates, $2 \log n$ of which are input gates, and there is a string x_1 of length $\log n$ such that for any string x_2 of length $\log n$, C on input $x_1 x_2$ outputs a 1.

The following examples include some *P*-complete problems [15]. Some of these are also complete for classes in the *SBH*. This does not prove that $P \subseteq SBH$, since the *SBH* is not known to be closed under the reductions, such as *LOGSPACE* or NC^1, for which these problems are *P*-complete.

MATRIX MULTIPLICATION TESTING is in $\tilde{\forall}P_1$: given three $\sqrt{n} \times \sqrt{n}$ matrices A, B, and C, is $A \cdot B = C$?

TOTAL UNARY ORDERED GENERATOR is complete for $\tilde{\forall}P_1$: given an ordered context-free grammar G and a set T of terminal symbols, does G generate a string in τ^* for each $\tau \in T$? (A grammar is *ordered* if, for each nonterminal A in G, each occurrence of A in the lefthand side of a production occurs before any occurrence of A on the righthand side of a production.) Buss and Goldsmith [8] show that the set of ordered unary generators is complete for $\tilde{\exists}P_1$; this variant follows easily from their proof.

The following five examples come from Fellows and Downey's work on fixed-parameter intractibility [12].

PERFECT CODE(k) is in $\tilde{\exists}^{k-1}P_1$: given a graph G, does G have a k-element perfect code; i.e., is there a set V' of k vertices such that for all vertices $v \in G$, v has a unique neighbor in V'? The algorithm nondeterministically chooses all but one of the vertices; the last must have a neighborhood equal to the vertices not in the neighborhoods of the chosen vertices.

WEIGHT(k) SAT is in $\tilde{\exists}^k P_1$: given a Boolean formula F in CNF, is there a weight-k satisfying assignment, i.e., a satisfying truth assignment with exactly k 1's?

WEIGHT(k) EXACT SAT is in $\tilde{\exists}^k P_1$: given a Boolean formula F in CNF, is there a weight-k satisfying assignment that assigns exactly one "true" to each clause?

UNIQUE WEIGHT(k) SAT is in $\tilde{\exists}^k \tilde{\forall}^k P_1$: given a Boolean formula F in CNF, is there exactly one weight-k satisfying assignment for F?

LEXICOGRAPHICALLY FIRST CSAT(k) is in $\tilde{\exists}^{k-1}\tilde{\forall}^k P_1$: given a topologically sorted circuit description of size n, with $k \log n$ input gates, and a string y of length $\log n$, does the lexicographically first satisfying assignment to the circuit include y as one of the k blocks of size $\log n$?

The *AW*[*]-complete problems SHORT GEOGRAPHY and SHORT NODE KAYLES [1] give examples of alternating quantifiers. Both model two-player games on a graph. In the geography game on a graph G, the players alternately pick nodes of a simple path; the first player unable to extend the path to a new node loses. A graph is in SHORT GEOGRAPHY(k) if the player moving first has a winning strategy of k moves or less. One can easily show that SHORT GEOGRAPHY($2k+1$) is in $(\tilde{\exists}\tilde{\forall})^k P_1$. Likewise, SHORT NODE KAYLES($2k$) is in $(\tilde{\exists}\tilde{\forall})^k P_1$. We refer the reader to Abrahamson, *et al.* [1] for more details on these problems and the class $AW[*]$.

In many cases, quantifiers in a problem specification can be eliminated by enhancing the quasilinear-time test. One such example is testing the Vapnik-Chervonenkis dimension of a set, which is defined as a Σ_3 property.

V-C DIMENSION(k) [21, 27]: given a family of subsets $\mathcal{C} = \{C_1, \ldots, C_r\}$, drawn from a finite universe $U = \{x_1, \ldots, x_m\}$, is the Vapnik-Chervonenkis (V-C) dimension of $\mathcal{C}$ at least k? That is, is there a set $S \subseteq U$, with $|S| = k$, such that for all subsets T of S, there is a $C_i \in \mathcal{C}$ such that $S \cap C_i = T$?

We assume that each subset C_i is specified in the input by an m-bit binary string, so that the total input size is $n = rm$. The V-C dimension of family $\mathcal{C}$ can be at most $\log r$.

(If the sets C_i are represented implicitly via circuits, rather than being explicitly listed, the problem is Σ_3^P-complete [25].)

A simple algorithm puts V-C DIMENSION(k) in $\tilde{\exists}^k P_1$: given $\mathcal{C}$ (of size rm), guess S using $k \log m \leq k \log n$ bits. Compute for each $i \leq r$ the set $D_i = C_i \cap S$. Sort the strings representing D_i and verify that there are 2^k distinct values.

Using a more sophisticated algorithm, one can show that V-C DIMENSION(k) is in $\tilde{\exists}^{k-1} P_1$.

Some P-complete problems do not appear to be in *SBH*. One example is the fundamental problem UNIT RESOLUTION [19], which is clearly in P_2, but not apparently in quasilinear time, even with the help of sharply bounded quantifiers.

2.3 Complete sets

Since the classes in *SBH* are closed under quasilinear-time functions, it is natural to define a notion of quasilinear-time, many-one complete sets for these classes [8, 16, 26]. Schnorr [26] showed that SAT is $\leq_m^{ql}$-complete for nondeterministic quasilinear time (*NQL*, a.k.a. NP_1). The same techniques apply to alternating quantifiers.

Theorem 1. *The set* QBF *of true quantified Boolean formulas is* $\leq_m^{ql}$*-hard for any class in SBH.*

Buss and Goldsmith [8] gave various problems that are complete for classes of the form $\tilde{\exists}^k P_1$, and Cai and Chen [9] have recently shown that the "weight $\leq k$ CSAT" problem is also complete for $\tilde{\exists}^k P_1$. Variants of some of these problems can be shown to be $\leq_m^{ql}$-complete for classes higher in *SBH*. For all S in $\{\tilde{\exists}, \tilde{\forall}\}^*$, we define the following two problems.

S-CSAT: Given a Boolean circuit C with $|S| \log |C|$ inputs, does $S\mathbf{x}[C$ on input $\mathbf{x}$ outputs a 1] hold?

S-SHORTSAT: Given a Boolean formula B with at least $|S| \log |C|$ variables, does $S\mathbf{x}$[the assignment of $\mathbf{x}$ to the first $|S| \log |C|$ variables causes B to unravel] hold? (A partial assignment of truth values to variables in the formula may induce values on other variables if the formula is to be true: if B is in 3CNF, and $(a_1 \vee a_2 \vee a_3)$ is a clause, and all but one of the a_i's are false, then the remaining variable is induced to assume the value true; if one variable a_i is true, then the clause is true, and no additional assignment is induced. If this procedure can be applied iteratively until all clauses evaluate to true, then we say the original partial assignment caused the formula to *unravel*.)

Theorem 2. *For all S in $\{\tilde{\exists}, \tilde{\forall}\}^*$, S-CSAT and S-SHORTSAT are complete for SP_1.*

2.4 Function classes

One can define function classes that correspond in a natural way to the levels of the *SBH*, or to the union of the whole hierarchy. In Section 2.1, we defined classes that correspond to the union of the hierarchies (over *QL* and *linear*). We demonstrate here that the recursion-theoretic characterizations give very closely related classes. The theorems in Section 5 reinforce this.

Let *FL* and *FQL* denote respectively the functions computable in linear and quasilinear deterministic time. Let $B(\cdot)$ denote the closure of a class under Boolean operations.

Theorem 3. *For any $S \in \{\tilde{\exists}, \tilde{\forall}\}^*$, a 0/1-valued function is in $FQL^{SP_1[1]}$ if and only if it is the characteristic function of a $B(SP_1)$ relation.*

Since $B(SBH) = SBH$, it follows immediately that

Corollary 4. *A 0/1-valued function is in $FQL^{SBH(QL)[1]}$ if and only if it is the characteristic function of an SBH relation.*

An analogous theorem and corollary hold with quasilinear time replaced by linear time throughout.

3 The structure of the classes

Like most hierarchies of complexity classes, *SBH* exhibits upward collapse and downward separation. Some of the collapses presented in this section concern the compression of like quantifiers (from $\tilde{\exists}\tilde{\exists}$ to $\tilde{\exists}$, for instance), as well as (or instead of) the quantifier-swapping we expect of collapsing results.

Theorem 5. *If C is a class in SBH, and $\tilde{\exists} C \subseteq C$, then for all $m \geq 0$, $\tilde{\exists}^m C \subseteq C$.*

The proof is a simple induction on m.

Theorem 6. *If C is a class in SBH, closed under complement (e.g. QL), and $\tilde{\exists}^k C \subseteq \tilde{\forall}^{k'} C$ for some $k > k'$, then for all $m > k$, $\tilde{\exists}^m C \subseteq \tilde{\exists}^{k'} C$.*

Proof. If $\tilde{\exists}^k C \subseteq \tilde{\forall}^{k'} C$, then dually $\tilde{\forall}^k \overline{C} \subseteq \tilde{\exists}^{k'} \overline{C}$. Since C is closed under complement by assumption, all four classes are equal. The result follows by Theorem 5.

Note that Theorem 6 and its proof do not allow the case $k = k'$; that is, we derive no consequences from the closure of $\tilde{\exists}^k C$ under complement. Unlike the case of unbounded quantifiers, closure under complement does not translate easily to a larger number of quantifiers.

If the classes $\tilde{\exists}^k P_1$ are all closed under complement, then *SBH* is contained in the Buss-Goldsmith existential hierarchy. This would certainly simplify *SBH*, but would it simplify it right down to *QL*? In the presence of an oracle, not necessarily. The following result is analogous to recent work on the β hierarchy [3]. We assume that a query to the oracle erases the query string from the tape.

Theorem 7. *There is an oracle A such that for all $k \geq 0$ and $h > 0$,*

- $\tilde{\exists}^k P_h^A \neq \tilde{\exists}^{k+1} P_h^A$ *and*
- $\tilde{\exists}^k P_h^A = \tilde{\forall}^k P_h^A$.

The proof relies on a delayed diagonalization. Coding conditions for the equalities are satisfied at each stage, while the inequalities are established (using diagonalization) only at those stages where sufficient room is available.

Placing the *SBH* with respect to deterministic computation would have significant consequences.

Theorem 8. *If for all l, $SBH \not\subseteq P_l$, then $P \neq PSPACE$.*

Proof. Suppose $P = PSPACE$. Then QBF $\in P$, and hence QBF $\in P_l$ for some $l \geq 1$. But then $SBH \subseteq P_l$.

Lemma 9. *If $P_r \subseteq SP_1$ for some string S of sharply-bounded quantifiers and some rational $r > 1$, then $P_{r^2} \subseteq SSP_1$.*

Proof. Let $A \in P_{r^2}$ and $B = \{ y \mid y = x10^{\lfloor |x|^r \rfloor - |x| - 1} \wedge x \in A \}$. Then $B \in P_r$, and thus in SP_1 by assumption. To decide $A(x)$, in time $Q(n^r)$, write $y = x10^{\lfloor |x|^r \rfloor - |x| - 1}$, and decide (in time $Q(|x|^r)$, with quantifiers S), if $y \in B$. In other words, $A \in SP_r \subseteq SSP_1$.

Theorem 10. *If for some rational $r > 1$, $P_r \subseteq SBH$, then $P \neq PSPACE$.*

Proof. Since P_r has a complete set [8], the inclusion $P_r \subseteq SBH$ implies that $P_r \subseteq SP_1$, for some string S of sharply bounded quantifiers.

Iterating the lemma, this implies $P \subseteq SBH \subseteq QLSPACE$ and hence $P \neq PSPACE$.

4 Characterizations of the classes

The definition of *SBH* in section 2.1 uses the apparently arbitrary bound of $\log n$ on the length of a quantified variable. There are several equivalent definitions, however, that do not depend on an explicit bound on nondeterminism.

4.1 An alternating machine model

Our first characterization comes from a modified alternating machine, which allows "blocks" of nondeterminism in a simple, natural way. The standard model of an alternating Turing machine [10] chooses (existentially or universally) one of two possible "next states" at each step. However, alternation can also be achieved by existentially or universally choosing a *head position*. This provides the desired $(\log n)$-bit blocks of nondeterminism.

Definition 11. A *length-alternating Turing machine* is a multi-tape Turing machine whose states are partitioned into universal, existential, and deterministic states. (We shall refer to universal and existential states collectively as indeterministic states.) Each state, whether deterministic or not, has a single next state. Each indeterministic state s has an associated tape k_s; when the machine reaches state s, it jumps the head on tape k_s to an arbitrary nonblank square, and changes to the next state. This transition counts as a single step of the computation.

Acceptance and rejection are determined for each possible configuration of a length-alternating Turing machine in the same way as for standard alternating machines [10].

The *indeterminism depth* of a configuration in an LATM computation tree is the number of its ancestor configurations in the tree which are indeterministic. The indeterminism depth of an entire LATM computation tree is the maximum indeterminism depth of all its configurations.

The *time bound* of an LATM computation tree is the depth of the tree, counting deterministic as well as indeterministic configurations.

Theorem 12. *A relation $R(\mathbf{x}) \subseteq (\Sigma^*)^k$ is in SBH(QL) if and only if there are a constant c, a quasilinear function q, and a length-alternating Turing machine such that for all $\mathbf{x} \in (\Sigma^*)^k$,*

- *the machine accepts on input $\mathbf{x}$ if and only if $R(\mathbf{x})$ holds,*
- *the indeterminism depth of the computation tree is at most c, and*
- *the time bound of the computation tree is at most $q(\max(|\mathbf{x}|))$.*

We leave the proof to the full version of this paper.

An exactly analogous characterization can be given of *SBH*(*linear*), the closure of *linear* time (on Turing machines with fixed but arbitrary numbers of tapes) under sharply bounded quantifiers.

4.2 First-order definability

The quantifiers of *SBH* have the same length as those of the log-time hierarchy, which is equivalent to logtime-uniform AC^0. One elegant characterization of uniform AC^0 (due to Immerman [2, 17, 18]) is in terms of first-order definability over a particular set of relations including $\leq$ and *Bit*. We have shown elsewhere [7] that the *SBH* can be characterized similarly, in terms of first-order definability over *QL*.

5 Function-algebraic characterizations

The *SBH* can also be characterized by function algebras, analogously to P. Early function-algebraic characterizations of P, such as Cobham's [11], relied on explicit polynomial bounds on recursion; more recent work [4, 5, 6, 20] has used the notion of *safe*, or *tiered*, recursion to replace such artificial bounds with more foundationally justifiable limitations. In this section, we present the relevant notions of safe recursion, give function-algebraic characterizations of both linear and quasilinear time, and modify these characterizations to characterize the *SBH* function classes defined earlier.

5.1 Safe parameters

Bellantoni and Cook [5] define a schema of "safe recursion on notation" that resembles Cobham's "bounded recursion on notation" [11] but replaces Cobham's growth-rate bounds by syntactic restrictions on the use of parameters. They consider each function to take two kinds of parameters, called "normal" and "safe"; the latter are subject to only a restricted set of operations. They distinguish the two kinds syntactically by listing a function's normal parameters first, separated from its safe parameters with a semicolon, *e.g.* $f(x_1, x_2; y, z_1, z_2)$. (A similar notion appeared, more or less independently, in Leivant [20].)

Safe and normal parameters differ in two important ways. First, a safe parameter may not be used to control the depth of a recursion; and second, a safe parameter to a function cannot be used as normal within the computation of that function, nor in any way affect a value used as normal within the computation of that function. Normal parameters, however, may be used as safe—there is a "one-way door" between the two.

To enforce these constraints, Bellantoni and Cook define function-constructor operations called "safe composition" and "safe recursion on notation", and show that closing a set of simple base functions under these constructors gives exactly polynomial time.

Characterizing linear or quasilinear time requires an additional technique. In particular, one cannot freely use as many copies of a parameter as one wishes; making copies of a linear-length value takes linear time and cannot be allowed inside any kind of loop. We therefore impose a third distinction between safe and normal parameters, one reminiscent of linear logic: a function may use only one copy of each safe parameter, but may make unlimited copies of normal parameters.

For technical reasons (again related to copying parameters), we use *tuple-valued* functions, which take j normal and k safe parameters and produce a tuple of m values (where j, k, and m are fixed for any given function).

We start with the following set of *BASE* functions.

- The constant 0 function;
- The function $Half(; y) = \lfloor y/2 \rfloor$;
- The function $s_0(; y) = 2y$;

- The function $s_1(;y) = 2y + 1$; and
- Any fixed permutation of any fixed subset of j normal and k safe parameters, for all j and k.

Note that the first four *BASE* functions return arity 1; only the last can return a nontrivial tuple.

The conditional function, as defined in [4, 5, 6, 20], uses two copies of its safe parameters, which we cannot allow. Therefore, this algebra uses a conditional *constructor* instead:

Definition 13. The function $f(\mathbf{x};\mathbf{y})$ is *defined by cases on the i-th parameter* from functions g_0, g_1, and g_2 if

$$f(\mathbf{x};\mathbf{y}) = \begin{cases} g_0(\mathbf{x};\mathbf{y}) & \text{if the } i\text{-th parameter of } \mathbf{x},\mathbf{y} \text{ is } 0 \\ g_1(\mathbf{x};\mathbf{y}) & \text{if the } i\text{-th parameter of } \mathbf{x},\mathbf{y} \text{ is odd} \\ g_2(\mathbf{x};\mathbf{y}) & \text{if the } i\text{-th parameter of } \mathbf{x},\mathbf{y} \text{ is positive and even.} \end{cases}$$

A function definable from the base functions using any fixed number of applications of definition by cases can be computed in constant time on a multi-tape Turing machine using the following representation for input and output. The machine has one tape for each input string (and possibly others); each has its head at the right (low-order) end of the string. The states of the machine are used to store, among other things, a permutation of the constantly many tapes; thus (for example) any fixed permutation can be computed in one step of the machine. Our proofs use this representation for intermediate computations; any reasonable input-output convention can be converted to and from this one in linear time at the start and end of a computation.

Definition 14. The function $f(\mathbf{x};\mathbf{y})$ is *defined by safe composition* from functions g, $u_1, \ldots, u_j$, and $v_1, \ldots, v_k$ if for some fixed partition of the safe parameters $\mathbf{y}$ into $\mathbf{y}_1, \mathbf{y}_2, \ldots, \mathbf{y}_k$,

$$f(\mathbf{x};\mathbf{y}) = g(u_1(\mathbf{x};), \ldots, u_j(\mathbf{x};); v_1(\mathbf{x};\mathbf{y}_1), \ldots, v_k(\mathbf{x};\mathbf{y}_k)).$$

Functions defined from *BASE* using only safe composition and cases are still very simple. If such a definition doesn't use multiple copies of its normal parameters (*e.g.* if it takes no normal parameters at all), then it can likewise be computed in constant time using the above input/output convention.

We define a restricted form of recursion as follows.

Definition 15. The function $f(z,\mathbf{x};\mathbf{y})$ is defined by *very safe recursion on notation* from functions $h(;\mathbf{w})$and $g(\mathbf{x};\mathbf{y})$ if

$$f(z,\mathbf{x};\mathbf{y}) = \begin{cases} g(\mathbf{x};\mathbf{y}) & \text{if } z = 0 \\ h\left(; f\left(\lfloor z/2 \rfloor, \mathbf{x};\mathbf{y}\right)\right) & \text{if } z > 0. \end{cases}$$

The variable of recursion is always a normal parameter to f, while the iterated function h takes no normal parameters.

Using very safe recursion and safe composition, one can define functions of any linear growth rate (measured as usual in terms of the length of the arguments and output). For example, let $Succ_k(;y)$ be the k-th iterate of $s_1(;y)$; then the function

$$f(x;) = \begin{cases} 0 & \text{if } x = 0 \\ Succ_k(;f(\lfloor x/2 \rfloor;)) & \text{if } x > 0 \end{cases}$$

satisfies $|f(x;)| = k \cdot |x|$. It follows from the results of the next section that no function with superlinear growth can be defined.

The following schema gives quasilinear growth rates.

Definition 16. The function $f(\mathbf{x};\mathbf{y})$ is *defined by length composition* from functions g, $u_1, \ldots u_j$, and $v_1, \ldots, v_k$ if for some fixed partition of the safe parameters $\mathbf{y}$ into $\mathbf{y}_1, \mathbf{y}_2, \ldots, \mathbf{y}_k$,

$$f(\mathbf{x};\mathbf{y}) = g(u_1(\mathbf{x},|\mathbf{y}|;), \ldots, u_j(\mathbf{x},|\mathbf{y}|;); v_1(\mathbf{x},|\mathbf{y}|;\mathbf{y}_1), \ldots, v_k(\mathbf{x},|\mathbf{y}|;\mathbf{y}_k)).$$

Combining length composition with very safe recursion can lead to superlinear growth rates. Let

$$q_0(z;y) = \begin{cases} y & \text{if } z = 0 \\ s_0(;q_0(\lfloor z/2 \rfloor;y)) & \text{if } z > 0, \end{cases}$$

and for all $k \geq 1$,

$$r_k(;y) = q_{k-1}(|y|;y)$$

and

$$q_k(z;y) = \begin{cases} y & \text{if } z = 0 \\ r_k(;q_k(\lfloor z/2 \rfloor;y)) & \text{if } z > 0. \end{cases}$$

For each k, the value of $q_k(z;z)$ has length $\Theta(|z| \log^{k-1} |z|)$.

5.2 Characterizing linear and quasilinear time

We now characterize linear and quasilinear time on multi-tape Turing machines by the use of safe parameters. Versions of Theorem 18 were conjectured simultaneously and independently by Bloch, by Bellantoni, and by Leivant, but to our knowledge neither Theorem 21 nor a precise statement or proof of Theorem 18 has appeared in the literature.[4]

Definition 17. The function class $\mathcal{D}$ is the closure of the set *BASE* under the operations of definition by cases, safe composition and very safe recursion on notation.

Theorem 18. *A function $f(\mathbf{x})$ is computable by a linear-time multi-tape Turing machine if and only if $f(\mathbf{x};) \in \mathcal{D}$.*

[4] Otto [23] has given a category-theoretic characterization of these classes inspired by an earlier version of this section.

Proof sketch. One direction uses the input-output convention described above to define the next-configuration function of a standard Turing machine. Iterating this function linearly many times gives the result of the machine.

The other direction uses Lemma 19 below. For any function and inputs $f(\mathbf{x};\mathbf{y})$, we denote by $Tf(\mathbf{x};\mathbf{y})$ the time required to compute f at $\mathbf{x};\mathbf{y}$ on a multi-tape Turing machine using the input-output convention defined in the previous section. If $f(\mathbf{x};) \in \mathcal{D}$, we bound both its growth rate and its required computation time by induction on its definition.

Lemma 19. *For all f in $\mathcal{D}$, there are constants l_f and c_f such that for all $\mathbf{x}$ and $\mathbf{y}$,*

$$|f(\mathbf{x};\mathbf{y})| \leq l_f \cdot (1 + \max(|\mathbf{x}|)) + \max(|\mathbf{y}|)$$

and

$$Tf(\mathbf{x};\mathbf{y}) \leq c_f \cdot (1 + \max(|\mathbf{x}|)).$$

(Thus some simple functions like $f(;x) = x + 1$ cannot be defined using only safe parameters.)

Definition 20. The function class $\mathcal{D}'$ is the closure of the set *BASE* under the operations of definition by cases, length composition and very safe recursion on notation.

Theorem 21. *A function $f(\mathbf{x})$ is computable by a quasilinear-time multi-tape Turing machine if and only if $f(\mathbf{x};) \in \mathcal{D}'$.*

Proof sketch. To state the analogue of Lemma 19 cleanly, we enhance the Turing machine model with a unit-time *length-replacement* operation. In one step, the length-replacement operation replaces the current contents of any one tape with the length of the non-blank string on another tape, written in binary.

A standard Turing machine can easily simulate a Turing machine with length replacement by doubling the number of tapes. Initially, the length of each input tape is computed and written in binary on a new tape. At each step, the lengths are updated to reflect head motion of the original machine. A length replacement copies the appropriate counter and computes its length. Each original step requires at most logarithmic time in the simulation; hence, quasilinear time is preserved.

Lemma 22. *For all f in $\mathcal{D}'$, there are polynomials l_f and p_f such that for all $\mathbf{x}$ and $\mathbf{y}$,*

$$|f(\mathbf{x};\mathbf{y})| \leq (1 + \max(|\mathbf{x}|)) \cdot l_f(||\mathbf{x}||, ||\mathbf{y}||) + \max(|\mathbf{y}|)$$

and

$$Tf(\mathbf{x};\mathbf{y}) \leq (\max(|\mathbf{x}|) + 1) \cdot p_f(||\mathbf{x}||, ||\mathbf{y}||)$$

on Turing machines with length replacement.

The proof is again an induction of the definition of a function.

For the reverse direction of Theorem 21, the machine encoding proceeds as in the linear case, but is iterated over a quasilinear time bound rather than a linear one, using the functions q_k of the previous section.

5.3 Characterizing *SBH*

We now characterize *SBH*(*QL*) and *SBH*(*linear*) by adding one additional schema to the tiered function algebras of Section 5.2. The 0/1-valued functions definable with the new schema will be the characteristic functions of *SBH*(*QL*) and *SBH*(*linear*) relations. Arbitrary (0/1-valued or not) definable functions are the functions in $FQL^{SBH(QL)[1]}$ and $FL^{SBH(linear)[1]}$.

Definition 23. A function f is defined by *normal-bounded zero-detection* from g if

$$f(z, \mathbf{x}; \mathbf{y}) = \begin{cases} 0 & \text{if } (\forall w < |z|)(g(w, \mathbf{x}; \mathbf{y}) = 0) \\ 1 & \text{if } (\exists w < |z|)(g(w, \mathbf{x}; \mathbf{y}) > 0) \end{cases}.$$

Definition 24.

- The function class $\mathcal{D}^{nzd}$ is the closure of the *BASE* functions under safe composition, very safe recursion on notation, and normal-bounded zero-detection.
- The function class $\mathcal{D}'^{nzd}$ is the closure of the *BASE* functions under length composition, very safe recursion on notation, and normal-bounded zero-detection.

Theorem 25.

- *A function* $f(\mathbf{x})$ *is in* $FL^{SBH(linear)[1]}$ *if and only if* $f(\mathbf{x};) \in \mathcal{D}^{nzd}$. *Furthermore, if* f *is 0/1-valued,* i.e., f *is the characteristic function of a relation* R_f, *then* $R_f \in SBH(linear)$ *if and only if* $f(\mathbf{x};) \in \mathcal{D}^{nzd}$.
- *A function* $f(\mathbf{x})$ *is in* $FQL^{SBH(QL)[1]}$ *if and only if* $f(\mathbf{x};) \in \mathcal{D}'^{nzd}$. *Furthermore, if* f *is 0/1-valued,* i.e., f *is the characteristic function of a relation* R_f, *then* $R_f \in SBH(QL)$ *if and only if* $f(\mathbf{x};) \in \mathcal{D}'^{nzd}$.

As in previous sections, we select a machine model suitable for reasoning about the definable functions. The natural model for $FQL^{SBH(QL)[1]}$ is an oracle machine: a quasilinear-time deterministic computation with a constant number of queries to an oracle in *SBH*. By Theorem 12, the oracle set is computable by an LATM in quasilinear time with constant indeterminism depth. In order to obtain clean time bounds in the inductive proofs below, we shall want to avoid copying strings to special oracle tapes. We therefore use a "self-query" machine, which combines the machine that queries the oracle with the machine that recognizes the oracle set.

A "self-query" state q designates three states q_t, q_y, and q_n, called "test", "yes", and "no" respectively. The test state q_t is the start of the computation of the oracle set. If the computation from q_t accepts, the answer to the query is yes, and the successor of query state q is q_y. If the computation from q_t rejects, then answer to the query is no, and the successor of q is q_n. Note that the self-query state has only one actual successor and is thus a deterministic state.

To treat this oracle machine and its query set as a single LATM with self-queries requires several restrictions. First, to keep the deterministic part that queries the oracle separate from the length-alternating part that decides the oracle, we require that query states may not appear as descendants of test states, while indeterministic states may *only* appear as descendants of test states. Second, the indeterminism depth of the whole computation tree must (as in Theorem 12) be constant.

Note that, since the time bound of an LATM is found by maximizing the length of *all* its branches, any linear or quasilinear time bound we impose will apply to both the deterministic and length-alternating parts.

With these restrictions, a self-querying LATM M using quasilinear time behaves essentially as a deterministic quasilinear-time oracle machine with an oracle in $SBH(QL)$, and thus computes a function in $FQL^{SBH(QL)[1]}$. The one difference is that M makes queries without copying its configuration to a special oracle tape. Because M makes only a constant number of queries, a standard oracle machine could compute the same result, with copying of query strings, also in quasilinear time. Similarly, linear time bounds on self-querying LATMs characterize exactly $FL^{SBH(linear)[1]}$.

Using self-querying LATMs, we can state simple bounds on the complexities of functions in $\mathcal{D}^{nzd}$ and in $\mathcal{D}'^{nzd}$. The following two lemmas can be proved by straightforward inductions on the definition of a function f.

Lemma 26. *If* $f(\mathbf{x};\mathbf{y}) \in \mathcal{D}^{nzd}$, *then*

- f *is computable on a self-querying multi-tape Turing machine in time* $O(|\mathbf{x}|)$ *with constantly many queries to an* $O(|\mathbf{x}|)$*-time, constant-depth LATM, and*
- *if* f *is 0/1-valued, then* R_f *is computable by an* $O(|\mathbf{x}|)$*-time, constant-depth LATM.*

In the analogous lemma for $\mathcal{D}'^{nzd}$, we strengthen the induction hypothesis by adding a statement about deterministic computation. The iterated function of a very safe recursion is calculated deterministically, giving a polylogarithmic growth rate.

Lemma 27. *If* $f(\mathbf{x};\mathbf{y}) \in \mathcal{D}'^{nzd}$, *then*

- f *is computable in time* $Q(|\mathbf{x}|, ||\mathbf{y}||)$ *on a self-querying LATM with length replacement,*
- $f(\mathbf{x};\mathbf{y})$ *is computable deterministically in time polynomial in* $|\mathbf{x}|$ *and* $||\mathbf{y}||$*, and*
- *if* f *is 0/1-valued, then* R_f *is computable by an* $Q(|\mathbf{x}|, ||\mathbf{y}||)$*-time, constant-depth LATM with length replacement.*

6 Further directions

We have characterized *SBH* in several ways; however, its relation to many other complexity classes remains unclear. Theorems 8 and 10 suggest limits on our ability to prove that *SBH* contains hard problems, but they allow some room to maneuver. Perhaps most significantly, we would like to know whether *SBH* contains all problems solvable in deterministic quasilinear time on a random-access machine (*NLT*). Gurevich and Shelah [16] show a complete problem for *NLT*; if this one problem is in *SBH*, then *SBH* is the same over random-access machines as over Turing machines. They show that all the machine models considered yield the same class with respect to unbounded nondeterminism (*i.e.*, $NQL = NNLT$).

Similarly, *SBH* might contain Grädel's [14] class $DTIME(n^{1+})$, which is strictly larger than *QL*, but is not large enough to apply our Lemma 9. Further, one could easily define an analogous hierarchy over $DTIME(n^{1+})$ using the model of Section 4.1; would the inclusion $DTIME(n^{1+}) \in SBH$ imply the collapse of the $DTIME(n^{1+})$ hierarchy?

Acknowledgements

The authors thank Peter Clote for suggesting that they work on this material, and making several helpful suggestions, and Michael Saks for pointing out that $\tilde{\forall}P_1$ contained interesting problems. Thanks to James Otto for pointing out an error in an earlier version of our safe-parameter definition.

References

1. K.A. Abrahamson, R.G. Downey, and M.R. Fellows, "Fixed parameter tractability and completeness IV: on completeness for $W[P]$ and PSPACE analogues," Technical report DCS-216-IR, University of Victoria, 1993.
2. D. Mix Barrington, N. Immerman, and H. Straubing, "On uniformity in NC^1, *J. Comput. System Sci.* 41 (1990) 274–306.
3. R. Beigel and J. Goldsmith, "Downward separation fails catastrophically for limited nondeterminism classes," in *Proceedings of the Ninth Annual Structure in Complexity Theory Conference*, IEEE Computer Society Press, 1994, pp. 134–138.
4. S. Bellantoni, *Predicative Recursion and Computational Complexity.* Ph.D. thesis, University of Toronto, 1992.
5. S. Bellantoni and S. Cook, "A new recursion-theoretic characterization of the polytime functions," *Computational Complexity* 2 (1992) 97–110.
6. S.A. Bloch, "Function-algebraic characterizations of log and polylog parallel time," *Computational Complexity* 4 (1994) 175–205. See also *Proceedings of the Seventh Annual Structure in Complexity Theory Conference*, IEEE Computer Society Press, 1992, pp. 193–206.
7. S.A. Bloch, J.F. Buss and J. Goldsmith, "Sharply-Bounded Quantifiers within *P*," technical report TR96-011, Electronic Colloquium on Computational Complexity, 1996. Available at http://ftp.eccc.uni-trier.de/pub/eccc/reports/1996/TR96-011/index.html.
8. J.F. Buss and J. Goldsmith, "Nondeterminism within P," *SIAM J. Comput.* 22 (1993) 560–572.
9. L. Cai and J. Chen, "On the amount of nondeterminism and the power of verifying," *SIAM J. Computing*, to appear. See also *Mathematical Foundations of*

Computer Science, Lecture Notes in Computer Science 711, Springer-Verlag, 1993, pp. 311–320.
10. A. Chandra, D. Kozen, and L. Stockmeyer, "Alternation," *J. Assoc. Comput. Mach.* 28 (1981) 114-133.
11. A. Cobham, "The intrinsic computational complexity of functions," in Y. Bar-Hillel, ed., *Logic, Methodology, and the Philosophy of Science: Proceedings of the 1964 International Congress*, North-Holland, 1965, pp. 24–30.
12. R.G. Downey and M.R. Fellows, "Fixed-parameter intractability," in *Seventh Annual IEEE Conference on Structure in Complexity Theory,* IEEE Computer Society Press, 1992, pp. 36–49.
13. N. Pippenger and M.J. Fischer, "Relations among complexity measures", *J. Assoc. Comput. Mach.* 26 (1979) 361–381.
14. E. Grädel, "On the notion of linear-time computability," *Internat. J. Foundat. Comput. Sci.* 1 (1990) 295–307.
15. R. Greenlaw, H.J. Hoover, and W.L. Ruzzo, *Limits to Parallel Computation: P-Completeness Theory,* Oxford University Press, 1995.
IEEE Computer Society Press, 1983, pp. 210–214.
16. Y. Gurevich and S. Shelah, "Nearly linear time," in A.R. Meyer and M.A. Taitslin, eds., *Logic at Botik '89,* Lecture Notes in Computer Science 363, Springer-Verlag, 1989, pp. 108–118.
17. N. Immerman, "Languages that capture complexity classes," *SIAM J. Comput.* 16 (1987) 760–778.
18. N. Immerman, "Expressibility and parallel complexity," *SIAM J. Comput.* 18 (1989) 625–638.
19. N.D. Jones and W.T. Laaser, "Complete problems for deterministic polynomial time," *Theoretical Computer Science* 3 (1976) 105–117.
20. D. Leivant, "Subrecursion and lambda representation over free algebras," in S.R. Buss and P. Scott, eds., *Feasible Mathematics*, Perspectives in Computer Science, Birkhäuser, 1990, pp. 281–291.
21. N. Linial, Y. Mansour, and R.L. Rivest, "Results on learnability and the Vapnik-Chervonenkis dimension," *Information and Control* 90 (1991) 33–49.
22. A.V. Naik, K.W. Regan, and D. Sivakumar, "Quasilinear-time complexity theory," *Theoretical Computer Science,* to appear.
23. J. Otto, "Tiers, tensor, and the linear-time hierarchy," in *Proceedings of Workshop on Logic and Computational Complexity*, Lecture Notes in Computer Science, Springer-Verlag, 1995.
24. N. Pippenger, "Fast simulation of combinational logic circuits by machines without random-access storage," in F.P. Preparata and M.B. Pursley, eds., *Fifteenth Allerton Conference on Communication, Control, and Computing,* 1977, pp. 25–33.
25. M. Schäfer, "Deciding the Vapnik-Červonenkis dimension is Σ_3^p-complete," in *Eleventh Annual IEEE Conference on Computational Complexity,* IEEE Computer Society Press, 1996, pp. 77–85.
26. C.P. Schnorr, "Satisfiability is quasilinear complete in *NQL*," *J. Assoc. Comput. Mach.* 25 (1978) 136–145.
27. V. Vapnik and A.Y. Chervonenkis, "On the uniform convergence of relative frequencies of events to their probabilities," *Theory of Probability and Its Applications* 16 (1971) 264–280.
28. C. Wrathall, "Rudimentary predicates and the linear-time hierarchy," *SIAM J. Comput.* 7 (1978) 194–209.

A Note on the Growth Rate of Planar Graphs

Neal Brand[1], Margaret Morton[2]

[1] Department of Mathematics , University of North Texas, Denton TX 76203, USA, brand@cas.unt.edu

[2] Department of Mathematics, University of Auckland, Auckland, New Zealand, morton@math.auckland.ac.nz

Abstract. If Γ is a planar, locally finite, vertex transitive, 1-ended graph, then there is a particular 'niceness' about the arrangement of the regions incident to a vertex in Γ. Using this feature, it can be shown that Γ can be embedded in either the Euclidean plane or the hyperbolic plane in such a way that every edge has the same length and every angle in an n-cycle bounding a region has the same measure. Moreover, there is a simple condition which tells whether Γ is embedded in the Euclidean plane or the hyperbolic plane. The geometry of these two planes is then exploited to show that Γ must have either quadratic growth or exponential growth depending on which plane it is embedded in.

1 Introduction

The graphs considered in this paper are simple, infinite, and planar. The symbols $V(\Gamma)$, $E(\Gamma)$ and $\mathrm{aut}(\Gamma)$ will denote respectively, the vertex set, the edge set, and the automorphism group of Γ. All graphs will be assumed to be *locally finite*; that is the degree of every vertex is finite. The graph Γ is said to be *transitive* if for every pair of vertices $\{u, v\}$ in Γ there exists an automorphism g of Γ such that $g(u) = v$. An *end* of a graph is an equivalence class of one way infinite paths where two one way infinite paths are equivalent if there are infinitely many pairwise disjoint paths in the graph connecting them. It is well known that every infinite, locally finite, vertex transitive graph has either one, two or infinitely many ends [4]. A useful alternative condition for a 1-ended graph is that if any finite subset T of $V(\Gamma)$ is removed then $\Gamma - T$ has just one infinite component.

Let Γ be a planar, locally finite, vertex transitive, 1-ended graph. Since Γ is 1-ended and vertex transitive each vertex has degree at least 3 and Γ is 3-connected [1]. By a generalization [3,4] of Whitney's result [5] we know that when an infinite planar graph is 3-connected, the cyclic order of the edges incident with each vertex when the graph is embedded in the plane becomes an intrinsic property of the graph and is the same for all planar embeddings. For a biconnected planar graph we define a proper embedding to be a planar embedding with the property that the boundary of each face is either an elementary circuit or a double ray. Using a result of Halin [2] it follows that every 1-ended, locally finite, planar graph has a proper embedding. In Bonnington *et al.* [1] it is shown that if Γ is a locally finite, planar, 3-connected, transitive graph, then it has no infinite faces in any proper embedding.

For the remainder of this paper we assume that Γ is a planar, locally finite, vertex transitive, 1-ended graph. Furthermore, we assume Γ is embedded in the plane and sometimes abuse notation by identifying vertices in Γ with their images in the plane. Suppose that each vertex in Γ has degree $k \geq 3$ and that the edges incident to a particular vertex v_0 in Γ, given in counterclockwise order, are $e_1, \ldots, e_k$ where $e_i = \{v_0, v_i\}, 1 \leq i \leq k$. Each consecutive pair of edges incident to v_0 form part of a cycle enclosing a region incident to v_0. Let R_i denote the closed region bounded by the cycle of n_i edges which includes e_i and e_{i+1}, $1 \leq i \leq k-1$, and R_k denote the closed region bounded by the cycle of n_k edges which includes e_k and e_1.

Lemma 1. *The regions $R_1, \ldots, R_k$ described above are distinct. Furthermore, if two regions share more than one vertex, they share exactly two vertices, v_0 and v_i, where v_i is adjacent with v_0.*

Proof. We first show the regions are distinct. Suppose that regions R_i and R_j, $1 \leq i < j \leq k$ are the same, let this be the region R. Then the edges $e_i, e_{i+1}, e_j, e_{j+1}$ are all part of the edge set of R. Now we can draw a simple closed curve C in R which intersects Γ only at the vertex v_0, and either v_{i+1} is interior to C while the vertex v_i is exterior to C or else v_{i+1} is exterior to C while v_i is interior to C. This means that the removal of the vertex v_0 disconnects Γ which contradicts the fact that Γ is 3-connected. Thus the regions are distinct.

We next show that no pair of the regions $R_1, \ldots, R_k$ have more than two vertices in common. Suppose the regions R_i and R_j, $1 \leq i < j \leq k$, have a vertex in common which is neither v_0 nor adjacent with v_0. (Note that if $j = i+1 \pmod{k}$ then R_i and R_j will have at least vertices v_0 and v_{i+1} in common.) Let u be a vertex common to the edge set of R_i and R_j where $u \neq v_i, v_{i+1}, v_j, v_{j+1}$. Let $R = R_i \cup R_j$. Now we can draw a simple closed curve C in R which intersects Γ only at the vertices v_0 and u; and either v_{i+1} is interior to C while the vertex v_i is exterior to C or vice versa. This means that the removal of the vertices v_0 and u disconnects Γ which contradicts the fact that Γ is 3-connected. Thus no pair of the regions $R_1, \ldots, R_k$ have more than two vertices in common.

2 Embeddings

Let M be a Riemannian 2-manifold. That is, M is a two-dimensonal manifold with a metric. The Riemannian manifolds considered here are the Euclidean plane ($\mathbf{R}^2$), the usual two-dimensional sphere of radius one ($\mathbf{S}^2$), and the hyperbolic plane ($\mathbf{H}^2$). (See [6] for more information about the hyperbolic plane.) These three manifolds are used because for any one of them given a pair of points x, y in the manifold there is an isometry mapping x to y. Furthermore, given any point, rotation about that point by any angle is an isometry. An embedding of a graph Γ into M is a *geometric embedding* if every edge of Γ in M has the same length ρ and there is a function $\alpha(n)$ such that every interior angle in every n-sided region has measure $\alpha(n)$.

Examples of geometric embeddings of a graph in the Euclidean plane include tesselations of the plane with squares or hexagons. The Platonic solids give examples of geometric embeddings of finite graphs in the sphere. A regular tessalation of the hyperbolic plane with heptagons is an example of a geometric embedding of a graph in the hyperbolic plane.

A *local geometric embedding* of Γ at v_0 into M is an embedding of a vertex v_0 of Γ and all the boundary vertices and edges of all the regions incident with v_0. Furthermore, it is required that all edges have the same length, ρ, and all angles in all n-gons have the same measure, $\alpha(n)$, as in the definition of geometric embedding.

A *geometric mapping* of Γ into M is a mapping of the vertices and edges into M which is a local geometric embedding at each vertex of Γ.

The 'niceness' of the arrangement of the regions incident to a vertex v in Γ suggests the possiblity of embedding Γ geometrically in either the Euclidean plane, the hyperbolic plane, or the sphere. The existence of such embeddings in the Euclidean and hyperbolic planes are then used to show that Γ has either quadratic or exponential growth. We give explicit conditions for when such embeddings exist. Let v_0 be a fixed vertex of Γ and $R_1, R_2, \ldots, R_k$ and $n_1, n_2, \ldots, n_k$ be as in Lemma 1. Consider the sum

$$S = \sum_{i=1}^{k}(1 - \frac{2}{n_i}).$$

Lemma 2. *The graph Γ has a local geometric embedding in either $\mathbf{R^2}$, $\mathbf{H^2}$ or $\mathbf{S^2}$ depending on whether S is two, greater than two, or less than two respectively.*

Proof. Note that each angle of a regular n-gon in the Euclidean plane has measure $\pi(1-2/n)$. Consequently, $S\pi$ is the sum of what the angles around v_0 would be if each region were regular and the graph were embedded in the Euclidean plane. Clearly if $S = 2$, then there is a local embedding in the Euclidean plane. We consider the remaining two cases:

Case A: $S < 2$. In this case we find a local embedding in the sphere $\mathbf{S^2}$. Each of the regions $R_1, \ldots, R_k$ can be mapped to a regular n_i-gon in $\mathbf{S^2}$ with sides of length ρ such that each of these n_i-gons is incident to the image point of v_0. We identify each R_i with its image n_i-gon. Let each n_i-gon, $1 \leq i \leq k$, have interior angle $\theta_{n_i,\rho}$. It is sufficient to show there is a value of ρ such that $\sum_{i=1}^{k} \theta_{n_i,\rho} = 2\pi$.

Let $n = \max\{n_i\}$, $1 \leq i \leq k$, then we consider values of ρ with $0 \leq \rho \leq 2\pi/n$. Note that the upper bound on ρ comes from the extreme case where the boundary of a region is a great circle, and each angle is π. If ρ is close to zero then an n-gon on the surface of a sphere is close to being 'flat' (as on the Euclidean plane) so $\lim_{\rho\to 0} \sum_{i=1}^{k} \theta_{n_i,\rho} = \sum_{i=1}^{k}(\pi - 2\pi/n_i) < 2\pi$ by the assumption that $\sum_{i=1}^{k}(1 - 2/n_i) < 2$.

We now show that $\sum_{i=1}^k \theta_{n_i,2\pi/n} > 2\pi$. Since the angles vary continuously with ρ, this is sufficient to show there is some $\rho < 2\pi/n$ where $\sum_{i=1}^k \theta_{n_i,\rho} > 2\pi$. First we consider a few special cases.

Subcase 1 (There is more than one region among $R_1, \ldots, R_k$ with n sides): Suppose R_i and R_j are two distinct polygons with n sides. Then $n = n_i = n_j$ and $\theta_{n_i,2\pi/n} = \theta_{n_j,2\pi/n} = \pi$, so since $k \geq 3$ it follows that $\sum_{i=1}^k \theta_{n_i,2\pi/n} > 2\pi$.

Subcase 2 ($k = 3$ and there is a unique region incident with v_0 with n sides): We break this subcase into three sub-subcases, based on the regions incident with v_0.

(i) ($k = 3$, an n-gon and two 3-gons where $n > 3$.)
Suppose that R_1 is an n-gon and R_2 and R_3 are both 3-gons. Now using the fact that each vertex in Γ must have degree 3 it follows that there is a vertex of R_1 which must have two further regions incident to it but cannot be adjacent with a vertex of R_2 or R_3. Since such a configuration is not possible in Γ the above situation cannot occur.

(ii) ($k = 3$, one 3-gon and an n-gon and an m-gon where $n > m > 3$.)
Suppose that R_1 is an n-gon, R_2 is an m-gon and R_3 is a 3-gon where $n > m > 3$. Let $u \neq v_0$ be the common vertex between R_2 and R_3 and $w \neq v_0$ be the common vertex between R_3 and R_1. Now u must also have an n-sided region (with edge $\{w, u\}$) incident to it and w must also have an m-sided region (with edge $\{w, u\}$) incident to it. This is not possible since w would be on the boundary of two distinct n-gons.

Observe that parts (i) and (ii) show that Γ cannot have a 3-sided region if $k = 3$.

(iii) ($k = 3$ and there is no 3-sided region.)
Suppose that R_1 is an n-gon, R_2 is an m-gon and R_3 is a t-gon where $n > m \geq t \geq 4$. Now $\theta_{n_3,2\pi/n} > \pi/2$ (since $\pi/2$ is the angle in the Euclidean case if $t = 4$), $\theta_{n_2,2\pi/n} > \pi/2$, and $\theta_{n_1,2\pi/n} = \pi$, therefore $\sum_{i=1}^k \theta_{n_i,2\pi/n} > 2\pi$.

Observe that we now need only consider $k \geq 4$.

Subcase 3 ($k \geq 4$): Let R_1, R_2, R_3 and R_4 be consecutive regions incident to the vertex v_0. Suppose that R_1 is an n_1-gon with $n = n_1$ and R_2, R_3 and R_4 have respectively n_2, n_3 and n_4 sides where equality may occur. Now $\theta_{n_1,2\pi/n} = \pi$ and $\theta_{n_i,2\pi/n} > \pi/3$ for $i = 2, 3, 4$ (since at worst these polygons could be triangles), therefore $\sum_{i=1}^k \theta_{n_i,2\pi/n} > 2\pi$.

Subcases 1 through 3 show that for those configurations of regions which can occur in Γ we always have $\sum_{i=1}^k \theta_{n_i,2\pi/n} > 2\pi$ where $n = \max\{n_1, \ldots, n_k\}$. Since $\sum_{i=1}^k \theta_{n_i,\rho}$ is a continuous function of ρ there is a value of ρ such that $\sum_{i=1}^k \theta_{n_i,\rho} = 2\pi$.

Case B: $S > 2$. In this case we find a local embedding in the hyperbolic plane $\mathbf{H}^2$. The idea is the same as in Case A, except that we note that as the side length of a regular n-gon approaches infinity, the angles in the n-gon approach 0. So, defining $\theta_{n_i,\rho}$ as an angle in a regular n-gon in $\mathbf{H^2}$ with sides of length ρ, we have

$$\lim_{\rho\to\infty}\sum_{i=1}^{k}\theta_{n_i,\rho}=0$$

and

$$\lim_{\rho\to 0}\sum_{i=1}^{k}\theta_{n_i,\rho}>2\pi.$$

The first limit holds since as the length of ρ increases the angle of the n-gon gets smaller and the second one corresponds to the Euclidean case. Therefore, there is a ρ such that $\sum_{i=1}^{k}\theta_{n_i,\rho}=2\pi$, which says that Γ locally embeds geometrically in $\mathbf{H^2}$.

Let $g:\Gamma\to M$ be a local geometric embedding where M is either $\mathbf{S^2}$, $\mathbf{R^2}$, or $\mathbf{H^2}$. With this embedding, we let $\alpha(n)$ denote the measure of each angle in an n-sided region incident with v_0. We define a mapping f from the walks starting at v_0 in Γ to walks in M starting at $g(v_0)$. Let $P=(v_0,v_1,v_2,\ldots,v_k)$ be a walk in Γ starting at v_0. Let $x_0=g(v_0)\in M$, and follow a geodesic from x_0 to $x_1=g(v_1)$. Then rotate in Γ from the edge $\{v_0,v_1\}$ to the edge $\{v_1,v_2\}$ counter clockwise and let $n_1,n_2,\ldots,n_h$ represent the number of sides in the regions we rotate through. In M rotate from the geodesic between x_0 and x_1 through an angle of $\sum_{i=1}^{h}\alpha(n_i)$. Then follow a geodesic from x_1 in this direction a distance of ρ. Continue this process from vertex to vertex along P until we reach the end vertex v_k. We define $f(P)$ to be the walk consisting of the union of the geodesic segments followed in this process. (See Figure 1.)

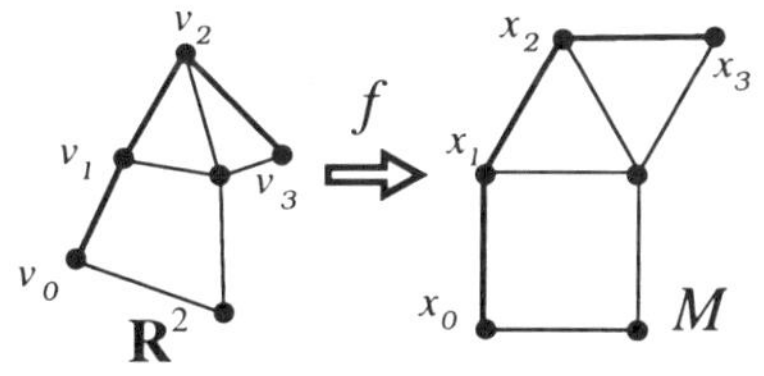

Fig. 1.

Our next goal is to show that the end point of $f(P)$ only depends on the end vertex of P. Consequently, we can think of f mapping the graph Γ into M. That is, we map a vertex v to the end vertex of $f(P)$ where P is any walk in Γ starting at v_0 and ending at v. An edge $\{v,w\}$ is mapped to the geodesic between $f(v)$ and $f(w)$. (In the sphere, we take the shortest geodesic between $f(v)$ and $f(w)$.)

We say a walk P starting at v_0 is *bad* if there is another walk starting at v_0, say P', such that P and P' have the same end vertex v, but $f(P)$ and $f(P')$ have different end points. Otherwise, the walk is said to be *good*.

Lemma 3. *A walk $P = (v_0, v_1, \ldots, v_i, w, v_i, v_{i+1}, \ldots, v_k)$ is good if and only if the walk $P' = (v_0, v_1, v_2, \ldots, v_{i-1}, v_i, v_{i+1}, \ldots, v_k)$ is good.*

Proof. This is easy to verify. The point is that the sum of the values for $a(n)$ around the vertex v_i is 2π.

Suppose that $P = (v_0, v_1, \ldots, v_{k-1}, v_k = v_0)$ is a closed walk. We say that P *ends well* if the last geodesic segment of $f(P)$ is the geodesic between $g(v_{k-1})$ and $g(v_0)$. In particular, if P ends well, then P is good.

Lemma 4. *If $w_0, \ldots, w_t$ is a cycle bounding a region in Γ then $P = (v_0, v_1, \ldots, v_k = w_0, w_1, w_2, \ldots, w_t = v_k, v_{k-1}, v_{k-2}, \ldots, v_0)$ ends well. Furthermore, the part of the path $f(P)$ corresponding to $v_k, v_{k-1}, \ldots, v_0$ retraces the part corresponding to $v_0, v_1, \ldots, v_k$.*

Proof. The cycle part of the walk P maps to a regular t-cycle in M by construction. Consequently, the next edge after the cycle maps to the same edge as the last edge before the cycle.

The above idea of a path retracing itself is used during the proof of the next few lemmas. We first attempt to clarify the concepts involved. Let $P = (v_0, v_1, \ldots, v_{k-1}, v_k = v_0)$ be a closed walk and Q be a walk which begins at v_0. We use PQ to denote the walk which starts at v_0 and first follows P and then follows Q. The path $f(Q)$ is defined to start with a certain geodesic segment based on the first edge of Q. On the other hand, when completing the closed walk P, we need to know that the next geodesic followed in $f(PQ)$ (corresponding to the first edge of Q) will be the same geodesic as the first one in $f(Q)$. This holds provided that P ends well.

Lemma 5. *Every cycle P ends well.*

Proof. We use induction on the number of regions enclosed by $P = (v_0, v_1, \ldots, v_{k-1}, v_k = v_0)$ to show that P ends well. If P encloses only one region then P ends well by Lemma 4.

Suppose that $P = (v_0, v_1, v_2, \ldots, v_k = v_0)$ encloses more than one region. Then by trimming one region as in Figure 2, we can break the area enclosed by P into one region and the union of the rest of the regions. That is, we have two cycles,

$$P' = (v_0, v_1, \ldots, v_t, w_1, w_2, \ldots, w_r, v_s, v_{s+1}, \ldots, v_k = v_0)$$

and

$$P'' = (v_t, w_1, w_2, \ldots, w_r, v_s, v_{s-1}, \ldots, v_t)$$

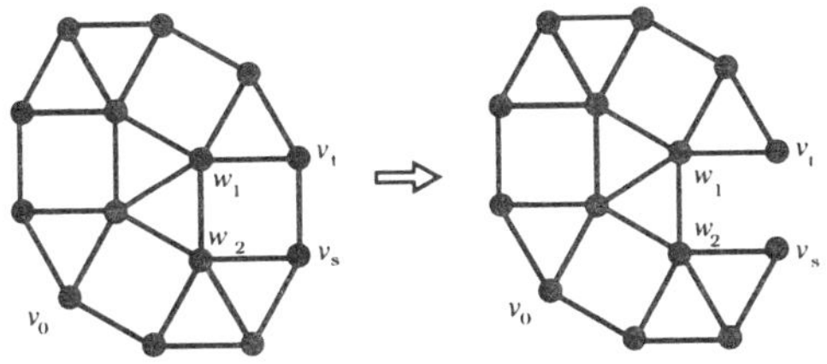

Fig. 2.

where P' encloses fewer regions than P encloses and P'' encloses one region. By induction, P' ends well. Furthermore, by Lemma 4 the closed walk

$$Q = (v_0, v_1, \ldots, v_t, \ldots, v_s, w_r, w_{r-1}, \ldots, w_1, v_t, v_{t-1}, v_{t-2}, \ldots, v_0)$$

ends well. But then, the walk QP' ends well since the path $f(Q)$ returns on the same geodesic segment as it begins. We deduce the cycle P is good by repeatedly applying Lemma 3 to QP. Furthermore, it is clear that the last geodesic segment of $f(P)$ is the same as the last geodesic segment of $f(P')$ in the case where P'' does not include $v_k = v_0$ and it is the same as the last geodesic segment of $f(P'')$ otherwise. In either case, P ends well.

If $P = (v_0, v_1, \ldots, v_k)$ is a walk, we denote the walk $(v_k, v_{k-1}, \ldots, v_0)$ by P^{-1}.

Lemma 6. *Let $C = (v_0 = u_0, u_1, \ldots, u_k = v_0)$ be a closed walk which ends well. Let ϕ be an automorphism of Γ with $\phi(v_0) = w_0$. Let $P = (v_0, v_1, \ldots, v_r = w_0)$ be a walk from v_0 to w_0. Then the walk $P\phi(C)P^{-1} = (v_0, v_1, \ldots, v_r = \phi(u_0), \phi(u_1), \ldots, \phi(u_k) = v_r, v_{r-1}, \ldots, v_1, v_0)$ ends well.*

Proof. The fact that C ends well implies that it would end well even if instead of starting at the point $g(v_0)$ we start at an arbitary point in M. Since in the construction of $f(P\phi(C)P^{-1})$, the part $\phi(C)$ is traced in the same way as C when computing $f(C)$ except that we start at a different point, it follows that the image of P^{-1} traces the path of the image of P backward.

Lemma 7. *Every closed walk P ends well.*

Proof. Suppose not. Let $P = (v_0, v_1, \ldots, v_k = v_0)$ be a closed walk of minimal length which does not end well. The closed walk P is not a cycle by Lemma 5.

Let i be the smallest positive integer such that there is a j with $v_i = v_j$. Let ϕ be an automorphism of Γ with $\phi(v_0) = v_i$. The closed walk $C = \phi^{-1}(v_i, v_{i+1}, \ldots, v_j)$ has length strictly smaller than the length of P. Therefore C ends well. By Lemma 6, the walk $W = (v_0, v_1, \ldots, v_j = v_i, v_{i-1}, v_{i-2}, \ldots, v_0)$ ends well. Furthermore, the closed walk $C' = (v_0, v_1, \ldots, v_i = v_j, v_{j+1}, v_{j+2}, \ldots, v_k = v_0)$ ends well as its length is less than k. Therefore, the closed walk WC' ends well. Now by applying Lemma 3, it follows that P ends well.

Lemma 8. *Let g be a local geometric embedding of Γ at v_0 into either $\mathbf{R}^2$, $\mathbf{S}^2$, or $\mathbf{H}^2$. Let P and P' be two paths in Γ each of which begins at the vertex v_0 and finishes at the vertex v. Then $f(P)$ and $f(P')$ have the same end point.*

Proof. Suppose that P and P' are both v_0v-walks. Then the walk $C = P'P^{-1}$ is a closed walk. By Lemma 7, C ends well. Since C is a closed walk, following $f(C)$ backward is the same as following $f(P)$ forward up until the end of P. But this is the end point of $f(P')$. Therefore, $f(P)$ and $f(P')$ have the same end point.

Theorem 9. *There is a geometric mapping of Γ into either $\mathbf{R}^2$, $\mathbf{H}^2$ or $\mathbf{S}^2$ depending on whether $S = \sum_{i=1}^{k}(1 - 2/n_i)$ is two, greater than two, or less than two respectively.*

Proof. By Lemma 2, Γ has a local geometric embedding into the indicated manifold. By construction, the mapping of Lemma 8 is a local geometric mapping at each vertex.

Lemma 10. *Let f be a geometric mapping from Γ to M. Then f extends to a covering projection map from $\mathbf{R}^2$ onto M.*

Proof. We first extend f by noting that Whitney's Theorem implies that if C is a cycle bounding a region of Γ, then $f(C)$ is a cycle bounding a region of $f(\Gamma)$. Since the regions bounded by C and $f(C)$ are homeomorphic, we extend f to all the regions of Γ by mapping each region by a homeomorphism to the corresponding region of $f(\Gamma)$. It is clear that f is continuous.

Let $x \in \mathbf{R}^2$. Either x is a vertex of Γ, x is interior to an edge of Γ, or else x is interior to a region of Γ. Since the mapping f is geometric, each region maps to a region consisting of a regular n-gon with side length ρ. For each of the three possibilities for x, there is an $\epsilon > 0$ and a neighborhood of x such that the neighborhood of x maps onto an ϵ neighborhood of $f(x)$. This implies that f is an open mapping, but even more, it implies that for some ϵ, if $y \in M$ is in the image of f then every point within ϵ of y is also in the image of f.

Suppose that f is not onto. Then there is a $y \in M$ such that $f(x) \neq y$ for every $x \in \mathbf{R}^2$. Let x_0 be any point such that $\text{dist}(f(x_0), y) < \inf\{\text{dist}(f(x), y) \mid x \in \mathbf{R}^2\} + \epsilon/2$. Then every point within ϵ of $f(x_0)$ is in the image of f, which is a contradiction. Therefore, f is onto.

Given a vertex $v \in V(\Gamma)$ let $N(v)$ be the union of the closed regions incident with v. The shaded region in Figure 3 represents $N(u_0)$. Since f is a local geometric mapping, at each vertex v, $f|_{N(v)}$ is a homeomorphism onto $f(N(v))$. Given a point x in $\mathbf{R}^2$, let $w \in V(\Gamma)$ be a vertex such that x is in the interior of $N(w)$. Now, for each vertex u in $N(w)$, let d_u be the distance in M between $f(u)$ and $f(x)$. Let u_0 be a vertex in $N(w)$ where this distance is minimized. Notice that there is a δ (independent of x) with the property that every point in M within δ of $f(x)$ is in $f(N(u_0))$. (See Figure 3.)

We now show the even covering property. Let $y \in M$ and let U be the disk centered at y with radius $\delta/2$. We show that $f^{-1}(U) = \cup U_i$ such that $U_i \cap U_j = \emptyset$

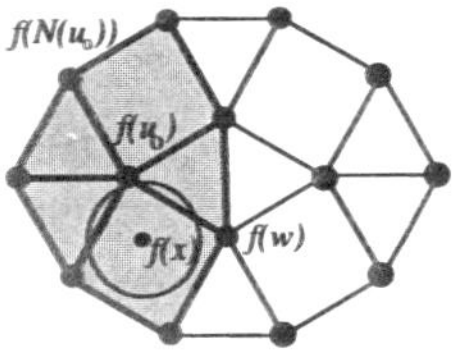

Fig. 3.

for $i \neq j$ and $f|_{U_i} : U_i \to U$ is a homeomorphism. Let $x \in f^{-1}(U)$ and u_0 be the vertex in $N(y)$ such that the distance between $f(u_0)$ and $f(x)$ in M is a minimum. Now it follows that $f|_{f^{-1}(U)\cap N(u_0)} : f^{-1}(U) \cap N(u_0) \to U$ is a homeomorphism as f is a geometric mapping. Therefore, f covers U evenly. So f is a covering projection map.

Theorem 11. *There is a geometric embedding of Γ into either $\mathbf{R}^2$ or $\mathbf{H}^2$. Let $S = \sum_{i=1}^{k}(1-2/n_i)$, then $S \geq 2$. When $S = 2$ the embedding is into $\mathbf{R}^2$, whereas for $S > 2$ the embedding is into $\mathbf{H}^2$.*

Proof. Recall that for any covering $f : X \to Y$, if Y is simply connected then f is a homeomorphism. Since $\mathbf{R}^2$ and $\mathbf{S}^2$ are not homeomorphic, there is no covering from $\mathbf{R}^2$ to $\mathbf{S}^2$. Consequently, by Lemma 10 and Theorem 9 it follows that $S \geq 2$. Also, by Lemma 10 and Theorem 9 there is a homeomorphism from $\mathbf{R}^2$ onto either $\mathbf{R}^2$ or $\mathbf{H}^2$ which, when restricted to Γ, gives a geometric embedding.

3 Growth rate

The *growth function* of a graph Γ with respect to any vertex v in Γ, is defined by

$$G_\Gamma(n) = |\{u \in V(\Gamma) \mid d(v,u) \leq n\}|, \quad \text{for all n} \in N.$$

Of course, if Γ is vertex transitive then the growth rate is independent of the choice of v. We say that Γ has *exponential growth* if there exist constants $c_1 > c_2 > 1$ such that $c_1^n \geq G_\Gamma(n) \geq c_2^n$ holds for all sufficiently large $n \in N$. Otherwise Γ has nonexponential growth. In particular, Γ has *quadratic growth* if $c_1 n^2 \leq G_\Gamma(n) \leq c_2 n^2$ holds for some positive constants c_1 and c_2.

We use the fact that any planar, locally finite, vertex transitive, 1-ended graph Γ can be geometrically embedded in either the Euclidean or hyperbolic plane to determine the growth rate of Γ. In each case we use the geometry of the plane in which Γ is embedded. In both cases we estimate the number of vertices within a circle of fixed radius and the result is what we would intuitively expect. In the Euclidean case the area of this circle is of course πr^2, where r is the radius, and we get quadratic growth. For the hyperbolic case the area is $2\pi(\cosh r - 1)$, where r is the hyperbolic radius, and we get exponential growth.

Theorem 12. *Let Γ be a planar, locally finite, vertex transitive, 1-ended graph with valence k. If $S = \sum_{i=1}^{k}(1 - 2/n_i) = 2$ then Γ has quadratic growth and if $S = \sum_{i=1}^{k}(1 - 2/n_i) > 2$ then Γ has exponential growth.*

Proof. Let v_0 be a vertex in Γ. In both the following cases we determine the radii (as linear functions of n) of two discs centered at v_0 such that all vertices within the smaller disk are within n steps of v_0 and all vertices within n steps of v_0 lie within the larger disk. The number of vertices within these respective discs give lower and upper bounds for the growth rate of Γ.

Case A: $S = 2$. In this case, we may assume by Theorem 11 that Γ is embedded geometrically in the Euclidean plane. Let $C(n)$ be the number of vertices within (graph) distance n of a fixed vertex v_0 in Γ. To determine an upper bound on $C(n)$ we count the number of vertices in Γ within a disc of radius approximately $n\rho$ centered at v_0 in the Euclidean plane. Let Γ^* be the dual graph of Γ in the Euclidean plane. In this dual graph each polygonal face represents a vertex of Γ. By vertex transitivity of Γ, all regions of Γ^* are congruent. Let β be the diameter of one of these regions. Consider a circle of radius $r = n\rho + \beta$ in $\mathbf{R}^2$ centered at v_0. Let a be the area of each region in Γ^*, then

$$\pi r^2/a \geq C(n).$$

This follows as a region of Γ^* corresponding to a vertex in Γ within n (graph distance) of v_0 is completely contained in the disk centered at v_0 and radius r. Therefore, $C(n) \leq c_1 n^2$ for some constant c_1.

Now we obtain a lower bound on $C(n)$. Let x_0 be a vertex of Γ with distance (in $\mathbf{R}^2$) r from v_0. Let L be a geodesic from x_0 to v_0. Let x be any point in $\mathbf{R}^2$ and x' be the closest vertex to x, then, as pointed out in Lemma 10, the disk centered at x with radius δ is contained in $N(x')$. Begin traversing L at x_0, since this is already a vertex a segment having length at least δ is contained in $N(x_0')$. Let x_1 be the point where L exits $N(x_0')$ and x_1' be the closest vertex to x_1. Again, starting at x_1 and following L toward v_0, one would move along L at least a distance δ before exiting $N(x_1')$. Continuing this process, it is clear that one would have to pass through at most r/δ sets of the form $N(x_i')$. Now, to get from any point in $N(x')$ to any other point in $N(x')$ takes at most b steps for some fixed b independent of x. Therefore, there is a path in Γ from x_0 to v_0 of (graph) length at most $(r/\delta)b$. Therefore, every vertex in a circle of radius $n\delta/b$ centered at v_0 is within n steps of v_0 in Γ. This gives a quadratic lower bound on $C(n)$. Therefore, Γ has quadratic growth.

Case B: $S > 2$. Here we can assume that Γ is geometrically embedded in the hyperbolic plane. The argument is the same as in the Euclidean case with just one essential difference. The area of a disk centered at v_0 in $\mathbf{H}^2$ grows exponentially as a function of r. Consequently, exponential upper and lower bounds follow just as quadratic bounds follow in the Euclidean case.

Acknowledgements

The first author was supported by a University of North Texas Faculty Research Grant and the second author by a University of Auckland Research Committee Grant.

References

1. C. Bonnington, W. Imrich and M.Watkins, *Separating double rays in locally finite planar graphs*, Discrete Mathematics **145** (1995) 61–72.
2. R. Halin, *Automorphisms and endomorphisms of infinite, locally finite graphs*, Abh. Math. Sem. Univ. Hamburg **39**(1973) 251–283.
3. W. Imrich, *On Whitney's theorem on the unique imbeddability of 3-connected planar graphs*, in: M. Fiedler, ed., Recent Advances in Graph Theory (Academia Praha, Prague, 1975) 303–306.
4. C. Thomasen, *Duality of infinite graphs*, Combinatorica **1**(1982) 285–288.
5. H. Whitney, *Non-separable and planar graphs*, Trans. Amer. Math. Soc. **34**(1932) 339–362.
6. H. Zeischang, *Finite Groups of Mapping Classes of Surfaces*, Lecture Notes in Mathematics No. 875, Springer-Verlag, New York, 1981

Secret Sharing Schemes Based on Room Squares

Ghulam Rasool Chaudhry and Jennifer Seberry

The Centre for Computer Security Research
Department of Computer Science
University of Wollongong
Wollongong, NSW 2522, AUSTRALIA

Abstract. In this paper, we describe secret sharing schemes. We discuss Room squares and their critical sets. We propose a model of secret sharing based on critical sets of Room squares.

1 Introduction

In information based systems, the integrity of the information is commonly provided for by requiring that certain operation(s) can be carried out only by one or more participants who have access rights. Access is gained by a key, password or token, and governed by a secure key management scheme. If the key or password is shared between several participants in such a way that it can be reconstructed only by a responsible group acting in agreement, then a high degree of security is attained. Shared security systems, of this sort, are also used in financial institutions, in communication networks, in computing systems serving educational institutions and distribution environments. However, the best known examples of shared security systems are in the military: for instance, in activating a nuclear weapon, several senior officers must concur before the necessary password can be reconstructed.

We describe another situation which motivates the subject of secret sharing:

> *The head of an organization keeps important documents in a safe of which only he or she knows the combination. However, the head is often absent for extended periods and occasionally information is needed from documents in the safe in order to maintain the day-to-day running of the organization. The head deems it undesirable for the combination to be trusted to any one of the five executive board members. What is regarded as acceptable, however, is a compromise situation whereby at least two of the executive board members acting together can gain access to the safe.*

Can such a system be devised?

Figure 1 provides a system to solve the above problem. The lines and points are chosen in projective space $PG(2,q)$ where $q \geq 5$. It is publicly known that the safe combination is a point on line l but the actual point is kept secret. Each of the five executives is privately given a point on line m and the safe

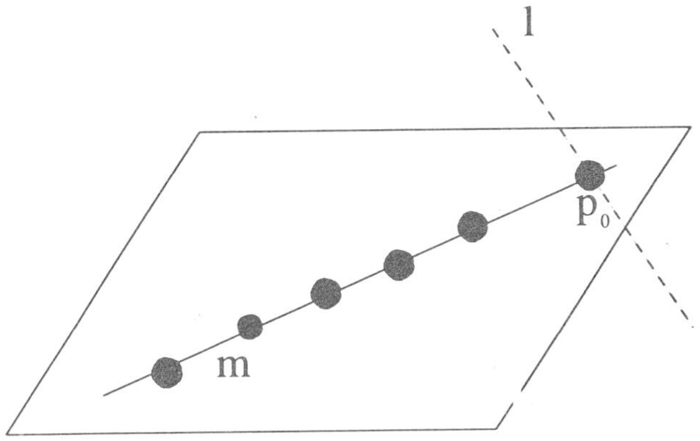

Fig. 1. A projective space

combination is chosen to be point p_0, the unique point of intersection of line m and line l. Any two executives can generate line m and hence evaluate p_0 by intersecting m and l. However, one executive acting alone knows only that the safe combination must be one of the points on line l—the actual point remains secret since for every point p on line l there is a unique line passing through the executive's point and p.

Secret sharing schemes are systems designed to solve problems of a similar type to the one we have just discussed. In general, there is a group of potential members of such a scheme and a collection of sets of these members which are desired to have access to some protected information.

The information is protected by distributing to each member of the scheme an amount of partial information which relates in some way to the protected information. This partial information is known only to the individual member to whom it is distributed and it is held secret by them. When any group of members of the scheme who are desired to have access to the protected information choose to do so, they can reconstruct it by pooling their pieces of partial information. Thus, in our opening problem, any two pieces of the partial information distributed to the executives must be sufficient to enable the combination of the safe to be determined.

Secret sharing schemes were first introduced by Blakley [2], Shamir [19] and Chaum [4] in 1979, and subsequently have been studied by numerous other authors. For a general discussion of shared secret schemes, see Simmons' paper [20]. A number of mathematical structures have been used to model shared secret schemes. Some of these are polynomials, geometric configurations, block designs, Reed-Solomon codes, vector spaces, matroids, near-right fields, complete multipartite graphs, orthogonal arrays and Latin squares. In this paper, we will model a new secret sharing scheme based on Room squares.

In most real-world applications there is also a need for a hierarchy to be built into the shared security system. That is, the key and password is shared between s individuals of rank $1, \ldots, r$ so that if a person of rank i is incapacitated, then a person of rank $j \geq i$, or a set of individuals of rank $l < i$, may replace the lost data. Brickell [3], Simmons [20] and Cooper, Donovan and Seberry [8] have adapted the basic schemes and constructed multilevel schemes.

We shall now introduce a more formal terminology for secret sharing schemes.

2 Secret sharing schemes

A *secret sharing scheme* is a method of sharing a secret S among a finite set of participants P in such a way that if the participants in $A \subseteq P$ are qualified to know the secret, then by pooling together their shares, they can reconstruct the secret S; but any set $B \subset P$, which is not qualified to know S, cannot reconstruct the secret. The key S is chosen by a special participant D called the *dealer* and it is usually assumed that $D \notin P$. The dealer gives partial information called a *share* to each participant to share the secret S.

Secret sharing schemes are useful in any important action that requires the concurrence of several designated people to be initiated as described in the examples in Section 1.

An *access structure* Γ is the family of all the subsets of participants that are able to reconstruct the secret. An access structure is said to be *monotone* if any set which contains a subset, that can recover the secret, can itself recover the secret, that is, if for any subsets B and C of P, where $B \subseteq C$ and $B \in \Gamma$, then $C \in \Gamma$. The subsets of P belonging to the access structure Γ are called *authorized sets* and those not belonging to the access structure are termed as *unauthorized sets.*

Example 1. In Figure 1, the secret sets are the points on line l and the secret is the point p_0. The shares are the five points on line m that are distributed to five participants. Finally, observing that P is the set of five participants, we have the access structure:

$$\Gamma = \{A \subseteq P : |A| \geq 2\}$$

One property of a monotone access structure is that it has a unique collection of authorized sets of minimal size. We define $B \in \Gamma$ to be a *minimal set* of Γ if $C \in \Gamma, C \subseteq B$ implies $C = B$. The collection of all minimal sets of Γ is denoted by Γ^-.

A *(k,n) threshold scheme* allows a secret to be shared among n participants in such a way that any k of them can recover the secret, but no group of $k-1$, or fewer participants, can do so.

A monotone access structure Γ defined on a participant set P such that $|P| = n$ and $\Gamma = \{X \subseteq P : |X| \geq k\}$ is known as (k,n) *threshold access structure.*

Example 2. A Latin square of order n is an $n \times n$ array of the integers 1, 2, ..., n such that each integer occurs precisely once in each row and each column. In this example, we take a Latin square L of order 3 as a (2,3) threshold scheme.

$$L = \begin{bmatrix} 1\,2\,3 \\ 2\,3\,1 \\ 3\,1\,2 \end{bmatrix}$$

Let $S = \{(2,1;2),(3,2;1),(1,3;3)\}$, a partial Latin square, be the union of some critical sets of L, where (2,1;2) means that integer 2 is at position (2,1), (3,2;1) means that integer 1 is at position (3,2) and (1,3;3) means that integer 3 is at position (1,3) in the Latin square L. We can construct a 2-out-of-3 secret sharing scheme on this set. The Latin square L is kept secret but its order is made public. When any two participants from S collaborate, they combine their shares and reconstruct the unique Latin square containing their shares.

A secret sharing scheme is said to be *perfect* provided the following two properties are satisfied:

(i) If an authorized subset of participants $A \subseteq P$ pool their shares, then they can determine value of the secret S.
(ii) If an unauthorized subset of participants $B \subset P$ pool their shares, then they can determine nothing more than any outsider about the value of the secret S.

The security of such a scheme is unconditional since we do not place any limit on the amount of computation that can be performed by a subset of participants.

The size of each share is defined as the number of bits s of non-redundant information in the share. The size of the secret t is the number of bits of non-redundant information in the secret. How to exactly determine s and t is not yet known.

In a ***multilevel scheme***, the participants are ranked and placed in levels $r_1, ..., r_w$. We assume that there are l_i participants in level r_i for $i = 1, ..., w$. So $\sum l_i = l$. A secret key S is chosen and l pieces of related information distributed, one piece to each participant. This is done in such a way that the secret can be recovered from the shares of t_i participants of rank r_i.

3 Room squares

A ***Room square*** R of order r is an $r \times r$ array each of whose cells may either be empty or contain an unordered pair of objects $0, 1, 2, ..., r$, subject to the following conditions :

(i) each of the objects $0, 1, 2, ..., r$ occurs precisely once in each row of R and precisely once in each column of R, and
(ii) every possible unordered pair of objects occurs precisely once in the whole array.

Theorem 1. (Mullin and Wallis [18]) *There exists a Room square of every odd integer order r greater than or equal to 7.*

We denote $N_r = 0, 1, 2, ...r$. A Room square of order r based on N_r is *standardized* if the ith diagonal cell, cell(i, i), contains $\{0, i\}$ for $1 \leq i \leq r$.

A *skew* Room square R is a standardized Room square with the property that when $i \neq j$, either the (i, j) or the (j, i) cell of R is occupied, but not both.

Two Room squares are said to be *isomorphic* if one can be obtained from the other by permuting the rows and columns and relabeling the elements. Two Room squares R and S are *equivalent* if R is isomorphic to S or to the transpose of S. All inequivalent Room squares of order 7 and 9 have been found by one-factorizations of graph, see [1, 11, 22] for details. There are exactly 6 inequivalent Room squares of order 7 and 257,630 inequivalent Room squares for order 9. No exact values of inequivalent Room squares are still known for higher order Room squares, although there are :

526,915,620 non-isomorphic one-factorizations of order 11,
9.876×10^{28} distinct one-factorizations of order 13,
1.148×10^{44} distinct one-factorizations of order 15,
1.520×10^{63} distinct one-factorizations of order 17,

and the number of all inequivalent Room squares for each of these orders are even much larger, see [10] for details.

A *partial Room square* R of order r is an $r \times r$ array each of whose cells may either be empty or contain an unordered pair of objects $0, 1, 2, ..., r$, subject to the following conditions :

(i) each of the objects $0, 1, 2, ..., r$ occurs at most once in each row of R and at most once in each column of R, and
(ii) every possible unordered pair of objects occurs at most once in the whole array.

A *critical set* $Q = \{Q_1, Q_2, Q_3, ..., Q_c\}$, $|Q| = c$, in a Room square R of order r, is a set of quadruples $Q_a = (i, j; k, l)$ such that if any Q_a is removed from the set, it can no longer be uniquely completed. In Q_a, (i, j) shows the position of the pair (k, l) in the square. That is, Q provides minimal information from which R can be reconstructed uniquely. In this paper, we consider that empty positions with '-' in the Room square are given.

Q^* is a *minimal critical set (min.cs)* of a Room square R if $|Q^*| = c$ is minimum for all critical sets Q of a Room square of order r whereas Q^* is a *maximal critical set (max.cs)* of a Room square R if $|Q^*| = c$ is the largest for all critical sets Q of a Room square of order r.

Example 3. Room square of order 7:

8,1	-	4,5	6,7	-	-	2,3
5,7	8,2	-	-	-	1,3	4,6
-	5,6	8,3	1,2	-	4,7	-
-	3,7	-	8,4	2,6	-	1,5
3,6	1,4	2,7	-	8,5	-	-
2,4	-	-	3,5	1,7	8,6	-
-	-	1,6	-	3,4	2,5	8,7

Critical set 1 of above square is:

**	–	4,5	**	–	–	2,3
**	**	–	–	–	1,3	4,6
–	**	8,3	1,2	–	4,7	–
–	**	–	**	2,6	–	1,5
**	**	**	–	**	–	–
**	–	–	**	**	**	–
–	–	**	–	**	**	**

Critical set 2 of above square is:

**	–	**	**	–	–	**
**	8,2	–	–	–	**	**
–	**	8,3	1,2	–	**	–
–	3,7	–	8,4	**	–	**
**	**	**	–	8,5	–	–
**	–	–	**	1,7	8,6	–
–	–	**	–	**	**	**

where "**" shows an unknown pair position and "–" shows an empty position in the square.

So the critical set 1 for the above Room square is: Q = {(1,3;4,5), (1,7;2,3), (2,6;1,3), (2,7;4,6), (3,3;8,3), (3,4;1,2), (3,6;4,7), (4,5;2,6), (4,7;1,5)}. That is, Q is the only Room square of order 7 with a pair 4,5 at position (1,3), pair 2,3 at position (1,7), pair 1,3 at position (2,6), pair 4,6 at position (2,7), pair 8,3 at position (3,3), pair 1,2 at position (3,4), pair 4,7 at position (3,6), pair 2,6 at position (4,5) and pair 1,5 at position (4,7).

There is very little known about critical sets in Room squares. Chaudhry and Seberry [5] studied critical sets in Room squares for orders 7, 9 and 11.

4 Proposed scheme

A secret sharing scheme can be constructed in which the secret key is a Room square R of order r. This scheme exhibits the following characteristics: The Room square is taken to be the secret key and therefore kept private. However, the order of the Room square is made public. The shares in the secret are based on a partial Room square $S = \{\cup A_i \mid A_i \text{ is a critical set in } R\}$. The union can be taken over all possible critical sets in R or over some subset of critical sets. The number of critical sets used will be dependent on the size of the Room square and the number of participants in the secret sharing scheme. The access structure will be the set $\Gamma = \{B \mid B \subseteq S \text{ and } B \supseteq A \text{ where } A \text{ is some critical set in } R\}$. One can easily see that Γ is monotone. The protocol for secret sharing scheme, involving l participants and based on a Room square is as follows.

- A Room square R of order r is chosen. The number r is made public, but the Room square R is kept secret to be the key.

- A set S which is the union of a number of critical sets in R is defined.
- For each $(i,j;k,l) \in S$, the share $(i,j;k,l)$ is distributed privately to a participant.
- When a group of participants whose shares constitute a critical set come together, they can reconstruct the Room square R and hence the secret key.

We demonstrate here how the scheme works on a small example and then give a more general construction.

Take a Room square of order 7 given in example 3. Let S be the partial Room square {(1,3;4,5), (1,7;2,3), (2,2;2,8), (2,6;1,3), (2,7;4,6), (3,3;8,3), (3,4;1,2), (3,6;4,7), (4,2;3,7), (4,4;4,8), (5,5;5,8), (4,5;2,6), (4,7;1,5), (6,5;1,7), (6,6;6,8) }. All the parties are told that the order of the Room square is 7. Each participant is given a share $(i,j;k,l)$, for one such element of S. In order to recover the secret, an authorized group of participants must place their shares in a partial Room square. They then reconstruct the unique Room square containing these entries. These authorized groups are based on the critical sets contained in S. Two of the critical sets contained in S are:

$$\begin{aligned} A_1 = \{&(1,3;4,5),(1,7;2,3),(2,6;1,3),(2,7;4,6),(3,3;8,3),\\ &(3,4;1,2),(3,6;4,7),(4,5;2,6),(4,7;1,5)\}\\ A_2 = \{&(2,2;2,8),(3,3;8,3),(3,4;1,2),(4,2;3,7),\\ &(4,4;4,8),(5,5;5,8),(6,5;1,7),(6,6;6,8)\} \end{aligned}$$

Note that $|A_1| = 9$ while $|A_2| = 8$.

Now for a more general example. Let R be a Room square of order r and Q be a critical set. Define $Q = \{Q' \mid Q' \text{ is the isotopic image of Q}\}$. Let $S' = \{Q' \mid Q' \in Q \text{ and } Q' \text{ is a critical set in } R\}$. We may use the protocol given above to construct a secret sharing scheme where the shares are drawn from the set S'.

The following points should be made about the secret sharing scheme.

- Since the authorized groups are based on critical sets in Room squares, the absence of one share implies that secret cannot be recovered uniquely.
- The scheme is obviously not perfect as an outsider can guess from the set of all possible Room squares of order r, whereas an unauthorized group of participants knows that the Room square must contain the partial Room square defined by their shares.
- The security of the scheme is based on the number of possible Room squares containing the partial Room square defined by an unauthorized group of participants.
- The number of inequivalent Room squares for higher order grows exponentially as mentioned earlier. The number of inequivalent Room squares of order greater than or equal to 11 are still unknown under the current computing resources, although lower bounds on these numbers are known. So the probability of guessing a secret key consisting of a Room square of order greater than or equal to 11 is very very small, and the scheme is very secure though not perfect.

5 Key management scheme

In this section we consider the situations where there are a number of secret sharing schemes all of which contain a common participant. This participant may be required to remember a number of shares. For example, a medical administrator (Registrar) may require access to several restricted files. These files may contain, say, patient data, hospital resources and organ bank data. Access to these files may be via a secret sharing scheme in which the registrar of the hospital always has a critical role. The registrar always has to remember several different shares. This obviously increases the complexity of the registrar's role and consequently reduces the security of the schemes.

We wish to model a key management scheme in which a secret key is common to a number of secret sharing schemes. The shares related to this key are such that a primary share is held by one participant and this share is a necessary part of the reconstruction process in each scheme. Each scheme will involve a number of minor shares which when combined with the primary share can be used to reconstruct the secret. It is also required that the secret can not be reconstructed uniquely from the combined information held by the minor shares.

Inequivalent critical sets in a Room square can be used to model a key management scheme of this nature. We illustrate this with an example. Take the Room square of order 7 given in example 3. Following are three distinct minimal critical sets of this Room square which have the common pairs (3,3;8,3) and (3,4;1,2).

$$\begin{aligned} A_1 = \{&(1,3;4,5),(1,7;2,3),(2,6;1,3),(2,7;4,6),(3,3;8,3),\\ &(3,4;1,2),(3,6;4,7),(4,5;2,6),(4,7;1,5)\}\\ A_2 = \{&(2,2;2,8),(3,3;8,3),(3,4;1,2),(4,2;3,7),\\ &(4,4;4,8),(5,5;5,8),(6,5;1,7),(6,6;6,8)\}\\ A_3 = \{&(1,1;1,8),(1,3;4,5),(1,4;6,7),(2,1;5,7),(2,7;4,6),\\ &(3,3;8,3),(3,4;1,2),(4,7;1,5),(1,5;3,6),(6,6;6,8)\} \end{aligned}$$

Note that $|A_1| = 9$, $|A_2| = 8$ and $|A_3| = 10$.

Each department is assigned a different critical set A_i with the same participant receiving a share which is common to each A_i. In this case the registrar will be given the common share (3,3;8,3) or (3,4;1,2). All departments will reconstruct the same secret, but each has a different set of keys to this secret. However if all participants in the minor (lower than registrar) level pool their shares, the secret cannot be reconstructed uniquely.

Another key management scheme can be developed to allow each department a different secret, but still have a common primary share using inequivalent Room squares of same order with some common pairs and same positions.

Multilevel schemes, based on critical sets of Room squares, can also be developed in which some participants are more important than others. In multilevel schemes, the share of a participant at level j can be replaced by two or more participants at level i, where $i < j$. Consider the case of an electronic transfer of funds between financial institutions. This transfer can only be initiated when an electronic signature is received. The signature will be reconstructed when the shares of two senior tellers and one vice-president or two vice-presidents, are entered. Critical sets for Room squares fulfilling these requirements are still not known.

6 Conclusion

In this paper, we have proposed a secret sharing scheme based on critical sets of Room squares. Since there is very little known about critical sets of Room squares, the implementation of this scheme is limited at the moment. However, the directions for future research are:

- Construct families of critical sets for Room squares.
- Construct families of smallest and largest critical sets for Room squares without a priori constraints on the Room square.
- Quantify the security of the scheme more effectively, that is, decide which critical sets are more secure than others.
- Find a general process that will start with an access structure and result in a Room square.
- Investigate the structure of critical sets from Room squares to see if possible to construct perfect secret sharing schemes.
- Devise multilevel schemes based on critical sets of Room squares.

References

1. D. S. Archdeacon, J. H. Dinitz and W. D. Wallis. *Sets of pairwise orthogonal one-factorizations of K_{10}.* Congr. Numer. 43 (1984), pp. 45-79.
2. G. R. Blakley. *Safeguarding cryptographic keys.* Proc. AFIPS 1979 Natl. Computer Conference, New York, 48, June (1979), pp. 313-317.
3. E.F. Brickel. *Some ideal secret sharing schemes.* J. Comb. Math. and Comb. Computing, 9, (1989), pp. 105-113.
4. D. Chaum. *Computer Systems established, maintained and trusted by mutually suspicious groups*, Memorandum No. UCB/ERL M179/10, University of California Berkley CA, 1979.
5. G. R. Chaudhry and J. Seberry. *Minimal and maximal critical sets in Room squares.* 7th Australasian Workshop on Combinatorial Algorithms (AWOCA'96), Magnetic Island, Australia, July 15-19 (1996), Technical Report 508, July 1996, Dept. of Computer Science, University of Sydney, Australia pp 75-86.
6. D. Chen and D.R. Stinson, *Recent results on combinatorial constructions for threshold schemes.* Aust. J. of Combin., 1 (1990), pp. 29-48.
7. J. Cooper, D. Donovan and J. Seberry. *Latin squares and critical sets of minimal size.* Aust. J. Combinatorics 4 (1991), pp. 113-120.
8. J. Cooper, D. Donovan and J. Seberry. *Secret sharing schemes arising from Latin squares.* Bull. of the Inst. of Combinatorics and its Applications, September (1994) pp. 33-43.
9. D. Curran and G.H.J. Van Rees. *Critical sets in Latin squares.* in Proc. Eighth Manitoba Conference on Numer. Math. and Computing, (1978), pp. 165-168.

10. J. H. Dinitz, D. K. Garnick and B. D. Mackay. *Non-isomorphic one-factorizations of K_{12}*. J. Comb. Design, Vol. 2, No. 4 (1994), pp. 273-285.
11. J. H. Dinitz and D. R. Stinson. *Room squares, Contemporary Design Theory: a Collection of Surveys*. John Wiley & Sons, Inc. (1992), pp. 137-204.
12. J. H. Dinitz and W. D. Wallis. *Four orthogonal one-factorizations on 10 points*. Ann. Disc. Math. 26 (1985), pp. 143-150.
13. M. R. Garey and D. S. Johnson. *Computers and Intractability: A Guide to the Theory of NP-Completeness*. W. H. Freeman and Company, San Francisco, 1979.
14. K. B. Gross. *Equivalence of Room designs I*. J. Comb. Theory A 16 (1974), pp. 264–265.
15. K. B. Gross. *Equivalence of Room designs II*. J. Comb. Theory A 17 (1974), pp. 299-316.
16. K. M. Martin, *Discrete Structures in the Theory of Secret Sharing*. PhD thesis, Royal Holloway and Bedford New College, University of London, 1991.
17. K. M. Martin, *New secret sharing schemes from old*. JCCMCC 14 (1993), pp. 65-77.
18. R. C. Mullin and W. D. Wallis. *The existence of Room squares*. Aequa. Math. 13 (1975), pp. 1-7.
19. A. Shamir *How to share a secret*. Comm. ACM, 22, No. 11, Nov. 1979, pp. 612-613.
20. G. J. Simmons. *An introduction to shared secret and/or shared control schemes and their applications*. in Contemporary Cryptology, the Science of Information Integrity, IEEE Press, Piscataway, 1991 , pp. 441-497.
21. D.R. Stinson, *An explication of secret sharing schemes*. Design, Codes and Cryptography, 2 (1992), pp. 357-390.
22. W.D. Wallis, A. P. Street and J. S. Wallis. *Combinatorics: Room Squares, Sum-free Sets, Hadamard Matrices*. Lect. Notes Math. 293, Springer-Verlag Berlin (1972).

Graph Classes between Parity and Distance-Hereditary Graphs*

Serafino Cicerone and Gabriele Di Stefano

Dipartimento di Ingegneria Elettrica, Università dell'Aquila
Monteluco di Roio, I-67040 L'Aquila – Italy.
Email: {cicerone,gabriele}@iinf02.ing.univaq.it

Abstract. Several graph problems (for example, Steiner tree, connected domination, Hamiltonian path, and isomorphism problem), which can be solved in polynomial time for distance-hereditary graphs, are NP-complete or have unknown complexity for parity graphs. Moreover, the metric characterizations of these two graph classes suggest an excessive gap between them. We introduce a family of classes forming an infinite lattice with respect to inclusion, whose extreme points are exactly parity and distance-hereditary classes. Then, we characterize these classes using Cunningham decomposition and use such a characterization in order to show efficient algorithms for the recognition and isomorphism problems.

1 Introduction

Parity graphs [5] and distance-hereditary graphs [4, 19] have been thoroughly studied because of their interesting metric properties. A graph has parity if and only if the lengths of any two induced paths joining the same pair of vertices have the same parity; on the other hand, a graph is distance-hereditary if and only if the lengths of such pair of paths are equal. Other important characterizations have been formed out for such graphs based on forbidden induced subgraphs and on generative operations for recursive definitions. With respect to the latter characterization, both parity [9] and distance-hereditary graphs [15] are generable, starting from a single vertex, using the following operations: extensions obtained by adding true and false twins or certain bipartite graphs. The latter operation appends an arbitrary bipartite graph for parity graphs, and the smallest non-trivial bipartite graph for distance-hereditary graphs.

Both metric and generative definitions stress the gap between the two classes: distance-hereditary graphs cover only a little part of parity graphs. Different computational properties of these classes are further confirmed by observing results about algorithmic studies on classic optimization problems. STEINER TREE, CONNECTED DOMINATION, and HAMILTONIAN PATH problems are NP-complete in parity graphs, because they are in bipartite graphs [22, 23], and are solvable in polynomial time for distance-hereditary graphs [15, 24]. Another fundamental

* This work has been partially supported by MPI on "Teoria dei Grafi ed Applicazioni in Informatica Medica".

problem in the algorithmic graph theory is the ISOMORPHISM PROBLEM. It is still open for general graphs [16] and in [6] the problem of bipartite graph isomorphism has been proved to be *isomorphism complete*, that is, polynomial-time equivalent to graph isomorphism in general. However, in the class of distance-hereditary graphs ISOMORPHISM PROBLEM is solvable in polynomial time [3].

A powerful tool for obtaining efficient solutions to graph problems is the divide-and-conquer paradigm, and one of its manifestations is graph decomposition. In [13] Cunningham introduced the *split decomposition* of graphs. Split decomposition has been used to characterize distance-hereditary, circle, and circular-arc graphs, giving efficient algorithms for the recognition [17, 25, 21] and the isomorphism [3, 21] problems.

The aim of this work is to investigate and to characterize graph classes between distance-hereditary and parity graphs in order to provide a reference framework for studying the computational complexity of fundamental graph problems. We define a family of graph classes where each class is generated by an operation that extends graphs by appending specific bipartite graphs. Hence, for any particular class of bipartite graphs, we have a corresponding class in the family. We prove that such a family of classes is an infinite lattice with respect to inclusion relation, whose bottom and top elements are the class of the distance-hereditary graphs and the class of the parity graphs, respectively.

Based on split decomposition, we give a characterization for each class in the lattice, defining the decomposition structure of graphs in an arbitrary class. A consequence of this result is the development of a polynomial time recognition algorithm for each class, assuming a polynomial recognition algorithm for the corresponding bipartite class. As far as the class of parity graphs is concerned, this general characterization coincides with that used in [11] in order to reduce the complexity of recognition and weighted clique number problems.

Since the proposed characterization provides a particular tree structure for graphs in the family classes, we also show that the ISOMORPHISM PROBLEM is solvable in polynomial time for some classes in the defined family.

Due to the lack of space, some proofs are omitted and we refer to [10] for the complete work. The paper is organized as follows. In the next section we give the notations used and recall the basic concepts regarding split decomposition. Section 3 introduces the proposed family of classes, and in Section 4 we characterize each graph classes by split decomposition. This characterization is used in Section 5 in order to show efficient algorithms for the recognition and isomorphism problems.

2 Notations and basic concepts

In this work we consider finite, simple loopless, undirected graphs $G = (V_G, E_G)$, and we use standard terminology from [18], some of which is briefly reviewed here.

Let S and T be two subsets of V_G. The *induced subgraph* $\langle S \rangle$ is the maximal subgraph of G with vertex set S. The removal of S from G results in the graph

$G - S$, that is, the induced subgraph $\langle V_G \setminus S \rangle$ of G. A vertex x of a connected graph G is an *articulation point* if $G - \{x\}$ is not connected. $|G|$ is the cardinality of V_G.

By $N_T(S)$ we denote the *neighborhood* of S in T, that is, the set of vertices in T that are adjacent to some vertex in S; moreover $N_T[S] = N_T(S) \cup S$. For the sake of simplicity, we shall omit T if T coincides with V_G, and we shall write $N(x)$ (or $N[x]$) when $S = \{x\}$. Two vertices x and y are *twins* if they have the same neighborhood; we distinguish between *false twins* iff $N(x) = N(y)$ and *true twins* iff $N[x] = N[y]$.

The symbol K_n is used to denote a *clique*, that is, the graph having every pair of its n vertices adjacent. The symbol $K_{m,n}$ denotes a *complete bipartite graph*, that is, a bipartite graph with mn edges when its partite sets V_1 and V_2 have m and n vertices, respectively. A *star* is a complete bipartite graph $K_{1,n}$ with $n > 1$. We call *center* and *pendants* of the star $K_{1,n}$ the vertex in V_1 and the vertices in V_2, respectively. The symbol C_n denotes the cycle graph with n vertices.

Now, we recall the split decomposition terminology defined by Cunningham [13] and also introduce some new terms.

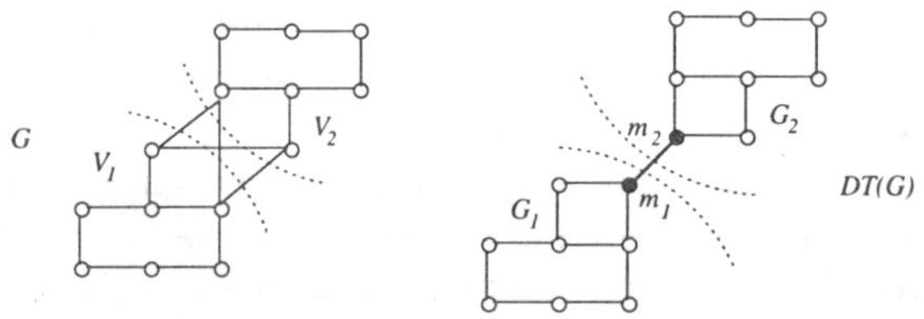

Fig. 1. *A graph G, a simple decomposition $\{G_1, G_2\}$, and, since the two components are prime, the split decomposition tree $DT(G)$.*

Let V_1, V_2 be a partition of the vertex set V_G with $|V_1|, |V_2| \geq 2$ such that there are sets $W_1 \subseteq V_1$, $W_2 \subseteq V_2$, for which all possible edges between W_1 and W_2 are in G and no other edges between V_1 and V_2 exist. We call $\{V_1, V_2\}$ a *split* of G, and if G has a split we say it is *decomposable.* The *simple decomposition* of G with respect to the split $\{V_1, V_2\}$ is the pair of graphs $\{G_1, G_2\}$, and it is obtained as follows: take the subgraphs of G induced by V_1 and V_2, add two new *marked vertices* m_1 and m_2 and join m_1 to all vertices in W_1 and m_2 to all vertices in W_2. Then, graphs G_1 and G_2 are $\langle V_1 \cup \{m_1\} \rangle$ and $\langle V_2 \cup \{m_2\} \rangle$, respectively. The *split decomposition* of a graph G is the set $D(G)$ of graphs obtained by the following recursive procedure: (i) if G has a split $\{V_1, V_2\}$, apply split decomposition to graphs G_1 and G_2 obtained by the simple decomposition $\{G_1, G_2\}$, (ii) if G does not have a split then G is called *prime.*

Elements of $D(G)$ are called *components.* An important property of simple decomposition is that any component in $D(G)$ is isomorphic to an induced subgraph of G. Notice that cliques and stars are not uniquely split decomposable.

But Cunningham proved the following uniqueness result.

Theorem 1 ([13]). *Each connected graph has a unique split decomposition into prime graphs, stars, and cliques with a minimum number of components.*

From now on, with $D(G)$ we denote the unique split decomposition of a graph G, and we call *trivial* each star and each clique component in $D(G)$. In this paper, we also use the version of Bouchet [7], where marked vertices m_1 and m_2 are joined by a marked edge. In this way, we can consider the *decomposition tree* $DT(G)$ of G, whose vertices are the components of $D(G)$ and whose edges are the marked edges (see Figure 1).

3 The family of graph classes

In this section we introduce a family of graph classes between the distance-hereditary and parity graph classes and show that every class in this family is hereditary, that is, it is closed under induced subgraphs. The following operators are introduced in order to define the classes of the proposed family by a generative characterization.

Definition 2. Let G be a graph, $x \in V_G$ and $\{x_1, x_2, \ldots, x_n\}$ a set of false twins of G. Let B be a bipartite graph with partite sets V_1 and V_2. We introduce the following three operations:

1. $\beta(G, x)$ is the operation that extends G by creating a true twin of x.
2. $\gamma(G, x)$ is the operation that extends G by creating a false twin of x.
3. $\phi(G, B, \{x_1, x_2, \ldots, x_n\})$ is the operation that extends G by the bipartite graph B, identifying certain vertices of the set V_1 with the set of false twins $\{x_1, x_2, \ldots, x_n\}$ (possibly $n = 1$).

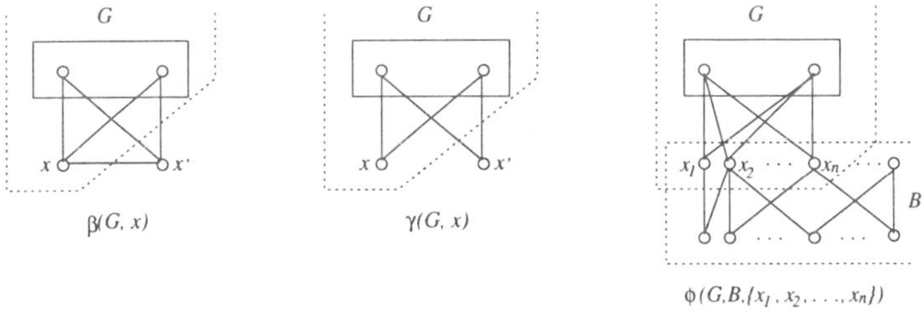

Fig. 2. *Extension of a graph G by β, γ and ϕ operations.*

When no ambiguity can arise, we omit some arguments of the above operations, and we write $\phi(x_1, x_2, \ldots, x_n)$ instead of $\phi(\{x_1, x_2, \ldots, x_n\})$.

These operations have been used to characterize the following graph classes: parity graphs [8], distance-hereditary graphs (where ϕ appends only K_2) [4], cographs (where ϕ is not used) [12], ptolemaic graphs (where ϕ appends K_2 and γ is applied only to vertices whose neighborhoods form a nonempty clique) [20] and $(6,2)$-chordal graphs (where ϕ appends K_2 and β is not used) [2].

If ϕ operation appends only graphs belonging to a specific subclass of bipartite graphs, then these three operations define an element of an entire family of new classes. Let $\mathcal{B}$ be the family containing all the hereditary subclasses of bipartite class, and $\mathbf{C}$ the generic element of $\mathcal{B}$. The following definition introduces the generic element $\Phi_{\mathbf{C}}$ of the new family $\mathcal{F}$ of classes induced by the bipartite subclass $\mathbf{C}$.

Definition 3. Let $\mathbf{C}$ be an element of $\mathcal{B}$ and G a connected graph. G belongs to $\Phi_{\mathbf{C}}$ if and only if it can be obtained from a single vertex applying successively and in any order the operations β, γ, and ϕ, where ϕ appends only non-trivial connected graphs in $\mathbf{C}$.

The smallest class in $\mathcal{B}$ contains only the graph K_2, whereas the biggest one is the class of bipartite graphs (denoted with $\mathbf{B}$). The former induces the class of distance-hereditary graphs and the latter the class of parity graphs. By Definition 3, these two classes are denoted by $\Phi_{\mathbf{K_2}}$ and $\Phi_{\mathbf{B}}$, respectively. Other elements of $\mathcal{B}$ induce classes in $\mathcal{F}$, such as the class of trees and the class of (6,2)-chordal bipartite graphs. The corresponding classes in $\mathcal{F}$ are $\Phi_{\mathbf{T}}$ and $\Phi_{(\mathbf{6,2})}$. It is easy to verify the proper inclusions $\Phi_{\mathbf{K_2}} \subset \Phi_{\mathbf{T}} \subset \Phi_{(\mathbf{6,2})}$. On the other side, there are different classes in $\mathcal{B}$ that induce the same class in $\mathcal{F}$. In fact, the class containing all the complete bipartite graphs induces the class of distance-hereditary graphs, that is, $\Phi_{\mathbf{K_2}} = \Phi_{\mathbf{K_{m,n}}}$. Now, two questions arise: "How many classes are in $\mathcal{F}$?", and "Are there classes in $\mathcal{F}$ for which graph problems can be efficiently solved?". The following theorem answers to the first question.

Theorem 4. *$(\mathcal{F}, \subseteq)$ is an infinite lattice whose top and bottom elements are $\Phi_{\mathbf{B}}$ and $\Phi_{\mathbf{K_2}}$, respectively.*

In the sequel we simply propose some argumentations and refer to [10] for the complete proof of this theorem. First, it is easy to verify that the inclusions $\Phi_{\mathbf{K_2}} \subseteq \Phi_{\mathbf{C}} \subseteq \Phi_{\mathbf{B}}$ holds for each $\mathbf{C} \in \mathcal{B}$. In order to prove that $\mathcal{F}$ contains an infinite number of element, we define a chain of classes starting from $\Phi_{\mathbf{K_2}}$ and bounded by $\Phi_{\mathbf{B}}$.

Definition 5. Let Φ^0 be the class of distance-hereditary graphs. Class Φ^i is recursively defined as $\Phi^i = \Phi_{\mathbf{B}^i}$ where $\mathbf{B}^i = \mathbf{B} \cap \Phi^{i-1}$ for each $i = 1, 2, \ldots$

This definition implies that $\Phi^i \in \mathcal{F}$ and $\Phi^i \subseteq \Phi^{i+1}$ for each $i = 0, 1, \ldots$ Now, we briefly show the elements of an induction on i showing that $\Phi^i \neq \Phi^{i+1}$ for each $i = 0, 1, \ldots$.

The graph C_6 does not belong to Φ^0, whereas it belongs to Φ^1. If there exists a graph $G_i = (V, E)$ such that $G_i \notin \Phi^{i-1}$ and $G_i \in \Phi^i$, then a graph G_{i+1} can be defined such that $G_{i+1} \notin \Phi^i$ and $G_{i+1} \in \Phi^{i+1}$. Let $V_1 = \{x_1 | \ x \in V\}$, $V_2 = \{x_2 | \ x \in V\}$, $E_1 = \{(x_1, y_1) | \ (x, y) \in E\}$ and $E_2 = \{(x_2, y_2) | \ (x, y) \in E\}$.

Then $G_{i+1} = (V_1 \cup V_2 \cup \{u,w\}, E_1 \cup E_2 \cup E')$ where $u, w \notin V$, and $E' = \{(x_1, u), (u, x_2), (y_1, w), (w, y_2)\}$ for an arbitrary edge $(x, y) \in E$.

In the following we report results proved in [10] showing that any class in $\mathcal{F}$ is hereditary. This property plays a fundamental role in order to achieve next results.

Theorem 6. *Let $G \in \Phi_\mathbf{C}$. If $G - \{x\}$ is connected, then $G - \{x\}$ belongs to $\Phi_\mathbf{C}$.*

According to Definition 3, if $G \in \Phi_\mathbf{C}$ then there must exist a sequence of operations $\sigma = (\sigma_1, \sigma_2, \ldots, \sigma_n)$ such that $G_0 = K_1$, $\sigma_i(G_{i-1}) = G_i$ with $\sigma_i \in \{\gamma, \beta, \phi\}$ for all $i = 1, 2, \ldots, n$, and $G_n = G$. This theorem can be proved by showing a sequence σ' to generate $G - \{x\}$. In both the cases, ϕ appends graphs from $\mathbf{C}$.

Corollary 7. *Any class belonging to $\mathcal{F}$ is a hereditary class.*

4 Characterization of graphs in the family

Now we can give a characterization of graphs in $\Phi_\mathbf{C}$ based on Cunningham decomposition. This aim will be achieved throughout two basic points: defining the structure of components in the decomposition (Theorem 12) and proving that certain decomposition trees fully qualify graphs in any class of $\mathcal{F}$ (Theorem 16).

The following three lemmas provide useful results to attain the first aim.

Lemma 8. *Let $G = (V_G, E_G)$ be a connected graph with $|V| \geq 4$. If $x \in V_G$ is an articulation point in G, or x has a twin, then G is decomposable.*

Proof. Let us suppose that A and B are the two connected components obtained by removing x from G. Then either V_A or V_B has at least 2 vertices. Let us assume that V_A has such a number of vertices. Then $\{V_A, V_B \cup \{x\}\}$ is a split. Moreover, if y is a twin of x, then $\{\{x, y\}, V_G - \{x, y\}\}$ is a split. □

Lemma 9. *Let G' be an element of $\Phi_\mathbf{C}$ having false twins $\{x_1, x_2, \ldots, x_n\}$. Let us suppose that G is obtained from G' by a $\phi(G', B, \{x_1, x_2, \ldots, x_n\})$ with $B \in \mathbf{C}$. If $|G' - \{x_1, x_2, \ldots, x_n\}| > 1$ then G is decomposable.*

Proof. A split is given by $\{G' - \{x_1, x_2, \ldots, x_n\}, B\}$. □

The following theorem shows that any non-trivial component of graphs in $\Phi_\mathbf{C}$ is a prime graph obtainable by a $\phi(K_{1,n}, B)$ which identifies the pendants of the star $K_{1,n}$ with some vertices of $B \in \mathbf{C}$. We first give the following definitions.

Definition 10. Let $G \in \Phi_\mathbf{C}$ be a graph obtainable by an operation $\phi(K_{1,n}, B)$ with $B \in \mathbf{C}$. If ϕ identifies the n pendants of the star $K_{1,n}$ with some vertices of B, and x is the center of $K_{1,n}$, we call x a *hinge* of G with respect to class $\mathbf{C}$.

Several induced stars of G generate different hinges for G with respect to class $\mathbf{C}$.

Definition 11. Let $G \in \Phi_{\mathbf{C}}$. The *hinge-set* of G with respect to the class $\mathbf{C}$ is

$$H_C(G) = \begin{cases} V_G & \text{if } G \text{ is clique or star} \\ \{x \in V_G | \ x \text{ is a hinge of } G \text{ w.r.t. } \mathbf{C}\} & \text{otherwise} \end{cases}$$

We say that a graph $G \in \Phi_{\mathbf{C}}$ is *hinged* with respect to class $\mathbf{C}$ if $H_C(G) \neq \emptyset$ (we also write $\mathbf{C}$-hinged).

Theorem 12. *Let $G \in \Phi_{\mathbf{C}}$. Any non-trivial component of $D(G)$ is $\mathbf{C}$-hinged.*

Proof. Let $D \in D(G)$ be a non-trivial component. By Corollary 7 $D \in \Phi_{\mathbf{C}}$, and by Lemma 8 D does not have any pair of twins. It follows that the last operation in any yielding sequence for D is $\phi(D', B, \{x_1, x_2, \ldots, x_n\})$ with $B \in \mathbf{C}$. By Lemma 9 it follows that $|D' - \{x_1, x_2, \ldots, x_n\}| = 1$, and Lemma 8 implies that $|\{x_1, x_2, \ldots, x_n\}| > 1$. Hence, D' must be a star $K_{1,n}$ whose pendants correspond to $\{x_1, x_2, \ldots, x_n\}$ and whose center belongs to $H_C(D)$. This proves the theorem. □

The previous theorem fully defines the structure of the components of a decomposition of graphs in $\Phi_{\mathbf{C}}$, but this is not enough to characterize the graph itself. In fact, the graph in Figure 1 has a decomposition with two components which are hinged with respect to (6,2)-chordal bipartite class (for example, nodes with degree 3 are in the hinge-set), but the graph is not in $\Phi_{(\mathbf{6},\mathbf{2})}$, because there are no sequences of operations to generate the graph. Hence, we need more information to decide whether a graph belongs to a specific class or not. Theorem 16 provides such information combining the structure of the components and generative sequences.

Lemma 13. *Let $\{G_1, G_2\}$ be a simple decomposition of a graph G such that $G_1 = \langle V_1 \cup \{m_1\}\rangle$, $G_2 = \langle V_2 \cup \{m_2\}\rangle$, and $G_1, G_2 \in \Phi_{\mathbf{C}}$. If G_2 is prime and $m_2 \in H_C(G_2)$, then $G \in \Phi_{\mathbf{C}}$.*

Proof. Theorem 12 implies that G_2 is trivial or $\mathbf{C}$-hinged.

If G_2 is trivial then G can be obtained from G_1 by true twin operations (if G_2 is a clique), by false twin operations (if G_2 is a star whose center is m_2), and by a ϕ operation that only appends a K_2 (if G_2 is a star and m_2 is a pendant). Hence, G is an element of $\Phi_{\mathbf{C}}$.

If G_2 is $\mathbf{C}$-hinged then the following facts held before the decomposition of G. Let us suppose that $N_{G_2}(m_2) = \{y_1, y_2, \ldots, y_m\}$. *(i)* $\langle V_1 \cup \{y_1, y_2, \ldots, y_m\}\rangle$ is an element of $\Phi_{\mathbf{C}}$ because it can be obtained from $\langle V_1 \cup \{y_1\}\rangle$ by $\gamma(y_1)$ applied $m-1$ times. In such a graph $\{y_1, y_2, \ldots, y_m\}$ are all false twins; *(ii)* $\langle V_2\rangle \in \mathbf{C}$ because G_2 is $\mathbf{C}$-hinged and $m_2 \in H_C(G_2)$. So, appending $\langle V_2\rangle$ to $\langle V_1 \cup \{y_1, y_2, \ldots, y_m\}\rangle$ by $\phi(y_1, y_2, \ldots, y_m)$, we obtain the graph $G \in \Phi_{\mathbf{C}}$. □

In order to better explain the main characterization of a graph G in a class $\Phi_{\mathbf{C}}$ we introduce a notation referring to marked vertices of $D(G)$. The symbol m_{ij} is used to represent the marked vertex in the component D_i that has the smallest distance from all the vertices in D_j. Notice that this notation implies that the marked vertex m_{ij_1} can be also referred as $m_{ij_2}, m_{ij_3} \ldots, m_{ij_k}$ if it is the vertex of D_i nearest to the components $D_{j_2}, D_{j_3} \ldots, D_{j_k}$, respectively. Such information can be easily deduced using $DT(G)$ and $D(G)$.

Definition 14. A decomposition tree $DT(G)$ of a graph G is $\mathbf{C}$-rooted if $D(G) = \{D_1, D_2, \ldots, D_m\}$, D_k is $\mathbf{C}$-hinged or trivial for all $k = 1, 2, \ldots, m$, and there exists a *root* $D_r \in D(G)$ such that $m_{jr} \in H_C(D_j)$ for each $D_j \in D(G) \setminus \{D_r\}$. A $\mathbf{C}$-rooted decomposition tree of a graph G is denoted by $RD_C(G)$.

Lemma 15. *Let $G = \phi(K_{1,n}, B)$ be a graph in $\Phi_{\mathbf{C}}$. If x is the center of $K_{1,n}$ then x is a hinge of a root of $RD_C(H)$.*

Proof. By definition of split decomposition, there is only one component D_r of $D(G)$ that contains the vertex x. Each component D_j different from D_r is an induced subgraph of B, and then any vertex of D_j is a hinge of D_j itself. Then either D_j is $\mathbf{C}$-hinged or it is trivial. To prove that G has a $\mathbf{C}$-rooted decomposition tree it remains to be shown that D_r is $\mathbf{C}$-hinged or trivial. In fact, if D_r is trivial the lemma holds, otherwise $D_r = \phi(\langle N_{D_r}[x]\rangle, \langle D_r - \{x\}\rangle)$ because $\langle N_{D_r}[x]\rangle$ is a star, D_r is bipartite and $\langle D_r - \{x\}\rangle$ is an induced subgraph of B. The latter fact also implies that x is a hinge of D_r. □

Theorem 16. *Let G be a graph. $G \in \Phi_{\mathbf{C}}$ if and only if G has a $\mathbf{C}$-rooted decomposition tree $RD_C(G)$.*

Proof ("only if" case). To prove that G belongs to $\Phi_{\mathbf{C}}$ we will present a sequence $\sigma = (\sigma_1, \sigma_2, \ldots, \sigma_m)$ that generates G itself. Let $D(G) = \{D_1, D_2, \ldots, D_n\}$ be the set of components of the $\mathbf{C}$-rooted decomposition tree of G, and let D_r its root. This sequence σ is given by the following actions. First, remove D_r from D and set $\sigma = (\sigma_1, \sigma_2, \ldots, \sigma_k)$ if such a subsequence $(\sigma_1, \sigma_2, \ldots, \sigma_k)$ generates the graph D_r belonging to $\Phi_{\mathbf{C}}$, and then set $G' = D_r$. Now, repeat the following steps until $D(G)$ is empty.

1. Choose $D_j \in D$ such that (m_1, m_2) is a marked edge between G' and D_j. By hypotheses it follows that $m_2 \in H_C(D_j)$;
2. Remove the split between G' and D_j, i.e., connect any vertex in $N_{G'}(m_1)$ to any vertex in $N_{D_j}(m_2)$ and then remove m_1 and m_2, to obtain G'', an induced subgraph of G. By Lemma 13 it follows that G'' belongs to $\Phi_{\mathbf{C}}$
3. Update σ adding all the operations shown in Lemma 13 to yield G'' from G' and D_j;
4. Remove D_j from D and set $G' = G''$.

It easy to see that σ applied to K_1 generates G.

Proof ("if" case). By induction on the number of operations necessary to obtain the graph G starting from K_1.

Basic step. Any graph G belonging to $\Phi_{\mathbf{C}}$ such that $|G| \leq 4$ verifies the theorem, because such a graph belongs to $\Phi_{\mathbf{K_2}}$ and any component of its split decomposition is trivial.

Inductive step. Let us suppose that a graph $G \in \Phi_{\mathbf{C}}$ generated by less than n operations has a **C**-rooted decomposition tree $RD_C(G)$. We have to prove that $\sigma_n(G)$ also has a **C**-rooted decomposition tree for any possible $\sigma_n \in \{\gamma, \beta, \phi\}$.

Case a. If $\sigma_n \in \{\gamma, \beta\}$ then, by Lemma 8, σ_n induces a split $\{V_1, V_2\}$. The corresponding simple decomposition $\{G_1, G_2\}$ implies that either $G_2 = K_3$ or $G_2 = K_{1,2}$. In both cases, by the inductive hypothesis, $\sigma_n(G)$ has a **C**-rooted decomposition tree because $m_2 \in H_C(G_2)$.

Case b. If $\sigma_n = \phi(G, B, \{x_1, x_2, \ldots, x_n\})$ appends a graph B to G, then two different cases can arise:

$|G - \{x_1, x_2, \ldots, x_n\}| = 1$. Then, by Lemma 15, it follows that the graph obtained by σ_n has a **C**-rooted decomposition tree.

$|G - \{x_1, x_2, \ldots, x_n\}| > 1$. Then, σ_n induces a split $\{V_1, V_2\}$. This split generates a component G_2 formed by B plus a vertex m_2 joined to all the vertices $x_1, x_2, \ldots, x_m$. The component G_2 is joined by the marked edge (m_1, m_2) to a graph isomorphic to $G - \{x_2, x_3, \ldots, x_m\}$. Obviously, this graph can be obtained from K_1 by a number of operations not grater than those needed to generate G, and hence, by the inductive hypothesis, it has a **C**-rooted decomposition tree.

By Lemma 15, G_2 has a **C**-rooted decomposition tree, the root component contains the vertex m_1, and m_1 is also a hinge of this component. It easily follows that $\sigma_n(G)$ has a **C**-rooted decomposition tree.

□

Let us consider the graph in Figure 1 again. Theorem 16 implies that such a graph does not belong to $\Phi_{(\mathbf{6,2})}$ because both the components of its decomposition cannot be considered roots, due to the fact that marked vertices m_1 and m_2 are not hinges with respect to (6,2)-chordal bipartite class. On the other hand, both components of decomposition of graph in Figure 3 are hinged with respect to (6,2)-chordal bipartite class and vertex m_2 is a hinge with respect to the same class. Hence, component with vertex m_1 is a root for the decomposition, and, by Theorem 16, the graph belongs to $\Phi_{(\mathbf{6,2})}$ class.

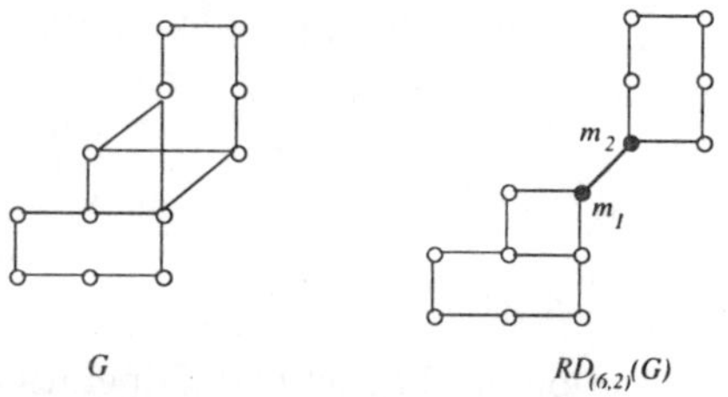

Fig. 3. *A graph G and its rooted decomposition with respect to the (6,2)-chordal bipartite class.*

Corollary 17. *Let $G \in \Phi_{\mathbf{B}}$. Then, any component of $D(G)$ is a root for $RD_C(G)$.*

Corollary 18. *Let G be a graph. $G \in \Phi_{\mathbf{B}}$ if and only if any component of $D(G)$ is trivial or bipartite.*

These results have been used in [11] in order to reduce the time complexity to linear time for recognition and weighted clique number problems for parity graphs.

5 Algorithmic problems

In this section we use the characterization of graphs in the class $\Phi_{\mathbf{C}}$ given in Theorem 16 in order to provide efficient algorithms to solve the recognition and the isomorphism problem.

5.1 Recognition problem

First of all we give a procedure `Test` that verifies whether a given graph G belongs to $\Phi_{\mathbf{C}}$ or not. Let us suppose that **C** is a class belonging to $\mathcal{B}$, and that a polynomial recognition algorithm exists for any graph in this class having a time bound $O(c)$. The following steps allows us to decide if a graph G belongs to the class $\Phi_{\mathbf{C}}$ with a polynomial time bound $O(nc)$, when $|G| = n$.

procedure `Test`;
input: G **output: true** iff $G \in \Phi_{\mathbf{C}}$
Compute $D(G)$
for each marked edge (m_1, m_2) in $DT(G)$
begin
Let $G_1, G_2 \in D(G)$ the components such that $m_1 \in G_1$ and $m_2 \in G_2$:
if $m_1 \in H_C(G_1)$ **then** add a direct edge from m_2 to m_1
if $m_2 \in H_C(G_2)$ **then** add a direct edge from m_1 to m_2
Remove the marked edge (m_1, m_2)
end
Find a vertex D_r of $DT(G)$ such that any other vertex of $DT(G)$ can be reached from D_r passing throughout directed edges
if a root D_r exists **then** compute $H_C(D_r)$ **else return false**
if $H_C(D_r) \neq \emptyset$ **then return true else return false**

We give an informal description of this procedure. Following the Theorem 16, the above procedure builds the decomposition $D(G)$ and then it tests if the decomposition tree $DT(G)$ is **C**-rooted. To the latter purpose, when a marked edge (m_1, m_2) joining the components G_1 and G_2 is considered, it can be (i) removed, (ii) replaced by a directed edge, or (iii) replaced by two opposite directed edges. The first case means that a root does not exist, the second one means that a root (if it exists) is in the subtree containing G_1 (G_2, resp.) if the inserted edge is directed from m_1 to m_2 (from m_2 to m_1, resp.), and the last one gives no information about the possible location oa a root. Notice that, for example, according to Corollary 17, any marked edge in the decomposition tree of a parity graph is replaced by two opposite directed edges. If a directed path exists from a component D_r to each other component, then D_r can be a root for $DT(G)$. Finally, $G \in \Phi_{\mathbf{C}}$ if and only if a root D_r exists and $H_C(D_r) \neq \emptyset$.

It is easy to see that the procedure works on $O(nc)$ time. In fact, step 1 requires a linear time [14] on the size of the graph, whereas the cycle at step 2 is bounded by $O(nc)$. Step 9 can be easily accomplished in linear time on the size of $DT(G)$. To perform the step 10, it is necessary to apply the recognition algorithm for graphs in the class **C** a number of times proportional to the size of the component D_r; this time, in the worst case, is $O(nc)$.

5.2 Isomorphism problem

We have already noted that the isomorphism problem can be solved in polynomial time in $\Phi_{\mathbf{K_2}}$ [3], and that its complexity is unknown in $\Phi_{\mathbf{B}}$. Now, we present a polynomial time procedure `Isomorphism` verifying if $G \cong G'$ for $G, G' \in \Phi_{\mathbf{C}}$. We assume the existence of polynomial algorithm A_C to test labelled isomorphism in **C**. A_C is used in order to test if an isomorphism between non-trivial hinged components such that a hinge corresponds to the other exists. This can be done by marking the neighborhood of the hinges, by removing the hinges, and by applying A_C to the remaining bipartite graphs.

As for the recognition problem, procedure `Isomorphism` is still based on the characterization of graphs in $\Phi_{\mathbf{C}}$ given in Theorem 16, and so it requires the **C**-rooted decompositions trees $RD_C(G)$ and $RD_C(G')$.

Fixed a root for both the decomposition trees, the following procedure returns a boolean variable `Iso` such that `Iso` $\Rightarrow G \cong G'$. To establish if graphs are really isomorphic it suffices to repeat the procedure for each root of one of them until either `Iso` is true or there are no more roots.

procedure `Isomorphism`;
input: $RD_C(G), RD_C(G')$ **output:** `Iso` (`Iso` $\Rightarrow G \cong G'$)
1. **if** $RD_C(G)$ and $RD_C(G')$ have different number of levels
2. **then** `Iso`=**false**; **HALT**
3. **else** Assume l is the number of levels when roots are at level 0.
4. Make the union of all the components of $RD_C(G)$ and $RD_C(G')$ at level l, and denote the corresponding set of components as C. Make the quotient set of C with respect to the isomorphism relation. Assign in a arbitrary way the integers 1,2, ... to the classes, and consequently assign the integer k to components belonging to the class k.
5. Let L_1 be the sorted list of integers assigned to components of $RD_C(G)$ at level l. Assume L_2 is the corresponding list for $RD_C(G')$.
6. **if** $L_1 \neq L_2$
7. **then** `Iso`=**false**; **HALT**
8. **else if** $l = 0$ **then** `Iso`=**true**; **HALT**
9. For each integer $k \in L_1$, if k is assigned to the component D_i at level l and D_j is its father at level $l-1$, then assign the label k to the marked vertex m_{ji}.
10. Repeat the previous step for L_2 and $RD_C(G')$.
11. $l = l - 1$.
12. goto step 5.

This algorithm follows the structure of the algorithm proposed in [1] to solve the isomorphism problem for labelled trees. We give an informal description.

The algorithm compares two **C**-rooted decomposition trees by levels, starting from the last one. A comparison is performed by assigning the same numerical label to isomorphic components of the same level, obtaining two lists of integers (one for each decomposition tree). Such lists are ordered and compared: if they are different the graphs are obviously not isomorphic. Before analyzing the other levels, it is necessary to record in each upper components all information which characterize (up to isomorphism) the subtrees having as root the components themselves. To this purpose, for each component D_i of the current level having associated label k and father D_j, we assign the label k to the marked vertex $m_{ji} \in D_j$. In this way two components belonging to the same level are isomorphic if and only if the subtrees of which they are roots are isomorphic. Finally, graphs G and G' are isomorphic if and only if the procedure assigns the same labels to the roots.

Theorem 19. *Let* $\mathbf{C} \in \mathcal{B}$ *be a hereditary subclass of bipartite class having a polynomial algorithm to test the labelled isomorphism. Then the problem of deciding if two graphs in* $\Phi_{\mathbf{C}}$ *are isomorphic can be solved in polynomial time.*

As the algorithm proposed in [1] solves the isomorphism problem for labelled trees, the following corollary can be stated.

Corollary 20. *The isomorphism problem for graphs in* $\Phi_{\mathbf{T}}$ *can be solved in polynomial time.*

In [3] the authors prove that the automorphism group of a distance-hereditary graph is that of a tree: this fact provides an injection, computable in polynomial time, from the class of distance-hereditary graphs to the class of trees and then the labelled isomorphism problem for distance-hereditary graphs can be solved in polynomial time. This result implies that labelled isomorphism in (6,2)-chordal bipartite class can be solved in polynomial time. Hence, the following corollary holds.

Corollary 21. *The isomorphism problem for graphs in* $\Phi_{(\mathbf{6},\mathbf{2})}$ *can be solved in polynomial time.*

Notice that $\Phi_{(\mathbf{6},\mathbf{2})} \equiv \Phi^1$ and that the classes Φ^i are built using the bipartite graphs of Φ^{i-1}. As the algorithm (modified as in [1]) can be used to solve the labeled isomorphism problem for Φ^1, a recursive application of this algorithm can be used for the isomorphism in a generic class Φ^i.

Corollary 22. *For a fixed* i*, the isomorphism problem for graphs in* Φ^i *can be solved in polynomial time.*

References

1. A. V. Aho, J. E. Hopcroft, and J. D. Ullman. *The Design and Analysis of Computer Algorithms.* Addison-Wesley, Reading, 1974.

2. G. Ausiello, A. D'Atri, and M. Moscarini. Chorality properties on graphs and minimal conceptual connection in semantic data models. *Jou. of Comp. System Sci.*, 33, 1986.
3. H. J. Bandelt, A. D'Atri, M. Moscarini, H. M. Mulder, and A. Schultze. Operations on distance hereditary graphs. Technical Report 226, CNR, Istituto di Analisi dei Sistemi e Informatica del CNR (Rome Italy), 1988.
4. H. J. Bandelt and M. Mulder. Distance-hereditary graphs. *Journal of Combinatorial Theory, Series B*, 41(2), 1986.
5. H. J. Bandelt and M. Mulder. Metric characterization of parity graphs. *DMATH: Discrete Mathematics*, 91:221–230, 1991.
6. K. S. Booth and C. J. Colbourn. Problems polynomially equivalent to graph isomorphism. Technical Report CS-77-04, Comp. Science Dept., Univ. of Waterloo, 1979.
7. A. Bouchet. Transforming trees by successive local complementations. *Journal of Graph Theory*, 4:196–207, 1988.
8. M. Burlet and J. P. Uhry. Parity graphs. *Annals of Discrete Math.*, 16:1–26, 1982.
9. M. Burlet and J. P. Uhry. Parity graphs. In C. Berge and V. Chvátal, editors, *Topics on Perfect Graphs*, number 21 in Annals of Discrete Math., pages 253–277, 1984.
10. S. Cicerone and G. Di Stefano. A family of graph classes, recognition and isomorphism problem. Technical Report R.96-1, Dip. di Ing. Elettrica, Univ. di L'Aquila - Italy, 1996.
11. S. Cicerone and G. Di Stefano. Linear algorithms for some problems on parity graphs. Technical Report R.96-16, Dip. di Ing. Elettrica, Univ. di L'Aquila - Italy, 1996.
12. D. G. Corneil, H. Lerchs, and L. Stewart-Burlingham. Complement reducible graphs. *Discrete Applied Mathematic*, 3:163–174, 1981.
13. W. H. Cunningham. Decomposition of directed graphs. *SIAM Journal on Alg. Disc. Meth.*, 3:214–228, 1982.
14. E. Dahlhaus. Efficient parallel and linear time sequential split decomposition. In *FSTTCS*. Lecture Notes in Computer Science, vol. 880, 1994.
15. A. D'Atri and M. Moscarini. Distance-hereditary graphs, steiner trees, and connected domination. *SIAM Journal on Computing*, 17:521–530, 1988.
16. M. R. Garey and D. S. Johnson. *Computers and intractability; a guide to the theory of NP-completeness*. W.H. Freeman, 1979.
17. P. L. Hammer and F. Maffray. Completely separable graphs. *DAMATH: Discrete Applied Math. and Combinatorial Operations Research and Comp. Sci.*, 27, 1990.
18. F. Harary. *Graph Theory*. Addison-Wesley, 1969.
19. E. Howorka. Distance hereditary graphs. *Quart. J. Math. Oxford*, 2(28):417–420, 1977.
20. E. Howorka. A characterization of ptolemaic graphs. *Jou. of Graph Theory*, 5:323–331, 1981.
21. W. L. Hsu. $O(mn)$ algorithms for the recognition and isomorphism problem on circular-arc graphs. *SIAM Journal on Computing*, 24:411–439, 1995.
22. Lai and Wei. The edge hamiltonian path problem is NP-complete for bipartite graphs. *Information Processing Letters*, 46(5), 1993.
23. H. Muller and A. Brandstädt. The NP-completeness of steiner tree and dominating set for chordal bipartite graphs. *TCS: Theoretical Comp. Sci.*, 53, 1987.
24. F. Nicolai. Hamiltonian problems on distance-hereditary graphs. Technical Report SM-DU-264, University Duisburg, 1994.

25. J. P. Spinrad. Recognition of circle graphs. *Jou. of Algorithms*, 16(1):264–282, 1994.

Sierpinski's Gasket, 2-Groups and Graph Symmetries

Marston D.E. Conder and Cameron G. Walker

Department of Mathematics, University of Auckland,
Private Bag 92019, Auckland, New Zealand

This is an extended abstract of a paper delivered at the DMTCS'96 conference. A full version of the paper will be published elsewhere under the title "Vertex-transitive graphs with arbitrarily large vertex-stabilizer".

1 Introduction

The following question is attributed to Chris Godsil: *Does there exist an infinite family of finite vertex-transitive graphs* $\{\Gamma_n\}$ *of fixed valency such that the order of a vertex-stabilizer in the smallest vertex-transitive group of automorphisms of* Γ_n *is a strictly increasing function of* n*?* We show the answer is "yes".

Without the condition on the *smallest* vertex-transitive group of automorphisms, the construction of vertex-transitive graphs with automorphism group having an arbitrarily large vertex-stabilizer is relatively easy. For example take the 4-valent graph with vertices $0, 1, 2, \ldots, 2n-1$, and edges joining each of $2i$ and $2i+1$ to each of $2i+2$ and $2i+3$ modulo $2n$ (for $0 \leq i < n$). This is a Cayley graph for the dihedral group D_n, and its full automorphism group is the wreath product $C_2 \operatorname{wr} D_n$ of order $2^n \times 2n$; in particular, the latter group is transitive on the $2n$ vertices, with vertex-stabilizer of order 2^n.

In this paper we show how to construct for each positive integer n a finite arc-transitive 4-valent graph Γ_n with full automorphism group the symmetric group S_N of degree N, where $N = p(n)$ is a prime greater than 2^{2^n+2} and congruent to 1 modulo 4. The stabilizer in S_N of any vertex v of Γ_n is a 2-group of order 2^{2^n+2}, defined with the help of Sierpinski's gasket, and due to the choice of N the smallest vertex-transitive group of group of automorphisms of Γ_n is the alternating group A_N, with vertex-stabilizer of order 2^{2^n+1}.

Before doing this in Section 2 (and verifying properties of Γ_n in Section 3), we first describe some background material on graph symmetries. If G is a group of automorphisms (symmetries) of the simple graph Γ which acts transitively on $V\Gamma$, then G is said to be vertex-transitive on Γ. Similarly if G acts transitively on the set of arcs (ordered edges) of Γ, then G is said to be *arc-transitive* on Γ. In the latter case, the stabilizer $G_v = \{g \in G : v^g = v\}$ of a vertex $v \in V\Gamma$ acts transitively on the set of arcs in Γ emanating from the vertex v. Further, if (v, w) is any one such arc, then by arc-transitivity there exists an automorphism $a \in G$ reversing (v, w), and then the structure of Γ may be defined completely in terms of a and G_v: vertices may be labelled with right cosets of G_v in G, and edges are the images under the action of G (by right multiplication) of the single edge $\{v, w\}$ labelled $\{G_v, G_v a\}$.

Conversely, given any group G containing a subgroup H and an element a such that $a^2 \in H$, we may construct a graph $\Gamma = \Gamma(G, H, a)$ on which G acts as an arc-transitive group of automorphisms, as follows: take as vertices of Γ the right cosets of H in G, and join two cosets Hx and Hy by an edge in Γ whenever $xy^{-1} \in HaH$. Defined in this way, Γ is a simple graph on which the group G acts as an arc-transitive group of automorphisms under the action $g : Hx \to Hxg$ for each $g \in G$ and each coset Hx in G.

The above construction is explained in more detail in [3] (and was used to answer similar questions in [1] and [2]). The graph $\Gamma = \Gamma(G, H, a)$ is connected if and only if G is generated by HaH, and is regular of degree d where d is the number of right cosets of H contained in the double coset HaH.

In what follows in this paper the group H will be a 2-group, generated by an element of order 4 and an increasing number of additional involutions. To define these involutions we use a modification of *Sierpinski's gasket*, or Pascal's triangle modulo 2; see §2.2 of [4].

2 The construction

First let ${}^nC_k$ be the standard binomial coefficient, and recall that these coefficients satisfy the symmetric identity ${}^nC_k = {}^nC_{n-k}$, and the additive identity ${}^nC_{k-1} + {}^nC_k = {}^{n+1}C_k$, which are fundamental properties of Pascal's triangle. Also ${}^nC_{n/2}$ is always even when n is even, while on the other hand, an easy exercise shows every coefficient ${}^nC_k$ is odd when $n+1$ is a power of 2. Next define $d_{rs} = {}^sC_{[r/2]}$ (where $[r/2]$ is the greatest integer not exceeding $r/2$), for $0 \leq r \leq 2s+1$. Then clearly $d_{0s} = d_{1s} = d_{2s,s} = d_{2s+1,s} = 1$ for all $s \geq 0$, and also whenever r is even they satisfy the following: $d_{rs} = d_{r+1,s}$, $d_{r-1,s} + d_{r,s} \equiv d_{r,s+1} \bmod 2$, and $d_{rr} \equiv 0 \bmod 2$. The proof is straightforward. A picture of the corresponding triangle of residues of these coefficients modulo 2 is given below for $s \leq 15$, with x's for 1's and blanks for 0's:

```
                              x x
                            x x x x
                          x x     x x
                        x x x x x x x x
                      x x             x x
                    x x x x         x x x x
                  x x     x x     x x     x x
                x x x x x x x x x x x x x x x x
              x x                             x x
            x x x x                         x x x x
          x x     x x                     x x     x x
        x x x x x x x x                 x x x x x x x x
      x x             x x             x x             x x
    x x x x         x x x x         x x x x         x x x x
  x x     x x     x x     x x     x x     x x     x x     x x
x x x x x x x x x x x x x x x x x x x x x x x x x x x x x x x x
```

Fig. 1. Modified Sierpinski Gasket

Now suppose n is any positive integer. Define $m = 2^n$, and let $N = p(n)$

be any prime such that $N \equiv 1 \bmod 4$ and $N > 2^{m+2}$. Let G be the symmetric group S_N, and define three elements a, b and c in S_N as follows:

$$\begin{aligned}
a = {} & (1,3)(2,4)(5,7)(6,8)\ldots \\
& (2m-7,2m-5)(2m-6,2m-4)(2m-3,2m-1)(2m-2,2m) \\
& (2m+1,N)(2m+2,2m+3)(2m+4,2m+5)\ldots \\
& (N-5,N-4)(N-3,N-2) \\
b = {} & (1,2m+1,2,2m+2)(3,5)(4,6)(7,9)(8,10)\ldots \\
& (2m-5,2m-3)(2m-4,2m-2) \\
& (2m+3,2m+4)(2m+5,2m+6)\ldots \\
& (N-4,N-3)(N-2,N-1) \\
c = {} & (1,2)(3,4)(5,6)(7,8)\ldots(2m-5,2m-4)(2m-3,2m-2)(2m-1,2m).
\end{aligned}$$

Within the alternating group A_N these permutations generate a subgroup which is transitive and indeed primitive (since N is prime). Also $c^{-1}ac = a$ while $c^{-1}bc = b^{-1}$. Next, note that c is the product of transpositions $t_j = (2j-1, 2j)$ for $1 \le j \le m$. Letting this product correspond to the first m entries of the mth row of the modified Sierpinski gasket, we define a sequence of additional permutations $c_1, c_2, \ldots, c_m$ in S_N as follows:

For $1 \le i \le m$ define $c_i = \prod_{j=i}^{m} t_j{}^{d_{j-i,m-i}}$, where t_j is the transposition $(2j-1, 2j)$, and $d_{j-i,m-i} = {}^{m-i}C_{[(j-i)/2]}$ (as defined earlier) for $i \le j \le m$.

Note that $c_1 = t_1 t_2 \ldots t_m = c$, while $c_m = t_m = (2m-1, 2m)$, and each of the c_i other than c_m is even. Also the choice of exponents implies

$$a^{-1}c_i a = \begin{cases} c_i & \text{if } i \text{ is odd} \\ c_{i-1}c_i & \text{if } i \text{ is even} \end{cases}$$

and

$$b^{-1}c_i b = \begin{cases} b^2 c_1 & \text{if } i = 1 \\ c_i & \text{if } i \in \{2,4,\ldots,m\} \\ c_{i-1}c_i & \text{if } i \in \{3,5,\ldots,m-1\}. \end{cases}$$

Definition. Let Γ_n be the graph $\Gamma(G, H, a)$, where $G = S_N$ and $a \in G$ are as defined earlier, and H is the subgroup of S_N generated by b and the m involutions $c_1, c_2, \ldots, c_m$.

Before discussing properties of this graph in Section 3, we make some observations about G, H and a. First, the subgroup generated by HaH contains b and a and is therefore primitive, and contains the single 2-cycle $c_m = (2m-1, 2m)$ and is therefore equal to S_N (by Jordan's theorem [5]). Next, the c_i are products of commuting transpositions t_j, and therefore generate an elementary abelian 2-group. Also for $1 \le k \le m$ the subgroup generated by $c_k, \ldots, c_m$ moves only the points $2k-1, 2k, \ldots, 2m-1, 2m$, and it follows that each such subgroup has order 2^{m-k+1}. From this we see that the subgroup H is a 2-group of order 2^{m+2}, and $H \cap a^{-1}Ha = \langle c_1, c_2, \ldots, c_m \rangle$ has index 4 in H.

3 Properties of the graphs

The graph Γ_n defined in Section 2 has $|G:H| = N!/2^{m+2}$ vertices, and is connected, 4-valent and arc-transitive, with S_N as an arc-transitive group of automorphisms (acting by right multiplication, with trivial kernel since A_N is simple). Also since the permutation $ab^2 = (1,3,2,4)(2m+1, N, 2m+2, 2m+3)$ has order 4, the graph $\Gamma_n = \Gamma(G,H,a)$ has a circuit of length 4 with vertices H, Ha, Hab^2a and Hab^2ab^2a, and thus Γ_n has girth 4.

The stabilizer in $G = S_N$ of the vertex H is the subgroup H itself, of order 2^{m+2}, and this group acts transitively as the dihedral group D_4 of order 8 on the neighbour-set $\Gamma_n(H)$. Also since every element $g \in G$ may be written in the form $g = hw$ where $h \in \langle c_1, c_2, \ldots, c_m\rangle$ and w is a word in the elements a and b, the subgroup $\langle a, b\rangle$ is transitive on vertices of Γ_n. On the other hand, the group S_N does not act transitively on 2-arcs (ordered paths of length 2) in Γ_n, because of the existence of circuits of length 4. Moreover, for $1 \le k \le m$ the subgroup of S_N generated by $c_k, \ldots, c_m$ is the stabilizer in S_N of the k-arc $(H(ab)^{[k/2]}, H(ab)^{[k/2]-1}, \ldots, Hab, H, Ha, Haba, \ldots, H(ab)^{[(k-1)/2]}a)$ of Γ_n, and this subgroup fixes all vertices at distance up to $[k/2]$ (but not all those at distance $[k/2]+1$) from H in Γ_n. These observations lead to the following:

Proposition. *The symmetric group S_N is the full automorphism group of Γ_n.*

Proof (Sketch). Assume the contrary, and let J be a subgroup of Aut Γ_n which properly contains S_N, and let K be the stabilizer in J of the vertex labelled H in Γ_n. By induction on the length of a stabilizer sequence, K is a 2-group, and local analysis shows $\theta \in K \setminus H$ may be chosen such that $\theta^{-1}a\theta = a$. It follows that θ normalizes $\langle H, a\rangle = G = S_N$, and then since Aut $S_N = S_N$ there must be a permutation $\sigma \in S_N$ corresponding to θ which normalizes H and centralizes a. But further local analysis shows $\theta^{-1}b\theta b^{-1} \in \langle c_m\rangle$, and therefore $\sigma^{-1}b\sigma = \theta^{-1}b\theta \in \{b, c_m b\}$. The case $\sigma^{-1}b\sigma = c_m b$ is impossible since b is even while $c_m b$ is odd, and therefore σ centralizes b. Consequently θ centralizes both a and b, from which it follows that θ fixes every vertex of Γ_n, a contradiction.

Since any vertex-transitive subgroup of Aut $\Gamma_n = S_N$ has index at most 2^{m+2} in S_N, but S_N has no proper subgroup of index less than N other than A_N, it follows that the smallest vertex-transitive group of automorphisms of Γ_n is the subgroup $\langle a, b\rangle = A_N$, in which the stabilizer of a vertex of Γ_n has order 2^{m+1}.

Acknowledgments

The first author gratefully acknowledges support from the Marsden Fund (grant number 95-UOA-MIS-0173), and the second author gratefully acknowledges support from a University of Auckland Postgraduate Scholarship.

References

1. Marston Conder: An infinite family of 5-arc-transitive cubic graphs, *Ars Combinatoria* **25A** (1988), 95–108.
2. Marston Conder and Peter Lorimer, Automorphism groups of symmetric graphs of valency 3, *J. Combinatorial Theory Ser. B* **47** (1989), 60–72.
3. R.C. Miller, The trivalent symmetric graphs of girth at most 6, *J. Combinatorial Theory Ser. B* **10** (1971), 163–182.
4. H.-O. Peitgen, H. Jürgens and D. Saupe, *Chaos and Fractals: New Frontiers of Science* (Springer-Verlag, 1992).
5. Helmut Wielandt, *Finite Permutation Groups* (Academic Press, New York, 1964).

A New Pumping Property of Context-Free Languages*

Pál Dömösi[1], Masami Ito[2], Masashi Katsura[2] and Chrystopher Nehaniv[3]

[1] Institute of Mathematics and Informatics, L. Kossuth University
Debrecen, Egyetem tér 1, H-4032, Hungary
e-mail: domosi@math.klte.hu

[2] Faculty of Science, Kyoto Sangyo University
Kyoto 603, Japan
e-mail: ito@ksuvx0.kyoto-su.ac.jp
katura@ksuvx0.kyoto-su.ac.jp

[3] School of Computer Science & Engineering
University of Aizu
Aizu-Wakamatsu City 965-80, Japan
e-mail: nehaniv@u-aizu.ac.jp

Abstract. A new *"pumping lemma"* is proven, providing a necessary condition for languages to be context-free.

1 Introduction

The following statement, a classical result on the class of context-free languages, is widely known as "Bar-Hillel's Lemma", or more precisely, "Bar-Hillel, Perles, and Shamir's Lemma" [2].

Theorem 1 (Bar-Hillel's Lemma [2]). *For every context-free language L we can give two natural numbers, p and q, such that each word in L that is longer than p has the form $uvwxy$ where $|vwx| \leq q, |vx| > 0$, and uv^iwx^iy is in L for all $i \geq 0$.* □

The above result is also called Classical Pumping Lemma, or simply, the Pumping Lemma and it has a number of extensions (see [1]-[4], [9]-[10], [12]-[14], [17]-[18], and [25]). It often can help to establish the fact that a language is not context-free but it is not guaranteed to do so. (Wise's result in [25] provides a necessary and sufficient condition for a language to be context-free.) Sometimes we shall use one of its above cited strengthened versions or other tools. (See, for example, [1, 7, 15, 16, 19, 20, 22, 24].)

* This work was supported by grants of the Hungarian National Foundation for Scientific Research (OTKA T019392, OTKA T7608), the Higher Education Research Foundation of Hungarian Ministry of Education (No. 222), Kyoto Sangyo University, and Grant-in-Aid for Scientific Research (No. 06640092), Ministry of Education, Science and Culture of Japan.

One can refine the Bar-Hillel Lemma by designating certain positions of the word z as *excluded*, i.e. forbidden for pumping in the sense the v and x may not contain any excluded positions, and by designating other positions as *distinguished*, where one is interested in the distribution of distinguished positions in the factorization $uvwxy$ of z. (Note here that the excluded and distinguished positions need *not* comprise disjoint sets). A great extension of the Bar-Hillel Lemma along these lines is the following result.

Theorem 2 (Strong Variant of Bader and Moura's Lemma [3, 13]). *Let L be a context-free language. There exists an integer $n \geq 2$ depending only on L such that, for any $z \in L$, if d positions are distinguished and e positions are excluded in z with $d > n^{e+1}$, then z can be written as $z = uvwxy$ so that*

1. *either each of u, v, w or each of w, x, y contains a distinguished position and vx contains no excluded position,*
2. *if d' and e' are the numbers of distinguished and excluded positions respectively in vwx, then $d' \leq n^{e'+1}$ and,*
3. *$uv^iwx^iy \in L$ for every $i \geq 0$.*

Using [1] or [3], it is easy to prove that the conclusion (2) can be slightly strengthened:

2'. if e' is the number of excluded positions in vwx, then $|vwx| \leq n^{e'+1}$.

In this paper we will demonstrate another pumping lemma providing a necessary condition for languages to be context-free.

For all notions and notations not defined here, see [3, 8, 9, 11, 21, 23]. An *alphabet* is a finite nonempty set. The elements of an alphabet are called *letters*. A *word* over an alphabet X is a finite string consisting of letters of X. For any alphabet X, let X^* denote the *free monoid* generated by X, i.e. the set of all words over X including the empty word λ and $X^+ = X^* \setminus \{\lambda\}$. The *length* of a word w, in symbols $|w|$, means the number of letters in w when each letter is counted as many times as it occurs. By definition, $|\lambda| = 0$. If u and v are words over an alphabet X, then their *catenation* uv is also a word over X. Especially, for any word uvw, we say that v is a *subword* of uvw. The set of all subwords of any word p is denoted by $Sub(p)$. A *language* over X is a set $L \subseteq X^*$. For any language L, we put $Sub(L) = \cup\{Sub(p)|p \in L\}$. We extend the concept of catenation for the class of languages as usual. Therefore, if L_1 and L_2 are languages, then $L_1L_2 = \{p_1p_2|p_1 \in L_1, p_2 \in L_2\}$. Let p be a word. We put $p^0 = \lambda$ and $p^n = p^{n-1}p(n > 0)$. Thus $p^k(k \geq 0)$ is the k-*th power* of p. If there is no danger of confusion, then sometimes we identify p with the singleton set $\{p\}$. Thus we will write p^* and p^+ instead of $\{p\}^*$ and $\{p\}^+$, respectively. An (*unrestricted generative*, or simply, *unrestricted*) *grammar* is an ordered quadruple $G = (V, X, S, P)$ where V and X are disjoint alphabets, $S \in V$, and P is a finite set of ordered pairs (W, Z) such that Z is a word over the alphabet $V \cup X$ and W is a word over $V \cup X$ containing at least one letter of V. The elements of V are called *variables* (or *nonterminals*) and those of X *terminals*. S is called the *start symbol*. Elements (W, Z) of P are called *productions* and are written $W \to Z$. If $W \to Z \in P$ implies $W \in V$ then G is called *context-free*. Especially, G is a *context-free grammar given in Chomsky normal form* if all productions are of the form $A \to BC$ or $A \to a$, where A, B, C are variables and a is a terminal. A word Q over $V \cup X$ *directly derives* a word R, in symbols, $Q \Rightarrow R$, if and only if there are words Q_1, Q_2, Q_3, R_1 such that $Q = Q_2Q_1Q_3, R = Q_2R_1Q_3$ and $Q_1 \to R_1$ belongs to P. Q *derives* R, or in symbols, $Q \overset{*}{\Rightarrow} R$ if and only if there is a finite sequence of words $W_0, \ldots, W_k (k \geq 0)$ over $V \cup X$ where $W_0 = Q, W_k = R$ and $W_i \Rightarrow W_{i+1}$ for $0 \leq i \leq k-1$. Thus for every $W \in (V \cup X)^*$ we have $W \overset{*}{\Rightarrow} W$. The *language* $L(G)$ *generated by* G is defined by $L(G) = \{w | w \in X^*, S \overset{*}{\Rightarrow} w\}$. $L \subseteq X^*$ is a *context-free language* if we have $L = L(G)$ for some context-free grammar G.

2 Main result

Theorem 3 (New Pumping Lemma). *For any context-free grammar G, there exists an effectively computable positive constant c depending only on G such that for any $z \in L(G)$, if $|z| \geq ce$ $(e > 0)$, and e positions of z are excluded, then z has the form $uvwxy$ where $|vx| > 0, vx$ does not contain any excluded positions, and uv^iwx^iy is in $L(G)$ for all $i \geq 0$.*

Proof. Consider a context-free grammar $G = (V, X, S, P)$ of the Chomsky normal form. If $|V| = 1$ then our statement holds trivially for $c > 1$. Assume $|V| > 1$ and let $c = |V|2^{|V|}$. Suppose that $|z| \geq ce$ for some $z \in L(G), e > 0$ and that e positions have been arbitrarily excluded in z. Let us denote T_z one of the derivation trees for z in G. If both children of a node in this tree have excluded descendants, call that node a *branch point*. The nodes and their children which have no excluded descendants are called *free*. Furthermore, a path is said to be *distinguished* if none of its intermediate nodes is a branch point (or it has no intermediate nodes), moreover, it starts with either a branch point or the root of the tree and ends with either a branch point or an excluded position. To prove our statement we will distinguish the following two possible cases:

Case I. Suppose that there exists a distinguished path with a repeated (nonterminal) label, say A, such that the upper one is not a branch point. (Then the upper one may be the initial node of the path only if it is the root of the tree, and simultaneously, this root is not a branch point.) Thus there exist two strings of terminals, v and x (which we call a pair of *pumping substrings*), and two nonterminals, B and C, such that $A \Rightarrow BC \overset{*}{\Rightarrow} vAx$. B and C cannot both dominate the lower A, therefore $|vx| > 0$. On the other hand, since there exists no intermediate branch point of the distinguished paths, we have that neither v nor x contains an excluded position. (Of course, the free children of the nodes of this path do not have excluded descendants.)

Case II. None of the distinguished paths has the above repetition of nonterminal labels. Since z has e excluded positions, T_z has exactly $e-1$ branch points.

Therefore we have not more than $2e-1$ distinguished paths and none of them has more than $|V|-1$ intermediate nodes. (Otherwise we could get our *Case I.*) Remember that only the root and these intermediate nodes of the distinguished paths may have free children. Thus the nodes of the distinguished paths have not more than $(2e-1)(|V|-1)$ free children altogether. But these free children are roots of the forest whose leaves cover all of the non-excluded positions. Therefore the free children in question have $(c-1)e$ non-excluded descendents at least. Since $\frac{(c-1)e}{(2e-1)(|V|-1)}$ $(=\frac{(|V|2^{|V|}-1)e}{(2e-1)(|V|-1)}) > 2^{|V|-1}$, at least one of the considered free children is a root of a tree containing more than $2^{|V|-1}$ leaves, where every leaf of this tree is a non-excluded position. But then this tree has a path of $|V|+1$ length (where the last node of this path has a terminal label). Therefore, there is a path in this tree with a repeated (nonterminal) label. Then we can take again an appropriate pair v, x of pumping strings as in the *Case I.*

The proof is complete. □

3 Further observations

Observe that by using a sufficiently large constant c in the proof of our new pumping lemma, we can obtain paths as in *case I* or *case II* having an arbitrarily given number of repetitions of nonterminal labels. Therefore, we have the following extension of our main result.

Proposition 4 (Extended version of our main result). *For every context-free grammar G, there exists an effectively computable positive constant c depending only on G such that for any $z \in L(G)$, if $|z| \geq ce$ $(e > 0)$, and e positions of z are excluded, then z has the form $uvwxy$ where u, v, w, x, y are nonempty words, vx does not contain any excluded position, and uv^iwx^iy is in $L(G)$ for all $i \geq 0$.* □

Take the grammar $G = (\{A, B, C, D, E, F, S\}, \{a, b, c\}, S, \{S \to AD, D \to EC, E \to AD, E \to FB, F \to FB, A \to a, B \to b, E \to b, F \to b, C \to c\})$. Obviously, $L(G) = \{a^m b^n c^m | m, n > 0\}$. For an arbitrary pair k, e of positive integers, consider the derivation tree T_z of the word $z = a^{ke} b^e c^{ke}$ of $L(G)$. (It is easy to see that our grammar is unambiguous. Therefore, T_z is determined unambiguously.) Consider an arbitrary pumping decomposition $uvwxy$ of z $(= a^{ke} b^e c^{ke})$ such that vx is a nonempty word which does not contain the letter b as subword. Since k and e are arbitrary positive integers, by taking $2k \geq c$ we obtain the following statement.

Proposition 5. *There exists a context-free grammar G such that, for any pair c, e of positive constants, there can be found a $z \in L(G)$ with $|z| \geq ce$ satisfying: we can exclude e positions of z in such a way that for every decomposition $uvwxy$ of z with $uv^iwx^iy \in L(G)$ and $|vx| > 0$ we have $|w| \geq e$ whenever vx does not contain any excluded position.*

Therefore, our pumping lemma cannot be extended to a generalization of Bar-Hillel's Lemma (in the sense of bounding the length of vwx by a constant). □

Proposition 6. *If G is given in Chomsky normal form then the length of vwx in Theorem 3 is bounded by $2^{max(e'|V|,\, 2|V|-1)}$, where e' denotes the number of excluded positions in vwx.*

Proof. If T_z has a free node with more than $2^{|V|-1}$ descendent leaves, then as in the usual proof of the Classical Pumping Lemma, the decomposition $uvwxy$ can be chosen so that vwx has no excluded positions and length of vwx can be bounded by $2^{|V|}$. (This includes *case II* of the proof of Theorem 3). Furthermore, if from any free node of T_z there is a path down of length $|V|+1$ or more, then the same is true.

Otherwise, we have *case I* and no such free nodes We may assume the upper node A of a distinguished path p as in *case I* to be of minimum height. We will bound the length of each path from A to a leaf, so we now consider only the subtree with root A (whose leaves are vwx).

By minimality, the length of each *case I* distinguished path starting at or below node A is not more than $|V|$ if its lowest node is a branch point (or $|V|+1$ if a leaf).

Now if a leaf letter a of vwx is not excluded, it is a descendant of a free node child of an intermediate node of some distinguished path. Hence by the condition on free nodes, the leaf a can be reached from node A by a path of length not more than $|V|+1+(|V|-1)=2|V|$. In particular this holds for all letters in v and x, and non-excluded letters of w.

If a is an excluded letter of w, then, again by minimality, among every $|V|+1$ consecutive variable nodes on the path from a up to the lower node in p there must be at least one branch point. Since there are only $e'-1$ branch points in the subtree of T_z under the upper node A, this path has length not more than $(e'-1)|V|+1$. Thus the path from excluded a to A has length no more than $|V|e'+1$.

It follows that the length of vwx is no more than $2^{max(e'|V|,\, 2|V|-1)}$. □

Problem 7. In fact, the above bound $2^{max(e'|V|,\, 2|V|-1)}$ can be replaced by the sharper bound $2^{e'|V|-1}+(|V|-1)2^{|V|-1}$ though we do not give the proof here. Could this upper bound for the length of the word vwx be improved? What is an optimal upper bound on $|vwx|$ in Proposition 6?

By Proposition 6, Theorem 3 and Bar-Hillel's Lemma, we have the following result.

Lemma 8 (Unified Pumping Lemma). *For every context-free grammar G given in Chomsky normal form, there exists an effectively computable positive constant c depending only on G and such that for any $z \in L(G)$, if $|z| \geq c \cdot max(e,1)$, and e positions of z are excluded then z has the form $uvwxy$ with $|vwx| \leq 2^{max(e',\, 2)\cdot|V|}$, where e' is the number of excluded positions of vwx, and $|vx|>0$, vx does not contain any excluded positions, and uv^iwx^iy is in $L(G)$ for all $i \geq 0$.*

Using our main result, we can give a new proof of the next result from [5]. (We also have a strengthened version of this statement [6], but it is not a straightforward consequence of our main result.)

Theorem 9 (see also [5]). *For any context-free language L there exists a positive integer k having the property that $p \in Sub(L)$ and $p \neq \lambda$ implies $qpr \in L$ for some pair q, r of words with $|qr| \leq k|p|$.*

Proof. Let us consider a context-free grammar $G = (V, X, S, P)$ with $L = L(G)$ and its constant c given in Theorem 3. Put $k = c - 1$ and let $qpr \in L$ for some $p \in Sub(L)$ and $q, r \in X^*$. If $|qr| \leq k|p|$ then we are done. Otherwise $qpr \geq c|p|$. Applying Theorem 3, exclude all positions of p in the word qpr. Then we can write the word qpr in the form uvw_1pw_2xy such that $|vx| > 0$ and $uv^iw_1pw_2x^iy \in L$ for all $i \geq 0$. Then we can consider the shorter word uw_1pw_2y instead of qpr. If $|uw_1w_2y| \leq k|p|$ then we are done. Otherwise use our previous treatment for uw_1 and w_2y instead of q and r. Continuing this procedure, we will get our result after some finite number of steps. □

Remark. The analogue of Theorem 9 does not hold true for the class of context-sensitive languages (see [6]).

Finally, combining our Unified Pumping Lemma with Bader and Moura's Lemma (and its modification (ii')), we have the following generalization of Bader and Moura's Lemma.

Theorem 10 (Generalized Bader-Moura Lemma). *Let L be a context-free language. There exists an integer $n \geq 2$ depending only on L such that, for any $z \in L$, if $|z| \geq n \cdot max(e, 1)$ and $e \geq 0$ positions of z are excluded, then z has the form $uvwxy$ with $|vwx| \leq n^{e'+1}$, where vwx has e' excluded positions, $|vx| > 0$, vx does not contain any excluded position, and uv^iwx^iy is in $L(G)$ for all $i \geq 0$.*

Moreover, if d positions have been distinguished in z, with $d > n^{e+1}$, then either each of u, v, w or each of w, x, y contains a distinguished position.

References

1. Bader, Ch., Moura, A., A generalization of Ogden's lemma, *Journ. of ACM,* **29** (1982), pp. 404–407.
2. Bar-Hillel, Y., Perles, M., and Shamir, E., On formal properties of simple phrase structure grammars, *Zeitschrift für Phonetik, Sprachwissenschaft, und Kommunikationsforschung,* **14** (1961), pp. 143–172.
3. Berstel, J., Boasson, L., Context-Free Languages, *in: Handbook of Theoretical Computer Sciences, Vol B: Formal Models & Semantics,* (J. van Leeuwen, ed.), Elsevier/MIT, 1994, pp. 60–102.
4. Boasson, L. and Horváth, S., On languages satisfying Odgen's lemma, *RAIRO Informatique Théorique,* **13** (1978), pp. 201–202.
5. Dömösi, P., Ito, M., On Subwords of Languages, *RIMS Proceedings, Kyoto Univ.,* 910 (1995), pp. 1–4.

6. Dömösi, P., Ito, M., Characterization of languages by lengths of their subwords, *Proc. Int. Conf. on Semigroups and Their Related Topics, Inst. of Math., Yunnan Univ., China* (submitted).
7. Ehrenfeucht, A., Rozenberg, G., On the separating power of EOL systems, *RAIRO Informatique Théorique* **17** (1983) pp. 13 -22.
8. Ginsburg, S., *The Mathematical Theory of Context-Free Languages,* McGraw-Hill, New York, 1966.
9. Harrison, M. A., *Introduction to Formal Language Theory,* Addison-Wesley Publishing Company, Reading, Massachusetts, 1978.
10. Hewett, R., Slutzki, G., Comparisons between some pumping conditions for context-free languages, *Math. Systems Theory,* **21** (1989), pp. 223–233.
11. Hopcroft, J.E. & Ullmann, J.D., *Introduction to Automata Theory, Languages, and Computation,* Addison-Wesley, Reading, Massachusetts, 1979.
12. Horváth, S., BHFL: the family of languages satisfying Bar-Hillel's lemma, *RAIRO Informatique Théorique,* **12** (1978), pp. 193–199.
13. Horváth, S., A comparison of iteration conditions on formal languages, *Proc. Colloq. Algebra, Combinatorics and Logic in Comp. Sci.,* Győr, Hungary, 1983, Coll. Math. Soc. J. Bolyai, **42**, J. Bolyai Math. Soc., Budapest, and North-Holland, Amsterdam, 1986, pp. 453–463.
14. Kløve, T., Pumping languages, *Int. Journ. of Comp. Math.,* **6** (1977), pp. 115–125.
15. van Leeuwen, J., A generalization of Parikh's theorem in formal language theory, *Proc. 2nd Colloq. on Automata, Languages and Programming,* Springer Lecture Notes in Computer Science, **14** (1974), pp. 17–26.
16. ijholt, A., A note on the sufficiency of Sokolowski's criterion for context-free languages, *Inf. Proc. Lett.,* **14** (1982), p. 207.
17. Nijholt, A., An annotated bibliography of pumping, *Bull. EATCS,* **17** (June, 1982), pp. 34–52.
18. Odgen, W., A helpful result for proving inherent ambiguity, *Math. Syst. Theory,* **2** (1968), pp. 191–194.
19. Ogden, W., Ross, R. J., Winklmann, K., An "Interchange Lemma" for context-free languages, *SIAM J. Comp.* **14** (1985) pp. 410–415.
20. Parikh, R. J., On context-free languages, *J. Assoc. Comput. Mach.* **13** (1966) pp. 570–581.
21. Révész, Gy. E., *Introduction to Formal Languages,* Dover, New York, 1991.
22. Ross, R., Winklmann, K., Repetitive strings are not context-free, *RAIRO Informatique Théorique,* **16** (1982) pp. 191–199.
23. Salomaa, A., *Formal Languages,* Academic Press, New York, London, 1973.
24. Sokolowski, S., A method for proving programming languages non context-free, *Inf. Proc. Lett.* **7** (1978), pp. 151–153.
25. Wise, D. S., A strong pumping lemma for context-free languages *Theor. Comp. Sci.,* **3** (1976), pp. 359–369.

The Parameterized Complexity of Relational Database Queries and an Improved Characterization of $W[1]$

Rod G. Downey[1]*
Michael R. Fellows[2]** and Udayan Taylor[2]

[1] Department of Mathematics,
Victoria University
P.O. Box 600
Wellington, New Zealand
rod.downey@vuw.ac.nz

[2] Department of Computer Science
University of Victoria
Victoria, B.C. V8W 3P6, Canada
{mfellows,udi}@csr.uvic.ca

Abstract. It is well known that for a fixed relational database query ϕ in m free variables, it can be determined in time polynomial in the size n of the database whether there exists an m-tuple x that belongs to the relation defined by the query. For the best known algorithms, however, the exponent of the polynomial is proportional to the size of the query. We study the data complexity of this problem parameterized by the size $k = |\phi|$ of the query, and answer a question recently raised by Yannakakis [Yan95]. Our main results show: (1) the general problem is complete for the parametric complexity class $AW[*]$, and (2) when restricted to monotone queries, the problem is complete for the fundamental parametric complexity class $W[1]$. The practical significance of these results is that unless the parameterized complexity hierarchy collapses, there are unlikely to be algorithms that solve this problem (even under the restriction to monotone queries) in time $f(k)n^c$ where f is an arbitrary function of k and c is a constant independent of k.
An important consequence of the proof of (2) is a significantly improved characterization of the parameterized complexity class $W[1]$. Previous results by Downey and Fellows characterize $W[1]$ in terms of the k-Weighted Circuit Satisfiability problem, for families of circuits that satisfy: (1) the depth of the circuits is bounded by a constant c, (2) on any input-output path there is at most one gate having unbounded fan-in (termed a *large* gate), with all other gates having fan-in bounded by c (that is, *small* gates). We show that the definition can be broadened by allowing circuits of depth bounded by an arbitrary function $f(k)$. If we denote this parameterized complexity class $W^*[1]$, then our corollary

* Research supported by the New Zealand Marsden Fund for Basic Science.
** Research supported by the National Science and Engineering Research Council of Canada.

is: $W^*[1] = W[1]$. As a consequence, the k-WEIGHTED SATISFIABILITY problem for Boolean expressions that are products of k-sums is complete for $W[1]$. Part of our argument establishes a general normalization lemma for $W^*[t]$ for all t.

1 Introduction

Many natural computational problems have input that consists of several elements of information. It is natural to consider some of these elements as a *parameter* and evaluate their relative contribution to the overall complexity of the problem. In many practical applications of computational problems only a small set of parameter values may be significant. In these cases, complexity that is exponential in the parameter, but otherwise a fixed polynomial, may be acceptable.

A framework in which to study this complexity issue has been developed in [DF95a, DF95b, DF95c, ADF95] and applied to problems in diverse areas of computer science [BDFW95, BDFHW95, BF95, BFH94, CCDF94, CW95, DEF93, DFHKW94, FK93]. The framework of parameterized computational complexity theory allows one to investigate how different aspects or different parts of the input to a problem may contribute differentially to overall problem complexity. For example, both of the well-known graph problems VERTEX COVER and DOMINATING SET take as input a graph G and a positive integer k. VERTEX COVER can be solved in time $O(f(k)n^c)$ where c is a constant independent of k, and n is the number of vertices of G, a complexity behavior termed *fixed-parameter tractability* in the theory. (For VERTEX COVER we may take $f(k) = (4/3)^k$ and $c = 1$ [BFR95].) The best that is known about DOMINATING SET is essentially that it can be solved by brute force (trying all k-subsets) in time $O(n^{k+1})$.

A parameter may also be some aggregate of information. For example, if for the above two problems the parameter describes not only the size of the vertex set, but also the treewidth of G, then both problems are fixed-parameter tractable. The theory has been shown to address many natural problems in diverse areas of computer science, and it must be regarded as one of the principal "applied" complexity theories [BDFW95, BDFHW95, BF95, BFH94, CCDF94, CW95, DEF93, DF95a, DFHKW94, FK93].

Part of the theory of parameterized complexity is concerned with general methodologies for demonstrating parametric tractability. On the other hand, as with classical complexity, if we cannot seem to be able to demonstrate parametric tractability, then we would like to prove, or at least give strong evidence for, parametric *intractability*. As we will see in the next section, to this end, Downey and Fellows developed a hierarchy of classes of parameterized languages called the W-hierarchy, and has the form

$$W[1] \subseteq W[2] \subseteq \ldots W[SAT] \subseteq W[P] \subseteq AW[1] = AW[*] \subseteq AW[P] \subseteq XP.$$

The parameterized complexity class $W[1]$ is particularly interesting and important, for several reasons. First of all, a demonstration of hardness for $W[1]$

can be regarded as the principal "working" reference point for likely fixed-parameter intractability for the reason that it provides a convincing "minimum" demonstration. Secondly, there is a substantial list of very natural and important problems that are precisely $W[1]$-complete, including: CLIQUE, INDEPENDENT SET, VAPNIK-CHERVONENKIS DIMENSION, MONOTONE DATA COMPLEXITY FOR RELATIONAL DATABASES (from this paper), SQUARE TILING, k-STEP DERIVATION FOR CONTEXT SENSITIVE GRAMMARS, m-LENGTH COMMON SUBSEQUENCE FOR k SEQUENCES, k-LENGTH POST CORRESPONDENCE, and the k-STEP HALTING PROBLEM FOR SINGLE-TAPE NONDETERMINISTIC TURING MACHINES (SHORT TURING MACHINE ACCEPTANCE [Ces96, DEF93, DF95b, DFHKW94, DFT96]. The last result is of particular significance since it is kind of a miniaturization of Cook's Theorem. The class $W[1]$ the the calls of parameterized languages which have the complexity of the problem below.
SHORT TURING MACHINE ACCEPTANCE
Input: A nondeterministic Turing machine M.
Parameter: A positive integer k.
Question: Does M have a computation path accepting the empty string in $\leq k$ steps?

However in the case of classical TURING MACHINE ACCEPTANCE, (i.e. the problem above with no parameter, and asking if the Turing Machine accepts in $\leq |M|^n$ steps for some fixed n) the point is that a nondeterministic Turing machine is such an opaque and generic object that it simply does not seem reasonable that we should be able to decide if a given one on a given input has some accepting path without looking at the paths. Cook's Theorem therefore gives powerful evidence that $P \neq NP$. It seems to us that if one accepts the philosophical argument that TURING MACHINE ACCEPTANCE is intractable then the same reasoning would suggest that SHORT TURING MACHINE ACCEPTANCE is *fixed parameter* intractable. Surely we would need to look at all the paths of length k. In this way, it is natural to regard the class $W[1]$ which has the problem SHORT TURING MACHINE ACCEPTANCE complete for it, as being a hardness measure for parametric intractability.

In this paper we report on applications of this perspective to problems of database theory, a line of investigation recently proposed by Yannakakis [Yan95]. We prove the following theorems.

Theorem 1. *The* QUERY NON-EMPTINESS *problem for relational databases is complete for the parameterized complexity class* $AW[*]$.

Theorem 2. *The* MONOTONE QUERY NON-EMPTINESS *problem for relational databases is* $W[1]$ *complete.*

The main Technical Lemma establishes a significantly broadened logical characterization of $W[1]$ than that proved in [DF95b]. The importance of this result is that it makes it easier to prove "upper bounds" on the complexity of problems with respect to the W hierarchy. In the present paper we use this characterization to prove $W[1]$ membership for the emptiness problem for monotone queries to relational databases, which together with the hardness arguments there, shows that this very natural problem is $W[1]$-complete.

Previous results by Downey and Fellows characterize $W[1]$ in terms of the k-WEIGHTED CIRCUIT SATISFIABILITY problem, for families of circuits that satisfy: (1) the depth of the circuits is bounded by a constant c, (2) on any input-output path there is at most one gate having unbounded fan-in (termed a *large* gate), with all other gates having fan-in bounded by c (that is, *small* gates). We show that the definition can be broadened by allowing circuits of depth bounded by an arbitrary function $f(k)$. If we denote this parameterized complexity class $W^*[1]$.

Theorem 3. $W[1] = W^*[1]$

As a consequence, the k-WEIGHTED SATISFIABILITY problem for Boolean expressions that are products of k-sums is complete for $W[1]$. Part of our argument establishes a general normalization lemma for $W^*[t]$ for all t. We suspect that Theorem 3 can be extended to the entire $W[t]$-hierarchy ($W^*[t] = W[t]$). As a step in the direction of this conjecture, we prove an important normalization theorem for $W^*[t]$, analogous to results concerning $W[t]$ in [DF95a].

In the next section we give a brief introduction to the formal framework of parameterized complexity theory. In §3 we introduce the terminology of relational database queries and prove Theorem 1. In §4 we prove Theorem 2, assuming Theorem 3. Finally in §5 and §6 we prove the extended "normalization" theorem, and prove the crucial technical Theorem 3. The final section discusses some related open problems.

2 Background on parameterized complexity

The formal framework is sketched as follows. We consider that input to a computational problem consists of two parts, one of which is expected to be relatively small. Thus we consider a *parameterized language* L to be a set of pairs of strings, $L \subseteq \Sigma^* \times \Sigma^*$. For notational convenience, we may equivalently consider a parameterized language L to be a subset of $\Sigma^* \times \mathbb{N}$, where for $(x, k) \in L$, k is the *parameter*. Note that a parameterized language can always be considered "classically" as well.

The fundamental concept of the theory is the notion of *fixed-parameter tractability*. We say that a parameterized language L is *fixed parameter tractable* (FPT) if it can be determined in time $f(k)n^c$ whether $(x, k) \in L$, where c is a constant independent of the parameter k and $n = |x|$ is the size of x. There are many examples of parameterized problems that are FPT, yet when considered classically are *NP*-complete, *PSPACE*-complete or even undecidable. A corresponding notion of problem reduction is defined as follows[3].

[3] Definition 4 is not the most general notion of parametric reduction, but is sufficient for our purposes. We refer the reader to [DF95a, DFta] for more details

Definition 4. Let A, B be parameterized decision problems. We say that A is (standardly) *reducible* to B if there is an algorithm Φ which transforms (x, k) into $(x', g(k))$ in time $f(k)|x|^\alpha$, where $f, g : N \to N$ are arbitrary functions and α is a constant independent of k, so that $(x, k) \in A$ if and only if $(x', g(k)) \in B$.

It is easy to see that if A reduces to B and B is fixed parameter tractable then so too is A.

Note that if $P = NP$ then a variety of natural parameterized *NP*-complete problems would be fixed-parameter tractable. Thus a completeness program is reasonable for establishing apparent fixed-parameter intractability. We remark that parametric reductions are much "finer" than classical ones, and hence instead of the classical observation that "normal" problems in *NP* being either in P or *NP* complete, there tend to be a large collection of hardness classes. The definitions to follow establish the basic parametric complexity classes. While they seem a little involved thc reader should keep in mind that the basic problems complete for the classes are simply analogues of the classical SATISFIABILITY problem. The parameterized version(s) of SATISFIABILITY is based upon the notion of *weight*, where the *weight* of a binary string is the number of 1's in it. We ask if a formula has a satisfying assignment of fixed weight k, which is the parameter. The w-hierarchy below is based on the belief that WEIGHTED CNF SATISFIABILITY and WEIGHTED 3SAT are of differing complexity, and the underlying reason for this differing complexity is the difference in logical depth of the formulae. Thus if CNF is thought of as products-of-sums of literals, then we can define a propositional formula to be *t-normalized* as follows.

Definition 5. We say that a propositional formula X is *t-normalized* if X is of the form products-of-sums-of-products... of literals with t-alternations.

Thus we see according to Definition 5, 2-normalized is the same as CNF. Now we believe that for all t, WEIGHTED t-NORMALIZED SATISFIABILITY, that is, satisfiability for t-normalized formulae, is strictly easier than WEIGHTED $t+1$-NORMALIZED SATISFIABILITY. This suggests a hierarchy based on the complexity of t-normalized formulae. We can consider t-normalized formula as special examples of certain classes of circuits, namely the weft t boolean circuits below.

Definition 6. A Boolean circuit is of *mixed type* if it consists of circuits having gates of the following kinds.

1. *Small gates*: $\neg$ gates, $\wedge$ gates and $\vee$ gates with bounded fan-in. We will assume that the bound on fan-in is 2 for *and* gates and *or* gates, and 1 for *not* gates.
2. *Large gates*: $\wedge$ gates and $\vee$ gates with unrestricted fan-in.

Definition 7. The *depth* of a circuit C is defined to be the maximum number of gates on an input-output path in C. The *weft* of a circuit C is the maximum number of large gates on an input-output path in C.

Definition 8. Let F be a family of decision circuits. We allow that F may have many different circuits with a given number of inputs. To F we associate the parameterized language

$$L_F = \{(C,k) : \exists \text{ an input vector } x \text{ of weight } k \text{ accepted by } C\}$$

Definition 9. A parameterized language L belongs to $W[t]$ if L reduces to $L_{F(t,h)}$ for the family $F(t,h)$ of mixed type decision circuits of weft at most t, and depth at most h, for some constant h.

In the above definition, the weight k vector x is governed by a single existential quantification. This can be generalized.

Definition 10. A k-structured circuit C is a circuit with inputs partitioned into k disjoint sets $I_1, ..., I_k$, each with an associated logical quantifier $Q_i \in \{\exists, \forall\}$ for $i = 1, ..., k$.

Definition 11. In a manner analogous to the definition of L_F we define

$$\begin{aligned} L_F^A = \{(C,k) &: C \text{ is a } k\text{-structured circuit} \\ &Q_1 i_1 \in I_1,\ Q_2 i_2 \in I_2,\ \cdots Q_k i_k \in I_k, \text{ such that} \\ &\text{the weight } k \text{ input } x \text{ with } x_{i_1} = x_{i_2} = \cdots = x_{i_k} = 1 \text{ is accepted by } C\} \end{aligned}$$

Definition 12. A parameterized language L belongs to $AW[t]$ if L reduces to $L^A_{F(h,t)}$ for the family $F(h,t)$ of structured circuits of weft at most t and depth at most h.

This gives us the following hierarchy of parameterized complexity classes:

$$FPT \subseteq W[1] \subseteq W[2] \subseteq \cdots \subseteq AW[1] \subseteq AW[2] \subseteq \cdots$$

In [ADF95] and [DFR96] it is shown that the $AW[t]$ hierarchy collapses to $AW[1]$. In [DF95a] the basic Normalization Theorem is proven:

Theorem (Basic Normalization Theorem). *For all* $t \geq 1$, WEIGHTED t-NORMALIZED SATISFIABILITY *is complete for* $W[t]$.

We will improve upon the Normalization Theorem in the present paper[4].

---

[4] The reader who is unfamiliar with parametric complexity might wonder if there are some connections with classical notions. There are two answers to this yes and no! In some sense, the theory is *orthogonal* to classical complexity since we are dealing with parameterized languages. Thus we are *definitely not* dealing with something inside of, say NP or $PSPACE$. For instance if A is *any* language however classically difficult, the language $A^P = \{(x,x) : x \in A\}$ is FPT but clearly has the same classical complexity as A. However, the event that, say, $W[1] = FPT$ does have classical consequences. We illustrate that for $W[P]$ and refer the reader to [ADF95] for further details. $NP[f(n)]$ denotes the class of decision problems that are solvable in P-time with an algorithm using only $f(n)$ bits of nondeterminism. Abrahamson, Downey and Fellows defined

3 Parametric data complexity of relational queries

We assume the reader is familiar with the basic ideas of relational database theory, as for example in [Ul88].

A relational database is essentially just a finite set of finite relations on a finite set U. Each relation can be viewed as a set of tuples listed in a table for the relation. The size of a database is the length of a description of all the relations by listing the contents of the relational tables.

Formally, there is a finite set U of attributes, each with an associated domain. A *relational schema* R is a subset of U. A *database schema* $D = (R_1, ..., R_n)$ is a set of relation schemas. A relation r_i over a relation schema R_i is a set of R_i-tuples, where a R_i-tuple is a mapping from the attributes of R_i to their domains. A database d over the schema D is a set $\{r_1, ..., r_n\}$ of finite relations.

We consider here the complexity of queries formulated in the *relational calculus*. Formulas of the calculus are built from atomic formulas of the form $x_1 = x_2$ and $R_i(x_1, ..., x_m)$, where the x_i are variables or constants of the appropriate domains, using Boolean connectives and quantifiers. A formula ϕ with free variables $y_1, ..., y_k$ specifies a query relation

$$r_\phi = \{(a_1, ..., a_k) : \phi(a_1, ..., a_k) \text{ is true in } d\}$$

A formula will be said to be in *normal form* if all quantifiers are up front and the body of the formula is in conjunctive normal form. For example, the query

$$\phi = \exists a \forall b \forall c\, (R_1(a, b, d) \vee \neg R_2(a, c) \vee \neg R_1(a, c, e)) \wedge (\neg R_1(a, d, e) \vee R_2(d, e))$$

is in normal form.

Various problems may be considered concerning the complexity of computations associated with a query ϕ. The following problem is simply stated, and a number of other problems can be easily reduced to it. (For example, the problem of determining whether a target tuple x belongs to the relation defined by the query is easily reduced to the non-emptiness problem.)

QUERY NON-EMPTINESS
Input: A database d, and a query ϕ.

---

the class $SUBEXPTIME(f(n))$ to be the set of languages L for which there is a polynomial $p(n)$ such that L is accepted in $DTIME(p(n)2^{g(n)})$ for some function g in $o(f(n))$. (That is, $\lim_{n\to\infty} f(n)/g(n) = \infty$.) They established the following result connecting parameterized and classical complexity.

Theorem (Abrahamson, Downey and Fellows).

1. *For the standard reducibility, $W[P] = FPT$ iff for every P-time function f with $f(n) \geq \log n$, there is a recursive function h such that for every $L \in NP[f(n)]$ $h(L)$ computes machine M which witnesses that $L \in SUBEXPTIME(f(n))$. (We say that $NP[f(n)]$ is constructively a subset of $SUBEXPTIME(f(n))$.)*
2. *Consequently, for the standard reducibility, $W[P] = FPT$ implies that for all P-time f, $NP[f(n)] \subseteq SUBEXPTIME(f(n))$.*

Parameter: $|\phi|$
Question: Is the query relation r_ϕ nonempty?

Proposition 13. QUERY NON-EMPTINESS *is in* $AW[1]$.

Proof. At the cost of a blow-up exponential in the size of ϕ we may assume that ϕ is in the normal form:

$$\phi = Q_1 x_1 Q_2 x_2 \cdots Q_r x_r E$$

where the the Q_i are logical quantifiers and E is in conjunctive normal form. Using multiplicative notation for convenience:

$$E = \prod_{i=1}^{l} \sum_{j=1}^{m} E_{ij}$$

where each E_{ij} is either a relational atomic formula or its negation. Let $y_1, ..., y_s$ denote the free variables of ϕ, and let U denote the base set over which the database d is defined, $|U| = n$.

We describe an appropriate k-structured circuit C to exhibit membership in $AW[4]$. Since the AW hierarchy collapses to $AW[1]$, this is sufficient for our purposes. The parameter k is obliged to be purely a function of ϕ. We may take $k = r + s$, the number of variables (both free and bound) occuring in ϕ.

The inputs to C are partitioned into k blocks of size n. Let J_i for $i = i, ..., r$ denote the blocks corresponding to the r bound variables of ϕ, and let J_i' for $i = 1, ..., s$ denote the blocks corresponding to the free variables of ϕ. Let

$$J_i = \{v[i, u]; u \in U\} \text{ for } i = 1, ..., r$$

and let

$$J_i' = \{w[i, u] : u \in U\} \text{ for } i = 1, ..., s$$

To each block J_i for $i = 1, ..., r$ we associate the quantifier Q_i. To each block J_i' for $i = 1, ..., s$ we associate the existential quantifier.

The circuit C for our reduction corresponds to the logical expression:

$$F = F_1 \cdot F_2 \cdot F_3$$

where

$$F_1 = \prod_{i=1}^{r} \prod_{u,u' \in U,\ u \neq u'} (\neg v[i, u] + \neg v[i, u'])$$

$$F_2 = \prod_{i=1}^{s} \prod_{u,u' \in U,\ u \neq u'} (\neg w[i, u] + \neg w[i, u'])$$

and

$$F_3 = \prod_{i=1}^{l} \sum_{j=1}^{m} F_{ij}$$

where the subexpressions F_{ij} correspond to the relational formulas (or negations of these) E_{ij} of E in the following way (which we will explain by example).

Suppose $E_{ij} = R(x_2, x_3, y_5)$. Then

$$F_{ij} = \sum_{(u,u',u'')\in R} (v[2,u] \cdot v[3,u'] \cdot w[5,u''])$$

The expression F_{ij} is thus simply a sum-of-products expression that evaluates R in the blocks corresponding to the variables occuring in the atomic relational formula (or its negation, as the case may be).

The reader can easily check that this construction works correctly, as it is little more than a notational translation. The circuit C has depth and therefore weft 4. Since $AW[4] = AW[1]$ we are done.

Proposition 14. QUERY NON-EMPTINESS *is hard for* $AW[*]$.

Proof. We reduce from the problem SHORT NODE KAYLES that is shown to be complete for $AW[*]$ in [ADF95]. This problem takes as input a graph $G = (V, E)$. The parameter is a positive integer k, and the question is whether the first player has a k-move winning strategy in the following game. Players alternate pebbling a vertex that is not adjacent to any previously pebbled vertex. The first player with no move loses.

The database d for our reduction has as a base set the vertex set of the graph, and it has two relations: (1) the binary adjacency relation E of the graph (where we consider also that a vertex is adjacent to itself), and (2) a unary padding relation $T = \{(u) : u \in V\}$.

It is straightforward to concoct a query that will define a non-empty relation if and only if the first player has a winning strategy. For example, the following query will serve when $k = 3$ and is easily generalized to arbitrary k.

$$\phi = \exists x_1(\forall y_1(E(x_1,y_1) \vee \exists x_2(\forall y_2(E(x_1,y_2) \vee E(x_2,y_2)\vee$$

$$\exists x_3(\forall y_3(E(x_1,y_3) \vee E(x_2,y_3) \vee E(x_3,y_3))))))) \wedge T(x)$$

The only free variable of ϕ is x. The unary relation defined by ϕ coincides with T when the first player has a 3-move winning strategy, and is otherwise empty.

Note that the size of ϕ depends only on k.

Together, Propositions 13 and 14 prove the theorem:

Theorem 1. QUERY NON-EMPTINESS *is* $AW[1]$*-complete.*

4 Monotone queries

In this section we consider the complexity of the restriction of the QUERY NON-EMPTINESS problem to monotone queries. A query ϕ is termed *monotone* if it has: (1) only existential quantifications, and (2) no negations.

Proposition 15. MONOTONE QUERY NON-EMPTINESS *belongs to* $W[1]$.

Proof. Use the same circuit construction as in the proof of Proposition 13. In this case, there are no blocks of input variables (as the circuit is now "ordinary" as opposed to k-structured). The subexpressions F_1 and F_2 enforce, however, that in any weight k input vector accepted by the circuit, there is exactly one input variable in each block with value 1, which is equivalent to an implicit existential quantification, as required for the correctness of the reduction from a monotone query.

An examination of the circuit C shows that on any input-output path there is at most one gate with unbounded fan-in, all other gates having fan-in bounded by a function of k. To complete the proof, we will need the following technical result:

Theorem 16 ($W[1]$ Characterization). *In the definition of* $W[1]$ *in terms of circuits, the definition of small gates can be relaxed to allow gates of fan-in* $f(k)$, *where* f *is an arbitrary function of the parameter* k.

(As we will see in Sections 5 and 6, Theorem 16 is just a restatement of Theorem 3, that $W[1] = W^*[1]$.)

By this result, we have established membership in $W^*[1] = W[1]$.

Proposition 17. MONOTONE QUERY NON-EMPTINESS *is hard for* $W[1]$.

Proof. A reduction from CLIQUE is straightforward and can be sketched as follows. Let $G = (V, E)$ be the graph for which we wish to decide on the presence of a k-clique. Let V be the base set of the database d that has binary relation R recording the edges of the graph. Order V arbitrarily (denoted $<$), and let:

$$R = \{(u, v) : u < v \text{ and } uv \in E\}$$

The database d also has a "padding" unary relation T:

$$T = \{(u) : u \in V\}$$

The query to determine whether there is a k-clique is just:

$$\phi = \exists x_1 \exists x_2 \cdots \exists x_k R(x_1, x_2) \wedge R(x_1, x_3) \wedge \cdots \wedge R(x_{k-1}, x_k) \wedge T(x)$$

Note that the size of ϕ is a function only of k.

The only free variable in ϕ is x. If there is a k-clique in G, then the unary relation R_ϕ defined by ϕ is:

$$R_\phi = \{(v) : v \in V\}$$

and if there is no k-clique in G then $R_\phi = \emptyset$. The ordering of V used in defining R implicitly forces a successful instantiation of the existentially quantified variables to represent k distinct vertices of G. Note that ϕ is monotone.

Remark. The expression of the CLIQUE problem as a database query problem is really quite straightforward. One can easily work out a similar representation for the DOMINATING SET problem — but not as a monotone query.

Propositions 15 and 17 together prove:

Theorem 2. MONOTONE QUERY NON-EMPTINESS *is complete for* $W[1]$.

5 W^* normalization

Now we turn to our improvements to the Normalization Theorem of [DF95a], and the proof of Theorem 3. We need a definition.

Definition 18. Fix a function f and consider a gate to be *small if it has fan-in $\leq f(k)$. The definition of *small allows us, as in §2 to define a notion of *weft. Then we say that a parameterized language L belongs to $W^*[t]$ if L reduces to $L_{F^*(t,h)}$ for the family $F^*(t,h)$ of mixed type decision circuits of *weft at most t, and depth at most $h(k)$, for some arbitrary function h. (Note that if we wish to consider families of circuits where *small* gates may have fan-in $f(k)$, then we can equivalently substitute a faster-growing function h.)

It is clear that $W^*[t]$ might well be a substantially larger class of parameterized languages than $W[t]$, and that trivially $W[t] \subseteq W^*[t]$.

In this section we sketch the proof of an important lemma concerning $W^*[t]$ that says essentially that the circuits can be put in a normal form that corresponds to a certain kind of Boolean formula. The normal form is defined recursively as follows.

Definition 19. A Boolean expression E is in 1-alternating normal form with respect to n and k if E is either an n-product of k-sums or an n-sum of k-products. A Boolean expression E is in t-alternating normal form with respect to n and k, for $t \geq 2$, if E either is of the form:

$$E = \prod_{i=1}^{n} \sum_{j=1}^{k} E_{ij}$$

or of the form:

$$E = \sum_{i=1}^{n} \prod_{j=1}^{k} E_{ij}$$

where each subexpression E_{ij} is $(t-1)$-alternating normal with respect to n and k.

Note that the above definition can be equivalently phrased in terms of circuits in a natural way. Define a tree circuit C to be in *t-alternating normal form with respect to n and k* if it has $2t$ layers of gates, with the fan-in alternating between n and k starting with fan-in n for the output gate.

Corresponding to this form we have the following parameterized problem.
t-ALTERNATING WEIGHTED SATISFIABILITY
Input: A Boolean expression E in t-alternating normal form with respect to n and k
Parameter: k
Question: Is there a weight k truth assignment that satisfies E?

Lemma 20 (Normalization). t-ALTERNATING WEIGHTED SATISFIABILITY *is complete for $W^*[t]$ for all $t \geq 1$.*

Proof (sketch). Assume that the given circuit C has an output large gate, since an output small gate subcircuit can be removed by employing additional nondeterminism in the manner used in [DF95a]. The idea is basically to progressively analyze and "copy" the circuit C into the required form, working progressively downward from the input level. In the first step, we locate the topmost large gates. Let g denote such a gate, and let $a_1, ..., a_r$ denote the inputs to g. Suppose g is a $\wedge$ gate. Each a_i is computed by a subcircuit consisting only of small gates, and therefore there the number of inputs to C on which a_i depends is bounded by a function of k. Thus, a_i can be re-expressed (at a cost that is bounded by a function of k) as a product-of-sums of total size bounded by a function of k. Thus g can be replaced by a product-of-sums "copy-gate" g', where there are at most $f(k)n$ sums (for some function f) and each sum has size bounded by some function $g(k)$. Make a copy of g' for each fan-out line of g.

For the case where g is a large $\vee$ gate, we replace each argument a_i with an equivalent sum-of-products, to obtain (for g') an $f(k)n$-sum of $g(k)$-products.

We next identify the topmost layer of large gates *below* the inputs to C and the copy gates created so far. Note that each copy gate has fan-out 1. We repeat the above recipe, creating two more layers of copy gates ($f'(k)n$-products of $g'(k)$-sums and $f'(k)$-sums of $g'(k)$-products, for new functions f' and g'). Eventually the entire circuit C is replaced by the copy gates, and the resulting copy gate circuit has the required form.

6 An improved characterization of $W[1]$

In this section we prove the main technical result, Theorem 3.

Theorem 3. $W^*[1] = W[1]$.

Proof. By the Normalization Lemma, we just have to show that 1-ALTERNATING WEIGHTED SATISFIABILITY is in $W[1]$. The only issue is for expressions that are n-products of k-sums, since the case of n-sums of k-products is in FPT. Let E denote the Boolean expression. We view E as consisting of a set $\mathcal{C}$ of clauses

$$\mathcal{C} = \{C_1, ..., C_n\}$$

where each clause is a set of k literals. Let X denote the set of variables of E,

$$X = \{x_1, \ldots, x_m\}$$

Our argument consists of a description of how to produce a Boolean expression E' over a set of variables X' that is a product of sums of size bounded by a constant, such that E' is satisfiable by a weight $k' = k + 2^{k+1}$ truth assignment if and only if E is satisfiable by a weight k truth assignment. By the results of [DF95b] this is sufficient to conclude the theorem.

Let $X_i \subseteq X$ denote the set of variables occurring in the clause C_i. Let X_i^+ denote the variables that occur positively in C_i and let X_i^- denote the set of variables that occur negated in C_i. (Below we will make the trivial assumption that $X_i^+ \cap X_i^- = \emptyset$).

Suppose $X_i^- = \{x_1, x_3, x_5, x_8\}$ The sequence (family) of increasingly larger sets beginning with the empty set, and progressively including the variable of least index among those remaining, we will call the ith *trail* $\mathcal{T}_i$. Thus for our example X_i^-, we would have

$$\mathcal{T}_i = \{\emptyset, \{x_1\}, \{x_1, x_3\}, \{x_1, x_3, x_5\}, \{x_1, x_3, x_5, x_8\}\}.$$

Let $\mathcal{T}$ be the union of the trails

$$\mathcal{T} = \bigcup_{i=1}^{n} \mathcal{T}_i$$

For $T \in \mathcal{T}$ the set X_T of *successors* of T is the set of variables $x = x_j$ such that:

1. $x_j \notin T$,
2. for all $x_i \in T$, $j > i$, and
3. $T' = T \cup \{x_j\} \in \mathcal{T}$.

Thus if x is a successor of T, then it will have the largest index in T'. Note also that the size of $\mathcal{T}$ is bounded by a quadratic polynomial in the size of E.

If $T \in \mathcal{T}$ then since T is a set of indexed variables, T has an ordering inherited from the indices. We will write $T(s)$ to denote the sth variable of T in this ordering.

For each $T \in \mathcal{T}$ define $\mathcal{C}_T$ to be the set of clauses for which:

1. the set of negated variables in the clause is contained in T, and
2. the set of positive variables of the clause is disjoint from T.

Formally:

$$\mathcal{C}_T = \{C_i \in \mathcal{C} : X_i^- \subseteq T \text{ and } X_i^+ \cap T = \emptyset\}$$

For example, $\mathcal{C}_\emptyset$ is the set of monotone clauses of E.

For each set of clauses $\mathcal{C}_T$ for $T \in \mathcal{T}$ we compute a rooted labeled tree Ω_T as follows. Each node of the tree is labeled by a pair $(\mathcal{A}, V)$ where $\mathcal{A}$ is a subset of $\mathcal{C}_T$, and V is a set of variables. The root of Ω_T is labeled by the pair $(\mathcal{C}_T, \emptyset)$. The tree is generated by beginning at the root and progressively expanding nodes, creating children in the following way.

In order for a node labeled $(\mathcal{A}, V)$ to be expandable, there are two conditions that must be met:

1. $\mathcal{A}$ must be nonempty, and
2. $|V| \leq k$.

The expansion of a node labeled $(\mathcal{A}, V)$ is achieved as follows. Suppose that the least indexed clause in $\mathcal{A}$ is C_j. One child is created for each variable x_s in X_j^+, that is, for each positive variable of the clause C_j. The child node is labeled $(\mathcal{A}', V')$ where

$$V' = V \cup \{x_s\}$$

and

$$\mathcal{A}' = \{C_i \in \mathcal{A} : X_i^+ \cap V' = \emptyset\}$$

The tree Ω_T is created by beginning with the root node, and expanding nodes of the tree until no more expansion is possible. Note that by condition (2), the height of Ω_T is at most k, and since the clauses of E have size k, the branching factor of Ω_T is also bounded by k. Consequently there are at most k^k leaves.

A leaf of Ω_T is *interesting* if it is labeled by a pair $(\mathcal{A}, V)$ with $\mathcal{A} = \emptyset$. Let $\mathcal{L}_T$ denote the family of sets of variables V that occur in the labels of interesting leaves of Ω_T:

$$\mathcal{L}_T = \{V : (\emptyset, V) \text{ is the label of an interesting leaf of } \Omega_T\}$$

Let $\mathcal{L}$ denote the union of the $\mathcal{L}_T$:

$$\mathcal{L} = \bigcup_{T \in \mathcal{T}} \mathcal{L}_T$$

The usefulness of the sets $\mathcal{L}_T$ is illustrated by the special case of $T = \emptyset$. In this case $\mathcal{C}_T$ is the set of monotone clauses of E, that is, the clauses having only positive literals. The following claim is easy to verify from the construction of $\Omega_\emptyset$.

Claim 21. *E is satisfiable by a weight k truth assignment only if $\exists L \in \mathcal{L}_\emptyset$ such that all of the variables in L are assigned the value true.*

For an arbitrary $T \in \mathcal{T}$, the set of clauses $\mathcal{C}_T$ must be satisfied "positively" in any satisfying truth assignment for E that assigns all of the variables of T to be *true*. By the construction of Ω_T the following claim holds.

Claim 22. *The clauses of $\mathcal{C}_T$ are positively satisfied by a weight k truth assignment that sets all of the variables of T to true, only if $\exists L \in \mathcal{L}_T$ such that all of the variables in L are assigned the value true.*

We next describe the set of variables X', which consists of three principal parts:

$$X' = X'[1] \cup X'[2] \cup X'[3]$$

For convenience, let $[k]$ denote the set $\{1, 2, \ldots, k\}$ and let $2^{[k]}$ (as usual) denote the set of all subsets of $[k]$. If $J \in 2^{[k]}$ then J is naturally ordered. We will write $J(s)$ to denote the sth element of J.

In the definition of $X'[2]$ and $X'[3]$ we employ a "dummy" index $\perp$. In order for our indexing scheme to work properly, we define

$$\mathcal{L}_\perp = \{\perp\}$$

The constituent sets of variables are:

$$X'[1] = \{a[i,x] : 1 \leq i \leq k,\ x \in X\}$$

$$X'[2] = \{b[J,T] : J \in 2^{[k]},\ T \in (\mathcal{T} \cup \{\perp\})\}$$

$$X'[3] = \{c[J,T,L,J'] : J, J' \in 2^{[k]},\ T \in (\mathcal{T} \cup \{\perp\}),\ L \in \mathcal{L}_T\}$$

The expression E' is the product of the following boolean expressions. To simplify the notation, the indices in the products implicitly range over the various important sets used to define the variables of X'. In particular: i and i' range over $[k]$; J, J' and J'' range over $2^{[k]}$; x and x' range over X; T and T' range over $\mathcal{T}$ (but not $\perp$); L and L' range over $\mathcal{L}$.

$$E'[1] = \prod_{i} \prod_{x \neq x'} (\neg a[i,x] \vee \neg a[i,x'])$$

$$E'[2] = \prod_{i \neq i'} \prod_{x} (\neg a[i,x] \vee \neg a[i',x])$$

$$E'[3] = \prod_{i < i'} \prod_{1 \leq j < j' \leq m} (\neg a[i,j'] \vee \neg a[i',j])$$

$$E'[4] = \prod_{J} \prod_{T \neq T'} (\neg b[J,T] \vee \neg b[J,T'])$$

$$E'[5] = \prod_{J,T:|J| \neq |T|} (\neg b[J,T])$$

$$E'[6] = b[\emptyset, \emptyset]$$

$$E'[7] = \prod_{J,T} \prod_{i > \max(J)} \prod_{x \in X_T} ((b[J,T] \wedge a[i,x]) \rightarrow b[J \cup \{i\}, T \cup \{x\}])$$

$$E'[8] = \prod_{J,T:|J|=|T|} \prod_{s=1}^{|J|} (b[J,T] \rightarrow a[J(s), T(s)])$$

$$E'[9] = \prod_{J''} \prod_{(T,L,J) \neq (T',L',J')} (\neg c[J'',T,L,J] \vee \neg c[J'',T',L',J'])$$

$$E'[10] = \prod_{J,T,L} \prod_{J':|J'|\neq|L|} (\neg c[J,T,L,J'])$$

$$E'[11] = \prod_{J,J'} \prod_{T\neq T'} \prod_{L} (\neg b[J,T] \vee \neg c[J,T',L,J'])$$

$$E'[12] = \prod_{J,T,J'} (\neg b[J,T] \vee \neg c[J,\perp,\perp,J'])$$

$$E'[13] = \prod_{J,T,L,J':|L|=|J'|} \prod_{s=1}^{|L|} (c[J,T,L,J'] \rightarrow a[J'(s),L(s)])$$

The variables of X' and the subexpressions of E' all have intuitive semantic interpretations that we next discuss, in order to clarify the argument.

Note that the variables of $X'[1]$ can be grouped into k *choice blocks* by the first index, each block having $|X|$ variables. Thus the variables of $X'[1]$ represent the k choices of a variable of X to set to *true* in a weight k truth assignment for E.

The subexpression $E'[1]$ enforces that in each block, at most one choice is indicated in any satisfying truth assignment for E'. The subexpression $E'[2]$ enforces that the indicated choices are distinct, and $E'[3]$ enforces that the choices of variables of X to be set *true* are made in increasing order according to the index of the variables.

The variables of $X'[2]$ pair sets of choice blocks with sets of variables in X. For example, the variable $b[\{2,5,7\},\{x_3,x_4,x_6\}]$ has the intended meaning: "x_3 is chosen in block 2, x_4 in block 5, and x_6 in block 7".

The subexpression $E'[4]$ enforces that for any subset of the k choice blocks, at most one corresponding "macro" set of variables is indicated. Note that it is not possible to supply all possible macros of size at most k since there are too many. Only those that occur in a certain way are available (or needed) as second indices for the variables of X'_2. When a macro corresponding to the choices made in a set of blocks is "not available", the set of blocks can be paired with the dummy index $\perp$. The subexpression $E'[5]$ enforces that in the pairing of a set of blocks with a set of variables, the sets have the same size, as required by the intended meaning.

The subexpression $E'[6]$ is essentially an artifact having to do with the availability of the dummy index. We require that the empty set of blocks is paired with the empty set of variables. Together with the subexpression $E'[7]$, this forces the variables of $X'[2]$ to do their intended job (the formal argument for this is made below). The subexpression $E'[8]$ forces any pairings asserted by the boolean variables of $X'[2]$ to be represented in the truth assignment for $X'[1]$.

The variables of $X'[3]$ each represent a fairly complicated predicate that is perhaps best illustrated by example. Suppose $J = \{2,3,5\}$ so that J represents the set of choice blocks indexed 2,3 and 5. By some variable of $X'[2]$ this is paired with (if it is available as an index, suppose this is so) the set of variables (of

X) chosen in these blocks. Suppose that this set of variables is $T = \{x_4, x_6, x_8\}$. There are then various clauses of E that are forced to be positively satisfied in any truth assignment that makes the variables of T *true*, and this situation is analyzed in the tree Ω_T. An index $L \in \mathcal{L}_T$ indicates a choice of further variables to set *true* in order to satisfy these clauses.

In summary, the variable $c[J, T, L, J']$ represents the assertion: "The variables of L are chosen to be *true* in the blocks indicated by J' in order to solve the positive-satisfaction problem created by the variables of T being set *true* in the blocks indicated by J."

The subexpression $E'[9]$ enforces that there is at most one such assertion made for any set of blocks J (by a satisfying truth assignment). $E'[10]$ insures that the right number of blocks are pointed to in the solution of the positive-satisfaction problem. $E'[11]$ and $E'[12]$ enforce that the solution indicated for the positive satisfaction problem created by T must be a solution computed in Ω_T. $E'[13]$ forces the indicated solution to be "implemented" in the variables of $X'[1]$.

We next argue more formally for the correctness of the reduction. Suppose τ is a weight k truth assignment that satisfies E. Let K denote the set of variables set to *true* and let $K(i)$ denote the ith variable of K (ordered by the indices of the variables in K).

For $J \subseteq [k]$, let $K(J) = \{x : \exists j \in J,\ K(j) = x\}$.

For $T \in \mathcal{T}$, let $L(T, K)$ denote the lexicographically smallest element of $S = \{K' \subseteq K : K' \in \mathcal{L}_T\}$.

Claim 23. *$L(T, K)$ is well-defined.*

Proof. It is enough to argue that the set S is nonempty. By Claim 22, any truth assignment for X that satisfies E must be a superset of at least one element of $\mathcal{L}_T$, otherwise a clause of $\mathcal{C}_T$ would not be satisfied.

Note that $L(T, K)$ is necessarily disjoint from T.

Let τ' be the weight k' truth assignment to the variables of X' that assigns the following variables the value *true*:

1. $\{a[i, K(i)] : 1 \le i \le k\}$
2. $\{b[J, K(J)] : J \in 2^{[k]},\ K(J) \in \mathcal{T}\}$
3. $\{b[J, \bot] : J \in 2^{[k]},\ K(J) \notin \mathcal{T}\}$
4. $\{c[J, K(J), L(K(J), K), J'] : J \in 2^{[k]},\ K(J) \in \mathcal{T},\ K(J') = L(K(J), K)\}$
5. $\{c[J, \bot, \bot, \emptyset] : J \in 2^{[k]},\ K(J) \notin \mathcal{T}\}$

Claim 24. *If τ satisfies E then τ' satisfies E'.*

Proof. To see that this is so, we just need to check that each of the subexpressions of E' is satisfied. The only case that is nontrivial is $E'[7]$. If one of the implications of this subexpression were false, then we would have a situation where τ' has assigned the value *true* to all possible true assertions represented by the variables of $X'[2]$ of the form: "the variables of T are chosen (in ascending order) in the blocks indicated by J", and where for one of these true assertions represented by $b[J,T]$, and for a further choice of a variable x set *true* by τ' in a block i (of larger index than any block in J), with x a successor of T, τ' has assigned the value *false* to $b[J \cup \{i\}, T \cup \{x\}]$. Since $T \cup \{x\} \in \mathcal{T}$, this is a contradiction, since this variable represents a statement that is clearly true.

That completes (the easier) half of the argument for the correctness of the reduction.

Now suppose that v' is a weight k' truth assignment to the variables of X' that satisfies E'. We argue that necessarily v has a certain restricted form, from which we can then describe a corresponding weight k truth assignment to the variables of X that satisfies E.

Claim 25. *The truth assignment v must assign exactly one variable of $X'[1]$ to be true in each choice block of $X'[1]$.*

Proof. Since $E'[1]$ is satisfied, there can at most one variable in each block assigned *true*. The satisfaction of $E'[4]$ and $E'[9]$ imply that that no more than 2^{k+1} variables are assigned *true* in $X'[2] \cup X'[3]$. This forces k to be assigned *true* in $X'[1]$ and the claim follows.

Let v denote the truth assignment for the variables of X that is indicated in the natural manner (as per Claim 25) by v'. Our goal is to show that the truth assignment v satisfies E. Part of our argument is based on the details of how v' satisfies E'.

Claim 26. *Necessarily v' assigns true to all possible variables of $X'[2]$ that represent true statements (about v) of the form: "the variables of T are chosen (in ascending order) in the blocks indicated by J".*

Proof. For any $T \in \mathcal{T}$ the proof is a simple induction along the trail leading to T, based on the satisfaction of $E'[6]$ (the base step), and $E'[7]$.

Claim 27. *The truth assignment v satisfies E.*

Proof. Suppose on the contrary that some clause C_i is not satisfied. Then all of the variables in C_i^- must be assigned the value *true* by v. Also, $C_i^- \in \mathcal{T}$. By Claim 26, there is some J such that the variable $b[J, C_i^-]$ is assigned the value *true* by v'. Since v' satisfies the subexpressions $E'[10]$, $E'[11]$ and $E'[12]$, it must assign *true* to some variable $c[J, C_i^-, L, J']$ with $|L| = |J'|$ and $L \in \mathcal{L}_{C_i^-}$. Since v' satisfies $E'[13]$, v assigns the value *true* to the variables of L. Since $C_i \in \mathcal{C}_{C_i^-}$, the clause C_i must be satisfied by v, a contradiction.

That completes the proof of the theorem.

7 Conclusions

The "concrete" results of this paper indicate that the size of a relational database query is not likely to be something that can be effectively exploited in isolation. It might be interesting to study various parameterizations of the *structure* of database queries in addition to their size to determine if there may be some opportunities for fixed-parameter tractability.

An interesting open problem concerning the W hierarchy is whether there are any *natural* degrees other than

$$FPT \subseteq W[1] \subseteq W[2] \subseteq \cdots \subseteq W[P]$$

(The four named in this tower are abundantly natural.) For example, are there natural problems that constitute a likely "$W[1/2]$" or "$W[3/2]$" degree? A promising place to look is in variations of the weighted t-normalized satisfiability problems that are shown to be complete for $W[t]$ in [DF95a]. Our main theorem here shows that one promising candidate ($W^*[1]$) for a representative of a "$W[3/2]$" degree is in fact no more than $W[1]$. It would be interesting to know if $W^*[t] = W[t]$ for all t.

References

[ADF95] K.A. Abrahamson, R.G. Downey and M.R. Fellows. Fixed parameter tractability and completeness IV: on completeness for W[P] and *PSPACE* analogs. *Annals of Pure and Applied Logic* 73 (1995), 235–276.

[BFR95] R. Balasubramanian, M. Fellows and V. Raman, An improved fixed parameter algorithm for vertex cover, To appear, *Information Processing Letters,* 1995.

[BDFW95] H. Bodlaender, R.G. Downey, M.R. Fellows and H.T. Wareham. The parameterized complexity of the longest common subsequence problem. *Theoretical Computer Science A* 147 (1995), 31–54.

[BDFHW95] H. Bodlaender, R.G. Downey, M.R. Fellows, M. Hallett and H.T. Wareham. Parameterized complexity analysis in computational biology. *Computer Applications in the Biosciences* 11 (1995), 49–57.

[BDFW94] H. Bodlaender, R.G. Downey, M.R. Fellows and H. T. Wareham. The parameterized complexity of sequence alignment and consensus. *Proceedings of the Fifth Symposium on Combinatorial Pattern Matching (CPM),* Springer-Verlag, Lecture Notes in Computer Science vol. 807 (1994), 15–30.

[BF95] H. Bodlaender and M.R. Fellows. On the complexity of k-processor scheduling. *Operations Research Letters* 18 (1995), 93–98.

[BFH94] H. Bodlaender, M.R. Fellows and M. Hallett. Beyond *NP*-completeness for problems of bounded width: hardness for the W hierarchy. *Proceedings of the ACM Symposium on the Theory of Computing (STOC)* (1994), 449–458.

[CCDF94] L. Cai, J. Chen, R.G. Downey and M.R. Fellows. On the parameterized complexity of short computation and factorization. To appear in *Arch. for Math. Logic.*

[Ces96] M. Cesati. Structural aspects of the parameterized complexity. Doctoral dissertation (Informatics). University of Rome, "La Sapienza", 1996.

[CW95] M. Cesati and H.T. Wareham. Parameterized complexity analysis in robot motion planning. In: *Proc. 25th IEEE Intl. Conf. on Systems, Man and Cybernetics.*

[DEF93] R.G. Downey, P.A. Evans and M.R. Fellows. Parameterized learning complexity. *Proceedings of the Sixth ACM Workshop on Computational Learning Theory (COLT'93)*, 51–57,

[DF95a] R.G. Downey and M.R. Fellows. Fixed-parameter tractability and completeness I: basic theory. *SIAM Journal of Computing* 24 (1995), 873–921.

[DF95b] R.G. Downey and M.R. Fellows. Fixed-parameter tractability and completeness II: completeness for W[1]. *Theoretical Computer Science A* 141 (1995), 109–131.

[DF95c] R.G. Downey and M.R. Fellows. Parameterized computational feasibility. *Proceedings of the Second Cornell Workshop on Feasible Mathematics*, Feasible Mathematics II, P. Clote and J. Remmel (eds.), Birkhauser Boston (1995), 219–244.

[DFta] R.G. Downey and M.R. Fellows. Parameterized Complexity. Monograph, Springer-Verlag, to appear.

[DFHKW94] R.G. Downey, M.R. Fellows, M.T. Hallett, B.M. Kapron, and H.T. Wareham. The parameterized complexity of some problems in logic and linguistics. *Proceedings Symposium on Logical Foundations of Computer Science (LFCS)*, Springer-Verlag, Lecture Notes in Computer Science vol. 813 (1994), 89–100.

[DFR96] R.G. Downey, M.R. Fellows and K. Regan. Parameterized circuit complexity and the W hierarchy. *Theoretical Computer Science*, to appear.

[DFT96] R.G.Downey, M.R. Fellows and U. Taylor. An improved characterization of $W[1]$. Manuscript. February 1996.

[FHW93] M.R. Fellows, M.T. Hallett and H.T. Wareham. DNA physical mapping: three ways difficult. In *Algorithms — ESA '93*, (Proceedings of the First European Symposium on Algorithms), Springer-Verlag, Berlin, Lecture Notes in Computer Science vol. 726 (1993), 157–168.

[FK93] M.R. Fellows and N. Koblitz. Fixed-parameter complexity and cryptography. *Proceedings of the Tenth International Symposium on Applied Algebra, Algebraic Algorithms and Error-Correcting Codes (AAECC'93)*, Springer-Verlag, Berlin, Lecture Notes in Computer Science vol. 673 (1993), 121–131.

[GJ79] M. Garey and D. Johnson. *Computers and Intractability: A Guide to the Theory of NP-completeness.* W.H. Freeman and Company, San Francisco, 1979.

[Ul88] J.D. Ullman. Database and Knowledge-base Systems. Computer Science Press, 1988.

[Yan95] M. Yannakakis. Perspectives on database theory. Proceedings FOCS 95, 224-246. Computer Science Press, 1988.

On the Logical Definability of Some Maximum Leaves Problems

Giulia Galbiati[1]*, Angelo Morzenti[2] and Francesco Maffioli[2]*

[1] Dipartimento di Informatica e Sistemistica, Università degli Studi di Pavia, Via Abbiategrasso 209, 27100 Pavia, Italia

[2] Dipartimento di Elettronica, Politecnico di Milano, Piazza Leonardo da Vinci 32, 20133 Milano, Italia

Abstract. This paper is about the logical definability of various combinatorial optimization problems aiming at computing the maximum number of leaves in a tree rooted at a given vertex ρ of an undirected graph. In the unbounded case there are no requirements on the depth of the tree and the graph is not necessarily connected. In the bounded case a tree of depth at most k is sought, for some $k \geq 2$, and the graph has every vertex at a distance less than or equal to k from ρ. Referring to the logical classes introduced by P. Kolaitis and M. Thakur we prove that, if the graphs are given as finite structures over the vocabulary of graph theory, then in the unbounded case the problem, called *MaxLeaves*, does not belong to the class MAX Π_1 but only to the biggest class MAX Π_2. In the bounded case we distinguish four problems: *k-MaxLeaves* and *k-MaxLeaves*$_{sp}$, where the bound k is fixed and the tree is required or not to be spanning, and their analogues *Var-MaxLeaves* and *Var-MaxLeaves*$_{sp}$, where the depth k is given in the instance. We prove that *k-MaxLeaves* does belong to MAX Π_1 whereas *Var-MaxLeaves*$_{sp}$ does not. The inclusion of *k-MaxLeaves*$_{sp}$ is open, but we show that another problem, called *k-MaxValue*, closely related to *k-MaxLeaves*$_{sp}$, is in MAX Π_1.

1 Introduction

The investigation of the relationship between structural complexity and the expressive power of logical languages is an important research topic. Well known results in this area are Fagin's [3] charactarization of NP in terms of definability in existential second order logic on finite structures, Immerman-Vardi's [9][12] and Grädel's [5] characterization of P in terms of definability in least fixpoint logic and in Horn existential second order logic with successor. A survey of the most important results in this field is [6].

More recently the interest has moved to the investigation of the logical definability of Non-deterministic Polynomial Optimization (NPO) problems [2]. Papadimitriou and Yannakakis [11] started the investigation by introducing two classes, called MAX SNP and MAX NP, of optimization problems which admit

* Partially supported by the M.U.R.S.T. National Research Project "Algoritmi, Modelli di Calcolo e Strutture Informative".

an L-reduction to problems whose optimum can be defined as the maximum number of tuples satisfying appropriate first order expressions. They show that the classes admit problems complete under L-reductions and that all problems in these classes are in APX, that is, are approximable in polynomial time within a constant factor, thus establishing a first relation between the structural complexity and the logical expressibility of an optimization problem. A subsequent step toward this goal has been the striking result of Arora, Lund, Motwani, Sudan, and Szegedy [1] showing that no MAX SNP hard problem can have a polynomial time approximation scheme, unless P=NP.

Later Panconesi and Ranjan [10] introduced a new class of problems, called MAX Π_1, which includes also problems which are not constant-approximable.

A more complete picture has been given by Kolaitis and Thakur [7], who extended the work of [10] by systematically investigating the logical definability of NPO problems. For maximization problems they defined four classes, called MAX Σ_0, MAX Σ_1, MAX Π_1 and MAX Π_2, proved that each class in the list is properly included in the following class, and proved that MAX Π_2 coincides with the class MAX NPO-PB of all maximization NPO problems that are polynomially bounded (that is, problems for which there exists a polynomial p such that, for all instances I, the optimum value of I is bounded above by $p(|I|)$). For minimization problems they also find a hierarchy of classes but this time only with two levels.

In this work we study various optimization problems on undirected graphs in order to classify some of them within Kolaitis and Thakur's logical classes. All problems aim at computing the maximum number of leaves in a tree rooted at a given vertex ρ of an undirected graph. Both the unbounded case, where there are no requirements on the depth of the tree, and the bounded case, where a tree of depth at most k is sought, for some integer $k \geq 2$, are analyzed. In the unbounded case the graph is not necessarily connected. In the bounded case the graph is connected and, without loss of generality, every vertex is assumed to be at a distance less than or equal to k from ρ. Referring to the logical classes introduced by P. Kolaitis and M. Thakur we prove that, if the graphs are given as finite structures over the vocabulary of graph theory, then in the unbounded case the problem, called *MaxLeaves*, does not belong to the class MAX Π_1 but only to the biggest class MAX Π_2. In the bounded case we distinguish four problems: *k-MaxLeaves* and *k-MaxLeaves*$_{sp}$, where the bound k on the depth of the tree is fixed and the tree is required or not to be spanning, and their analogues *Var-MaxLeaves* and *Var-MaxLeaves*$_{sp}$, where the depth of the tree is given in the instance. We prove that *k-MaxLeaves* is in MAX Π_1 whereas *Var-MaxLeaves*$_{sp}$ cannot belong to this class. The inclusion of *k-MaxLeaves*$_{sp}$ is open, but we show that another problem, called *k-MaxValue*, closely related to *k-MaxLeaves*$_{sp}$ since the optimum value of its objective function differs only by an instance-dependent constant from the optimum of *k-MaxLeaves*$_{sp}$, belongs to MAX Π_1.

We conclude this section by noticing that the inclusion in, or exclusion from, a logically defined class for a problem strictly depends on the encoding of its instances as finite structures over a specific vocabulary; changing the vocabulary may modify the result. As complexity results depend on data structures, so logical results depend on the chosen vocabularies. We think that finding logical classifications of NPO problems independent from the encoding of the instances is an open research challenge.

2 The maximum leaves problems

Before describing all problems that we will examine we need a few definitions.

Throughout the rest of the paper k will be an integer greater or equal to 2, $G = (V, E)$ an undirected graph having a specified vertex ρ called root.

Definition 1. G is k-bounded if every vertex has a distance less than or equal to k from ρ.

Definition 2. A tree T of G, rooted at ρ, is k-bounded if its depth is at most k.

Definition 3. A path p of G, rooted at ρ, is k-bounded if its length is at most k.

Now all problems that we address in this paper for establishing to which logical class they belong can be described as follows:

Instance : an undirected graph $G = (V, E)$ with vertex set V, $|V| = n$, and edge set E; a specified vertex $\rho \in V$ called root.
Objective function : the number of leaves of a tree T of G rooted at ρ.
Optimization criterion : maximize.

As it is formulated this problem is called *MaxLeaves*. Depending on additional requirements on the instance and on tree T we distinguish five other problems which we name:

i) $MaxLeaves_{sp}$ if G is connected, T required to be spanning,
ii) *k-MaxLeaves* if G is k-bounded, T required to be k-bounded,
iii) $k\text{-}MaxLeaves_{sp}$ as above, with T also required to be spanning,
iv) *Var-MaxLeaves* as in ii), for some k,
v) $Var\text{-}MaxLeaves_{sp}$. as in iii), for some k.

MaxLeaves and $MaxLeaves_{sp}$ deal with the unbounded case, the other four with the bounded one. Notice that a k-bounded graph is connected, so all problems except *MaxLeaves* deal with connected graphs.

We remind the reader that in [8] Lu and Ravi show that the NPO problem of finding a spanning tree having a maximum number of leaves, and not only of computing the maximum number of leaves of a spanning tree as $MaxLeaves_{sp}$ instead does, is approximable within 3. In [4] we prove that the same problem is MAX SNP hard and therefore cannot have a polynomial time approximation scheme, unless P=NP. We observe here that the proof given in [4], slightly modified, also allows to conclude that the problem of finding a spanning tree of depth bounded by k, having a maximum number of leaves, is also MAX SNP hard, for each fixed $k \geq 2$.

We now turn our attention to the logical definability of the above problems, starting with the unbounded case.

To make the paper self-contained, in the next section we present some definitions and logic preliminaries as given in Kolaitis and Thakur [7], together with some explanatory examples.

3 Preliminaries

We will be interested mostly in the vocabulary of graph theory, but another vocabulary will be utilized in the sequel as well.

Definition 4. A *vocabulary* (also known as *similarity type* or *signature*) $\sigma = (\tilde{R}_1, \ldots, \tilde{R}_s)$ is a finite set of predicate symbols. Each predicate symbol $\tilde{R}_i$ has a positive integer r_i as its assigned arity.

The vocabulary of graph theory is $\sigma_{\mathbf{g}} = (\tilde{E})$ with $\tilde{E}$ of arity 2.

We will be interested in describing the instances of optimization problems as finite structures over some vocabulary σ.

Definition 5. A *finite structure* $\mathbf{A} = (A, R_1, ..., R_s)$ over the vocabulary $\sigma = (\tilde{R}_1, \ldots, \tilde{R}_s)$ consists of a finite set A, called the *universe* of $\mathbf{A}$, and relations $R_1, ..., R_s$ on A of arities $r_1, ..., r_s$.

As a first example, the finite structures $\mathbf{G} = (V, E)$ over the vocabulary $\sigma_{\mathbf{g}} = (\tilde{E})$ of graph theory describe the finite graphs, if V is the set of vertices of a graph and $E(v_1, v_2)$ is true if vertices v_1 and v_2 are adjacent in the graph.

Instead the finite structures $\mathbf{B} = (X, C, P, N)$ over $\sigma = (\tilde{C}, \tilde{P}, \tilde{N})$, with predicates of arities 1, 2 and 2 respectively, describe the boolean expressions in conjunctive normal form, if X is the set of variables and clauses of an expression, $C(x)$ is true if x is a clause, $P(v, c)$ is true if variable v is positive in clause c, $N(v, c)$ is true if variable v is negative in clause c.

Now let σ be a vocabulary, $\mathbf{A}$ a finite structure over σ, $\mathbf{w}$ a finite sequence of fixed arity of variables.

With $\phi(\mathbf{w})$ we denote a formula of first order logic over σ having $\mathbf{w}$ as its free variables.

With $\{\mathbf{w} : \mathbf{A} \models \phi(\mathbf{w})\}$ we denote the set of all tuples $\mathbf{w}$ from the universe A of $\mathbf{A}$ for which ϕ becomes true, that is, the set of all tuples $\mathbf{w}$ from A that satisfy $\phi(\mathbf{w})$.

Now let $\mathbf{S} = (S_1, \ldots, S_m)$ be a sequence of predicate symbols of arities $s_1, \ldots, s_m$ not in σ.

With $\phi(\mathbf{w}, \mathbf{S})$ we denote a formula of first order logic over the vocabulary $\sigma \cup \mathbf{S}$ having $\mathbf{w}$ as its free variables.

With $(\mathbf{A}, \mathbf{S})$ we denote the expanded structure $(A, R_1, \ldots, R_s, S_1, \ldots, S_m)$, where we intend that $S_1, \ldots, S_m$ are relations over A of arities $s_1, \ldots, s_m$.

Finally with $\{\mathbf{w} : (\mathbf{A}, \mathbf{S}) \models \phi(\mathbf{w}, \mathbf{S})\}$ we denote the set of all tuples $\mathbf{w}$ from A for which the formula $\phi(\mathbf{w}, \mathbf{S})$ becomes true on the expanded structure $(\mathbf{A}, \mathbf{S})$.

It is well known that every formula of first order logic is equivalent to a formula in *prenex normal form*, that is, to a formula in which all quantifiers are to the left of all other symbols. We write Σ_n (respectively, Π_n), $n \geq 1$, for the class of first-order formulae in prenex normal form that have $n-1$ alternations of quantifiers and start with a block of existential (respectively, universal) quantifiers. The class of quantifier-free formulae is denoted by Σ_0 (equivalently Π_0).

Definition 6. Let Q be a maximization problem whose instances are given as finite structures $\mathbf{A}$ over σ. We say that Q is in MAX Π_n, (respectively, MAX Σ_n), $n \geq 0$, iff there is a Π_n (respectively, Σ_n) formula $\phi(\mathbf{w}, \mathbf{S})$ with predicate symbols among those in σ and $\mathbf{S}$ such that for every instance $\mathbf{A}$ of Q we have

$$opt_Q(\mathbf{A}) = \max_{\mathbf{S}} |\{\mathbf{w} : (\mathbf{A}, \mathbf{S}) \models \phi(\mathbf{w}, \mathbf{S})\}|.$$

As already said in the introduction, Kolaitis and Thakur have proved that

$$\text{MAX } \Sigma_0 \subset \text{MAX } \Sigma_1 \subset \text{MAX } \Pi_1 \subset \text{MAX } \Pi_2 = \text{MAX NPO-PB}$$

and that all inclusions are strict.

As an example, Problem MAXSAT, asking for the maximum number of clauses that can be satisfied in the conjunctive normal form of a boolean expression, given as a finite structure over $\sigma = (\tilde{C}, \tilde{P}, \tilde{N})$, is in MAX Σ_1 since

$$opt_{\text{MAXSAT}}(\mathbf{B}) = \max_{\mathbf{S}} |\{\mathbf{c} : (\mathbf{B}, \mathbf{S}) \models \exists x\ \psi(\mathbf{c}, \mathbf{S}, x)\}|$$

where $\mathbf{S} = (S)$, $\mathbf{c} = (c)$ and

$$\psi(\mathbf{c}, \mathbf{S}, x) = C(c) \wedge ((P(x, c) \wedge S(x)) \vee (N(x, c) \wedge \neg S(x))).$$

Another example is Problem MAXCLIQUE, which is not in MAX Σ_1 [10], but is in MAX Π_1, with the usual encoding of a graph, since

$$opt_{\text{MAXCLIQUE}}(\mathbf{G}) = \max_{\mathbf{S}} |\{\mathbf{w} : (\mathbf{G}, \mathbf{S}) \models \forall y_1\ \forall y_2\ \psi(\mathbf{w}, \mathbf{S}, y_1, y_2)\}|$$

where $\mathbf{S} = (S)$, $\mathbf{w} = (w)$ and

$$\psi(\mathbf{w}, \mathbf{S}, y_1, y_2) = S(w) \wedge (S(y_1) \wedge S(y_2) \wedge (y_1 \neq y_2) \rightarrow E(y_1, y_2)).$$

We conclude this section observing that, unless explicitly said, graphs will be given as finite structures over the vocabulary of graph theory.

4 The unbounded case

In this section we prove that Problem *MaxLeaves* does not belong to MAX Π_1. As a consequence, since it is polynomially bounded and NP-hard, it belongs to the widest class MAX Π_2, which coincides with the class MAX NPO-PB of all polynomially bounded maximization NPO problems, as proved in [7]. The next theorem was inspired by Theorem 2 of [7].

Theorem 7. *Problem MaxLeaves does not belong to* MAX Π_1.

Proof. Assume, by contradiction, that for each structure $\mathbf{G} = (V, E)$

$$opt_{MaxLeaves}(\mathbf{G}) = \max_{\mathbf{S}} |\{\mathbf{w} : (\mathbf{G}, \mathbf{S}) \models \forall \mathbf{y} \psi(\mathbf{w}, \mathbf{S}, \mathbf{y})\}| , \tag{1}$$

where ψ is quantifier free and $\mathbf{w}$ ranges over tuples of arity r of vertices of V. Let $\mathbf{T}$ be a complete balanced binary tree of depth m, with $m > r$, and hence with 2^m leaves. Call $v_{i,j}$ the vertex of $\mathbf{T}$ at level i and position j (counting from the left, for instance) and by $\mathbf{T}_{i,j}$ the subgraph of $\mathbf{T}$ obtained by removing $v_{i,j}$ and all edges incident to it, for all i, j, $1 \leq i \leq m$, $1 \leq j \leq 2^i$. Assume that the optimum value in (1) for $\mathbf{T}$, equal to 2^m, is achieved for some $\mathbf{S}^*$, and let $\mathbf{S}^*_{i,j}$ be the restriction of $\mathbf{S}^*$ to the vertex set $V_{i,j} = V - \{v_{i,j}\}$ of $\mathbf{T}_{i,j}$.

The next two equations, where we implicitly define sets W and $W_{i,j}$, for each i,j, follow from the preceding considerations and from (1) applied to graph $\mathbf{T}_{i,j}$.

$$|W| = |\{\mathbf{w} : (\mathbf{T}, \mathbf{S}^*) \models \forall \mathbf{y} \psi(\mathbf{w}, \mathbf{S}^*, \mathbf{y})\}| = 2^m , \tag{2}$$

$$|W_{i,j}| = |\{\mathbf{w} : (\mathbf{T}_{i,j}, \mathbf{S}^*_{i,j}) \models \forall \mathbf{y} \psi(\mathbf{w}, \mathbf{S}^*_{i,j}, \mathbf{y})\}| \leq 2^m - 2^{m-i} . \tag{3}$$

Notice that, in (2), $\mathbf{w}$ and $\mathbf{y}$ range over tuples of elements of V, while in (3) they range over tuples of elements of $V_{i,j}$. Now for each i, j we partition W into two disjoint sets called $I_{i,j}$ and $N_{i,j}$ consisting respectively of those r-tuples that include or do not include vertex $v_{i,j}$. Since the formula in (1) is a universal one, it follows that $N_{i.j} \subseteq W_{i.j}$. Hence $|N_{i,j}| \leq 2^m - 2^{m-i}$ and therefore $|I_{i,j}| \geq 2^{m-i}$, so vertex $v_{i,j}$ occurs in at least 2^{m-i} tuples of W. Since $\mathbf{T}$ has 2^i vertices at level i, the total number of occurrences of any vertex of level i in the tuples of W is therefore at least 2^m. Since the tree has m levels, the total number of occurrences in the tuples of W of any vertex of level i, for all $i = 1, \ldots, m$, is at least $m2^m$, a contradition because the total number of occurrences is $r2^m$, and $m > r$. □

We now observe that Problem *MaxLeaves*, when restricted to connected graphs, coincides with Problem $MaxLeaves_{sp}$ because of the following result.

Theorem 8. *For any undirected, connected graph* $\mathbf{G} = (V, E)$, *having root* ρ,

$$opt_{MaxLeaves}(\mathbf{G}) = opt_{MaxLeaves_{sp}}(\mathbf{G}) .$$

Proof. It is enough to prove that for each tree T rooted at ρ having a maximum number of leaves there exists a spanning tree with at least the same number of leaves. Since $\mathbf{G}$ is connected, this is simply done by connecting to T external vertices, and noticing that this process never decreases the number of leaves. □

However, since in the proof of Theorem 7 graph $\mathbf{T}_{i,j}$ is not connected, the exclusion of $MaxLeaves_{sp}$ from MAX Π_1 remains open.

5 The bounded case

We begin this section by remembering that in all problems of the bounded case the input graph is k-bounded, hence connected. Notice now that, if $k \geq 3$, the analogue of Theorem 8 does not hold, since there are graphs, as those illustrated in Figure 1(a), whose maximum number of leaves among the 3-bounded trees is strictly greater that the same maximum among the 3-bounded spanning trees. It is easy to see however that for $k = 2$ the two maxima coincide.

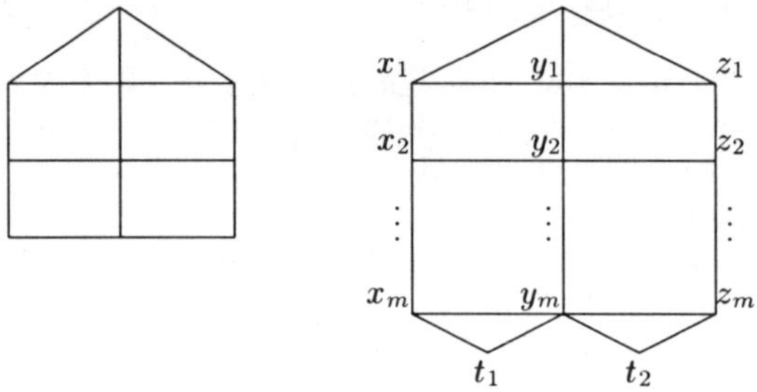

(a) (b)

Fig. 1.

The first two results we state in this section are one in the negative and one in the positive.

Theorem 9. *Problem Var-MaxLeaves$_{sp}$ does not belong to* MAX Π_1.

Proof. (*Sketch*) Assume, by contradiction, that for each graph $\mathbf{G}$, k-bounded for some k,

$$opt_{VarMaxLeaves_{sp}}(\mathbf{G}) = \max_{\mathbf{S}} |\{\mathbf{w} : (\mathbf{G}, \mathbf{S}) \models \forall \mathbf{y} \psi(\mathbf{w}, \mathbf{S}, \mathbf{y})\}|,$$

where ψ is quantifier free and $\mathbf{w}$ ranges over tuples of arity r of vertices of $\mathbf{G}$. The proof follows that of Theorem 7, where the role of graph $\mathbf{T}$ is played by the graph $\mathbf{H}$ of Figure 1(b), k-bounded with $k = m + 1$, $m > 2r$, and the role of the $\mathbf{T}_{i,j}$ is played by the subgraphs $\mathbf{H}_i$ obtained by removing vertex y_i, $i = 1, \ldots, m$, considered k-bounded with the same value of k. We have that $|W| = 2(m + 1)$, since an $(m+1)$-bounded spanning tree for $\mathbf{H}$ with a maximum number of leaves connects both t_1 and t_2 through the y_i. We also have that $|W_i| \leq m+1$, since an $(m + 1)$-bounded spanning tree for $\mathbf{H}_i$ with a maximum number of leaves must connect t_1 through the x_i and t_2 through the z_i. So, for each i, vertex y_i occurs in at least $m + 1$ tuples of W and the total number of occurrences of the y_i is at least $m(m + 1)$. This is again a contradiction, since the allowed number of total occurrences is $2r(m + 1)$ and $m > 2$. $\square$

Theorem 10. *For each $k \geq 2$, Problem k-MaxLeaves belongs to* MAX Π_1.

Proof. We begin by observing that there is a one-to-one correspondence between the leaves of a rooted tree and the paths from the root to them. Therefore maximizing the number of leaves in a k-bounded tree is equivalent to maximizing the number of such paths. We proceed in two steps: first, we write an informal formula where each (bold) variable is intended to represent a k-bounded path, then we rewrite the formula, by substituting each variable with a k-tuple of variables representing vertices, according to the fact that the graph is given as a finite structure over $\sigma_{\mathbf{g}}$.

We use the following predicates: $Path(\mathbf{p})$ holds if $\mathbf{p}$ is a k-bounded path; $C(\mathbf{p}, \mathbf{q})$ holds if paths $\mathbf{p}$ and $\mathbf{q}$ are compatible, that is, their union is a k-bounded tree; $I(\mathbf{p}, \mathbf{q})$ holds if path $\mathbf{p}$ is properly included in (proper prefix of) path $\mathbf{q}$. Now the optimum value computed by *k-MaxLeaves* on an input structure $\mathbf{G}$ is given by the following formula

$$opt_{k\text{-}MaxLeaves}(\mathbf{G}) = \max_{\mathbf{S}} |\{\mathbf{p} : (\mathbf{G}, \mathbf{S}) \models \forall \mathbf{x} \forall \mathbf{y}\ \psi(\mathbf{p}, \mathbf{S}, \mathbf{x}, \mathbf{y})\}| \tag{4}$$

where $\mathbf{S} = (S)$ and :

$$\begin{aligned}\psi(\mathbf{p}, \mathbf{S}, \mathbf{x}, \mathbf{y}) = {} & S(\mathbf{p}) \wedge \\ & (S(\mathbf{x}) \rightarrow Path(\mathbf{x})) \wedge \\ & (S(\mathbf{x}) \wedge S(\mathbf{y}) \rightarrow C(\mathbf{x}, \mathbf{y})) \wedge \\ & (S(\mathbf{x}) \wedge S(\mathbf{y}) \rightarrow \neg I(\mathbf{x}, \mathbf{y}) \wedge \neg I(\mathbf{y}, \mathbf{x})) .\end{aligned}$$

The second conjunct asserts that predicate S identifies a set of paths rooted at ρ, the third conjunct says that only compatible paths belong to the set, the last conjunct ensures that no initial part of a path in the set also belongs to the set.

Let us now code each k-bounded path $\mathbf{p}$ with the tuple $(p_1, \ldots, p_k)$ made by the vertices following ρ in the path; if the length of $\mathbf{p}$ is less that k repeat the last vertex $k-h$ times. We show now that all predicates in (4) may be characterized by quantifier free, first order formulas on k-tuples of variables ranging over vertices of the graph. We have the following.

$$\begin{aligned} & Path(x_1, \ldots, x_k) = \\ & \quad (E(\rho, x_1) \wedge (x_1 = x_2) \wedge \ldots \wedge (x_{k-1} = x_k)) \vee && /*\text{ of length } 1 */ \\ & \quad (E(\rho, x_1) \wedge E(x_1, x_2) \wedge (x_2 = x_3) \wedge \ldots \wedge (x_{k-1} = x_k)) \vee && /*\text{ of length } 2 */ \\ & \quad \ldots\ldots \\ & \quad (E(\rho, x_1) \wedge E(x_1, x_2) \wedge \ldots \wedge E(x_{k-1}, x_k)) && /*\text{ of length } k */ \end{aligned}$$

Since two paths are compatible if and only if they share an initial, possibly empty, portion, so that their tuples coincide in the first d elements, for some $0 \leq d \leq k$, and each pair of successive vertices differ, we also have that

$$C(x_1, \ldots, x_k, y_1, \ldots, y_k) = \bigvee_{d=0}^{k} C_d(x_1, \ldots, x_k, y_1, \ldots, y_k)$$

where:

$$C_d(x_1, \ldots, x_k, y_1, \ldots, y_k) = \bigwedge_{j \leq d} (x_j = y_j) \wedge \bigwedge_{d<h\leq k} \bigwedge_{d<l\leq k} (x_h \neq y_l).$$

We leave to the reader the characterization of predicate $I(x_1, \ldots, x_k, y_1, \ldots, y_k)$. □

Now we observe that the inclusions in MAX Π_1 of *Var-MaxLeaves* and *k-MaxLeaves*$_{sp}$ are open. As already anticipated in the introduction, we are instead able to include in MAX Π_1 a problem closely related to *k-MaxLeaves*$_{sp}$, called *k-MaxValue*. The next subsection is devoted to this goal.

5.1 The *k-MaxValue* problem

The problem is defined as follows.

Instance : an undirected, k-bounded graph $G = (V, E)$ with vertex set V, $|V| = n$, and edge set E; a specified vertex $\rho \in V$ called root.
Objective function : the value $v(T)$ of a k-bounded tree T of G .
Optimization criterion : maximize,

where:

$$v(T) = \sum_{e\in T} w(e)\,,$$

and, considering root ρ always an internal node,

$$w(e) = \begin{cases} k & \text{if } e \text{ is an edge having a leaf as an endpoint} \\ k-1 & \text{otherwise.} \end{cases}$$

With the next two theorems we prove that any k-bounded tree T maximizing the value function $v(T)$ is a spanning tree maximizing the number of leaves.

Theorem 11. *If a k-bounded tree T maximizes $v(T)$ among all k-bounded trees of G, then it is spanning.*

Proof. Without loss of generality, suppose that there exists a k-bounded path $p = \rho v_{i_1} v_{i_2} \ldots v_{i_j}$, $1 < j \leq k$, where all nodes except v_{i_j} are in T, and suppose that it is a shortest k-bounded path with this property. This impies that, for each $l = 1, \ldots, j-1$, vertex v_{i_l} is in T at a distance from ρ greater or equal to l. We show that T cannot maximize $v(T)$. Gradually modify T by first connecting ρ to v_{i_1} and removing the edge incident to v_{i_1}, then by connecting v_{i_1} to v_{i_2} and removing the edge incident to v_{i_2}, and so on until v_{i_j} is connected to $v_{i_{j-1}}$. Because in the original T every vertex v_{i_l} was at a distance from ρ greater or equal to l, this process does not violate the property of T being k-bounded. At the end of the process, T has one more vertex, namely vertex v_{i_j}, which is a leaf, whereas all vertices $v_{i_1} v_{i_2} \ldots v_{i_{j-1}}$ have become internal. It is now simple to see that the value of $v(T)$ for this new tree is bigger than the value for the original tree. In fact the gain in value due to p is $(j-1)(k-1)+k$. The loss depends on the values of the edges removed to break the cycles, which depend on the roles in the original T of the vertices v_{i_l}, $l = 1, \ldots, j-1$. If all vertices are leaves the loss is maximum and equal to $(j-1)k$, otherwise it is less. Even in the worst case $(j-1)(k-1)+k > (j-1)k$. □

Theorem 12. *If a k-bounded tree T maximizes $v(T)$ among all k-bounded trees of G, then it also maximizes the number of leaves among all k-bounded spanning trees of G.*

Proof. If we denote by f_T and i_T the number of leaves and of internal vertices of T it follows by definition that $v(T) = kf_T + (k-1)(i_T - 1) = (k-1)(f_T + i_T - 1) + f_T$. Because of Theorem 11 the value $f_T + i_T$ is n, hence maximizing $v(T)$ also maximizes f_T among all k-bounded spanning trees of G. □

The following corollary, a direct consequence of the proof of the preceding theorem, establishes the relation between the optimum computed by *k-MaxLeaves*$_{sp}$ and that computed by *k-MaxValue*. It shows that the two optima differ only by an instance-dependent constant.

Corollary 13. *For any undirected, k-bounded graph* $\mathbf{G}$ *having root* ρ

$$opt_{k\text{-}MaxLeaves_{sp}}(\mathbf{G}) = opt_{k\text{-}MaxValue}(\mathbf{G}) - (k-1)(n-1).$$

In order to reach the goal of showing that *k-MaxValue* is in MAX Π_1 we introduce an auxiliary problem, called *k-MaxPaths*. It is nothing but a different face of *k-MaxValue*, since for each graph it computes the same optimum; but an instance graph is given to *k-MaxValue* as a finite structure over the usual vocabulary of graph theory, whereas it is given to *k-MaxPaths* as a finite structure $\mathbf{G_p}$ over a new vocabulary $\sigma_\mathbf{p}$. Then we find the logical expression that first assigns *k-MaxPaths* and then *k-MaxValue* to MAX Π_1.

The *k-MaxPaths* problem

Let $\sigma_\mathbf{p} = (\tilde{P}, \tilde{O}^1, \ldots, \tilde{O}^{k-1}, \tilde{C}, \tilde{I})$ be a signature with all predicates having arity 2, except for $\tilde{P}$ which has arity 1.

Then a k-bounded graph G having root ρ can be described as a finite structure $\mathbf{G_p} = \{U, P, O^1, \ldots, O^{k-1}, C, I\}$ over $\sigma_\mathbf{p}$ if the following hold.

The universe is the set

$$U = \{p_1, p_1^1, \ldots, p_1^{k-1}, p_2, p_2^1, \ldots, p_2^{k-1}, \ldots, p_m, p_m^1, \ldots, p_m^{k-1}\}$$

that includes all k-bounded paths $p_1, \ldots, p_m$ of G, called *proper paths*, and moreover, for each proper path p_i, also includes $k-1$ copies, denoted $p_i^1, \ldots, p_i^{k-1}$, called its *shadow paths*.

P is a unary predicate: $P(p)$ holds iff p is a proper path.

All O^j, $1 \le j \le k-1$, are binary predicates: $O^j(q,p)$ holds iff q is the j-th shadow of p, which is a proper path.

C and I are binary predicates expressing the compatibility and proper inclusion relations among paths (proper or shadow), as defined in Theorem 10.

Now Problem *k-MaxPaths* and the optimum value it computes on the input structure $\mathbf{G_p}$ can be defined operatively by the following formula

$$opt_{k\text{-}MaxPaths}(\mathbf{G_p}) = \max_{\mathbf{S}} |\{\mathbf{w} : (\mathbf{G_p}, \mathbf{S}) \models \forall p \forall q \forall r \ \psi(\mathbf{w}, \mathbf{S}, p, q, r)\}|, \quad (5)$$

where $\mathbf{S} = (S)$, $\mathbf{w} = $(w) and

$$\begin{aligned}\psi(\mathbf{w}, \mathbf{S}, p, q, r) = S(w) \wedge \\ &(S(p) \wedge S(q) \to C(p,q)) \wedge \\ &(S(p) \wedge S(q) \wedge P(p) \wedge P(q) \wedge I(p,q) \wedge O^{k-1}(r,p) \to \neg S(r)) \wedge \\ &(P(p) \wedge S(r) \wedge (O^1(r,p) \vee O^2(r,p) \vee \ldots \vee O^{k-1}(r,p)) \to S(p)).\end{aligned}$$

The first conjunct of ψ shows that S is a 1-arity predicate identifying the elements of U that satisfy a Π_1 first order formula on the expanded structure $(\mathbf{G_p}, \mathbf{S})$. The second conjunct asserts that only compatible paths can belong to S, so that the paths of S actually identify a k-bounded tree of graph G. The third conjunct says that if a proper path in S is included in another proper path in S then its $(k-1)^{st}$ copy is not included in S. The last conjunct imposes that any shadow path can be in S only if its corresponding proper path is also in S.

The next result follows immediately from (5).

Theorem 14. *For each $k \geq 2$, Problem k-MaxPaths is in* MAX Π_1.

The following theorem shows that problems *k-MaxPaths* and *k-MaxValue* are two different faces of the same medal, since for each graph they compute the same optimum value, but the first problem receives in input a finite structure over $\sigma_\mathbf{p}$, the second over $\sigma_\mathbf{g}$.

Theorem 15. *For any undirected, k-bounded graph G having root ρ, described as a finite structure $\mathbf{G}$ over $\sigma_\mathbf{g}$ and as a finite structure $\mathbf{G_p}$ over $\sigma_\mathbf{p}$*

$$opt_{k\text{-}MaxValue}(\mathbf{G}) = opt_{k\text{-}MaxPaths}(\mathbf{G_p}).$$

Proof. Observe that in a maximal set S for (5), which identifies a k-bounded tree T because of the second conjunct, every shadow path appears with its proper path and any proper path with all the allowed shadows. These shadows are $k-1$ for each path ending in a leaf, but only $k-2$ for paths ending in an internal node, because of the third conjunct. Thus each internal node of T accounts for $k-1$ elements of S, 1 proper plus $k-2$ shadows, while each leaf accounts for k elements, 1 proper plus $k-1$ shadows. Obviously $|S| = v(T)$ and the conclusion follows in a straigthforward way. □

Now the goal pursued in Subsection 5.1 is achieved.

Theorem 16. *For each $k \geq 2$, Problem k-MaxValue is in* MAX Π_1.

Proof. (*Sketch*) View each element of U as a $(k+1)$-tuple of appropriate elements of V and express, using only Π_1 first order logic expressions over the stucture $(\mathbf{G}, \mathbf{S})$, all predicates P, I, C and O^j as predicates over V^{k+1}. □

6 Final remarks

In the present work we show that the problem of finding the maximum number of leaves in a rooted tree of an undirected graph does not belong to the class MAX Π_1 but only to the biggest class MAX Π_2. The same result holds for the problem of finding the maximum number of leaves in a k-bounded spanning tree of an undirected graph, k-bounded for some $k \geq 3$. Instead the problem of finding the maximum number of leaves in a k-bounded tree of an undirected k-bounded graph does belong to the class MAX Π_1, for each fixed $k \geq 2$.

Various questions remain open. The most challenging of these are the unbounded problem $MaxLeaves_{sp}$ on connected graphs and the bounded problem k-$MaxLeaves_{sp}$ whose related problem k-*MaxValue* is instead in MAX Π_1. We finally observe that the problems that we showed not to belong to MAX Π_1 by means of a counting argument that refers only to the usual vocabulary of graph theory may instead be included in MAX Π_1 if rephrased into (formally distinct) problems, with the graphs given as finite structures over a different vocabulary. In our view, it is an open reserch challenge to find results of inclusion or non-inclusion of a problem in a class independent from a specific signature.

References

1. Arora, S., Lund, C., Motwani, R., Sudan, M., and Szegedy, M.: Proof verification and hardness of approximation problems. *Proc. 33rd Ann. IEEE Symp. on Foundations of Computer Science* (1992) 14–23
2. Crescenzi, P., Panconesi, A.: Completeness in approximation classes. *Information and Computation* **2** (1993) 241–262
3. Fagin, R.: Generalized first-order spectra and polynomial-time recognizable sets. In *Complexity of Computation*, Karp, R. M. ed., SIAM-AMS Proc. **7** (1974) 43–73
4. Galbiati, G., Maffioli, F., and Morzenti, A.: A Short Note on The Approximability of The Maximum Leaves Spanning Tree Problem. *Information Processing Letters* **52** (1994) 45–49
5. Grädel, E.: The expressive power of second-order Horn logic. *Proc. 8th Symp. on Theor. Aspects of Comp. Sci.*, Lecture Notes in Computer Science **480** (1991) 466–477
6. Gurevich, Y.: Logic and the challenge of Computer Science. In *Current trends in theoretical computer science*, E. Borger, ed., Computer Science Press (1988) 1–57
7. Kolaitis, P., and Thakur, M.: Logical Definability of NP Optimization Problems. *Information and Computation* **115** (1994) 321–353
8. Lu, Hsueh-I, and Ravi, R.: The Power of Local Optimization: Approximation Algorithms for Maximum-Leaf Spanning Tree. *Proc. Allerton Conf.* (1992) 533–542
9. Immerman, N.: Relational queries computable in polynomial time. *Information and Control* **68** (1986) 86–104
10. Panconesi A., and Ranjan, D.: Quantifiers and Approximation, *Proc. 22th Annual ACM Symposium on the Theory of Computing* (1990) 446–456
11. Papadimitriou, C. H., and Yannakakis, M.: Optimization, approximation, and complexity classes. *J. Comput. and Syst. Sci.* **43** (1991) 425–440
12. Vardi, M. Y.: The complexity of relational query languages. *Proc. 14th ACM Symp. on the Theory of Computing* (1962) 137–146

The Functional Lambda Abstraction Algebras Form a Variety

Robert Goldblatt

Mathematics Department, Victoria University
P. O. Box 600, Wellington, New Zealand
Rob.Goldblatt@vuw.ac.nz

Abstract. The variety (equational class) of *lambda abstraction algebras* (LAAs) was introduced by Pigozzi and Salibra [PS93, PS95] to provide an algebraic version of the the untyped lambda calculus. The main examples are the *functional* algebras (FLAs) consisting of suitable functions based on environment models of lambda calculus, and *restricted* functional algebras (RFAs) obtained by allowing only those environments that agree almost everywhere (cofinitely) with some fixed environment.
Pigozzi and Salibra showed that the RFAs form a variety that is generated by the FLAs. In this paper a proof is given that every RFA is isomorphic to an FLA, so this very natural class of functional models of lambda calculus (FLAs) is itself a variety.
The proof given here is based on an ultrapower construction. At the end of the paper an indication is given of how an alternative approach can be developed by using methods from nonstandard analysis.

1 Functional lambda abstraction algebras

What kind of algebra stands to the untyped lambda calculus as Boolean algebra stands to propositional logic and cylindric algebra stands to quantificational logic? To consider this question we first examine the syntax of lambda calculus, which provides for the formation of *terms* $M, N, \ldots$ which are intended to denote functions. These terms are formed inductively from a set I of variables $x, y, \ldots$ by two operations:

- *Application*: the term MN is intended to denote the result of applying the function denoted by term M to the entity denoted by N.
- *Abstraction*: the term $\lambda x.M$ is intended to denote the function which assigns to an entity v the denotation that the term M receives when the variable x is given value v. (Thus in a real analysis context, $\lambda x.x^2$ would denote the squaring function, $\lambda x.e^x$ the exponential function, and so on.)

Note that *self-application* MM is allowed, suggesting that the denotation of M be thought of as a function that belongs to its own domain. But if V is the set of possible denotations of terms, then the set V^V of *all* functions from V to V is of larger cardinality than V (by Cantor's diagonalisation argument, provided V has more than one element) and so V^V cannot be identified with V itself. So to obtain a mathematical semantics for lambda calculus the best we can do is identify V with some subclass of V^V.

This leads to the notion of a *functional domain* as a structure of the form

$$\mathbf{V} = \langle V, \bullet^{\mathbf{V}}, \lambda^{\mathbf{V}} \rangle ,$$

where V is a non-empty set, $\bullet^{\mathbf{V}}$ a binary operation on V ("application"), and $\lambda^{\mathbf{V}}$ a *partial* function from V^V to V ("abstraction") such that the condition

$$\lambda^{\mathbf{V}}(f) \bullet^{\mathbf{V}} v = f(v)$$

holds for all f in the domain of $\lambda^{\mathbf{V}}$, and all $v \in V$.

The operation $\bullet^{\mathbf{V}}$ induces a map $\Phi^{\mathbf{V}}$ from V to V^V, identifying each $u \in V$ with the function $\Phi^{\mathbf{V}}(u) : V \to V$ that acts by "left multiplication":

$$\Phi^{\mathbf{V}}(u)(v) = u \bullet^{\mathbf{V}} v.$$

Then the previous equation amounts to the requirement that

$$\Phi^{\mathbf{V}}\left(\lambda^{\mathbf{V}}(f)\right) = f.$$

It follows that $\lambda^{\mathbf{V}}$ is injective, and therefore associates a unique element of V with each function in the domain of $\lambda^{\mathbf{V}}$.

Suppose now we have a fixed set I of variables for constructing lambda calculus terms. An element p of V^I, assigning *values* in V to variables, will be called an *environment*. The value assigned to x by p will be denoted p_x, rather than $p(x)$. For $v \in V$ and $x \in I$, $p\{v/x\} \in V^I$ is the new environment that updates p by assigning v to x. Thus

$$p\{v/x\}_y = \begin{cases} p_y & \text{if } y \neq x \\ v & \text{if } y = x. \end{cases}$$

Let

$$V_I = \left\{ f : V^I \xrightarrow{f} V \right\}$$

be the set of all functions from environments to values in V. The element $x^{\mathbf{V}_I}$ of V_I is defined by putting

$$x^{\mathbf{V}_I}(p) = p_x$$

for any $x \in I$ and $p \in V^I$, while a binary operation $\bullet^{\mathbf{V}_I}$ on V_I is defined by

$$\left(f \bullet^{\mathbf{V}_I} g\right)(p) = f(p) \bullet^{\mathbf{V}} g(p).$$

A function $f \in V_I$ will be called $\mathbf{V}$-*admissible* if for each $p \in V^I$ and each $x \in I$ the function

$$\langle f(p\{v/x\}) : v \in V \rangle$$

belongs to the domain of $\lambda^{\mathbf{V}}$. In particular the functions $\{x^{\mathbf{V}_I} : x \in I\}$ are all $\mathbf{V}$-admissible iff the domain of $\lambda^{\mathbf{V}}$ contains all the constant functions $\langle u : v \in V \rangle$ for $u \in V$ as well as the identity function $\langle v : v \in V \rangle$.

If f is $\mathbf{V}$-admissible then we can define $(\lambda x^{\mathbf{V}_I}.f) \in V_I$ by

$$(\lambda x^{\mathbf{V}_I}.f)(p) = \lambda^{\mathbf{V}}(\langle f(p\{v/x\}) : v \in V\rangle).$$

Then $\lambda x^{\mathbf{V}_I}$ is an operation on V_I that algebraically models lambda abstraction of the variable x (see below). If $A \subseteq V_I$ is a set of $\mathbf{V}$-admissible functions that includes $\{x^{\mathbf{V}_I} : x \in I\}$ and is closed under the binary operation $\bullet^{\mathbf{V}_I}$ and the unary operations $\lambda x^{\mathbf{V}_I}$, then the algebra

$$\mathbf{A} = \left\langle A, \bullet^{\mathbf{V}_I}, \{\lambda x^{\mathbf{V}_I} : x \in I\}, \{x^{\mathbf{V}_I} : x \in I\}\right\rangle$$

is a *functional lambda abstraction algebra of dimension* I. We use the acronym FLA_I for the class of all functional lambda abstraction algebras of dimension I, and refer to a member of this class as "an FLA_I".

Using FLA_I-operations, the denotation of a lambda calculus term M can be given as a function

$$\mathbf{V}(M) : V^I \to V$$

from environments to values, i.e. $\mathbf{V}(M) \in V_I$. For each $p \in V^I$, the element $\mathbf{V}(M)(p)$ of V is the *value of* M *in environment* p. Inductively,

$$\begin{aligned} \mathbf{V}(x) &= x^{\mathbf{V}_I} \\ \mathbf{V}(MN) &= \mathbf{V}(M) \bullet^{\mathbf{V}_I} \mathbf{V}(N) \\ \mathbf{V}(\lambda x.M) &= \lambda x^{\mathbf{V}_I}.\mathbf{V}(M). \end{aligned}$$

The last equation specifies that for any $p \in V^I$,

$$\mathbf{V}(\lambda x.M)(p) = \lambda^{\mathbf{V}}(\langle\, \mathbf{V}(M)(p\{v/x\}) : v \in V\,\rangle).$$

For this to make sense, the function $\mathbf{V}(M)$ has to be $\mathbf{V}$-admissible, which requires that the function

$$\langle\, \mathbf{V}(M)(p\{v/x\}) : v \in V\,\rangle$$

actually belongs to the domain of $\lambda^{\mathbf{V}}$ for any p. An *environment model* [Mey82, page 96] is a functional domain in which this requirement is always satisfied, so that the denotation of every term is defined. Meyer expresses the view that "environment models are the natural, most general formulation of what might be meant by mathematical models for the untyped lambda calculus".

Axiomatically, the lambda calculus is a theory about equations between terms. In a term of the form $\lambda x.M$, the lambda abstraction operator λx acts like a quantifier and binds all occurrences of x in M. An occurrence of x in a term is *free* if it is not within the scope of the operator λx. $M[N/x]$ is the term obtained by substituting N for all free occurrences of x in M. N is *free for* x *in* M if no free occurrence of a variable in N becomes bound in $M[N/x]$.

The two principal axiom schemes of the lambda calculus are

(α) $\lambda x.M = \lambda y.M[y/x]$ if y does not occur free in M.

(β) $(\lambda x.M)N = M[N/x]$ if N is free for x in M.

Here (α) states that the names of bound variables do not matter, just as $\int f(x)dx = \int f(y)dy$ in integration theory. (β) states that functional application can be computed by syntactic substitution. Thus in the language of arithmetic, if f is the function $\lambda x.(x^2+1)$ then $f(2)$ is computed as $(x^2+1)[2/x] = 2^2+1$ etc.

The rules for deriving equations from instances of (α) and (β) are the usual ones from equational logic asserting that equality is a congruence for application and abstraction. An equation $M = N$ is *valid* in an environment model $\mathbf{V}$ if M and N have the same denotation in $\mathbf{V}$, i.e. the functions $\mathbf{V}(M)$ and $\mathbf{V}(N)$ are identical. Any equation derivable in the lambda calculus is valid in all environment models [Mey82, page 113].

There are several known "syntax-free" characterisations of environment models ([Sco80, Mey82, Koy82, Koy84] et. alia. Cf. footnote 4 of [Mey82] for references). For instance, [Mey82, page 89] defines a *combinatory model* as a structure of the form

$$\langle V, \bullet, \mathbf{s}, \mathbf{k}, \mathbf{1} \rangle$$

with $\bullet$ a binary operation and $\mathbf{k}, \mathbf{s}, \mathbf{1}$ distinguished elements satisfying the equations

$$(\mathbf{k}{\bullet}u){\bullet}v = u$$
$$((\mathbf{s}{\bullet}u){\bullet}v){\bullet}w = (u{\bullet}w){\bullet}(v{\bullet}w)$$
$$(\mathbf{1}{\bullet}u){\bullet}v = u{\bullet}v$$
$$\mathbf{1}{\bullet}\mathbf{1} = \mathbf{1}$$

and the *Meyer-Scott axiom*

$$\text{If} \quad (\forall w \in V)\,(u{\bullet}w = v{\bullet}w) \quad \text{then} \quad \mathbf{1}{\bullet}u = \mathbf{1}{\bullet}v.$$

Associated with such a combinatory model is the environment model $\mathbf{V}$ in which

(i) $\bullet^{\mathbf{V}}$ is the given operation $\bullet$;
(ii) the domain of $\lambda^{\mathbf{V}}$ is the collection

$$\{\Phi^{\mathbf{V}}(u) : u \in V\} \subseteq V^V$$

of functions induced by left multiplication by $\bullet$, i.e. $\Phi^{\mathbf{V}}(u)(w) = u{\bullet}w$; and
(iii) the action of $\lambda^{\mathbf{V}}$ is given by $\lambda^{\mathbf{V}}(\Phi^{\mathbf{V}}(u)) = \mathbf{1}{\bullet}u$, which is well defined by the Meyer-Scott axiom, since the latter asserts that $\Phi^{\mathbf{V}}(u) = \Phi^{\mathbf{V}}(v)$ only if $\mathbf{1}{\bullet}u = \mathbf{1}{\bullet}v$.

Conversely, every environment model becomes a combinatorial model when $\mathbf{k}$, $\mathbf{s}$, and $\mathbf{1}$ are taken as the values of the terms $\lambda x.(\lambda y.x)$, $\lambda x.(\lambda y.(\lambda z.((xz)yz)))$ and $\lambda x.(\lambda y.xy)$. These terms get the same value in any environment. This will be discussed further in Section 4 when we make use of the Meyer-Scott axiom to lift the functional domain structure to an ultrapower of $\mathbf{V}$.

2 Relativised functional algebras

On the set V^I of environments of a functional domain, define an equivalence relation $\equiv_\omega$ by putting

$$p \equiv_\omega q \quad \text{iff} \quad \{x \in I : p_x \neq q_x\} \text{ is finite.}$$

So $p \equiv_\omega q$ when p and q agree on a *cofinite* subset of I. For instance, $p \equiv_\omega p\{v/x\}$ in general. For a particular environment $r \in V^I$, let

$$V_r^I = \{q \in V^I : r \equiv_\omega q\}$$

be the $\equiv_\omega$-equivalence class of r. Note that V_r^I can be much smaller than V^I: if both V and I are countably infinite then so is V_r^I, while V^I is uncountable. On the other hand, if I is finite then $V_r^I = V^I$.

The construction of FLA_Is can be relativised in this context, by replacing V^I everywhere by V_r^I. We write $V_{I,r}$ for the set of all functions from V_r^I to V. Then $x^{\mathbf{V}_{I,r}} \in V_{I,r}$ is the restriction of $x^{\mathbf{V}_I}$ to V_r^I, while $\bullet^{\mathbf{V}_{I,r}}$ is the binary operation on $V_{I,r}$ defined in the same manner as $\bullet^{\mathbf{V}_I}$. Thus

$$x^{\mathbf{V}_{I,r}}(p) = p_x$$
$$(f \bullet^{\mathbf{V}_{I,r}} g)(p) = f(p) \bullet^{\mathbf{V}} g(p)$$

for $p \in V_r^I$ and $f, g \in V_{I,r}$. A function $f \in V_{I,r}$ will be called $\mathbf{V}_r$-*admissible* if the function $\langle f(p\{v/x\}) : v \in V\rangle$ belongs to the domain of $\lambda^{\mathbf{V}}$ whenever $p \in V_r^I$. If f is $\mathbf{V}_r$-admissible then we have $(\lambda x^{\mathbf{V}_{I,r}}.f) \in V_{I,r}$ defined by

$$(\lambda x^{\mathbf{V}_{I,r}}.f)(p) = \lambda^{\mathbf{V}}(\langle f(p\{v/x\}) : v \in V\rangle).$$

If $A \subseteq V_{I,r}$ is a set of $\mathbf{V}_r$-admissible functions that includes $\{x^{\mathbf{V}_{I,r}} : x \in I\}$ and is closed under the operations $\bullet^{\mathbf{V}_{I,r}}$ and $\lambda x^{\mathbf{V}_{I,r}}$, then the algebra

$$\mathbf{A} = \left\langle A, \bullet^{\mathbf{V}_{I,r}}, \{\lambda x^{\mathbf{V}_{I,r}} : x \in I\}, \{x^{\mathbf{V}_{I,r}} : x \in I\}\right\rangle$$

is a *point-relativised functional lambda abstraction algebra of dimension* I, or RFA_I.

Every FLA_I can be decomposed as a subdirect product of RFA_Is with the same underlying functional domain [PS95, Proposition 3.13]. To see how, let $\mathbf{A}$ be an FLA_I with functional domain $\mathbf{V}$. Then for each $r \in V^I$ and $f \in A$ let $f_r \in V_{I,r}$ be the restriction of f to V_r^I. The map $\theta_r : f \mapsto f_r$ is a homomorphism from $\mathbf{A}$ onto an RFA_I $\mathbf{A}_r$. The product map $\theta : f \mapsto \langle f_r : r \in V^I\rangle$ then gives the desired subdirect embedding

$$\theta : \mathbf{A} \longrightarrow \prod_{r \in V^I} \mathbf{A}_r.$$

It was shown in [PS95, Corollary 7.5] that the class $\mathbf{I}\mathsf{RFA}_I$ of algebras isomorphic to RFA_Is is an equational class, or variety, by proving that it is closed under homomorphic images, subalgebras, and direct products. From the subdirect decomposition just described it then follows that $\mathsf{FLA}_I \subseteq \mathbf{I}\mathsf{RFA}_I$. Furthermore it was shown that every RFA_I is a homomorphic image of an FLA_I [PS95, Proposition 5.13], implying that the variety $\mathbf{I}\mathsf{RFA}_I$ is generated by FLA_I.

The purpose of this paper is to set out a proof that every RFA_I is isomorphic to an FLA_I, thereby showing that $\mathbf{I}\mathsf{FLA}_I = \mathbf{I}\mathsf{RFA}_I$, so the algebras isomorphic to functional lambda abstraction algebras form a variety. This will be demonstrated by showing that an RFA_I $\mathbf{A}$ is isomorphic to an FLA_I whose functional domain is an ultrapower of the domain underlying $\mathbf{A}$. (We are only interested in the case of infinite I here, for if I is finite then every RFA_I is an FLA_I, because in that case $V_r^I = V^I$ in general.)

3 The ultrapower of V

From now on let I be an infinite set, and $\mathbf{A}$ a fixed RFA_I over a functional domain $\mathbf{V} = \langle V, \bullet^{\mathbf{V}}, \lambda^{\mathbf{V}} \rangle$, with the underlying set A of $\mathbf{A}$ contained in $V_{I,r}$ for some environment $r \in V^I$.

The fact that $\mathbf{V}$ is a functional domain means that $\Phi^{\mathbf{V}}\left(\lambda^{\mathbf{V}}(f)\right) = f$ whenever $\lambda^{\mathbf{V}}(f)$ is defined. We are going to define a functional domain structure on an ultrapower of $\mathbf{V}$ and motivation for this will come from the following characterisation of the domain of the function $\lambda^{\mathbf{V}}$. The result follows from the existence of an RFA_I based on $\mathbf{V}$.

Lemma 1. *The domain* $\operatorname{dom}\lambda^{\mathbf{V}}$ *of* $\lambda^{\mathbf{V}}$ *is the collection of functions*

$$\{\Phi^{\mathbf{V}}(u) : u \in V\}.$$

Proof. If $f \in \operatorname{dom}\lambda^{\mathbf{V}}$, put $u = \lambda^{\mathbf{V}}(f)$. The functional domain axiom then gives $f = \Phi^{\mathbf{V}}(u)$.

Conversely, take any $u \in V$, let x, y be distinct members of I and put $p = r\{u/x\} \in V_r^I$. The entity $\left(x^{\mathbf{V}_{I,r}} \bullet^{\mathbf{V}_{I,r}} y^{\mathbf{V}_{I,r}}\right)$ belongs to the RFA_I $\mathbf{A}$, so is $\mathbf{V}_r$-admissible. If follows that the function

$$h = \left\langle \left(x^{\mathbf{V}_{I,r}} \bullet^{\mathbf{V}_{I,r}} y^{\mathbf{V}_{I,r}}\right)(p\{v/y\}) : v \in V \right\rangle$$

belongs to the domain of $\lambda^{\mathbf{V}}$. But

$$\left(x^{\mathbf{V}_{I,r}} \bullet^{\mathbf{V}_{I,r}} y^{\mathbf{V}_{I,r}}\right)(p\{v/y\}) = \left(x^{\mathbf{V}_{I,r}}\right)(p\{v/y\}) \bullet^{\mathbf{V}} \left(y^{\mathbf{V}_{I,r}}\right)(p\{v/y\}) = u \bullet^{\mathbf{V}} v,$$

(because $p\{v/y\}_x = p_x = u$) so $h = \langle u \bullet^{\mathbf{V}} v : v \in V\rangle = \Phi^{\mathbf{V}}(u)$. Hence $\Phi^{\mathbf{V}}(u)$ belongs to $\operatorname{dom}\lambda^{\mathbf{V}}$.

□

Another consequence of the existence of $\mathbf{A}$ is that $\mathbf{V}$ satisfies the *Meyer-Scott axiom* referred to at the end of Section 1. To see this, take distinct $x, y \in I$ and let

$$h = \lambda x^{\mathbf{V}_{I,r}}.\left(\lambda y^{\mathbf{V}_{I,r}}.\left(x^{\mathbf{V}_{I,r}} \bullet^{\mathbf{V}_{I,r}} y^{\mathbf{V}_{I,r}}\right)\right) \in \mathbf{A}.$$

Then h is the interpretation in $\mathbf{A}$ of the lambda calculus term $\lambda x.(\lambda y.xy)$, and for any $p \in V_r^I$ we calculate

$$\begin{aligned} h(p) &= \lambda^{\mathbf{V}}\Big(\Big\langle \lambda^{\mathbf{V}}(\langle u \bullet^{\mathbf{V}} v : v \in V\rangle) : u \in V\Big\rangle\Big) \\ &= \lambda^{\mathbf{V}}\Big(\Big\langle \lambda^{\mathbf{V}}(\Phi^{\mathbf{V}}(u)) : u \in V\Big\rangle\Big). \end{aligned}$$

Let $\mathbf{1}^{\mathbf{V}}$ be this value $h(p) \in V$, which is the same for all $p \in V_r^I$.

Lemma 2. *The structure* $\langle V, \bullet^{\mathbf{V}}, \mathbf{1}^{\mathbf{V}}\rangle$ *satisfies*

(i) *The Meyer-Scott axiom*

$$\big[\forall w \in V\, (u \bullet w = v \bullet w)\big] \longrightarrow \mathbf{1} \bullet u = \mathbf{1} \bullet v.$$

(ii) *The equation*

$$(\mathbf{1} \bullet u) \bullet v = u \bullet v.$$

Proof.

(i) For any $u \in V$,

$$\begin{aligned} \mathbf{1}^{\mathbf{V}} \bullet u &= h(p) \bullet u \\ &= \lambda^{\mathbf{V}}\Big(\Big\langle \lambda^{\mathbf{V}}(\Phi^{\mathbf{V}}(u)) : u \in V\Big\rangle\Big) \bullet u \\ &= \lambda^{\mathbf{V}}(\Phi^{\mathbf{V}}(u)), \end{aligned}$$

the last step being by the functional domain axiom. It follows that if $\Phi^{\mathbf{V}}(u) = \Phi^{\mathbf{V}}(v)$ then $\mathbf{1}^{\mathbf{V}} \bullet u = \mathbf{1}^{\mathbf{V}} \bullet v$, which is what the Scott-Meyer axiom asserts.

(ii)

$$\begin{aligned} (\mathbf{1}^{\mathbf{V}} \bullet u) \bullet v &= \lambda^{\mathbf{V}}(\Phi^{\mathbf{V}}(u)) \bullet v && \text{by the proof of (i)} \\ &= \Phi^{\mathbf{V}}(u)(v) && \text{by the functional domain axiom} \\ &= u \bullet v && \text{definition of } \Phi^{\mathbf{V}}. \end{aligned}$$

□

Now take J to be the set of all finite subsets of I, and $\mathcal{F}$ an ultrafilter on J that contains the set

$$J_i = \{j \in J : i \subseteq j\}$$

for each $i \in J$. Hence $\mathcal{F}$ contains $J_x = \{j \in J : x \in j\}$ for each $x \in I$.

$\mathcal{F}$ induces an equivalence relation $=_{\mathcal{F}}$ on the set V^J of all function from J to V by

$$a =_{\mathcal{F}} b \quad \text{iff} \quad \{j \in J : a(j) = b(j)\} \in \mathcal{F}.$$

Let $a/\mathcal{F} = \{b \in V^J : a =_{\mathcal{F}} b\}$ be the *$\mathcal{F}$-equivalence class of f* and

$$V^J/\mathcal{F} = \{a/\mathcal{F} : a \in V^J\}$$

the associated quotient set, known as the *ultrapower of* V *by* $\mathcal{F}$.

Put $^*V = V^J/\mathcal{F}$. To make this into a functional domain

$$^*\mathbf{V} = \left\langle {}^*V, \bullet^{^*\mathbf{V}}, \lambda^{^*\mathbf{V}} \right\rangle,$$

begin by defining

$$(a/\mathcal{F}) \bullet^{^*\mathbf{V}} (b/\mathcal{F}) = \langle a(j) \bullet^{\mathbf{V}} b(j) : j \in J \rangle/\mathcal{F}$$

for any $a, b \in V^J$. Use $\bullet^{^*\mathbf{V}}$ to define $\Phi^{^*\mathbf{V}} : {}^*V \to \left({}^*V\right)^{^*V}$, i.e.

$$\Phi^{^*\mathbf{V}}(d)(e) = d \bullet^{^*\mathbf{V}} e.$$

Next let $\mathbf{1}^{^*\mathbf{V}}$ be the element $\overline{\mathbf{1}^{\mathbf{V}}}/\mathcal{F}$ of $^*V$, where $\overline{\mathbf{1}^{\mathbf{V}}}$ is the function on J constantly equal to $\mathbf{1}^{\mathbf{V}}$. The structure

$$\left\langle {}^*V, \bullet^{^*\mathbf{V}}, \mathbf{1}^{^*\mathbf{V}} \right\rangle$$

is, by definition, the ultrapower of the structure $\left\langle V, \bullet^{\mathbf{V}}, \mathbf{1}^{\mathbf{V}} \right\rangle$, so they both satisfy the same elementary properties (Łoś's Theorem). In particular, by Lemma 2(i) the new structure satisfies the Meyer-Scott axiom, and so

$$\Phi^{^*\mathbf{V}}(d) = \Phi^{^*\mathbf{V}}(e) \quad \text{implies} \quad \mathbf{1}^{^*\mathbf{V}} \bullet^{^*\mathbf{V}} d = \mathbf{1}^{^*\mathbf{V}} \bullet^{^*\mathbf{V}} e.$$

Hence (with motivation also from Lemma 1) we can put

$$\operatorname{dom} \lambda^{^*\mathbf{V}} = \left\{ \Phi^{^*\mathbf{V}}(d) : d \in {}^*V \right\}$$

and

$$\lambda^{^*\mathbf{V}}\left(\Phi^{^*\mathbf{V}}(d)\right) = \mathbf{1}^{^*\mathbf{V}} \bullet^{^*\mathbf{V}} d$$

to get a well defined function $\lambda^{^*\mathbf{V}}$.

The ultrapower also satisfies the equation of Lemma 2(ii), which means that if $d \in {}^*V$, then

$$\left(\mathbf{1}^{^*\mathbf{V}} \bullet^{^*\mathbf{V}} d\right) \bullet^{^*\mathbf{V}} e = d \bullet^{^*\mathbf{V}} e$$

for all $e \in {}^*V$. This entails that

$$\Phi^{^*\mathbf{V}}\left(\lambda^{^*\mathbf{V}}\left(\Phi^{^*\mathbf{V}}(d)\right)\right) = \Phi^{^*\mathbf{V}}(d),$$

which is the condition required for

$$^*\mathbf{V} = \left\langle {}^*V, \bullet^{^*\mathbf{V}}, \lambda^{^*\mathbf{V}} \right\rangle$$

to be a functional domain.

4 Choice functions

Consider a map of the form

$$\mathrm{ch} : V^J/\mathcal{F} \longrightarrow \left(V^J\right)^I,$$

assigning to each $\mathcal{F}$-equivalence class $S \in V^J/\mathcal{F}$ an "I-tuple"

$$\mathrm{ch}(S) = \langle \mathrm{ch}(S)_x : x \in I \rangle$$

of functions $\mathrm{ch}(S)_x \in V^J$. Such a map will be called a *choice function* if for all $x \in I$, $\mathrm{ch}(S)_x \in S$ and hence $\mathrm{ch}(S)_x/\mathcal{F} = S$.

Any choice function induces a map

$$\mathrm{ch}^+ : \left(V^J/\mathcal{F}\right)^I \longrightarrow \left(V^I\right)^J$$

such that if $p = \langle p_x : x \in I \rangle \in \left(V^J/\mathcal{F}\right)^I$, then the J-tuple

$$\mathrm{ch}^+(p) = \langle \mathrm{ch}^+(p)(j) : j \in J \rangle$$

is given by

$$\mathrm{ch}^+(p)(j) = \langle \mathrm{ch}(p_x)_x(j) : x \in I \rangle.$$

In other words, ch^+ is determined by the requirement that

$$\mathrm{ch}^+(p)(j)_x = \mathrm{ch}(p_x)_x(j).$$

For technical reasons that will emerge later, we need a choice function with very special properties. These are summarised in the following result.

Choice Lemma. *There exists a choice function*

$$\mathrm{ch} : V^J/\mathcal{F} \longrightarrow \left(V^J\right)^I,$$

such that

(i) $\mathrm{ch}^+(p)(j) \in V_r^I$ *for all* $p \in \left(V^J/\mathcal{F}\right)^I$ *and all* $j \in J$;

(ii) *for each* $s \in V_r^I$ *there exists a* $p^s \in \left(V^J/\mathcal{F}\right)^I$ *such that* $\mathrm{ch}^+(p^s)(j) = s$ *for* $\mathcal{F}$*-almost all* j, *i.e.* $\{j \in J : \mathrm{ch}^+(p^s)(j) = s\} \in \mathcal{F}$.

Proof. For any object v, let $\overline{v} = \langle v : j \in J \rangle$ be the function on J with constant value v. For each $S \in V^J/\mathcal{F}$ and $x \in I$, choose $\mathrm{ch}(S)_x \in S$ to satisfy the following for all $x \in I$:

(a) $\mathrm{ch}(S)_x(j) = r_x$ if $x \notin j$, i.e. if $j \notin J_x$.

(b) for each $v \in V$, if $\overline{v} \in S$, then $\mathrm{ch}(S)_x(j) = v$ for all $j \in J_x$.

To see that such a choice is possible, observe that since S is closed under $=_{\mathcal{F}}$, if we alter the values of a member of S only on a set that is not in $\mathcal{F}$, the new function still belongs to S. In particular we can modify any member of S on the complement of J_x arbitrarily and still have a member of S. So (a) can always be arranged.

Now there can be at most one v with $\overline{v} \in S$, as distinct constant functions cannot be $\mathcal{F}$-equivalent. If such a $\overline{v}$ exists, modify it to be equal to r_x outside of J_x to define $\mathrm{ch}(S)_x$ satisfying (a) and (b). If there is no $\overline{v} \in S$, then only (a) has to be fulfilled. Thus $\mathrm{ch}(S)_x$ can indeed always be chosen to satisfy (a) and (b).

(i) Let $p \in \left(V^J/\mathcal{F}\right)^I$. Then for any $j \in J$,

$$\begin{aligned} \mathrm{ch}^+(p)(j)_x &= \mathrm{ch}(p_x)_x(j) && \text{by definition of ch}^+ \\ &= r_x \text{ if } x \notin j && \text{by (a).} \end{aligned}$$

Hence $\mathrm{ch}^+(p)(j)$ agrees with r_x outside the *finite* set j, so $\mathrm{ch}^+(p)(j) \equiv_\omega r$, giving $\mathrm{ch}^+(p)(j) \in V_r^I$.

(ii) Let $s = \langle s_x : x \in I \rangle \in V_r^I$. Define $p^s = \langle p_x^s : x \in I \rangle$ by

$$p_x^s = \overline{s_x}/\mathcal{F} \in V^J/\mathcal{F}.$$

Then for any $j \in J$,

$$\begin{aligned} \mathrm{ch}^+(p^s)(j)_x &= \mathrm{ch}(p_x^s)_x(j) && \text{by definition of ch}^+ \\ &= \mathrm{ch}(\overline{s_x}/\mathcal{F})_x(j) && \text{by definition of } p^s \\ &= \begin{cases} r_x & \text{if } x \notin j, \\ s_x & \text{if } x \in j. \end{cases} && \text{by (a) and (b).} \end{aligned}$$

Hence

(c) $\mathrm{ch}^+(p^s)(j)$ agrees with r outside j; and

(d) $\mathrm{ch}^+(p^s)(j)$ agrees with s on j.

But $s \in V_r^I$, so there is some finite $k \subseteq I$ such that

(e) r agrees with s outside k.

Then if $j \in J_k$, we have $k \subseteq j$, so r agrees with s outside j by (e). Therefore $\mathrm{ch}^+(p^s)(j)$ agrees with s outside j by (c), and agrees with s on j by (d), showing $\mathrm{ch}^+(p^s)(j) = s$. Thus we have

$$\{j \in J : \mathrm{ch}^+(p^s)(j) = s\} \supseteq J_k \in \mathcal{F},$$

proving (ii). □

5 The isomorphism

The set A underlying the RFA_I $\mathbf{A}$ is a collection of functions of the form $f : V_r^I \to V$. To represent $\mathbf{A}$ as an FLA_I over ${}^*\mathbf{V}$ we construct a map

$$\theta : A \longrightarrow {}^*V_I,$$

where ${}^*V_I$ is the collection of functions of the form ${}^*V^I \to {}^*V$. For this purpose we take a choice function fulfilling the Choice Lemma, and then for each $f \in A$ define $\theta(f) \in {}^*V_I$ by putting

$$\theta(f)(p) = \langle f\big(\mathrm{ch}^+(p)(j)\big) : j \in J\rangle/\mathcal{F}$$

for all $p \in {}^*V^I$. Note that part (i) of the Choice Lemma is needed here to ensure that $\mathrm{ch}^+(p)(j)$ is in the domain V_r^I of f, so that $f\big(\mathrm{ch}^+(p)(j)\big)$ is always defined as an element of V, making $\theta(f)(p)$ defined as an element of $V^J/\mathcal{F} = {}^*V$.

Our task now is to show that $\theta(A)$ is a collection of ${}^*\mathbf{V}$-admissible functions, and that θ is an injection which is an algebraic homomorphism:

$$\theta\big(x^{\mathbf{V}_{I,r}}\big) = x^{{}^*\mathbf{V}_I} \tag{1}$$
$$\theta\big(f \bullet^{\mathbf{V}_{I,r}} g\big) = \theta(f) \bullet^{{}^*\mathbf{V}_I} \theta(g) \tag{2}$$
$$\theta\big(\lambda x^{\mathbf{V}_{I,r}}.f\big) = \lambda x^{{}^*\mathbf{V}_I}.\theta(f)\ . \tag{3}$$

This will establish that $\theta(A)$ contains the distinguished elements $x^{{}^*\mathbf{V}_I}$ and is closed under the operations $\bullet^{{}^*\mathbf{V}_I}$ and $\lambda x^{{}^*\mathbf{V}_I}$, so that

$$\theta(\mathbf{A}) = \Big\langle \theta(A),\ \bullet^{{}^*\mathbf{V}_I}, \Big\{\lambda x^{{}^*\mathbf{V}_I} : x \in I\Big\}, \Big\{x^{{}^*\mathbf{V}_I} : x \in I\Big\}\Big\rangle$$

is an FLA_I isomorphic to the RFA_I $\mathbf{A}$.

Proof that θ is injective

Let $f, g \in A$ with $f \neq g$. Then for some $s \in V_r^I$, $f(s) \neq g(s)$. By part (ii) of the Choice Lemma there exists some $p^s \in {}^*V^I$ such that the set

$$K = \{j \in J : \mathrm{ch}^+(p^s)(j) = s\}$$

belongs to $\mathcal{F}$. But for $j \in K$, $f\big(\mathrm{ch}^+(p^s)(j)\big) = f(s)$, so it follows that

$$\langle f\big(\mathrm{ch}^+(p^s)(j)\big) : j \in J\rangle/\mathcal{F} = \langle f(s) : j \in J\rangle/\mathcal{F},$$

which means that $\theta(f)(p^s) = \overline{f(s)}/\mathcal{F}$. Likewise $\theta(g)(p^s) = \overline{g(s)}/\mathcal{F}$. Hence as $f(s) \neq g(s)$, $\theta(f)(p^s) \neq \theta(g)(p^s)$, showing $\theta(f) \neq \theta(g)$.

Proof of Equation (1)

For any $p \in {}^*V^I$,

$$\begin{aligned}
\theta(x^{\mathbf{V}_{I,r}})(p) &= \langle x^{\mathbf{V}_{I,r}}(\mathrm{ch}^+(p)(j)) : j \in J\rangle/\mathcal{F} && \text{definition of } \theta\\
&= \langle \mathrm{ch}^+(p)(j)_x : j \in J\rangle/\mathcal{F} && \text{definition of } x^{\mathbf{V}_{I,r}}\\
&= \langle \mathrm{ch}(p_x)_x(j) : j \in J\rangle/\mathcal{F} && \text{definition of } \mathrm{ch}^+\\
&= \mathrm{ch}(p_x)_x/\mathcal{F}\\
&= p_x && \text{as } \mathrm{ch}(p_x)_x \in p_x\\
&= x^{{}^*\mathbf{V}_I}(p) && \text{definition of } x^{{}^*\mathbf{V}_I},
\end{aligned}$$

so $\theta(x^{\mathbf{V}_{I,r}}) = x^{{}^*\mathbf{V}_I}$.

Proof of Equation (2)

$$\begin{aligned}
&\theta(f \bullet^{\mathbf{V}_{I,r}} g)(p)\\
&= \langle f \bullet^{\mathbf{V}_{I,r}} g(\mathrm{ch}^+(p)(j)) : j \in J\rangle/\mathcal{F} && \text{definition of } \theta\\
&= \langle f(\mathrm{ch}^+(p)(j)) \bullet^{\mathbf{V}} g(\mathrm{ch}^+(p)(j)) : j \in J\rangle/\mathcal{F} && \text{definition of } \bullet^{\mathbf{V}_{I,r}}\\
&= \langle f(\mathrm{ch}^+(p)(j)) : j \in J\rangle/\mathcal{F} \bullet^{{}^*\mathbf{V}} \langle g(\mathrm{ch}^+(p)(j)) : j \in J\rangle/\mathcal{F} && \text{definition of } \bullet^{{}^*\mathbf{V}}\\
&= \theta(f)(p) \bullet^{{}^*\mathbf{V}} \theta(g)(p) && \text{definition of } \theta\\
&= [\theta(f) \bullet^{{}^*\mathbf{V}_I} \theta(g)]\,(p) && \text{definition of } \bullet^{{}^*\mathbf{V}_I},
\end{aligned}$$

so $\theta(f \bullet^{\mathbf{V}_{I,r}} g) = \theta(f) \bullet^{{}^*\mathbf{V}_I} \theta(g)$.

Proof of Equation (3)

This requires a lengthy and intricate preliminary analysis. If $p \in {}^*V^I$ and $a \in V^J$, then for $x \in I$ the function (${}^*V$-environment) $p\{(a/\mathcal{F})/x\}$ also belongs to ${}^*V^I$. It differs from p only in assigning $a/\mathcal{F}$ to x. By the Choice Lemma (i), for each $j \in J$ the two V-environments

$$s(j) = \mathrm{ch}^+(p\{(a/\mathcal{F})/x\})(j),$$
$$t(j) = (\mathrm{ch}^+(p)(j))\{a(j)/x\}$$

both belong to V_r^I.

Lemma 3.

(i) $\mathrm{ch}^+(p\{(a/\mathcal{F})/x\})(j) = \mathrm{ch}^+(p)(j)\{a(j)/x\}$ *for $\mathcal{F}$-almost all* $j \in J$.

(ii) $\langle f(\mathrm{ch}^+(p\{(a/\mathcal{F})/x\})(j)) : j \in J\rangle/\mathcal{F} =$
$$\langle f(\mathrm{ch}^+(p)(j)\{a(j)/x\}) : j \in J\rangle/\mathcal{F}.$$

(iii) $\theta(f)\big(p\{(a/\mathcal{F})/x\}\big) = \langle f\big(\mathrm{ch}^+(p)(j)\{a(j)/x\}\big) : j \in J\rangle/\mathcal{F}$.

Proof. If $y \neq x$, then

$$\begin{aligned}
s(j)_y &= \mathrm{ch}^+\big(p\{(a/\mathcal{F})/x\}\big)(j)_y && \text{definition of } s(j)\\
&= \mathrm{ch}(p\{(a/\mathcal{F})/x\}_y)_y(j) && \text{definition of ch}^+\\
&= \mathrm{ch}(p_y)_y(j) && \text{as } y \neq x\\
&= \mathrm{ch}^+(p)(j)_y && \text{definition of ch}^+\\
&= \big[\big(\mathrm{ch}^+(p)(j)\big)\{a(j)/x\}\big]_y && \text{as } x \neq y\\
&= t(j)_y.
\end{aligned}$$

Hence for any j the functions $s(j)$ and $t(j)$ differ at most at x, implying

$$s(j) = t(j) \quad \text{iff} \quad s(j)_x = t(j)_x. \tag{4}$$

But

$$\begin{aligned}
s(j)_x &= \mathrm{ch}^+\big(p\{(a/\mathcal{F})/x\}\big)(j)_x && \text{definition of } s(j)\\
&= \mathrm{ch}(p\{(a/\mathcal{F})/x\}_x)_x(j) && \text{definition of ch}^+\\
&= \mathrm{ch}(a/\mathcal{F})_x(j),
\end{aligned}$$

and

$$\begin{aligned}
t(j)_x &= \big[\mathrm{ch}^+(p)(j)\{a(j)/x\}\big]_x\\
&= a(j),
\end{aligned}$$

so from (4) follows

$$s(j) = t(j) \quad \text{iff} \quad \mathrm{ch}(a/\mathcal{F})_x(j) = a(j). \tag{5}$$

Now $\mathrm{ch}(a/\mathcal{F})_x$ and a both belong to $a/\mathcal{F}$, so the set

$$K = \{j \in J : \mathrm{ch}(a/\mathcal{F})_x(j) = a(j)\}$$

belongs to the ultrafilter $\mathcal{F}$. But $K = \{j \in J : s(j) = t(j)\}$ by (5) , so this proves part (i) of the Lemma.

Next, observe that $K \subseteq \{j \in J : f(s(j)) = f(t(j))\}$, so

$$\langle f(s(j)) : j \in J\rangle/\mathcal{F} = \langle f(t(j)) : j \in J\rangle/\mathcal{F}$$

which is (ii).

Finally $\theta(f)\big(p\{(a/\mathcal{F})/x\}\big)$ is equal to the left side of (ii), by definition of $\theta(f)$, so (iii) follows from (ii). □

Lemma 4. *Let $x \in I$ and $f \in A$. For any $p \in {}^*V^I$ there is a function $a_p \in V^J$ such that*

(i) $\Phi^{\mathbf{V}}(a_p(j)) = \langle f\big(\mathrm{ch}^+(p)(j)\{v/x\}\big) : v \in V\rangle$ *for all $j \in J$;*

(ii) $\left(\lambda x^{\mathbf{V}_{I,r}}.f\right)\left(\mathrm{ch}^{+}(p)(j)\right) = \lambda^{\mathbf{V}}\left(\Phi^{\mathbf{V}}\left(a_p(j)\right)\right)$;

(iii) $\left\langle \theta(f)(p\{u/x\}) : u \in {}^{*}V \right\rangle = \Phi^{{}^{*}\mathbf{V}}(a_p/\mathcal{F})$.

Hence $\theta(f)$ is a ${}^{*}\mathbf{V}$-admissible function.

Proof. f is $\mathbf{V}_r$-admissible because it belongs to the RFA_I $\mathbf{A}$. Thus for any $j \in J$, since $\mathrm{ch}^{+}(p)(j)$ belongs to V_r^I the function

$$h_j = \left\langle f\left(\mathrm{ch}^{+}(p)(j)\{v/x\}\right) : v \in V \right\rangle$$

belongs to $\mathrm{dom}\,\lambda^{\mathbf{V}}$. By definition of $\lambda x^{\mathbf{V}_{I,r}}$,

$$\left(\lambda x^{\mathbf{V}_{I,r}}.f\right)\left(\mathrm{ch}^{+}(p)(j)\right) = \lambda^{\mathbf{V}}(h_j).$$

Now by Lemma 1, $h_j = \Phi^{\mathbf{V}}(u_j)$ for some $u_j \in V$. Put $a_p(j) = u_j$. This defines a_p as a function satisfying (i) and (ii). For (iii), let $u \in {}^{*}V$ with $u = a/\mathcal{F}$ for some $a \in V^J$. Then

$$\begin{aligned}
\Phi^{{}^{*}\mathbf{V}}(a_p/\mathcal{F})(u) &= \Phi^{{}^{*}\mathbf{V}}(a_p/\mathcal{F})(a/\mathcal{F}) && \\
&= (a_p/\mathcal{F}) \bullet^{{}^{*}\mathbf{V}} (a/\mathcal{F}) && \text{definition of } \Phi^{{}^{*}\mathbf{V}} \\
&= \left\langle a_p(j) \bullet^{\mathbf{V}} a(j) : j \in J \right\rangle/\mathcal{F} && \text{definition of } \bullet^{{}^{*}\mathbf{V}} \\
&= \left\langle \Phi^{\mathbf{V}}\left(a_p(j)\right)\left(a(j)\right) : j \in J \right\rangle/\mathcal{F} && \text{definition of } \Phi^{\mathbf{V}} \\
&= \left\langle f\left(\mathrm{ch}^{+}(p)(j)\{a(j)/x\}\right) : j \in J \right\rangle/\mathcal{F} && \text{by (i)} \\
&= \theta(f)\left(p\{(a/\mathcal{F})/x\}\right) && \text{by Lemma 3(iii)} \\
&= \theta(f)\left(p\{u/x\}\right),
\end{aligned}$$

which proves (iii). But (iii) itself asserts that for any $x \in I$ and any $p \in {}^{*}V^I$ the function

$$\left\langle \theta(f)(p\{u/x\}) : u \in {}^{*}V \right\rangle$$

is of the form $\Phi^{{}^{*}\mathbf{V}}(d)$ for some $d \in {}^{*}V$ and so belongs to $\mathrm{dom}\,\lambda^{{}^{*}\mathbf{V}}$. This means that $\theta(f)$ is ${}^{*}\mathbf{V}$-admissible. □

Lemma 5. *For any $b \in V^J$,*

$$\lambda^{{}^{*}\mathbf{V}}\left(\Phi^{{}^{*}\mathbf{V}}(b/\mathcal{F})\right) = \left\langle \lambda^{\mathbf{V}}\left(\Phi^{\mathbf{V}}(b(j))\right) : j \in J \right\rangle/\mathcal{F}.$$

Proof.

$$\begin{aligned}
\lambda^{{}^{*}\mathbf{V}}\left(\Phi^{{}^{*}\mathbf{V}}(b/\mathcal{F})\right) &= \mathbf{1}^{{}^{*}\mathbf{V}} \bullet^{{}^{*}\mathbf{V}} (b/\mathcal{F}) && \text{definition of } \lambda^{{}^{*}\mathbf{V}} \\
&= \left\langle \mathbf{1}^{\mathbf{V}} \bullet^{\mathbf{V}} b(j) : j \in J \right\rangle/\mathcal{F} && \text{definition of } \bullet^{{}^{*}\mathbf{V}},\ \mathbf{1}^{{}^{*}\mathbf{V}} \\
&= \left\langle \lambda^{\mathbf{V}}\left(\Phi^{\mathbf{V}}(b(j))\right) : j \in J \right\rangle/\mathcal{F} && \text{proof of Lemma 2(i).}
\end{aligned}$$

□

At last we are ready for the proof of equation (3). Let $f \in A$, and take any $p \in {}^*V^I$. Let a_p be as in Lemma 4. That Lemma showed $\theta(f)$ to be $^*\mathbf{V}$-admissible, so $\lambda x^{{}^*\mathbf{V}_I}.\theta(f)$ is defined. We calculate

$$\begin{aligned}
\theta\big(\lambda x^{\mathbf{V}_{I,r}}.f\big)(p) &= \big\langle\big(\lambda x^{\mathbf{V}_{I,r}}.f\big)\big(\mathrm{ch}^+(p)(j)\big) : j \in J\big\rangle/\mathcal{F} && \text{definition of } \theta \\
&= \big\langle\lambda^{\mathbf{V}}\big(\Phi^{\mathbf{V}}\big(a_p(j)\big)\big) : j \in J\big\rangle/\mathcal{F} && \text{by Lemma 4(ii)} \\
&= \lambda^{{}^*\mathbf{V}}\big(\Phi^{{}^*\mathbf{V}}(a_p/\mathcal{F})\big) && \text{by Lemma 5} \\
&= \lambda^{{}^*\mathbf{V}}\big(\langle\theta(f)(p\{u/x\}) : u \in {}^*V\rangle\big) && \text{by Lemma 4(iii)} \\
&= \big(\lambda x^{{}^*\mathbf{V}_I}.\theta(f)\big)(p) && \text{definition of } \lambda x^{{}^*\mathbf{V}_I},
\end{aligned}$$

so $\theta\big(\lambda x^{\mathbf{V}_{I,r}}.f\big) = \lambda x^{{}^*\mathbf{V}_I}.\theta(f)$.

6 Finite axiomatisation

We have now completed the proof that any RFA_I $\mathbf{A}$ is isomorphic to the FLA_I $\theta(\mathbf{A})$. This proof was found by the present author in June 1995. Later in joint work with Salibra [SG96] it was used to help show that the variety $\mathbf{I}\mathsf{FLA}_I$ of algebras isomorphic to functional lambda abstraction algebras is axiomatisable by finitely many equational schemes. Pigozzi and Salibra [PS93, PS95] had earlier introduced the class LAA_I of *lambda abstraction algebras of dimension* I, which are algebras of the form

$$\mathbf{A} = \Big\langle A,\ \bullet^{\mathbf{A}},\ \{\lambda x^{\mathbf{A}} : x \in I\},\ \{x^{\mathbf{A}} : x \in I\}\Big\rangle$$

satisfying the following equations for $x, y, z \in I$ and $\xi, \mu, \nu \in A$ (we simplify notation by suppressing the $\mathbf{A}$-superscripts and abbreviating $\lambda x(\xi)$ to $\lambda x.\xi$, $\lambda x(\lambda y(\xi))$ to $\lambda xy.\xi$, and $\xi \bullet \mu$ to $\xi\mu$):

- $(\lambda x.x)\xi = \xi$
- $(\lambda x.y)\xi = y, \quad x \neq y$
- $(\lambda x.\xi)x = \xi$
- $(\lambda xx.\xi)\mu = \lambda x.\xi$
- $(\lambda x.\xi\mu)\nu = (\lambda x.\xi)\nu\big((\lambda x.\mu)\nu\big)$
- $\big(\lambda xy.\xi\big)\big((\lambda y.\mu)z\big) = \lambda y.\big(\lambda x.\xi\big)\big((\lambda y.\mu)z\big), \qquad x \neq y,\ z \neq y$
- $\lambda x.(\lambda y.\xi)z = \lambda y.\big(\lambda x.(\lambda y.\xi)z\big)y, \qquad z \neq y.$

These equations express algebraically various instances of the (α) and (β) axioms of lambda calculus. They can be rendered more comprehensible by reformulating them in terms of operators $S^x_b(a)$ that provide an algebraic version of the syntactic procedure of substitution of a term for a variable. In line with the (β) axiom scheme, these operators are defined by $S^x_b(a) := (\lambda x^{\mathbf{A}}.a)b$, which may be thought of as "a with b substituted for free x". See [PS95, Section 1] for details.

Every FLA_I is an LAA_I. In [SG96] the converse is established by showing that every LAA_I is isomorphic to an RFA_I, and hence to an FLA_I. This shows $\mathbf{I}\mathsf{FLA}_I = \mathsf{LAA}_I$, giving the asserted finite axiomatisation of $\mathbf{I}\mathsf{FLA}_I$.

The constructions in this paper involving choice functions are motivated by work on ultrapowers in the representation theory of cylindric set algebras [HMT$^+$81, I.7.13],[HMT85, 3.1.102]. Our conclusions provide a notable divergence between the algebra of the lambda calculus and cylindric algebra theory. For the latter, the analogue of FLA_Is are the *representable cylindric set algebras*, which are algebras of sets of sequences constructible from models of first-order logic. They too form a variety, but one that is *not* axiomatisable by finitely many equational schemes [Mon69].

7 A nonstandard approach

An alternative proof that an RFA_I $\mathbf{A}$ is isomorphic to an FLA_I can be developed by using the conceptual framework of nonstandard analysis. For readers familiar with that framework we sketch the main ideas.

Let $\mathbb{U}$ be a set-theoretic universe that contains the sets I and V, and in which all members of $I \cup V$ are regarded as individuals. Let $\mathbb{U}'$ be a second universe which is an enlargement of $\mathbb{U}$ by a transfer map $\mathbb{U} \overset{*}{\to} \mathbb{U}'$ taking each $b \in \mathbb{U}$ to an entity ${}^*b \in \mathbb{U}'$, and satisfying ${}^*b = b$, for all $b \in I \cup V$. Members of $\mathbb{U}'$ of the form ${}^*b$ with $b \in \mathbb{U}$ are called *standard*, while the *internal* members of $\mathbb{U}'$ are those that belong to standard sets.

${}^*I$ and ${}^*V$ are sets containing I and V, respectively. The environment $r : I \to V$ transfers to a map ${}^*r : {}^*I \to {}^*V$. ${}^*r$ is a member of ${}^*(V^I)$, which is itself the set of all *internal* functions from ${}^*I$ to ${}^*V$.

The binary operation $\bullet^{\mathbf{V}}$ on V transfers to a binary operation on ${}^*V$ which we will denote $\bullet^{{}^*\mathbf{V}}$. The operator $\lambda^{\mathbf{V}}$, defined on a subset of V^V, transfers to a function ${}^*\lambda^{\mathbf{V}}$ whose domain is a set of internal functions from ${}^*V$ to ${}^*V$. The structure

$$ {}^*\mathbf{V} = \langle {}^*V, \bullet^{{}^*\mathbf{V}}, {}^*\lambda^{\mathbf{V}} \rangle $$

then proves to be a functional domain.

The set $\mathcal{P}_{fin}(I)$ of all finite subsets of I transfers to a set ${}^*\mathcal{P}_{fin}(I)$ of internal subsets of ${}^*I$. These internal subsets are called *hyperfinite* subsets of ${}^*I$. All finite subsets of ${}^*I$ prove to be hyperfinite, but some infinite subsets may be hyperfinite as well. In particular the enlargement principle ensures that

$$ \textit{there exists a hyperfinite set } J \in {}^*\mathcal{P}_{fin}(I) \textit{ with } I \subseteq J \subseteq {}^*I. \tag{6} $$

The set V^I_r of environments agreeing cofinitely with r transfers to a set

$$ {}^*(V^I_r) \subseteq {}^*(V^I) \subseteq ({}^*V)^{{}^*I} $$

of internal functions from ${}^*I$ to ${}^*V$. Transferring the definition of V^I_r shows that the members of ${}^*(V^I_r)$ are precisely those internal functions from ${}^*I$ to ${}^*V$ that agree *co-hyperfinitely* with ${}^*r : {}^*I \to {}^*V$, i.e. agree with ${}^*r$ outside some hyperfinite subset of ${}^*I$.

Next we need to assume that the enlargement is *comprehensive*, which implies that each function of the type $p : I \to {}^*V$ can be extended to an *internal* function ${}^+p : {}^*I \to {}^*V$. This allows us to define an embedding

$$(\text{-})_r : {}^*V^I \longrightarrow {}^*(V_r^I) \tag{7}$$

taking each $p : I \to {}^*V$ to an element p_r of ${}^*(V_r^I)$ which extends p. Using the hyperfinite set J given by (6), we specify p_r to be that function on ${}^*I$ which agrees with ${}^+p$ on J and with ${}^*r$ on ${}^*I - J$. p_r is thus the disjoint union of the two functions ${}^+p \upharpoonright J$ and ${}^*r \upharpoonright ({}^*I - J)$, each of which is internal as it is the restriction of an internal function to an internal subset of its domain. Hence p_r is internal, and moreover is defined to agree with ${}^*r$ outside the hyperfinite set J, so belongs to ${}^*(V_r^I)$.

Each element f of the RFA_I $\mathbf{A}$ is a function of the form $V_r^I \to V$ and so transfers to a function ${}^*f : {}^*(V_r^I) \to {}^*V$. Composing this with the embedding (7) then gives a function ${}^*V^I \to {}^*V$ that will be denoted $\theta(f)$:

$$\begin{array}{ccc} {}^*V^I & \xrightarrow{(\text{-})_r} & {}^*(V_r^I) \\ \theta(f)\searrow & & \swarrow {}^*f \\ & {}^*V & \end{array}$$

Thus $\theta(f)(p) = {}^*f(p_r)$ for all $p \in {}^*V^I$. This construction defines a map

$$\theta : A \longrightarrow {}^*V_I$$

where ${}^*V_I$ is the set of all maps of the form ${}^*V^I \to {}^*V$. At this point we are in a similar situation to that at the beginning of Section 5 in this paper, and it remains to show that $\theta(A)$ supports an FLA_I isomorphic to $\mathbf{A}$. The proof is quite direct and in essence reduces to a transfer of the definitions of the LAA_I operations of $\mathbf{A}$.

The full details of this nonstandard argument are available in [Gol96]. The existence of the two proofs provides an interesting contrast between the non-standard and standard methodologies.

References

[Gol96] Robert Goldblatt. Enlargements of Functional Algebras for the Lambda Calculus. Research Report 96–187, Mathematics Department, Victoria University of Wellington, June 1996.

[HMT+81] L. Henkin, J. D. Monk, A. Tarski, H. Andréka, and I. Németi. *Cylindric Set Algebras*, volume 883 of *Lecture Notes in Mathematics*. Springer-Verlag, 1981.

[HMT85] L. Henkin, J. D. Monk, and A. Tarski. *Cylindric Algebras II*. North-Holland, 1985.

[Koy82] C. P. J. Koymans. Models of the Lambda Calculus. *Information and Control*, 52:306–332, 1982.

[Koy84] C. P. J. Koymans. *Models of the Lambda Calculus*, volume 9 of *CWI Tracts*. Centrum voor Wiskunde en Informatica, Amsterdam, 1984.

[Mey82] Albert R. Meyer. What is a Model of the Lambda Calculus? *Information and Control*, 52:87–122, 1982.

[Mon69] J. D. Monk. Nonfinitisability of Classes of Representable Cylindric Algebras. *Journal of Symbolic Logic*, 34:331–343, 1969.

[PS93] D. Pigozzi and A. Salibra. An Introduction to Lambda Abstraction Algebras. In *IX Simposio Latinoamericano de Logica Matematica*, volume 38 of *Notas de Logica Matematica*, pages 93–112. Universidad Nacional del Sur, Bahia Blanca, Argentina, 1993.

[PS95] Don Pigozzi and Antonino Salibra. Lambda Abstraction Algebras: Representation Theorems. *Theoretical Computer Science*, 140:5–52, 1995.

[Sco80] Dana Scott. Relating Theories of the λ Calculus. In J. P. Seldin and J. R. Hindley, editors, *To H. B. Curry: Essays on Combinatory Logic, Lambda Calculus and Formalism*, pages 403–450. Academic Press, 1980.

[SG96] Antonino Salibra and Robert Goldblatt. A Finite Equational Axiomatisation of the Functional Algebras for the Lambda Calculus. Research Report 96–174, Mathematics Department, Victoria University of Wellington, January 1996.

A Formalization of Anytime Constraint Satisfaction*

Hans W. Guesgen**

Computer Science Department, University of Auckland, New Zealand

Abstract. Anytime algorithms play a significant role in solving real world problems, as these problems often require a solution within limited time (for example, planning the path of a robot to retrieve explosive materials from a site or diagnosing the injuries of a human after an accident). A class of algorithms that has received more and more attention over recent years is the class of constraint satisfaction algorithms, i.e., algorithms that solve problems stated as a set of constraints over a set of variables. In this paper, we look at some constraint satisfaction algorithms and apply the concept of anytime algorithms to these algorithms.

1 Introduction and motivation

An important issue in the area of knowledge-based systems has been the development of efficient methods to reason about the knowledge represented by the system. Constraints are one way to represent knowledge, and constraint satisfaction algorithms one way to reason about that knowledge. A constraint satisfaction algorithm is a search algorithm for problems that can be stated in the following way: Given a set of variables over some finite domains and constraints on subset of these variables, find values for the variables such that all constraints are satisfied.

Constraint satisfaction has been applied successfully in various areas, for example:

- Computer vision [20]
- Circuit analysis [18]
- Planning experiments in molecular biology [19]
- Job-shop scheduling [5]
- Diagnosis [2, 3, 7]
- Temporal and spatial reasoning [1, 6, 10, 11, 16]

* A preliminary version of this paper has been presented at the Florida Artificial Intelligence Research Symposium (FLAIRS-96) at the beginning of this year [8]. Feedback from discussions at FLAIRS-96 has led to this revised version.

** The author performed part of this work while on leave at the International Computer Science Institute (ICSI), Berkeley, California, where he has been supported by the University of Auckland Research Fund under the grant number A18/XXXXX/62090/F3414025.

- Logic programming [4, 13]

Unfortunately, constraint satisfaction in general is NP-complete. Thus it is very unlikely that any deterministic complete constraint satisfaction algorithm has polynomial worst-case time complexity. On the other hand, exponential time complexity is unacceptable for most knowledge-based systems. We would rather accept an approximation of the solution computed in polynomial time than a solution of the problem computed in exponential time. More than that, we often require an algorithm to produce some result in a predefined period of time.

An algorithm that determines or approximates the solution of a problem in a given time period is called an anytime algorithm. Usually, the quality of its result gradually improves as available runtime increases. Anytime algorithms can be classified as follows:

1. If the algorithm knows in advance how much computation time is available for computing a result, the algorithm is called an *informed* anytime algorithm. In the context of AI planning, the notion of an informed anytime algorithm is remotely related to the notion of a contract algorithm [12, 21].
2. If the algorithm does not know in advance how much computation time is available for computing a result, the algorithm is called an *uninformed* anytime algorithm.

We will formalize these notions in the following.

2 Formalization of anytime constraint satisfaction

We will start the formalization with standard definitions for constraints, constraint networks, etc. as they can be found in the constraint literature. For a more elaborate introduction of these concepts, see [9] or similar work.

The basis of our formal framework is the notion of constraints and constraint networks. Informally, a constraint network is a set of constraints interconnected by shared variables, each constraint restricting the values that may be assigned to its variables. Formally, constraints and constraint networks can be defined as follows:

Definition 1 (Constraint). A k-ary constraint C consists of variables $V_1, \ldots, V_k$ over domains $D_1, \ldots, D_k$ and a decidable relation $R \subseteq D_1 \times \cdots \times D_k$, i.e., a k-ary relation the i-th place of which is symbolized by V_i.

Definition 2 (Constraint Network). A constraint network N is given by a set of constraints with shared variables. This means a constraint network on variables $V_1, \ldots, V_m$ consists of constraints $C_1, \ldots, C_n$, the variables of each C_i being a subset of $V_1, \ldots, V_m$.

We consider a constraint to be satisfied if the assignment of values to its variables is admissible with respect to the constraint relation. If all constraints in the network are satisfied, the assignment of values to the variables in the network is called a solution. The problem is to find such an assignment. Formally:

Definition 3 (Satisfaction). A tuple $(d_1, \ldots, d_m) \in D_1 \times \cdots \times D_m$ satisfies a constraint C_i if the subsequence of $(d_1, \ldots, d_m)$ that corresponds to the variables of C_i is an element of the relation of C_i.

Definition 4 (Solution). A solution of a constraint network that is given by constraints $C_1, \ldots, C_n$ on variables $V_1, \ldots, V_m$ over domains $D_1, \ldots, D_m$ is a tuple $(d_1, \ldots, d_m) \in D_1 \times \cdots \times D_m$ that satisfies $C_1, \ldots, C_n$.

Definition 5 (CSP). The task of finding one, several, or all solutions of a constraint network is called the constraint satisfaction problem (CSP).

There are various algorithms that can be used to either solve a given CSP or to transform it into an equivalent but easier-to-solve CSP. (In this context, a CSP is equivalent to another CSP if their solution spaces are identical.) Since we want to look at some of these algorithms from the viewpoint of anytime algorithms, we have to introduce notions that help us to assess the result and also the intermediate states of the algorithms.

Let us assume that $\mathcal{A}$ is a constraint satisfaction algorithm to be applied to a constraint network N. Instead of running the algorithm until it terminates, we interrupt it from time to time and look at the intermediate results of the algorithm. For example, if $\mathcal{A}$ is a backtracking algorithm [17], the intermediate result is the current assignment of values to (some of) the variables; if $\mathcal{A}$ is a filtering algorithm like Waltz filtering [20], the intermediate result is the current assignment of restricted domains to the variables; and so on.

Generally speaking, the intermediate results of an algorithm $\mathcal{A}$ applied to a constraint network N can be extracted directly from the states which N assumes during the application of $\mathcal{A}$. These states are usually defined by the data structures that are used by $\mathcal{A}$ and thus differ from algorithm to algorithm. In the following, $\mathcal{A}(N)$ denotes the set of states that N assumes when $\mathcal{A}$ is applied to N, and $\mathcal{A}(N)_{|t}$ the state of N at time t.

To compare different states of a constraint network with each other, we need some measure which assesses these states:

Definition 6 (State Measure). Let N be a constraint network and $\mathcal{A}$ a constraint satisfaction algorithm, then a state measure $\mu : \mathcal{A}(N) \to [0, 1]$ for N with respect to $\mathcal{A}$ is a function from $\mathcal{A}(N)$ to the interval of real numbers between 0 and 1.

Based on the above definition, we can now define the notion of an anytime constraint satisfaction algorithm:

Definition 7 (Anytime Algorithm). Given a constraint satisfaction algorithm $\mathcal{A}$, let N be a constraint network and μ a state measure of N with respect to $\mathcal{A}$. Then $\mathcal{A}$ is an anytime constraint satisfaction algorithm if:

$$\forall t, t' \in [0, \infty] : t \leq t' \Longrightarrow \mu(\mathcal{A}(N)_{|t}) \leq \mu(\mathcal{A}(N)_{|t'})$$

Furthermore, $\mathcal{A}$ is an informed anytime constraint satisfaction algorithm, if t and t' are input of $\mathcal{A}$:

$$\forall t, t' \in [0, \infty] : t \leq t' \Longrightarrow \mu(\mathcal{A}(N, t)_{|t}) \leq \mu(\mathcal{A}(N, t')_{|t'})$$

Otherwise, $\mathcal{A}$ is called an uninformed anytime constraint satisfaction algorithm.

Note that in the case of an informed constraint satisfaction algorithm, it may be possible that some successor states are worse than its predecessors during the run of the algorithm, i.e., the following may hold for an informed constraint satisfaction algorithm:

$$\exists t, t', t'' \in [0, \infty] : t \leq t' \land \mu(\mathcal{A}(N, t'')_{|t}) > \mu(\mathcal{A}(N, t'')_{|t'})$$

It is only required that, given two different contract times t and t' with $t \leq t'$, the result of applying $\mathcal{A}$ to N and t' must be at least as good as the result of applying $\mathcal{A}$ to N and t.

Obviously, not every constraint satisfaction algorithm is an anytime algorithm. Backtracking, for example, shows a very bad anytime behavior. On the other hand, there are some constraint satisfaction algorithms with perfect anytime behavior (i.e., that are anytime constraint satisfaction algorithms according to the above definition) or at least with good-on-average anytime behavior.

In the following, we will restrict our discussion to uninformed anytime constraint satisfaction algorithms. We will first introduce a class of constraint satisfaction algorithms that satisfy the condition of an anytime algorithm. We will then look at some algorithms that don't show anytime behavior but can be converted into anytime algorithms.

3 Discussion of constraint satisfaction algorithms

There are constraint satisfaction algorithms that show perfect anytime behavior. Among these algorithms are filtering algorithms like the Waltz algorithm [20] and the arc and path consistency algorithms [14]. As a representative of the filtering algorithms, we will discuss here the filtering algorithm that was introduced in [9]. The algorithm is shown in Figure 1.

The idea of a filtering algorithm is to iterate through the constraints of the network, comparing each constraint relation with the current domains of the variables and deleting incompatible values from the domains of the variables. When no more values can be deleted from the domains, the list of constraints to iterate through finally becomes empty and the algorithm terminates.

The control flow of the algorithm is influenced by two components:

1. M, the list of constraints to iterate through.
2. $D_1, \ldots, D_m$, the domains of the variables of N.

1. Initialize M with the constraints of a given constraint network N.
2. While $M \neq \emptyset$ do:
   (a) Select a constraint $C \in M$.
   Let $D_1, \ldots, D_k$ be the domains of the variables of C.
   (b) Remove C from M.
   (c) For each D_i, $i \in \{1, \ldots, k\}$, do:
      i. Delete each value d_i from D_i for which the following holds:
      There exist no
      $$\begin{array}{l} d_1 \in D_1 \\ d_2 \in D_2 \\ \vdots \\ d_{i-1} \in D_{i-1} \\ d_{i+1} \in D_{i+1} \\ \vdots \\ d_k \in D_k \end{array}$$
      such that $(d_1, \ldots, d_k)$ is an element of the relation of C.
      ii. If D_i has been changed, then add to M all other constraints that share a variable with C.

Fig. 1. A filtering algorithm for constraint networks.

Since M gets finally empty, it does not contribute to the result of the algorithm, and thus we can neglect it. $D_1, \ldots, D_m$, on the other hand, have an impact on the result of the algorithm; actually, they are the result. Thus we use $D_1, \ldots, D_m$ to represent the states that the algorithm assumes when applied to N.

Since the aim of filtering algorithms is to make the domains of the variables tighter (thereby restricting the search space), an adequate state measure for a constraint network N with respect to filtering algorithms is a measure that takes into account the number of values in the domains of the variables. There are several ways to do so, one of which is to calculate the difference between the total number of values in the current variable domains and the number of values in the initial domains, and to normalize the result:[3]

$$\mu(\mathcal{A}(N)_{|t}) = \frac{\sum_{i=1}^{m} |D_{i\,|0}| - \sum_{i=1}^{m} |D_{i\,|t}|}{\sum_{i=1}^{m} |D_{i\,|0}|}$$

During the execution of a filtering algorithm, values are deleted from the domains of the variables. As a result, the number of values in the domains does not increase over time, which means:

$$\forall t, t' \in [0, \infty] : t \leq t' \Longrightarrow \sum_{i=1}^{m} |D_{i\,|t}| \geq \sum_{i=1}^{m} |D_{i\,|t'}|$$

[3] $|D_{i\,|t}|$ denotes the number of values in the domain D_i at time t.

From this we can conclude the property that must hold for anytime constraint satisfaction algorithms:

$$\forall t, t' \in [0, \infty] : t \leq t' \Longrightarrow \mu(\mathcal{A}(N)_{|t}) \leq \mu(\mathcal{A}(N)_{|t'})$$

Although filtering algorithms have become a substantial part of constraint satisfaction research, they are not the only algorithms under investigation in this area, the reason being that filtering algorithms in general compute only an approximation of the solution of a given constraint satisfaction problem rather than the solution itself. Backtracking, on the other hand, does compute a solution (if one exists). However, backtracking algorithms in general are *not* anytime constraint satisfaction algorithms.

The best we can hope for in the case of a backtracking algorithm is a behavior that is close to that of an anytime constraint satisfaction algorithm. It turned out that some backtracking algorithms are closer to anytime behavior than others. We performed empirical studies on a variety of backtracking approaches (see [17] for information on these approaches) and measured the quality of the intermediate results.

In particular, we performed our experiments with the algorithms shown in Figure 2. Beyond that, we tested simple backtracking and backjumping with various variable and value ordering heuristics.[4] As state measure for the algorithms, we used the number of correctly instantiated variables over the total number of

1. Simple backtracking (BT)
2. Backjumping (BJ)
3. Conflict-directed backjumping (CBJ)
4. Graph-based backjumping (GBJ)
5. Backmarking with simple backtracking (BM)
6. Backmarking with backjumping (BM_BJ)
7. Backmarking with conflict-directed backjumping (BM_CBJ)
8. Backmarking with graph-based backjumping (BM_GBJ)
9. Forward checking with simple backtracking (FC)
10. Forward checking with backjumping (FC_BJ)
11. Forward checking with conflict-directed backjumping (FC_CBJ)
12. Forward checking with graph-based backjumping (FC_GBJ)
13. Full arc consistency forward checking with simple backtracking (FCarc)
14. Full path consistency forward checking with simple backtracking (FCpath)

Fig. 2. Various backtracking algorithms for constraint satisfaction.

[4] We used the constraint satisfaction system CSP [15] for our experiments, which provides all the algorithms mentioned here.

variables of the constraint network, i.e.:[5]

$$\mu(\mathcal{A}(N)_{|t}) = \frac{|S_{|t}|}{m}$$

where $S_{|t}$ denotes the set of variable instantiations at time t.

As expected, none of the above algorithms are anytime constraint satisfaction algorithms, the reason being that any backtracking algorithm retracts values from time to time, which causes the algorithm to assume a state that is worse than the previous state. We can overcome this problem by caching the best instantiation so far and returning this instantiation when the algorithm is interrupted.

From this point of view, every backtracking algorithm can be converted into an anytime algorithm. However, the question that arises then is: How fast do the above algorithms make progress, i.e., how quickly does the quality of the states improve over time? We found that the more informed an algorithm is, the faster the improvement over time.

Another way of getting an indication for how fast a particular algorithm makes progress is to look at the following:

1. How often does the algorithm assume a state that is worse than the previous one?
2. What is the standard deviation of the intermediate states?

Table 1 summarizes our findings for the various backtracking approaches. We checked 30 intermediate states for each algorithm and calculated the standard deviation of the number of correctly instantiated variables in each state (see 'Standard deviation' in Table 1). Beyond that, we counted how many of the 30 intermediate states were worse than previous states (see 'Degradations' in Table 1).

The table shows that BM_GBJ is the most likely candidate for an algorithm that makes good progress over time and therefore is worthwhile being converted into an anytime algorithm. Only 7 out of the 30 intermediate states in our experiment were states that were worse than previous states.

4 Summary

In this paper, we applied the concept of anytime algorithms to constraint satisfaction. We introduced the concept of anytime constraint satisfaction, provided a formalization, and discussed filtering algorithms with respect to their anytime behavior. We also showed that backtracking algorithms in general are not anytime algorithms, but can be converted into anytime constraint satisfaction algorithms very easily.

[5] It is arguable whether this is an appropriate measure for backtracking algorithms. Sometimes it is quite easy to instantiate the first $m-k$ variables (for some small k), but very hard to instantiate the k remaining ones. This observations is not reflected in the measure.

Algorithm	Standard deviation	Degradations
BT	1.26	12
BJ	1.65	13
CBJ	1.84	9
GBJ	1.14	9
BM	1.11	13
BM_BJ	2.16	10
BM_CBJ	1.99	12
BM_GBJ	1.04	7
FC	1.20	11
FC_BJ	1.43	11
FC_CBJ	1.96	14
FC_GBJ	2.08	12

Table 1. Standard deviation of the intermediate states and number of degradations for various backtracking approaches.

References

1. J.F. Allen. Maintaining knowledge about temporal intervals. *Communications of the ACM*, 26:832–843, 1983.
2. R. Davis. Diagnostic reasoning based on structure and behavior. *Artificial Intelligence*, 24:347–410, 1984.
3. J. de Kleer and B.C. Williams. Diagnosing multiple faults. In *Proc. AAAI-86*, pages 132–139, Philadelphia, Pennsylvania, 1986.
4. M. Dincbas, H. Simonis, and P. van Hentenryck. Extending equation solving and constraint handling in logic programming. In *Proc. Colloquium on Resolution of Equations in Algebraic Structures*, Austin, Texas, 1987.
5. M.S. Fox, B. Allen, and G. Strohm. Job-shop scheduling: An investigation in constraint-directed reasoning. In *Proc. AAAI-82*, pages 155–158, Pittsburgh, Pennsylvania, 1982.
6. C. Freksa. Qualitative spatial reasoning. In *Proc. Workshop RAUM*, pages 21–36, Koblenz, Germany, 1990.
7. H. Geffner and J. Pearl. An improved constraint-propagation algorithm for diagnosis. In *Proc. IJCAI-87*, pages 1105–1111, Milan, Italy, 1987.
8. H.W. Guesgen. Towards anytime constraint satisfaction. In *Proc. FLAIRS-96*, pages 20–24, Key West, Florida, 1996.
9. H.W. Guesgen and J. Hertzberg. *A Perspective of Constraint-Based Reasoning*. Lecture Notes in Artificial Intelligence 597. Springer, Berlin, Germany, 1992.

10. H.W. Guesgen and J. Hertzberg. A constraint-based approach to spatiotemporal reasoning. *Applied Intelligence (Special Issue on Applications of Temporal Models)*, 3:71–90, 1993.
11. D. Hernández. *Qualitative Representation of Spatial Knowledge*. Lecture Notes in Artificial Intelligence 804. Springer, Berlin, Germany, 1994.
12. J. Hertzberg. Anytime-Algorithmen. *KI*, 1/95:40, 1995.
13. J. Jaffar and J.L. Lassez. Constraint logic programming. In *Conference Record of the 14 th Annual ACM Symposium on Principles of Programming Languages*, pages 111–119, Munich, Germany, 1987.
14. A.K. Mackworth. Consistency in networks of relations. *Artificial Intelligence*, 8:99–118, 1977.
15. D. Manchak and P. van Beek. A c-library of constraint satisfaction techniques. Technical report, Available by anonymous ftp from ftp.cs.ualberta.ca:pub/ai/csp, 1994.
16. A. Mukerjee and G. Joe. A qualitative model for space. In *Proc. AAAI-90*, pages 721–727, Boston, Massachusetts, 1990.
17. P. Prosser. Hybrid algorithms for the constraint satisfaction problem. *Computational Intelligence*, 9:268–299, 1993.
18. R.M. Stallman and G.J. Sussman. Forward reasoning and dependency-directed backtracking in a system for computer-aided circuit analysis. *Artificial Intelligence*, 9:135–196, 1977.
19. M. Stefik. Planning with constraints (MOLGEN: Part 1). *Artificial Intelligence*, 16:111–140, 1981.
20. D.L. Waltz. Generating semantic descriptions from drawings of scenes with shadows. Technical Report AI-TR-271, MIT, Cambridge, Massachusetts, 1972.
21. S. Zilberstein and S.J. Russell. Efficient resource-bounded reasoning in at-ralph. In *Artificial Intelligence Planning Systems: Proc. First International Conference*, pages 260–266, College Park, Maryland, 1992.

The Generating Functions of Lambda Terms (Extended Abstract)*

Ryu Hasegawa

Oxford University Computing Laboratory, Wolfson Building,
Parks Road, Oxford OX1 3QD, England**

Abstract. The canonical normal functor models of the lambda calculus admit interpretations in the shape of formal power series in non-negative integer coefficients. We give an analytic description of the interpretation of looping combinators by the Lagrange-Good inversion formula.

1 Introduction

We apply the ideas of enumerative combinatorics to analyze the normal functor models [6] of the lambda calculus [2]. In this model, every lambda term is interpreted as a normal functor, which is regarded as a formal power series where the coefficiets are sets, possibly being infinite sets. So a natural question is whether the interpretations of lambda terms have finite coefficients [6]. We show that all coefficients are finite in the canonical models obtained via initial algebras, while this assertion fails in general models.

One of the principal vehicles in enumerative combinatorics is generating functions, which are formal power series in non-negative integer coefficients. The reason generating functions are useful is that operations on formal power series often correspond to natural graphical manipulations that are necessary to tackle the problems of enumeration. We employ this observation to the normal functor models. Namely we consider the operations on formal power series induced by two primitives of lambda calculus, viz., lambda abstraction and application, and we associate graphical configurations to those operations. Thence the finiteness of coefficients is verified by showing that the number of the structures matching the configurations is always finite.

One of the most important lambda terms is the looping combinator $\mathsf{Y} = \lambda y.\,(\lambda x.\,y(xx))(\lambda x.\,y(xx))$, which is the form of self-application widely used in mathematical logic such as Kleene's recursion theorem and Gödel's incompleteness theorem. This term Y realizes an iteration process, that has much coincidence with the idea behind the Lagrange-Good inversion formula in analysis. In fact, we show in §4 that the interpretation of the looping combinator Y has an analytic description by a direct application of the inversion formula.

---

* This work is supported by EPSRC GR1J97366.

** Correspondence (after August 1996): Graduate School of Mathematical Sciences, The University of Tokyo, Komaba 3-8-1, Meguro-ku, Tokyo 153, Japan

2 Preliminaries

In this section, we summarize basic properties of normal functors.

A *C-monoid* in a cartesian closed category is an object X satisfying an isomorphism $X^X \cong X$. Such an object serves as a model of the lambda calculus. Terms M in free variables included in $\{x_1, x_2, \ldots, x_n\}$ $(n \geq 0)$ are interpreted as morphisms $[\![M]\!]_{\mathbf{x}}$ from X^n to X. Moreover, the $\beta\eta$-equality is sound in the model. Namely, if $M =_{\beta\eta} N$ then $[\![M]\!]_{\mathbf{x}} = [\![N]\!]_{\mathbf{x}}$ holds. See [10] for the details.

Normal functors are certain functors (see below) between the categories of the form $\mathbf{Set}^A$ where A is a discrete category, that is, a set. One may consider $\mathbf{Set}^A$ to be the category of presheaves from A, or the cartesian product of $\operatorname{card} A$ copies of the category $\mathbf{Set}$ of sets. By the latter point of view, a presheaf z in $\mathbf{Set}^A$ is regarded as a vector of sets indexed by the members of A.

Convention.

1. For a presheaf H in $\mathbf{Set}^A$, the square bracket $H[a]$ denotes the set that is the value of H at $a \in A$. We reserve the brace $F(\text{-})$ for the application of functors, accommodated to the notation in [9].
2. If F is a functor from $\mathbf{Set}^A$ to $\mathbf{Set}^B$, then we have a functor $F_b : \mathbf{Set}^A \to \mathbf{Set}$ for each $b \in B$, defined by $F_b(z) = F(z)[b]$.

For a functor F from some category $\mathbf{C}$ to $\mathbf{Set}$, let $\mathrm{el}(F)$ denote the category of elements [11]. The objects of $\mathrm{el}(F)$ are the pairs (c, X) of objects X of $\mathbf{C}$ and a member c in $F(X)$.

For a set A, let $\exp(A)$ denote the collection of all finite multisets of members of A. Every member γ in $\exp(A)$ is regarded as the object of $\mathbf{Set}^A$ sending $a \in A$ to the cardinality of occurrences of a in γ. Then γ is a finitely presentable object [1] of $\mathbf{Set}^A$, and conversely every finitely presentable object of $\mathbf{Set}^A$ is isomorphic to a unique γ in $\exp(A)$. The function exp can be lifted to a functor from $\mathbf{Set}$ to $\mathbf{Set}$: For a function $f : A \to B$, we define $\exp(f)$ to send multisets $\{a_1, a_2, \ldots, a_n\}$ in $\exp(A)$ to multisets $\{f(a_1), f(a_2), \ldots, f(a_n)\}$ in $\exp(B)$. In fact, exp turns out to be an analytic functor in the sense of Joyal corresponding to the uniform species [8, 9]. Let us note that, for a singleton $\{a\}$, it holds that $\mathbf{Set}^A(\{a\}, z) = z[a]$.

Theorem 1. *Let F be a functor from $\mathbf{Set}^A$ to $\mathbf{Set}^B$.*
The following are equivalent:

1. *F_b is isomorphic to a small coproduct of the functors representable by finitely presentable objects of $\mathbf{Set}^A$, for every b in B.*
2. *Every slice category $\mathrm{el}(F_b)/(c, Y)$ has an initial object $(d, X) \to (c, Y)$ where X is a finitely presentable object of $\mathbf{Set}^A$, for all $b \in B$.*
3. *F preserves filtered colimits and connected limits (that is, the limits of the diagrams whose underlying graphs are connected).*

Definition 2. A *normal functor* $F \in [\mathbf{Set}^A, \mathbf{Set}^B]_{\mathrm{NF}}$ is a functor from $\mathbf{Set}^A$ to $\mathbf{Set}^B$ satisfying one of the equivalent conditions in Theorem 1. In particular, the condition (1) says that there is a family of sets $\tilde{F}_b[\alpha] = \tilde{F}[\alpha, b]$ indexed by $(\alpha, b) \in \exp(A) \times B$ such that

$$F_b(z) \;\cong\; \sum_{\alpha \in \exp(A)} \tilde{F}_b[\alpha]\, \mathbf{Set}^A(\alpha, z)$$

for all z in $\mathbf{Set}^A$ and all b in B. Here the symbol $\sum$ signifies coproduct in **Set**, and the cartesian product is written simply by concatenation.

If we write $\mathbf{Set}^A(\alpha, z) = z^\alpha$, then the summation form in Definition 2 is regarded as a formal power series with coefficients in **Set**. Therefore we call the summation form the *power series representation* of a normal functor F, though the coefficients $\tilde{F}[\alpha, b]$ are not necessarily finite.

Convention. For a normal functor F, we use the notation $\tilde{F}[\text{-}]$ to indicate the coefficients, or sometimes $F[\text{-}]^{\sim}$ by aesthetic reasons.

Remark. Normal functors are a special case of analytic functors in the sense of Joyal [9]. Namely, $F : \mathbf{Set}^A \to \mathbf{Set}^B$ is a normal functor iff F is an analytic functor where all involved actions of groups G are free, that is, the actions of elements of G save the unit have no fixpoints. In this case, the representation of analytic functors analogous to exponential generating functions turns out to be equivalent to the power series representation of normal functors analogous to ordinary generating functions.

We can make the collection $[\mathbf{Set}^A, \mathbf{Set}^B]_{\mathrm{NF}}$ of normal functors into a category by taking cartesian natural transformations [6] as morphisms. The equivalence of categories $[\mathbf{Set}^A, \mathbf{Set}^B]_{\mathrm{NF}} \cong \mathbf{Set}^{\exp(A)\times B}$ holds, as suggested by the correspondence of $\tilde{F}$ to F in Definition 2.

Definition 3 (of $\mathbf{CAAcc}_{\mathrm{NF}}$). $\mathbf{CAAcc}_{\mathrm{NF}}$ is the category where the objects are $\mathbf{Set}^A$ for sets A and the morphisms are normal functors. The justification of this name will be given in the full paper.

Proposition 4. *The category* $\mathbf{CAAcc}_{\mathrm{NF}}$ *is cartesian closed.*

3 Normal functor models and enumeration

Finding a C-monoid $\mathbf{Set}^{\exp(A)\times A} \cong \mathbf{Set}^A$ in the category $\mathbf{CAAcc}_{\mathrm{NF}}$ amounts to finding a set A having a bijection $\psi : \exp(A) \times A \xrightarrow{\sim} A$. If we find such a set A, then $\mathbf{Set}^A$ provides a model of the lambda calculus with $\beta\eta$-reduction. The interpretation $[\![M]\!]_{\mathbf{x}}$ is a normal functor in $[(\mathbf{Set}^A)^n, \mathbf{Set}^A]_{\mathrm{NF}}$ if M is a lambda term in at most n free variables. Specifically

$$\begin{aligned}
[\![x_i]\!]_{\mathbf{x}}(z_1, \ldots, z_n) &= z_i \\
[\![MN]\!]_{\mathbf{x}}(z_1, \ldots, z_n) &= \mathrm{ev}'([\![M]\!]_{\mathbf{x}}(z_1, \ldots, z_n), [\![N]\!]_{\mathbf{x}}(z_1, \ldots, z_n)) \\
[\![\lambda y.\, M]\!]_{\mathbf{x}}(z_1, \ldots, z_n) &= \mathrm{lam}'([\![M]\!]_{\mathbf{x},y}(z_1, \ldots, z_n, \text{-}))
\end{aligned}$$

where ev′ is the composition $\mathbf{Set}^A \times \mathbf{Set}^A \xrightarrow{\sim} \mathbf{Set}^{\exp(A)\times A} \times \mathbf{Set}^A \xrightarrow{\mathrm{ev}} \mathbf{Set}^A$, and lam′ is the isomorphism $[\mathbf{Set}^A, \mathbf{Set}^A]_{\mathrm{NF}} \xrightarrow{\sim} \mathbf{Set}^{\exp(A)\times A} \xrightarrow{\sim} \mathbf{Set}^A$.

We are particularly interested in the model obtained by initial algebra construction. Let A_0 be an arbitrary non-empty set, and let A_{n+1} be defined by $\exp(A_n) \times A_n$. A one-to-one function $d_0 : A_0 \to A_1$ is defined by $a \mapsto (\emptyset, a)$, and one-to-one functions $d_{n+1} : A_{n+1} \to A_{n+2}$ are defined by $\exp(d_n) \times d_n$. Let A_∞ be a colimit of the ω-diagram $A_0 \xrightarrow{d_0} A_1 \xrightarrow{d_1} A_2 \rightarrow \cdots$ in the category **Set**. Obviously we have a bijection ψ from $\exp(A_\infty) \times A_\infty$ to A_∞.

Definition 5. The *canonical model* is the C-monoid $\mathbf{Set}^{A_\infty}$ in the category $\mathbf{CAAcc}_{\mathrm{NF}}$.

We will find it convenient later to write the elements of A_∞ (and of other sets A with bijection $\psi : \exp(A) \times A \xrightarrow{\sim} A$) in flat tree forms: In some cases,

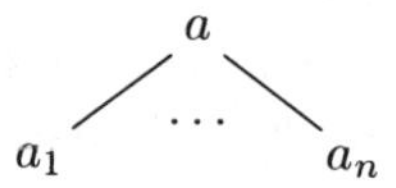

if the element is equal to $\psi(\{a_1, \ldots, a_n\}, a)$; in other cases,

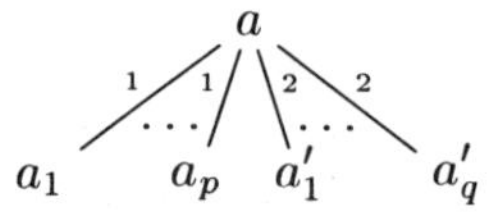

if the element is equal to $\psi^2(\{a_1, \ldots, a_p\}, \{a'_1, \ldots, a'_q\}, a)$ where $\psi^2 : \exp(A) \times \exp(A) \times A \xrightarrow{\sim} A$ is an obvious bijection induced by ψ. In the first form, the order of $a_1, \ldots, a_n$ is irrelevant, but for later use it is more convenient to fix an arbitrary planar representation. Namely we give a linear order on $\{a_1, \ldots, a_n\}$, this order being taken arbitrarily but having to be fixed throughout the manipulation. In the second form, we fix arbitrary orders on $\{a_1, \ldots, a_p\}$ and $\{a'_1, \ldots, a'_q\}$ as well as we indicate the distinction between the first and the second arguments.

In the sequel, we analyse the canonical model $\mathbf{Set}^{A_\infty}$ by the following plan: First we show that the interpretation $F = [\![M]\!]_{\mathbf{x}}$ depends on only the Böhm tree of M. Next we associate configurations $\tilde{C}_M(\alpha_1, \ldots, \alpha_n, a)$ to the Böhm tree of M. Then the coefficients $\tilde{F}[\alpha_1, \ldots, \alpha_n, a]$ are computed as the number of the structures matching the configurations.

It is by the familiar method of labeled reduction that we show that the interpretation depends on only Böhm trees. [12, 13]. What we want is the following theorem, by which we can assume that every lambda term is in head normal form. We leave a proof to the reader.

Theorem 6. *In the canonical model, every unsolvable term M is interpreted as empty presheaves. Namely $[\![M]\!]_{\mathbf{x}}(z_1, \ldots, z_n) = \emptyset$.*

Configurations are graphical descriptions that help to count the number of structures in problems of enumerative combinatorics. Joyal [8] gives the configurations associated to various operations on species, so that the coefficients in exponential generating functions are exactly the number of the structures matching the configuration. Here we need configurations for counting the *types* of the structures to obtain the coefficients of the ordinary generating function.

Suppose that G is a normal functor from $\mathbf{Set}^A$ to $\mathbf{Set}^B$ and γ is a member of $\exp(B)$. We define a functor $G^\gamma : \mathbf{Set}^A \to \mathbf{Set}$ by $G^\gamma(z) \triangleq \mathbf{Set}^B(\gamma, G(z))$. Since γ is finitely presentable, G^γ is a normal functor by (3) in Theorem 1. The coefficients $G^\gamma[\alpha]^{\tilde{}}$ is computed by $\sum_g(\prod_{b\in\gamma}\tilde{G}_b[g^{-1}(b)])$ where $b \in \gamma$ takes multiplicity into account and the summation of g ranges over the quotient set $\mathbf{Set}(\Sigma\alpha, \Sigma\gamma)/\mathrm{Aut}_{\mathbf{Set}^A}(\alpha)$. Here $\Sigma\alpha$ is a set of n elements if α is a multiset of n elements taking multiplicity into account. Furthermore $\Sigma\alpha$ is acted obviously by the group $\mathrm{Aut}_{\mathbf{Set}^A}(\alpha)$ of all isomorphisms on α in the category $\mathbf{Set}^A$.

Let M be a term in its Böhm tree and in free variables included in $\{x_1, \ldots, x_n\}$. The interpretation $F = [\![M]\!]_{\mathbf{x}}$ in a model $\mathbf{Set}^A$ is a normal functor in $[(\mathbf{Set}^A)^n, \mathbf{Set}^A]_{\mathrm{NF}}$. So the association $\tilde{F}[\cdot]$ of coefficients is a presheaf in $\mathbf{Set}^{\exp(A)^n\times A}$. Let $\alpha_1, \ldots, \alpha_n$ be members of $\exp(A)$ and a a member of A. To make it easy to compute the coefficient $\tilde{F}[\alpha_1, \ldots, \alpha_n, a]$, we employ a graphical presentation called a *configuration* $\tilde{C}_M(\alpha_1, \ldots, \alpha_n, a)$. The coefficient is computed as the number of the structures matching the configuration.

If $F = [\![\lambda y.\, M]\!]_{\mathbf{x}}$, then $\tilde{F}[\alpha_1, \ldots, \alpha_n, \psi(\gamma, a)]$ equals $\tilde{G}[\alpha_1, \ldots, \alpha_n, \gamma, a]$ where G is $[\![M]\!]_{\mathbf{x},y}$, thus we can put $\tilde{C}_{\lambda y.M}(\alpha_1, \ldots, \alpha_n, \psi(\gamma, a)) = \tilde{C}_M(\alpha_1, \ldots, \alpha_n, \gamma, a)$. So it suffices to give the configuration for terms of the form $xM_1 \cdots M_k$ without lambda bindings in front.

We give a detailed description of the configuration $\tilde{C}_M(\alpha_1, \ldots, \alpha_n, a)$ for the case $k = 2$, from which the general cases are readily induced. So let us suppose that $M = x_iNN'$ and put $G(z_1, \ldots, z_n) \triangleq [\![N]\!]_{\mathbf{x}}(z_1, \ldots, z_n)$ and $G'(z_1, \ldots, z_n) \triangleq [\![N']\!]_{\mathbf{x}}(z_1, \ldots, z_n)$. Then the interpretation $F(z_1, \ldots, z_n) = [\![M]\!]_{\mathbf{x}}(z_1, \ldots, z_n)$ is given by

$$\sum_{\gamma,\gamma'\in\exp(A)} \mathbf{Set}^A(\{\psi^2(\gamma, \gamma', a)\}, z_i)\; G^\gamma(z_1, \ldots, z_n)\; G'^{\gamma'}(z_1, \ldots, z_n)$$

where $\psi^2 : \exp(A) \times \exp(A) \times A \xrightarrow{\sim} A$ is a bijection induced by ψ. Hence the coefficient $\tilde{F}[\alpha_1, \ldots, \alpha_n, a]$ equals $\sum_{\gamma,\gamma'}\sum_{\beta_j,\beta'_j} G^\gamma[\beta_1, \ldots, \beta_n]^{\tilde{}}\; G'^{\gamma'}[\beta'_1, \ldots, \beta'_n]^{\tilde{}}$ where the second summation is over all ordered partitions of $\alpha_1, \ldots, \alpha_n$ satisfying that $\alpha_j = \beta_j + \beta'_j$ if $j \neq i$ and $\alpha_i = \beta_i + \beta'_i + \{\psi^2(\gamma, \gamma', a)\}$, allowing empty classes and taking into account of only the types of ordered partitions.

We define the configuration $\tilde{C}_M(\alpha_1, \ldots, \alpha_n, a)$ of a term $M = x_iNN'$ over $(\alpha_1, \ldots, \alpha_n, a)$ where $\alpha_1, \ldots, \alpha_n \in \exp(A)$ and $a \in A$, as in Figure 1. There the numbers 1 and 2 on the edges signify the first and second arguments respectively, nd $p, q \geq 0$. The coefficients $\tilde{F}[\alpha_1, \ldots, \alpha_n, a]$ for $F = [\![M]\!]_{\mathbf{x}}$ is computed as the nber of the structures matching the configuration. Namely, we are engaged in unting all possible ways to choose an element from α_i, this element having root a if it is expanded by $(\psi^2)^{-1}$, to partition α_i minus this element and all other α_j into $p+q$ classes, this number determined by the element chosen first, to form p $\tilde{G}$-structures over $(\delta_{11}, \ldots, \delta_{n1}, a_1)$ etc. and q $\tilde{G}'$-structures over $(\delta'_{11}, \ldots, \delta'_{nq}, a'_1)$

etc. (where $G = [\![N]\!]_{\mathbf{x}}$ and $G' = [\![N']\!]_{\mathbf{x}}$), and to plug these structures into a_1 through a'_q.

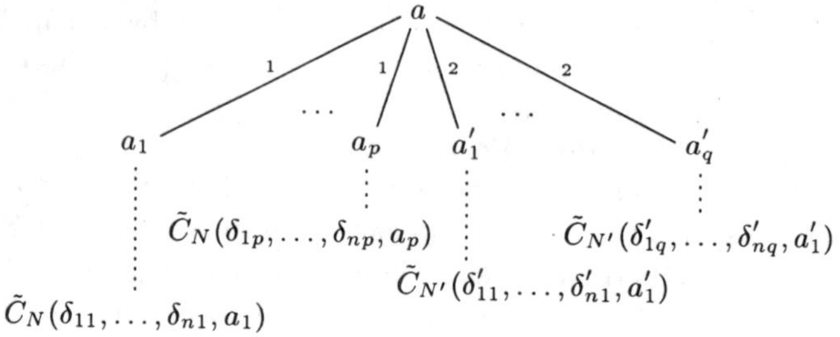

$$\begin{aligned} \text{with constraint } \alpha_j &= \delta_{j1} + \cdots + \delta_{jp} + \delta'_{j1} + \cdots + \delta'_{jq} \qquad (j \neq i) \\ \alpha_i &= \delta_{i1} + \cdots + \delta_{ip} + \delta'_{i1} + \cdots + \delta'_{iq} \\ &\quad + \{\psi^2(\{a_1, \ldots, a_p\}, \{a'_1, \ldots, a'_q\}, a)\} \end{aligned}$$

Fig. 1. Configuration $\tilde{C}_M(\alpha_1, \ldots, \alpha_n, a)$

Remark.

1. For a term of the form $M = xM_1 \cdots M_k$, the configuration $\tilde{C}_M$ refers to configurations $\tilde{C}_{M_i}$. However, $\tilde{C}_M$ is not a recursive definition, since the Böhm tree of M_i is not necessarily simpler than that of M.
2. The definition of the configurations is valid for all models, not only for the canonical model.

Example 1. Let Y be the normal functor in $[\mathbf{Set}^A, \mathbf{Set}^A]_{\mathrm{NF}}$ interpreting the term $\mathsf{Y}x$ in free variable x, where Y is Curry's looping combinator. The Böhm tree of $\mathsf{Y}x$ satisfies the following recursive equation

$$\mathsf{Y}x \;=\; \begin{array}{c} x \\ | \\ \mathsf{Y}x \end{array} .$$

The following is valid for every term having the same recursive equation, as does Θx where Θ is Turing's looping combinator. The configuration $\tilde{C}_{\mathsf{Y}x}(\alpha, a)$ is given as in Figure 2. Unfolding the recursive calls of $\tilde{C}_{\mathsf{Y}x}$, the configuration $\tilde{C}_{\mathsf{Y}x}(\alpha, a)$ turns out to form a tree in Figure 3. We remark that the tree is finite, for the given multiset α is finite.

Remark. One may think of

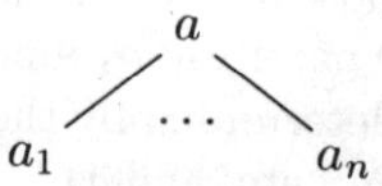

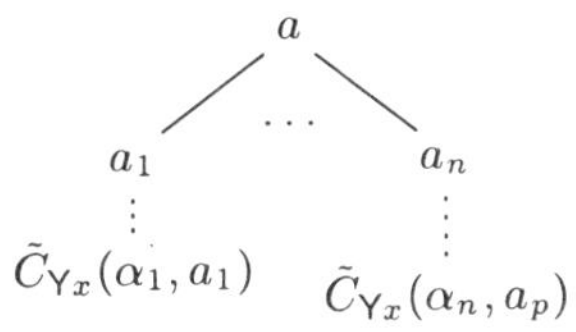

with constraint $\alpha = \{\psi(\{a_1, \ldots, a_n\}, a)\} + \alpha_1 + \cdots + \alpha_n$.

Fig. 2. Configuration $\tilde{C}_{\mathsf{Y}x}(\alpha, a)$

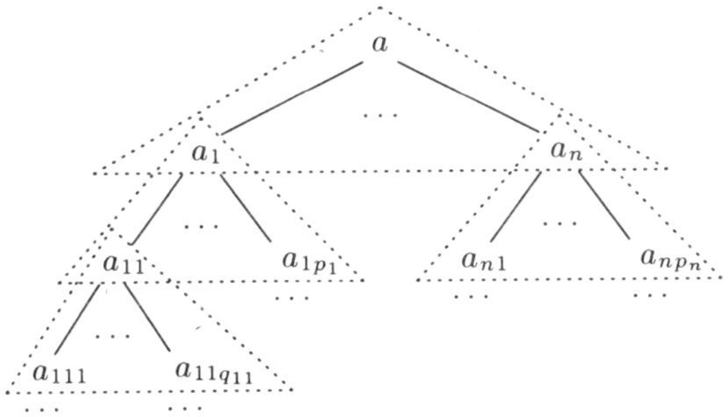

$$\begin{aligned}\text{with } \alpha \;=\; & \{\psi(\{a_1, \ldots, a_n\}, a),\\ & \psi(\{a_{11}, \ldots, a_{1p_1}\}, a_1),\\ & \cdots\\ & \psi(\{a_{n1}, \ldots, a_{np_n}\}, a_n),\\ & \psi(\{a_{111}, \ldots, a_{11q_{11}}\}, a_{11}),\\ & \cdots\}\end{aligned}$$

Fig. 3. Unfolding of $\tilde{C}_{\mathsf{Y}x}(\alpha, a)$

as a finite piece of function that provides the following instruction: If n tokens $a_1, \ldots, a_n$ are given as input, then a single token a is produced as output. Hence the unfolded configuration of $\mathsf{Y}x$ is regarded to display a flow of computation from bottom of the figure to top. The coefficient $\tilde{Y}[\alpha, a]$ is the number of ways to form the flows of computation producing output a from a given multiset α of finite pieces of function.

Example 2.

1. Let us consider a multiset $\alpha = \{a_1, a_2, a_2, a_3, a_4\}$ in $\exp(A)$ where

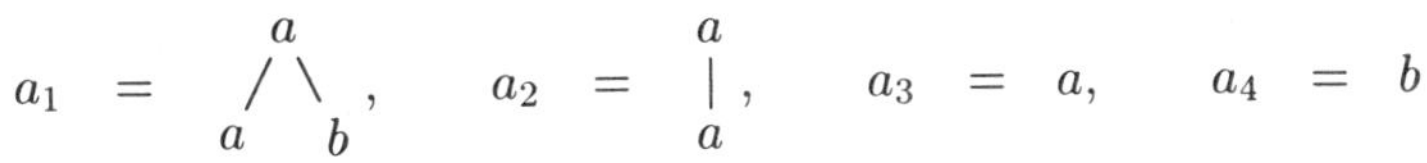

with $a \neq b$. There are three ways to form the flows of computation with output a as in Figure 4(i). Hence the coefficient $\tilde{Y}[\alpha, a]$ equals 3.

2. If b equals a, we have the multiset $\alpha' = \{a_5, a_2, a_2, a_3, a_3\}$ where

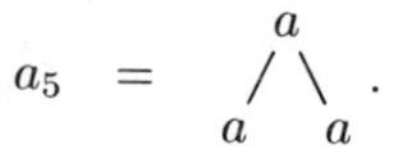

We have six ways to form the flows of computation with output a as shown in Figure 4(ii). Hence the coefficient $\tilde{Y}[\alpha', a]$ is 6.

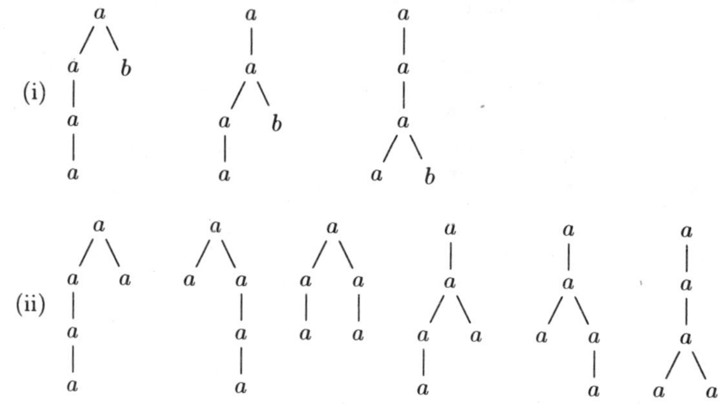

Fig. 4. Trees matching $\tilde{C}_{\mathsf{Y}x}(\alpha, a)$ and $\tilde{C}_{\mathsf{Y}x}(\alpha', a)$

If we suppose that the element a_1 corresponds to variable z_1 etc., then the power series representation of the normal functor $Y = [\![\mathsf{Y}x]\!]_x$ has the form $Y(z) = \cdots + 3z_1{z_2}^2z_3z_4 + \cdots + 6{z_2}^2{z_3}^2z_5 + \cdots$.

Remark. Note that the first two trees of Figure 4(ii) are counted distinctly. We recall that, in the flat tree representation of $\psi(\{a_1, \ldots, a_n\}, a)$, the order of $a_1, \ldots, a_n$ is irrelevant, but we fix an arbitrary planar representation. This phenomenon may be considered as follows: The order to provide the inputs is irrelevant, but it matters which slot each input is provided from.

In [6], Girard posed a question whether the coefficients $\tilde{F}[\alpha_1, \ldots, \alpha_n, a]$ for $F = [\![M]\!]_{\mathbf{x}}$ are all finite in a certain class of models for every term M. A model $\mathbf{Set}^A$ is called *regular* if A is a set with bijection $\psi : \exp(A) \times A \xrightarrow{\sim} A$ subject to the following condition: There is a function $l : A \to \mathrm{N}$ such that $l(\psi(\{a_1, \ldots, a_n\}, a)) = \max_{1 \le i \le n} l(a_i) + 1$. The canonical model $\mathbf{Set}^{A_\infty}$ is regular. The problem is whether the question just mentioned holds for regular models. In other words, it asks whether the power series representation of $F = [\![M]\!]_{\mathbf{x}}$ has non-negative integer coefficients as usual generating functions. We have a counter-example for general regular models, though not having enough space to include it. But the answer is affirmative for the canonical model:

Theorem 7. *Let* $F = [\![M]\!]_{\mathbf{x}}$ *be the interpretation of a lambda term* M *as a normal functor in* $[(\mathbf{Set}^{A_\infty})^n, \mathbf{Set}^{A_\infty}]_{\mathrm{NF}}$.

The coefficient $\tilde{F}[\alpha_1, \ldots, \alpha_n, a]$ *is finite, for all* $\alpha_1, \ldots, \alpha_n$ *in* $\exp(A_\infty)$ *and all* a *in* A_∞.

4 The generating function of looping combinators

The Lagrange-Good inversion formula [7, 3, 5] is a standard method to count the number of trees in enumerative combinatorics. Since the unfolded configuration $\tilde{C}_{\mathsf{Y}x}(\alpha, a)$ in Figure 3 has the form of tree, we naturally expect to apply the formula to compute the generating function associated to looping combinators. We need a version of the inversion formula in infinitely many variables [4]. In the next theorem, $[z^\alpha]f(z)$ denotes the coefficient of z^α in the formal power series $f(z)$.

Theorem 8 (Lagrange-Good inversion formula). *Let* $\boldsymbol{z} = (z_i)_{i \in A}$ *be a sequence of possibly infinitely many variables. Suppose that* $\boldsymbol{f}(\boldsymbol{z})$ *is a family of formal power series in* $\mathbb{C}[[\boldsymbol{z}]]$ *indexed by the members of* A.

There is a unique solution $\boldsymbol{w} = \boldsymbol{h}(\boldsymbol{z})$ *for the equation* $\boldsymbol{w} = \boldsymbol{z} \times \boldsymbol{f}(\boldsymbol{w})$ *and for this solution* $[\boldsymbol{z}^\alpha]\, q(\boldsymbol{h}(\boldsymbol{z})) = [\boldsymbol{w}^\alpha]\, q(\boldsymbol{w}) D_{f,\alpha}(\boldsymbol{w})$ *holds for every formal power series* $q(\boldsymbol{z})$ *in* $\mathbb{C}[[\boldsymbol{z}]]$ *and every* α *in* $\exp(A)$. *Here* $D_{f,\alpha}(\boldsymbol{w})$ *is the determinant*

$$\det\left(\delta_{ik}(f_i(\boldsymbol{w}))^{\alpha[i]} - w_k \frac{\partial f_i(\boldsymbol{w})}{\partial w_k}(f_i(\boldsymbol{w}))^{\alpha[i]-1}\right)_{ik}$$

where $\alpha[i]$ *is the number of the occurrences of* i *in* α *and* i, k *range over the carrier* $\{i \in A \mid \alpha[i] \neq \emptyset\}$ *of* α.

The following form of the Lagrange-Good inversion is standard: $[\boldsymbol{z}^\alpha]q(\boldsymbol{h}(\boldsymbol{z})) = [\boldsymbol{w}^\alpha]\, q(\boldsymbol{w})(\boldsymbol{f}(\boldsymbol{w}))^\alpha D'_{f,\alpha}(\boldsymbol{w}, \, \boldsymbol{w}/\boldsymbol{f}(\boldsymbol{w}))$ where the determinant $D'_{f,\alpha}(\boldsymbol{w}, \boldsymbol{z})$ is given by $\det\,(\delta_{ik} - z_i(\partial f_i(\boldsymbol{w})/\partial w_k))_{ik}$. We can substitute z_k for z_i without changing the value of the determinant. So $(\boldsymbol{f}(\boldsymbol{w}))^\alpha D'_{f,\alpha}(\boldsymbol{w}, \boldsymbol{w}/\boldsymbol{f}(\boldsymbol{w}))$ is equal to $D_{f,\alpha}(\boldsymbol{w})$.

Turning to the normal functors (or species), the solution of the equation in the theorem above gives rise to the species of certain trees:

Definition 9 (of A_M). Let M be a normal functor in $[\mathbf{Set}^A, \mathbf{Set}^A]_{\mathrm{NF}}$.

A_M is defined as a solution $\boldsymbol{w} = A_M(\boldsymbol{z})$ of the equation $\boldsymbol{w} = \boldsymbol{z} \times M(\boldsymbol{w})$. The solution uniquely exists and is a normal functor in $[\mathbf{Set}^A, \mathbf{Set}^A]_{\mathrm{NF}}$, having finite coefficients if so does M. In fact, A_M is the multivariate species of M-enriched trees [4].

Let $\mathbf{Set}^A$ be a model of the lambda calculus with bijection $\psi : \exp(A) \times A \xrightarrow{\sim} A$, and let $\mathrm{ev}' : \mathbf{Set}^A \times \mathbf{Set}^A \to \mathbf{Set}^A$ be the composition of the evaluation functor ev preceded by a natural isomorphism, as given in section 3. We want to reformulate ev′ in terms of more elementary operations.

Definition 10 (of S and P). Let a range over A and $\boldsymbol{z}$ over $\mathbf{Set}^A$.

1. Let S be a normal functor from $\mathbf{Set}^A$ to $\mathbf{Set}^A$, this functor S defined by $S(z)[a] = \sum z[b^\circ]$ where the summation is over those $b^\circ \in A$ which have the form $b^\circ = \psi(\gamma, a)$ for some γ in $\exp(A)$.
2. Let P be a normal functor from $\mathbf{Set}^A$ to $\mathbf{Set}^A$, this functor P defined by $P(z)[a] = \mathbf{Set}^A(\gamma, z)$ if a has the form $\psi(\gamma, a_0)$ for some $a_0 \in A$.

Lemma 11. *It holds that* $\mathrm{ev}'(G, F) = S(G \times P(F))$ *for all presheaves* G *and* F *in* $\mathbf{Set}^A$.

Let Y be the interpretation $[\![\mathsf{Y}x]\!]_x$ in $[\mathbf{Set}^A, \mathbf{Set}^A]_{\mathrm{NF}}$, or as the same thing $[\![\Theta x]\!]_x$. The normal functors Y and A_M satisfy the following equations respectively:

$$Y(z) = \mathrm{ev}'(z, Y(z))$$
$$A_M(z) = z \times M(A_M(z)).$$

We want to write Y in terms of A_M.

Proposition 12. *It holds that* $Y = S \circ A_M$ *where* M *is defined by the composition* $P \circ S$.

In the next corollary, we use the following notations: Suppose that α in $\exp(A)$ has the carrier $\{a_1, a_2, \ldots, a_n\}$. Let us put $w_{a_j} = w[a_j]$ and $p_j = \alpha[a_j]$. Then $\alpha!$ equals $p_1!p_2!\cdots p_n!$, that is, the order of the group $\mathrm{Aut}_{\mathbf{Set}^A}(\alpha)$, and $(\frac{\partial}{\partial w})^\alpha$ equals $(\frac{\partial}{\partial w_{a_1}})^{p_1}(\frac{\partial}{\partial w_{a_2}})^{p_2}\cdots(\frac{\partial}{\partial w_{a_n}})^{p_n}$.

Corollary 13. *Let* $D_{M,\alpha}(w)$ *be the determinant*

$$\det\left(\delta_{ik}(M(w))^{\alpha[i]} - w_k \frac{\partial M_i(w)}{\partial w_k}(M_i(w))^{\alpha[i]-1}\right)_{ik}$$

for i, k *ranging over the carrier of* α, *where* $M = P \circ S$.

The interpretation $Y = [\![\mathsf{Y}x]\!]_x$ *of looping combinators in* $[\mathbf{Set}^A, \mathbf{Set}^A]_{\mathrm{NF}}$ *is given by* $\tilde{Y}_a[\alpha] = (S_a \times D_{M,\alpha})[\alpha]^{\sim}$ *for every* $a \in A$ *and every* $\alpha \in \exp(A)$. *In other words,*

$$Y_a(z) = \sum_{\alpha \in \exp(A)} \frac{z^\alpha}{\alpha!}\left[\left(\frac{\partial}{\partial w}\right)^\alpha S_a(w) D_{M,\alpha}(w)\right]_{w:=0}.$$

Example 3. We compute coefficient $\tilde{Y}[\alpha, a]$ in Example 2 (1), using the Lagrange-Good inversion formula. We let variable z_1 correspond to a_1 etc., and we write $M_1 = M_{a_1}$ and so on. Then $M_1(z) = P(S(z))[a_1]$ equals $S_a(z) \times S_b(z)$, that is, $(z_1 + z_2 + z_3)z_4$. Note: Although $S_a(z)$ and $S_b(z)$ are infinite sums by definition, we can omit the terms that are irrelevant of a given multiset α. Likewise $M_2(z) = z_1 + z_2 + z_3$ and $M_3(z) = M_4(z) = 1$. The determinant $D_{M,\alpha}(z)$ is computed as

$$\det\begin{pmatrix} z_2z_4 + z_3z_4 & -z_2z_4 & \cdots & \cdots \\ -(z_1{}^2 + z_1z_2 + z_3z_1) & z_1{}^2 + z_3{}^2 + z_1z_2 + z_2z_3 + 2z_3z_1 & \cdots & \cdots \\ 0 & 0 & 1 & 0 \\ 0 & 0 & 0 & 1 \end{pmatrix}.$$

Hence $S_a(z)D_{M,\alpha}(z) = (z_1 + z_2 + z_3)(\cdots + 1z_2{}^2z_3z_4 + \cdots + 2z_1z_2z_3z_4 + \cdots + 0z_1z_2{}^2z_4 + \cdots)$, which equals $\cdots + 3z_1z_2{}^2z_3z_4 + \cdots$. Therefore the coefficient $\tilde{Y}_a[\alpha]$ equals 3. Note that the coefficient 1 of $z_2{}^2z_3z_4 = z_1z_2{}^2z_3z_4/z_1$ corresponds to the first tree in Figure 4(i) starting with a_1. The coefficient 2 of $z_1z_2z_3z_4 = z_1z_2{}^2z_3z_4/z_2$ corresponds to the last two trees starting with a_2, and there is no tree starting with a_3, reflected by the coefficient 0 of $z_1z_2{}^2z_4 = z_1z_2{}^2z_3z_4/z_3$.

References

1. J. Adámek and J. Rosický, *Locally Presentable and Accessible Categories*, London Math. Soc. Lecture Note Ser. 189, (Cambridge University Press, 1994).
2. H. P. Barendregt, *The Lambda Calculus, Its Syntax and Semantics*, Revised Edition, Studies in Logic and the Foundations of Mathematics, Volume 103, (North-Holland, 1984).
3. N. G. de Bruijn, The Lagrange-Good inversion formula and its applications to integral equations, *J. Math. Analysis Applications* **92** (1983) 397–409.
4. R. Ehrenborg and M. Méndez, A bijective proof of infinite variated Good's inversion, *Advances Math.* **103** (1994) 221–259.
5. I. M. Gessel, A combinatorial proof of the multivariable Lagrange inversion formula, *J. Combinatorial Th., Ser. A* **45** (1987) 178–195.
6. J.-Y. Girard, Normal functors, power series and λ-calculus, *Ann. Pure Applied Logic* **37** (1988) 129–177.
7. I. J. Good, Generalization to several variables of Lagrange's expansion, with applications to stochastic processes, *Proc. Cambridge Philosophical Soc.* **56** (1960) 367–380.
8. A. Joyal, Une théorie combinatoire des séries formelles, *Advances Math.* **42** (1981) 1–82.
9. A. Joyal, Foncteurs analytiques et espèces de structures, *Combinatoire Enumérative*, Proceedings, Montreal, Québec, Canada, 1985, G. Labelle, P. Leroux, eds., Lecture Notes in Mathematics 1234, (Springer, 1986) pp. 126–159.
10. J. Lambek and P. J. Scott, *Introduction to Higher Order Categorical Logic*, Cambridge Studies in Advanced Mathematics 7 (Cambridge University Press, 1986).
11. S. Mac Lane and I. Moerdijk, *Sheaves in Geometry and Logic, A First Introduction to Topos Theory*, (Springer, 1992).
12. C. P. Wadsworth, The relation between computational and denotational properties for Scott's D_∞-models of the lambda calculus, *SIAM J. Computing* **5** (1976) 488–521.
13. C. P. Wadsworth, Approximate reduction and lambda calculus models, *SIAM J. Computing* **7** (1978) 337–356.

Effectivity and Effective Continuity of Functions between Computable Metric Spaces

Peter Hertling

Theoretische Informatik I, FernUniversität, D-58084 Hagen, Germany.
Email: peter.hertling@fernuni-hagen.de

Abstract. Based on the Turing machine model there are essentially two different notions of computable functions over the real numbers. The *effective* functions are defined only on computable real numbers and are Type 1 computable with respect to a numbering of the computable real numbers. The *effectively continuous* functions may be defined on arbitrary real nunbers. They are exactly those functions which are Type 2 computable with respect to an appropriate representation of the real numbers.
We characterize the Type 2 computable functions on computable real numbers as exactly those Type 1 computable functions which satisfy a certain additional condition concerning their domain of definition. Our result is a sharp strengthening of the well-known continuity result of Tseitin and Moschovakis for effective functions. The result is presented for arbitrary computable metric spaces.

1 Introduction

In this paper we compare two approaches for defining computability of functions between computable metric spaces. The first approach can be described as effectivity or computability with respect to numberings, the second as effective continuity or computability with respect to representations. We shall give special attention to the space of real numbers.

In the first approach one considers countable metric spaces M, M' with numberings. One of the most important examples is the space $\mathbb{R}_c$ of computable real numbers. A numbering of this space is derived from a numbering of the algorithms or machines that compute such a number. A function $f :\subseteq M \to M'$ [1] is called *Type 1 computable* or *effective* if there is partial computable function $g \in P^{(1)}$ which transforms any index of an element $x \in \operatorname{dom} f$ into an index of the result $f(x) \in M'$. This approach, with some variations, has been considered in the Russian school of constructive analysis, represented by Markov [10], Shanin [15], Tseitin [21], Kushner [9], and others. It has also been considered by Moschovakis [11], Aberth [1] and by many others. Closely related is the constructive approach to analyis presented by Bishop and Bridges [3].

In the second approach one considers metric spaces M, M' with a numbering of a dense subset, respectively. Consider for example the space $\mathbb{R}$ of real numbers with a standard numbering of the rational numbers. A representation of M

[1] We use the notation $f :\subseteq X \to Y$ for a (partial) function f with domain $\operatorname{dom} f \subseteq X$.

(respectively of M') is defined via the numbering of the dense subset. One can represent an element x by sequences p in the Baire space $\mathbb{B} = \mathbb{N}^{\mathbb{N}} = \{p : \mathbb{N} \to \mathbb{N}\}$ which describe normed Cauchy sequences consisting of elements in the enumerated dense subset with limit x. A function $f :\subseteq M \to M'$ is *Type 2 computable* if there is a partial recursive operator in the sense of Rogers [14] on the Baire space $\mathbb{B}$ that transforms any name $p \in \mathbb{B}$ for an element $x \in \operatorname{dom} f$ into a name for $f(x)$. This is equivalent to *effective continuity* in the following sense: there is an algorithm which, given a desired output precision 2^{-n}, needs only a finite approximation of an element $x \in \operatorname{dom} f$ in order to compute an approximation of the result $f(x)$ up to precision 2^{-n}, and which is able to compute a sufficient input precision. For real functions this leads to the notion of computability considered by Grzegorczyk [4], Klaua [6], Pour-El and Richards [13], Weihrauch [22], Ko [7], and by many others.

How are the two computability notions related? From the normed Cauchy representation of a metric space M (with a numbering of a dense subset) one can derive a numbering of the subspace $M_c \subseteq M$ of the computable elements: a number i is an index for a computable element x if it is an index of a name $p \in R^{(1)} = \{p : \mathbb{N} \to \mathbb{N} \mid p \text{ recursive}\}$ for a recursive normed Cauchy sequence with limit x. In the following we consider Type 2 computability with respect to normed Cauchy representations and Type 1 computability with respect to the derived numberings of the computable elements. We assume that the metric spaces M and M' are computable and recursively separable; see Section 3.1.

It is easy to see that the restriction to M_c of a Type 2 computable function $f :\subseteq M \to M'$ is Type 1 computable. Note that this implies $f(M_c) \subseteq {M'}_c$. Hence, for a function $f :\subseteq M_c \to M'$ Type 1 computability is a necessary condition for Type 2 computability. A simple example of Myhill (see Remark 2 in [8]) shows that it is not a sufficient condition. But Weihrauch's [22] version of the well-known Tseitin/Moschovakis continuity result shows that a Type 1 computable function with a recursively separable domain must be Type 2 computable. Yet, in general, recursive separability of the domain is a too strong additional condition: there is a Type 2 computable function $f :\subseteq M_c \to M'$ which cannot be extended to a Type 1 computable function with a recursively separable domain. It is the aim of this paper to give a sharp condition. We show that a function $f :\subseteq M_c \to M'$ is Type 2 computable if and only if it is Type 1 computable and satisfies a certain additional condition which in general is weaker than recursive separability of the domain (to be precise: it is weaker than the possibility to extend the function to a Type 1 computable function with a recursively separable domain).

This result is a sharp strengthening of the continuity result of Tseitin [20, 21] and Moschovakis [11]. It seems to be new even for functions between the space $R^{(1)}$ of total computable functions or the space $\mathbb{R}_c$ of computable real numbers. Note that a special version of the continuity result had been obtained by Kreisel, Lacombe, and Shoenfield [8]. Related results can be found in [12, 19, 22, 2, 17, 18]. But the problem of giving a characterization of effective continuity as above, seems to be still open in the general case. For a discussion of the two approaches for defining computability the reader is referred to Kushner [9], pp. 304–306, and to Weihrauch [22], Ch. 3.6.

First we formulate a corollary of the main result for the special case of operators on the space of total recursive functions, which explains already the kind of characterization. In the following section the notions concerning computable metric spaces, Type 1 computable functions (effective functions), and Type 2 computable functions (effectively continuous functions) are introduced. Then we discuss the classical results more precisely. In the last section the new condition is introduced and the main results are stated. Actually we give two conditions, which are equivalent in all interesting cases. Due to lack of space the proofs are omitted.

Notation

$\mathbb{N} = \{0, 1, 2, \ldots\}$ is the set of nonnegative integers, $\mathbb{Q}$ is the space of rational numbers, $\mathbb{R}$ is the space of real numbers, $\mathbb{B} = \{p : \mathbb{N} \to \mathbb{N}\}$ is the Baire space (see Example 3), $\langle .,. \rangle : \mathbb{N}^2 \to \mathbb{N}$ is a standard pairing function, and the maps $\pi_i : \mathbb{N} \to \mathbb{N}$ for $1 \le i \le 2$ are the corresponding inverse maps with $\pi_j\langle n_1, n_2\rangle = n_j$. By $P^{(1)} := \{p :\subseteq \mathbb{N} \to \mathbb{N} \,|\, p \text{ computable}\}$ we denote the set of all (partial) computable number functions, by $R^{(1)} := \{p : \mathbb{N} \to \mathbb{N} \,|\, p \text{ computable}\}$ the space of total recursive functions.

A *numbering* of a denumerable set S is a (partial) surjective function $\nu :\subseteq \mathbb{N} \to S$. For $i \in \operatorname{dom} \nu$ we denote the corresponding element in S either by ν_i or by $\nu(i)$. A numbering ν is *reducible* to a numbering ν' (written $\nu \le \nu'$) if there is a $g \in P^{(1)}$ such that $\nu(i) = \nu' \circ g(i)$, for all $i \in \operatorname{dom} \nu$. If $\nu \le \nu'$ and $\nu' \le \nu$ then the numberings ν, ν' are *equivalent* (written $\nu \equiv \nu'$).

The map $\nu_\mathbb{Q} : \mathbb{N} \to \mathbb{Q}$ is the standard total numbering of $\mathbb{Q}$ defined by $\nu_\mathbb{Q}\langle i, j, k\rangle = (i - j)/(k + 1)$. $\varphi : \mathbb{N} \to P^{(1)}$ is a total standard numbering of $P^{(1)}$ and $W_i := \operatorname{dom} \varphi_i$ for all $i \in \mathbb{N}$. Hence, W is a total numbering of all *recursively enumerable (r.e.)* sets.

2 Operators on the space of total recursive functions

In order to give an impression of our main result we formulate a corollary for the important special case of an operator mapping a subset of the total recursive functions $R^{(1)}$ to $R^{(1)}$.

The restriction $\varphi' := \varphi|_{\varphi^{-1}(R^{(1)})}$ of φ is a partial numbering of $R^{(1)}$. A function $f :\subseteq R^{(1)} \to R^{(1)}$ is called (φ', φ')*-computable* or *effective* if there is a function $g \in P^{(1)}$ such that for all $i \in \varphi^{-1}(\operatorname{dom} f)$ we have $i \in \operatorname{dom} g$, $\varphi_{g(i)} \in R^{(1)}$, and $f(\varphi_i) = \varphi_{g(i)}$. The function f is called *effectively continuous* if there is a (partial) recursive operator F in the sense of Rogers [14] with $F|_{\operatorname{dom} f} = f$; compare Section 3.3. Kreisel, Lacombe and Shoenfield [8] proved that an effective function $f :\subseteq R^{(1)} \to R^{(1)}$ with a recursively separable domain is effectively continuous (see Theorem 1 in [8] and the remark preceding Theorem 3 in [8]).

The domain dom f is *recursively separable* if there is an r.e. set $E \subseteq \mathbb{N}$ with

$$(1)\ E \subseteq \varphi^{-1}(\mathrm{dom}\, f), \quad (2)\ \mathrm{dom}\, f \subseteq \overline{\varphi(E)}\,.$$

If $g \in P^{(1)}$ is a function which "computes" f, i.e. $f(\varphi_i) = \varphi_{g(i)}$ for all $i \in \mathrm{dom}\, f$, then dom f and hence also $\varphi(E)$ must be contained in the set

$$X(g) := \{p \in R^{(1)} \mid (\forall i, j \in \varphi^{-1}\{p\})\, g(i) \text{ and } g(j) \text{ are defined and } \varphi_{g(i)} = \varphi_{g(j)} \in R^{(1)}\}$$

of elements $p \in R^{(1)}$ at which g "defines a value in $R^{(1)}$". Now the main point is that it is not necessary that g really "defines a value in $R^{(1)}$" at the elements in $\varphi(E)$. Instead it is sufficient to demand that for each $n \in \mathbb{N}$ one can find an r.e. set of elements whose closure contains dom f and at which g "defines a value up to precision n". Therefore we define the set

$$X_n(g) := \{p \in R^{(1)} \mid (\forall i, j \in \varphi^{-1}\{p\})\ g(i) \text{ and } g(j) \text{ are defined and } (\forall k \leq n)\ \varphi_{g(i)}(k) = \varphi_{g(j)}(k) \text{ are defined}\}$$

of elements p at which g *defines a value up to precision* n. We call a function $f :\subseteq R^{(1)} \to R^{(1)}$ *D′-effective* if there is a function $g \in P^{(1)}$ with $f(\varphi_i) = \varphi_{g(i)}$ for all $i \in \varphi^{-1}(\mathrm{dom}\, f)$ and if additionally there is a function $e \in R^{(1)}$ such that for all $n \in \mathbb{N}$

$$(1)\ W_{e(n)} \subseteq \varphi^{-1}(X_n(g)), \quad (2)\ \mathrm{dom}\, f \subseteq \overline{\varphi(W_{e(n)})}\,.$$

Obviously, if a function $f :\subseteq R^{(1)} \to R^{(1)}$ is effective and has a recursively separable domain, then it is D′-effective.

Theorem 1. *For a function $f :\subseteq R^{(1)} \to R^{(1)}$ the following two conditions are equivalent:*

1. *f is D′-effective,*
2. *f is effectively continuous.*

3 Basic notions

In this section the basic notions are introduced and simple statements connecting them are deduced. First we consider computable metric spaces. Then effective functions are defined, that is, functions that are (Type 1) computable with respect to numberings. Finally we introduce effectively continuous functions and functions that are (Type 2) computable with respect to normed Cauchy representations.

3.1 Computable metric spaces

The various approaches for introducing effectivity on a metric space M can be summarized by saying that one imposes effectivity conditions on the metric with respect to a numbering of a dense subset. Hence effectivity for metric spaces has been introduced only for separable metric spaces.

Under a metric space we always understand a triple (M, d, α) where M is a nonempty set, $d : M \times M \to \mathbb{R}$ is a metric, and $\alpha :\subseteq \mathbb{N} \to M$ is a numbering of a dense subset of M. Sometimes we write only (M, α), leaving aside d.

Definition 2. A metric space (M, d, α) (or just (M, α)) is called *semicomputable* if there is an r.e. set $D^{\alpha}_{<} \subseteq \mathbb{N}$ such that for all $i, j \in \operatorname{dom} \alpha$ and all $n \in \mathbb{N}$

$$\langle i, j, n \rangle \in D^{\alpha}_{<} \iff d(\alpha_i, \alpha_j) < \nu_{\mathbb{Q}}(n) .$$

It is called a *computable metric space* if furthermore there is an r.e. set $D^{\alpha}_{>} \subseteq \mathbb{N}$ such that for all $i, j \in \operatorname{dom} \alpha$ and all $n \in \mathbb{N}$

$$\langle i, j, n \rangle \in D^{\alpha}_{>} \iff d(\alpha_i, \alpha_j) > \nu_{\mathbb{Q}}(n) .$$

Computationally interesting spaces are usually not just separable but recursively separable.

Definition 3. A subset X of a metric space (M, α) is called *α-recursively separable* if there is an r.e. set $E \subseteq \alpha^{-1}(X)$ such that $X \subseteq \overline{\alpha(E)}$.

Often we simply write "recursively separable" instead of "α-recursively separable" if it is clear which numbering α is meant. Note that in general a subspace of a recursively separable metric space is of course separable, but it does not have to be recursively separable; compare Slisenko [16].

The effectivity properties of a metric space depend only on the equivalence class of the numbering of a dense subset, as the following easily proved lemma shows.

Lemma 4. *Let (M, α) be a metric space, and let α' be a numbering of a dense subset of M with $\alpha' \leq \alpha$. If (M, α) is semicomputable (computable), then (M, α') is semicomputable (computable). If (M, α') is recursively separable, then (M, α) is recursively separable.*

We remark that, if a metric space (M, α) is recursively separable, then one can easily construct a total numbering α' of a dense subset of M with $\alpha' \leq \alpha$.

We wish to consider computations over the real numbers that are based on approximations. The idea is to describe the real numbers by Cauchy sequences of rational numbers that converge recursively. Among these it is sufficient to consider the Cauchy sequences with a normed rate of convergence. This can be done also for arbitrary metric spaces with a numbering of a dense subset. A *normed Cauchy sequence* is a sequence of points $(x_i)_{i \in \mathbb{N}}$ in a metric space M with $d(x_i, x_j) < 2^{-\min\{i,j\}}$. Using normed Cauchy sequences and a numbering of a dense subset one can define a natural representation of the metric space, define its *computable elements* and derive a numbering of its computable elements.

Definition 5. Let (M, α) be a metric space. The *Cauchy representation* $\rho_\alpha :\subseteq \mathbb{B} \to M$ of (M, α) is defined by

$$\rho_\alpha(p) = x \quad \text{iff} \quad (\alpha_{p(n)})_{n \in \mathbb{N}} \text{ is a normed Cauchy sequence with limit } x \in M\,.$$

By $\nu_\alpha(i) := \rho_\alpha \varphi_i$ for $i \in \operatorname{dom} \nu_\alpha := \{i \in \mathbb{N} \mid \varphi_i \in \operatorname{dom} \rho_\alpha\}$ we define a partial numbering ν_α of the set $M_c := \operatorname{range} \nu_\alpha$ of the *computable elements* of M.

The normed Cauchy representations are special cases of the admissible representations of T_0-spaces with countable bases considered in the Type 2 Theory of Effectivity; see Weihrauch [22]. In Section 3.3 we will define the computability of functions with respect to representations. Here we are interested in the numbering ν_α.

Lemma 6. *If (M, α) is a semicomputable (computable, recursively separable) metric space then also (M, ν_α) and (M_c, ν_α) are semicomputable (computable, recursively separable) metric spaces.*

Furthermore one easily checks $\alpha \leq \nu_\alpha$ for any metric space (M, α). A metric space (M, α) is called *weakly complete* if the converse relation is true also, that is, given an index of a recursive normed Cauchy sequence converging to an enumerated point, one can compute an index of the point:

$$\nu_\alpha|_{\nu_\alpha^{-1}(\text{range } \alpha)} \leq \alpha.$$

The most important special property of the derived numbering ν_α is that the spaces (M, ν_α) and (M_c, ν_α) are weakly complete. The expression "weakly complete" is taken from Kushner [9]. For computable metric spaces with a surjective numbering the condition above is equivalent to the Condition (A) introduced by Moschovakis [11].

Example 1. One of the simplest examples of a computable, recursively separable, and weakly complete metric space is the discrete space $\mathbb{N}$ with the trivial total numbering $\mathrm{id}_\mathbb{N}$ (and the distance $d(n, m) = 1$ for $n \neq m$).

Example 2. The space $(\mathbb{R}, \nu_\mathbb{Q})$ is a computable recursively separable metric space, but it is not weakly complete. The space $(\mathbb{R}_c, \nu_{\nu_\mathbb{Q}})$ is the standard space of the *computable real numbers*. It is computable, recursively separable, and weakly complete.

Example 3. Finally let $\mu^* : \mathbb{N} \to \mathbb{N}^*$ be a total bijective standard numbering of the finite words over $\mathbb{N}$, With $\mu(i) := \mu^*(i)0^\omega$ the space $(\mathbb{B}, \mu)$ is a computable recursively separable metric space. We use the metric $d : \mathbb{B} \times \mathbb{B} \to \mathbb{R}$ defined by

$$d(p, q) := \begin{cases} 2^{-\min\{i \mid p(i) \neq q(i)\}} & \text{if } p \neq q \\ 0 & \text{else} \end{cases}$$

for any $p = p(0)p(1)p(2)\ldots \in \mathbb{B}$, $q = q(0)q(1)q(2)\ldots \in \mathbb{B}$. Because of range $\nu_\mu = R^{(1)}$, the computable elements of $(\mathbb{B}, \mu)$ are just the total recursive functions. The restriction $\varphi' := \varphi|_{\varphi^{-1}(R^{(1)})}$ is also a partial numbering of $R^{(1)}$. One easily checks $\nu_\mu \equiv \varphi'$. The spaces $(R^{(1)}, \nu_\mu)$ and $(R^{(1)}, \varphi')$ are computable, recursively separable, and weakly complete metric spaces.

3.2 Effective functions

In this section (M, d, α) and (M', d', β) are two fixed metric spaces. We introduce *effective* or *Type 1 computable* functions between M and M'.

Definition 7. A function $f :\subseteq M \to M'$ is called (α, β)*-computable* if $\mathrm{dom}\, f \subseteq \mathrm{range}\, \alpha$ and there is a $g \in P^{(1)}$ with

$$(\forall i \in \alpha^{-1}(\mathrm{dom}\, f))\ f(\alpha_i) = \beta_{g(i)}.$$

The first property to be mentioned is that the computability of a function depends only on the equivalence class of the numbering. This follows immediately from the following obvious lemma.

Lemma 8. *Let $\tilde{\alpha} :\subseteq \mathbb{N} \to M$ and $\tilde{\beta} :\subseteq \mathbb{N} \to M'$ be additional numberings of dense subsets with $\tilde{\alpha} \leq \alpha$ and $\beta \leq \tilde{\beta}$. If $f :\subseteq M \to M'$ is (α, β)-computable, then f is $(\tilde{\alpha}, \tilde{\beta})$-computable.*

We are mainly interested in (ν_α, ν_β)-computable functions. If it is clear which numberings α and β are meant then we will call the (ν_α, ν_β)-computable functions *effective* functions.

Remark. The above notion of computability with respect to numberings is the notion which is used for example in the early paper [10] by Markov, by Helm [5], by Weihrauch [22] and by many other authors. In constructive analysis (see for example Kushner [9]) a more restrictive definition of an effective function $f :\subseteq M \to M'$ between enumerated sets M and M' with numberings $\alpha :\subseteq \mathbb{N} \to M$ and $\beta :\subseteq \mathbb{N} \to M'$ (that is, α and β are surjective) is often used. One demands that there is a partial recursive function $g \in P^{(1)}$ with $f(\alpha_i) = \beta_{g(i)}$ for all $i \in \alpha^{-1}(\mathrm{dom}\, f)$ and additionally with $\alpha^{-1}(\mathrm{dom}\, f) = \mathrm{dom}\, \alpha \cap \mathrm{dom}\, g$. If a function f can be computed in this way, Weihrauch [22] calls it *strongly* (α, β)*-computable.* For example Tseitin [21] and Moschovakis [11] have formulated their continuity results for effective operators in this stronger sense.

3.3 Computable and effectively continuous functions

Let (M, α) and (M', β) be metric spaces. We define *Type 2 computable* functions and two forms of *effective continuity* for functions $f :\subseteq M \to M'$.

In order to define effective continuity we use a numbering B of a base of (M, α) defined by

$$B_i := B(\alpha_{\pi_1(i)}, 2^{-\pi_2(i)}) = \{x \in M \mid d(\alpha_{\pi_1(i)}, x) < 2^{-\pi_2(i)}\}$$

for $i \in \mathrm{dom}\, B := \{i \mid \pi_1(i) \in \mathrm{dom}\, \alpha\}$. The corresponding numbering of a base of (M', β) will be denoted by B' in this section.

The first of the following three definitions of global effective continuity expresses that the preimage of a basic open set is open in an effective way. The second is based on ε-δ continuity. The third expresses the computability used in Type 2 Theory of Effectivity [22].

Definition 9. A function $f :\subseteq M \to M'$ is called

1. *weakly effectively continuous* if there is a function $g \in P^{(1)}$ with
$$(\forall j \in \mathrm{dom}\, B')(W_{g(j)} \subseteq \mathrm{dom}\, B \text{ and } f^{-1}(B'_j) = \mathrm{dom}\, f \cap \bigcup\{B_i \mid i \in W_{g(j)}\}),$$
2. *effectively continuous* if there is an r.e. set $V \subseteq \mathbb{N}$ such that
$$(\forall \langle i,j\rangle \in V)(i \in \mathrm{dom}\, B \wedge j \in \mathrm{dom}\, B' \wedge f(B_i) \subseteq B'_j)$$
$$\text{and } (\forall x \in \mathrm{dom}\, f)(\forall \varepsilon > 0)(\exists \langle i,j\rangle \in V)(x \in B_i \wedge 2^{-\pi_2(j)} < \varepsilon),$$
3. $(\rho_\alpha, \rho_\beta)$*-computable* if there is an effectively continuous function $F :\subseteq \mathbb{B} \to \mathbb{B}$ such that for all $p \in \rho_\alpha^{-1}(\mathrm{dom}\, f)$
$$f\rho_\alpha(p) = \rho_\beta F(p)\,.$$

A subset X of M is called a *Lacombe set* if there is an r.e. set $A \subseteq \mathrm{dom}\, B$ such that $X = \bigcup\{B_i \mid i \in A\}$. Hence a function is weakly effectively continuous if the preimage of any ball B'_j is a Lacombe set and given an index j of such a ball one can compute an index of this Lacombe set. The first condition in Part (2) (effective continuity) of the definition states that V enumerates only pairs of balls B_i, B'_j such that B_i is mapped to B'_j. By the second condition sufficiently many such pairs are enumerated. Finally note that for a function $F :\subseteq \mathbb{B} \to \mathbb{B}$ the following conditions are equivalent: (1) F is weakly effectively continuous, (2) F is effectively continuous, (3) F can be extended to a partial recursive operator $G :\subseteq \{p :\subseteq \mathbb{N} \to \mathbb{N}\} \to \{p :\subseteq \mathbb{N} \to \mathbb{N}\}$ in the sense of Rogers [14], §9.8, (4) F can be extended to a total recursive operator G in the same sense.

The metric spaces considered in Type 2 Theory of Effectivity (Weihrauch [22, 23]) are usually semicomputable metric spaces with a total numbering of a dense subset, hence semicomputable, recursively separable metric spaces. A more effective version of the next proposition has been given by Weihrauch [23].

Proposition 10. *If (M, α) and (M', β) are semicomputable and recursively separable, then for a function $f :\subseteq M \to M'$ the conditions (1), (2), and (3) in Definition 9 are equivalent.*

4 Classical results

We discuss shortly the known results on the relation between Type 1 computability and Type 2 computability for functions between computable metric spaces.

The following proposition is well-known and easy to prove.

Proposition 11. *Let (M, α) be a semicomputable metric space and (M', β) be a metric space. If $f :\subseteq M \to M'$ is effectively continuous then $f|_{M_c}$ is (ν_α, ν_β)-computable.*

On the other hand Remark 2 in [8] contains a simple example, due to Myhill, of an effective function $f :\subseteq R^{(1)} \to \mathbb{N}$ which is discontinuous and, hence, cannot be Type 2 computable. Thus, effectivity alone does not imply continuity. But the famous result of Tseitin [20, 21] implies that a strongly (ν_α, ν_β)-computable function $f :\subseteq M_c \to M'_c$ is effectively continuous if (M, α) is computable and recursively separable and (M, β) is computable. In this case the domain of f must itself be recursively separable; compare Kushner [9], Ch. 9. §2, Lemma 5. In fact one can prove the following result; see Weihrauch [22].

Proposition 12. *Let (M, α) be a semicomputable metric space and (M', β) be a computable, recursively separable metric space. If a function $f :\subseteq M_c \to M'$ is (ν_α, ν_β)–computable and has a ν_α-recursively separable domain, then it is effectively continuous.*

Actually Tseitin formulated his result not for the derived numberings ν_α, ν_β, but more generally for weakly complete numberings. Moschovakis gave a similar but slightly weaker formulation ([11], Theorem 12)

Remark. Moschovakis also gave another continuity result in [11], Theorem 3. Using slightly weaker assumptions he derived effective "pointwise" continuity of effective functions. Helm [5] has given an example of an effective and effectively pointwise continuous function which is not effectively continuous.

The question arises whether the last proposition is sharp, that is, whether it covers all effectively continuous functions. The answer is Yes if the image space is discrete.

Theorem 13. *Let (M, α) be a computable, recursively separable metric space and (M', β) be a discrete, computable, recursively separable metric space. Then for a function $f :\subseteq M_c \to M'$ the following three conditions are equivalent.*

1. *f can be extended to a strongly (ν_α, ν_β)-computable function.*
2. *f can be extended to a (ν_α, ν_β)-computable function with a ν_α-recursively separable domain.*
3. *f is effectively continuous.*

But in the general case the answer is No. There is even a counterexample with a discrete domain.

Theorem 14. *There is an effectively continuous function $f :\subseteq \mathbb{N} \to R^{(1)}$ which cannot be extended to an effective function $g :\subseteq \mathbb{N} \to R^{(1)}$ with a recursively separable domain.*

This poses the problem to characterize the relation between effectivity and effective continuity precisely, which is the aim of the paper.

5 Characterization of effective continuity by effectivity and an additional condition

In this section (M, d, α) and (M', d', β) are two fixed metric spaces. We have seen that an effective function with a recursively separable domain is effectively continuous. We shall show that a weaker condition than recursive separability suffices. In fact we give two conditions, which turn out to be equivalent under suitable assumptions.

Let us assume that a function $f :\subseteq M \to M'$ is (ν_α, ν_β)–computable via a function $g \in P^{(1)}$, that is, $\operatorname{dom} f \subseteq \operatorname{range} \nu_\alpha$ and $f(\nu_\alpha(i)) = \nu_\beta(g(i))$ for all $i \in \nu_\alpha^{-1}(\operatorname{dom} f)$. Then $\operatorname{dom} f$ is ν_α-recursively separable if and only if there is an r.e. set $E \subseteq \mathbb{N}$ such that

$$(1)\ E \subseteq \nu_\alpha^{-1}(\operatorname{dom} f), \quad (2)\ \operatorname{dom} f \subseteq \overline{\nu_\alpha(E)}.$$

The domain $\operatorname{dom} f$ must be contained in the set

$$CX(g) := \{x \in M \mid (\forall i, j \in \nu_\alpha^{-1}\{x\})\ (\beta(\varphi_{g(i)}(n)))_{n\in\mathbb{N}} \text{ and } (\beta(\varphi_{g(j)}(n)))_{n\in\mathbb{N}} \text{ are normed Cauchy sequences with the same limit}\}$$

of points in which g "defines a value" with respect to ν_α and ν_β. For any $n \in \mathbb{N}$ we define the set

$$CX_n(g) := \{x \in M \mid (\forall i, j \in \nu_\alpha^{-1}\{x\})\ \varphi_{g(i)}(n) \in \operatorname{dom} \beta,\ \varphi_{g(j)}(n) \in \operatorname{dom} \beta,\ d(\beta(\varphi_{g(i)}(n)), \beta(\varphi_{g(j)}(n))) < 3 \cdot 2^{-n}\}$$

of points in which g *defines a value up to precision n*. A slightly stronger version is given by the sequence of sets

$$SCX_n(g) := \{x \in M \mid (\forall i, j \in \nu_\alpha^{-1}\{x\})\ (\forall k \le n)\ \varphi_{g(i)}(k) \in \operatorname{dom} \beta,\ \varphi_{g(j)}(k) \in \operatorname{dom} \beta,\ d(\beta(\varphi_{g(i)}(k)), \beta(\varphi_{g(j)}(k))) < 3 \cdot 2^{-k}\}$$

of points in which g *strongly defines a value up to precision n*.

Obviously $CX(g) \subseteq SCX_n(g) \subseteq CX_n(g)$ for all $n \in \mathbb{N}$.

Definition 15.

1. A function $f :\subseteq M \to M'$ is called *D-(ν_α, ν_β)–computable* or *D-effective* if it is (ν_α, ν_β)–computable via a function $g \in P^{(1)}$ and additionally there is a function $e \in R^{(1)}$ such that for all $n \in \mathbb{N}$ we have

$$(1)\ W_{e(n)} \subseteq \nu_\alpha^{-1}(CX_n(g)), \quad (2)\ \operatorname{dom} f \subseteq \overline{\nu_\alpha(W_{e(n)})}.$$

2. A function $f :\subseteq M \to M'$ is called *SD-(ν_α, ν_β)–computable* or *strongly D-effective* if it is (ν_α, ν_β)–computable via a function $g \in P^{(1)}$ and additionally there is a function $e \in R^{(1)}$ such that for all $n \in \mathbb{N}$ we have

$$(1)\ W_{e(n)} \subseteq \nu_\alpha^{-1}(SCX_n(g)), \quad (2)\ \operatorname{dom} f \subseteq \overline{\nu_\alpha(W_{e(n)})}.$$

Obviously a strongly D-effective function is D-effective. We shall see that under suitable conditions on the spaces M and M' a D-effective function is effectively continuous and, vice versa, an effectively continuous function is even strongly D-effective.

Proposition 16. *Let $f :\subseteq M \to M'$ be (ν_α, ν_β)–computable via a function $g \in P^{(1)}$ and let* dom f *be recursively separable. Then there is a function $e \in R^{(1)}$ such that f is SD-(ν_α, ν_β)–computable via g and e.*

Proof. Let $E \subseteq \mathbb{N}$ be an r.e. set such that $E \subseteq \nu_\alpha^{-1}(\text{dom } f)$ and dom $f \subseteq \overline{\nu_\alpha(E)}$. The statement is proved by a function $e \in R^{(1)}$ with $W_{e(n)} = E$ for all $n \in \mathbb{N}$. □

By the last proposition and by Theorem 14 the following result is a strengthening of the Tseitin/Moschovakis result (for a recursively separable image space).

Theorem 17. *Let (M, α) be a semicomputable metric space, and let (M', β) be a computable recursively separable metric space. If $f :\subseteq M \to M'$ is D-(ν_α, ν_β)–computable then it is $(\rho_\alpha, \rho_\beta)$–computable.*

We have copied from Weihrauch [22] the observation that the domain does not need to be computable but only semicomputable. The theorem is a sharp result since the converse is also true.

Theorem 18. *Let (M, α) be a computable recursively separable metric space and let (M', β) be a semicomputable metric space. If $f :\subseteq M \to M'$ is effectively continuous then $f|_{M_c}$ is SD-(ν_α, ν_β)–computable.*

Note that here we do not need that the image space (M', β) is computable. Instead semicomputability suffices. Summarizing Theorem 17 and Theorem 18 we obtain our main result.

Theorem 19. *Let (M, α) and (M', β) be computable recursively separable metric spaces. Then for a function $f :\subseteq M_c \to M'$ the following conditions are equivalent.*

1. *f is D-(ν_α, ν_β)–computable.*
2. *f is SD-(ν_α, ν_β)–computable.*
3. *f is weakly effectively continuous.*
4. *f is effectively continuous.*
5. *f is $(\rho_\alpha, \rho_\beta)$–computable.*

Proof. The equivalences (3) $\iff$ (4) $\iff$ (5) have been stated already in Proposition 10. (4) $\Rightarrow$ (2) is the content of Theorem 18, (2) $\Rightarrow$ (1) is trivial, and (1) $\Rightarrow$ (5) is the content of Theorem 17. □

Theorem 19 describes precisely the difference between the Type 1 computability notion for real functions based on the numbering $\nu_{\nu_\mathbb{Q}}$ of the computable real numbers and the Type 2 computability notion based on representations like the Cauchy representation $\rho_{\nu_\mathbb{Q}}$; see the introduction.

References

1. O. Aberth. *Computable analysis.* McGraw–Hill, New York, 1980.
2. U. Berger. Total sets and objects in domain theory. *Annals of Pure and Applied Logic*, 60:91–117, 1993.
3. E. Bishop and D. S. Bridges. *Constructive Analysis.* Springer-Verlag, Berlin, Heidelberg, 1985.
4. A. Grzegorczyk. On the definitions of computable real functions. *Fund. Math.*, 44:61–71, 1957.
5. J. P. Helm. On effectively computable operators. *Zeitschrift f. math. Logik und Grundlagen d. Math.*, 17:231–244, 1971.
6. D. Klaua. *Konstruktive Analysis.* VEB Deutscher Verlag der Wiss., Berlin, 1961.
7. K.-I Ko. *Complexity Theory of Real Functions.* Birkhäuser, Boston, 1991.
8. G. Kreisel, D. Lacombe, and J. R. Shoenfield. Partial recursive functionals and effective operations. In A. Heyting, editor, *Constructivity in Mathematics, Proceedings of the Colloquium held at Amsterdam 1957*, pages 290–297, Amsterdam, 1959. North–Holland Publishing Company.
9. B. A. Kushner. *Lectures on Constructive Mathematical Analysis*, volume 60 of *Translations of Mathematical Monographs.* American Math. Soc., Providence, Rhode Island, 1984.
10. A. A. Markov. On the continuity of constructive functions. *Uspekhi Mat. Nauk*, 9(3 (61)):226–230, 1954. (Russian).
11. Y. N. Moschovakis. Recursive metric spaces. *Fund. Math.*, 15:215–238, 1964.
12. J. Myhill and J. C. Shepherdson. Effective operations on partial recursive functions. *Zeitschrift f. math. Logik und Grundlagen d. Math.*, 1:310–317, 1955.
13. M. B. Pour-El and J. I. Richards. *Computability in Analysis and Physics.* Springer–Verlag, Berlin, Heidelberg, 1989.
14. H. Rogers, Jr. *Theory of Recursive Functions and Effective Computability.* McGraw–Hill Book Company, New York, 1967.
15. N. A. Shanin. *Constructive real numbers and constructive function spaces*, volume 21 of *Translations of Math. Monographs.* Amer. Math. Soc., Providence, R. I., 1968.
16. A. O. Slisenko. On constructive nonseparable spaces. *Amer. Math. Soc Transl.*, 2(100):195–199, 1972. (Trudy Mat. Inst. Steklov. 72 (1964), 533–536, Russian).
17. D. Spreen. On effective topological spaces. Technical Report 94–01, Universität Siegen, 1994. 41 pages.
18. D. Spreen. Representations versus numberings. In this volume.
19. D. Spreen and P. Young. Effective operators in a topological setting. In M. M. Richter et al., editors, *Computation and Proof Theory, Proc. Logic Colloquium Aachen 1983, Part II*, number 1104 in Lect. Notes Math., pages 437–451, Berlin, 1984. Springer–Verlag.
20. G. S. Tseitin. Algorithmic operators in constructive complete separable metric spaces. *Dokl. Akad. Nauk. SSSR*, 128:49–52, 1959. (Russian).
21. G. S. Tseitin. Algorithmic operators in constructive metric spaces. *Amer. Math. Soc. Transl.*, 2(64):1–80, 1967. (Trudy Mat. Inst. Steklov. 67 (1962), 259–361, Russian).
22. K. Weihrauch. *Computability*, Springer–Verlag, Berlin, 1987.
23. K. Weihrauch. Computability on computable metric spaces. *Theoretical Computer Science*, 113:191–210, 1993.

On Shuffle Closures of Commutative Regular Languages

B. Imreh[1]*, M. Ito[2]**, M. Katsura[2]

[1] Dept. of Informatics, József Attila University, Árpád tér 2, H-6720 Szeged, Hungary
[2] Faculty of Science, Kyoto Sangyo University, Kyoto 603, Japan

Abstract. In this paper, we present two characterizations of those commutative regular languages whose shuffle closures are regular as well. In particular, from our characterizations it follows that it is decidable whether the shuffle closure of a commutative regular language is regular.
Keywords: formal languages, shuffle closures, decidability

1 Introduction

The operation of the shuffle product appears in different contexts in formal language theory and in algebra (see for example [1], [2], [3], [6], [7], [8], [9], [10], [11], [12], [13], [14] and [4], [5], respectively). In this paper, we study the shuffle closures of regular languages. The paper is organized as follows. After the preliminaries and general remarks of Section 2, we present in Section 3 the first characterization based on the fact that if the shuffle closure of a commutative regular language is regular, then it can be expressed by a shuffle product of a union of finite shuffle powers of the considered language and a regular language. Section 4 gives another characterization based on the additive decomposition of a commutative regular language. Both characterizations provide the decidability.

2 Preliminaries

Let X^* denote the free monoid generated by the non-empty finite alphabet X and let $X^+ = X^* \setminus \{1\}$ where 1 denotes the empty word of X^*. If $A \subseteq X^*$, then A^+ denotes the set of all concatenations of words in A and $A^* = A^+ \cup \{1\}$. In particular, if $A = \{w\}$, then we write w^+ and w^* instead of $\{w\}^+$ and $\{w\}^*$, respectively. If $u \in X^*$, then $|u|$ denotes the length of u. Moreover, if $a \in X$, then the number of the occurences of the letter a in the word u is denoted by $|u|_a$. Let $A \subseteq X^*$. By $\mathrm{alph}(A)$ we denote the alphabet of A, that is, $a \in \mathrm{alph}(A)$ if and only if a occurs in at least one word of A. We shall apply this notation to words as well, so $\mathrm{alph}(u)$ denotes the set of those letters which occur in the

* This work has been supported by the Japan Society for the Promotion of Science, the Ministry of Culture and Education of Hungary Grant 223/95.

** It has been also supported by the Ministry of Education, Science and Culture of Japan, Grant-in-Aid No. 08640074.

word u. Let $N = \{0, 1, \ldots\}$ and $X = \{a_1, \ldots, a_n\}$. The mapping Ψ of X^* into the set N^n defined by $\Psi(u) = (|u|_{a_1}, \ldots, |u|_{a_n})$, $u \in X^*$, is called the *Parikh mapping*. For a language $A \subseteq X^*$, we define $\Psi(A) = \{\Psi(u) : u \in A\}$. Moreover, if $S \subseteq N^n$, then $\Psi^{-1}(S) = \{u : u \in X^* \ \&\ \Psi(u) \in S\}$. The language A is called *commutative* if $A = \Psi^{-1}\Psi(A)$. The *syntactic congruence* P_A of the commutative language A is defined as follows. For any $u, v \in X^*$, $u \equiv v(P_A)$ iff $(uw \in A \rightarrow vw \in A \ \&\ vw \in A \rightarrow uw \in A)$ is valid for all $w \in X^*$. By the *shuffle product* of words $u, v \in X^*$ we mean the set

$$u \diamond v = \{w : w = u_1 v_1 u_2 v_2 \cdots u_k v_k, u = u_1 \cdots u_k, v = v_1 \cdots v_k, u_i, v_j \in X^*\}.$$

Obviously, $u \diamond v = v \diamond u$ and $(u \diamond v) \diamond w = u \diamond (v \diamond w)$ hold for all $u, v, w \in X^*$.

Now, let $U, V \subseteq X^*$. By the shuffle product of U and V we mean the set $\cup\{u \diamond v : u \in U, v \in V\}$ which is denoted by $U \diamond V$. It is easy to see that $U \diamond V = V \diamond U$ and $(U \diamond V) \diamond W = U \diamond (V \diamond W)$ for all $U, V, W \subseteq X^*$. It is worth noting that the shuffle product of commutative languages is also a commutative language.

Let $A \subseteq X^*$. The language $A^\diamond$ determined as the union of all the finite shuffle products of words in A is called the *shuffle closure* of A.

The language obtained by adjoining the empty word to $A^\diamond$ is denoted by A°. It is obvious that if $A = \emptyset$, then $A^\diamond = \emptyset$ and $A^\circ = \{1\}$. Moreover, if $A = \{1\}$, then $A^\diamond = A^\circ = \{1\}$. For the sake of simplicity, if $A = \{u\}$ for some word $u \in X^*$, then we write $u^\diamond$ and u° instead of $\{u\}^\diamond$ and $\{u\}^\circ$, respectively.

We shall use the following notations. Let $m \geq 2$ be an arbitrarily fixed integer. We denote the language $A^{(m-1)\diamond} \diamond A$ by $A^{m\diamond}$ where $A^{1\diamond} = A$. Moreover, the language $\cup\{A^{i\diamond} : i = 1, \ldots, m\}$ is denoted by $A^{m\diamond\downarrow}$. By the definitions, we have $A^\diamond = \cup\{A^{i\diamond} : i = 1, 2, \ldots\}$. For a finite index set $I = \{i_1, i_2, \ldots, i_k\}$ and languages F_{i_t}, $t = 1, \ldots, k$, the language $F_{i_1} \diamond F_{i_2} \diamond \cdots \diamond F_{i_k}$ is denoted by $\prod^\diamond_{j \in I} F_j$. In particular, if $I = \emptyset$, then let $\prod^\diamond_{j \in I} F_j = \{1\}$. The following statement is well-known (see for example [7]).

Lemma 1. *If $A_1, A_2 \subseteq X^*$ are regular languages, then $A_1 \diamond A_2$ is regular as well.*

3 Characterization by finite shuffle power

To prove our main result, we need some preparation. First of all, observe that if $A = \emptyset$ or $A = \{1\}$, then both A and $A^\diamond$ are regular. Therefore, in what follows, throughout this paper, $A \subseteq X^*$ will denote a language with $A \neq \emptyset$ and $A \neq \{1\}$.

Lemma 2. *If A is commutative and $A^\diamond$ is regular, then $1 \neq u \in A$, $a \in \mathrm{alph}(u)$ and $a^+ \cap A = \emptyset$ imply $a^t X^* \cap (A \cap \mathrm{alph}(u)^*) \neq \emptyset$ for all positive integer t.*

Proof. Let $1 \neq u \in A$, $a \in \mathrm{alph}(u)$, $a^+ \cap A = \emptyset$ and let us suppose that the statement is not valid. Since $1 \neq u \in A$ and $a \in \mathrm{alph}(u)$, there exists at least one positive integer t such that $a^t X^* \cap (A \cap \mathrm{alph}(u)^*) \neq \emptyset$. On the other hand, observe that $a^t X^* \cap (A \cap \mathrm{alph}(u)^*) \neq \emptyset$ implies $a^j X^* \cap (A \cap \mathrm{alph}(u)^*) \neq \emptyset$

for all j, $j = 1, \ldots, t$. Thus, from our assumption it follows that there exists a greatest integer $t_0 > 0$ such that $a^{t_0}X^* \cap (A \cap \mathrm{alph}(u)^*) \neq \emptyset$. Consequently, by the commutativity of A, we have $|q|_a \leq t_0$ for all $q \in A \cap \mathrm{alph}(u)^*$. Now, let $a^{t_0}w \in A$ where $w \in (\mathrm{alph}(u) \setminus \{a\})^*$. By $a^{t_0}X^* \cap (A \cap \mathrm{alph}(u)^*) \neq \emptyset$, such a word exists. Then $a^{ht_0}w^h \in A^\diamond$ for any positive integer h. On the other hand, $A^\diamond$ is regular and thus there are positive integers k, p with $a^k \equiv a^{k+p}(P_{A^\diamond})$. Choosing an h_0 satisfying $h_0t_0 \geq k$, by the above relation, we obtain that $a^{h_0t_0+mp}w^{h_0} \in A^\diamond$ for all nonnegative integer m. Choose a large enough integer m_0 for which $m_0p \geq h_0t_0|w|$. Since $v_0 = a^{h_0t_0+m_0p}w^{h_0} \in A^\diamond$ and $\mathrm{alph}(v_0) \subseteq \mathrm{alph}(u)$, there are $u_1, \ldots, u_s \in A \cap \mathrm{alph}(u)^+$ such that $v_0 = a^{h_0t_0+m_0p}w^{h_0} \in u_1 \diamond \cdots \diamond u_s$. From $a^+ \cap A = \emptyset$ it follows that any word in $A \cap \mathrm{alph}(u)^+$ contains at least one letter from $\mathrm{alph}(u) \setminus \{a\}$. Therefore, by the above relation, we obtain that $s \leq h_0|w|$. On the other hand, $|q|_a \leq t_0$ for all $q \in A \cap \mathrm{alph}(u)^+$. Consequently, $|u_j|_a \leq t_0$, $j = 1, \ldots, s$ and thus we have $|v|_a \leq st_0 \leq h_0t_0|w|$ for all $v \in u_1 \diamond \cdots \diamond u_s$. But $|v_0|_a = h_0t_0 + m_0p > h_0t_0|w|$ which is a contradiction. This completes the proof of Lemma 2.

Lemma 3. *If $A \subseteq X^*$ is a commutative regular language, then there exists a non-empty finite commutative language $F \subseteq A$ such that $A^\diamond = A^{|X|\diamond\downarrow} \diamond F^\diamond$.*

Proof. Let $X = \{a_1, \ldots, a_n\}$. Without loss of generality, we may assume that $\mathrm{alph}(A) = \{a_1, \ldots, a_r\}$ for some $1 \leq r \leq n$. First, we define a suitable set F. Since A is a regular language, there exist positive integers k, p such that $a_j^k \equiv a_j^{k+p}(P_A)$ for all j, $j = 1, \ldots, r$. Now, define the set F by $F = \{u : u \in A \ \&\ |u|_{a_j} < k + p$ for all j, $j = 1, \ldots, r\}$. Then $F \subseteq A$ and F is a finite commutative language. Moreover, it is easy to see that $F \neq \emptyset$. Let $B = A^{n\diamond\downarrow} \diamond F^\diamond$. It is obvious that $A^\diamond \supseteq B$. To prove the converse inclusion, let $u \in A^\diamond$ be an arbitrary word. If $u = 1$, then $u \in B$ is valid as well. Assume that $u \neq 1$. Then u can be decomposed into the form $u \in u_1 \diamond \cdots \diamond u_m$ where $m \geq 1$ and $u_j \in A$, $j = 1, \ldots, m$. If $m \leq n$, then we immediately obtain that $u \in B$. Now, assume that $m > n$. Let $|u_i|_{a_j} = t_{ij}$, $i = 1, \ldots, m$, $j = 1, \ldots, r$. Since A and B are commutative, we may assume that $u_i = a_1^{t_{i1}} \cdots a_r^{t_{ir}}$, $i = 1, \ldots, m$. Using the relations $a_j^k \equiv a_j^{k+p}(P_A)$, $j = 1, \ldots, r$, we define a new shuffle decomposition for u as follows.

If $t_{i1} < k+p$ for all i, $i = 1, \ldots, m$, then let $t'_{i1} = t_{i1}$ for all i, $i = 1, \ldots, m$. In the opposite case, there exists at least one $i \in \{1, \ldots, m\}$ satisfying $t_{i1} \geq k + p$. Without loss of generality, we may assume that $t_{11} \geq k + p$. Now, for every $i \in \{2, \ldots, m\}$, let $n_i = 0$ if $t_{i1} < k + p$ and let n_i such an integer for which $k \leq t_{i1} - n_ip < k + p$ is valid if $t_{i1} \geq k + p$. Moreover, let $t'_{i1} = t_{i1} - n_ip$. Finally, define t'_{11} by $t'_{11} = t_{11} + (n_2 + \ldots + n_m)p$. Consider the words: $u_i^{(1)} = a_1^{t'_{i1}}a_2^{t_{i2}} \cdots a_r^{t_{ir}}$, $i = 1, \ldots, m$. Then $u_i^{(1)} \in A$, $i = 1, \ldots, m$ and $u \in u_1^{(1)} \diamond \cdots \diamond u_m^{(1)}$ where $t'_{i1} < k + p$ for all i, $i = 2, \ldots, m$.

Using the same idea as above, we can construct a further sequence of words $u_i^{(2)} = a_1^{s_{i1}} \cdots a_r^{s_{ir}}$, $i = 1, \ldots, m$, such that $u_i^{(2)} \in A$, $i = 1, \ldots, m$ and $u \in u_1^{(2)} \diamond \cdots \diamond u_m^{(2)}$ where $s_{i1} < k+p$ if $i \in \{2, \ldots, m\}$ and $s_{i2} < k+p$ if $i \in \{3, \ldots, m\}$.

Continuing this procedure, finally we obtain such a sequence of words $u_i^{(r)} = a_1^{p_{i1}} \cdots a_r^{p_{ir}}$, $i = 1, \ldots, m$, for which $u_i^{(r)} \in A$, $i = 1, \ldots, m$ with $u \in u_1^{(r)} \diamond \cdots \diamond u_m^{(r)}$ and, for all $t \in \{1, \ldots, r\}$, $p_{it} < k + p$ is valid if $i \in \{1 + t, \ldots, m\}$. Now, observe that $u_{r+1}^{(r)}, \ldots, u_m^{(r)} \in F$ and thus we obtain that $u \in A^{r\diamond} \diamond F^\diamond \subseteq A^{n\diamond\downarrow} \diamond F^\diamond$. Consequently, $u \in B$ which completes the proof of Lemma 3.

Lemma 4. *Let A be a commutative regular language and $Z = \{a : a \in X \ \& \ a^+ \cap A \neq \emptyset\}$. If $1 \neq u \in A$, $a \in \mathrm{alph}(u)$ and $a^+ \cap A = \emptyset$ imply $a^t X^* \cap (A \cap \mathrm{alph}(u)^*) \neq \emptyset$ for all positive integers t, then there exists a positive integer M such that $w^\diamond \subseteq A^{M\diamond\downarrow} \diamond \prod_{a \in Z}{}^\diamond (a^* \cap A^*)$ is valid for all $w \in A$.*

Proof. Denote the elements of Z by $b_1, \ldots, b_s$ and let $Y = X \setminus Z = \{a_1, \ldots, a_r\}$. We note that $Z \cup Y \neq \emptyset$. By the regularity of A, there exist positive integers k, p' such that $a_j^k \equiv a_j^{k+p'} (P_A)$ for all j, $j = 1, \ldots, r$ and $b_i^k \equiv b_i^{k+p'} (P_A)$ for all i, $i = 1, \ldots, s$. On the other hand, for every $b_i \in \{b_1, \ldots, b_s\}$, we have $b_i^+ \cap A \neq \emptyset$ and thus there exists a smallest positive integer p_i with $b_i^{p_i} \in A$. Let $t = \mathrm{l.c.m.}\{p_1, \ldots, p_s\}$ provided that $Z \neq \emptyset$. Then $b_i^t \in A^*$ for all i, $i = 1, \ldots, s$.

Now, let

$$p = \begin{cases} p' & \text{if } Z = \emptyset, \\ \mathrm{l.c.m.}\{p', t\} & \text{otherwise.} \end{cases}$$

Then we have

$$\begin{aligned} &a_j^k \equiv a_j^{k+p} (P_A), \quad j = 1, \ldots, r, \\ &b_i^k \equiv b_i^{k+p} (P_A), \quad i = 1, \ldots, s, \\ &b_i^p \in A^*, \quad i = 1, \ldots, s. \end{aligned} \tag{1}$$

Let f be the smallest positive integer for which $fp \geq (k + p)n$ is valid where $n = |X|$. We define the integers q and M by $q = k + fp$ and $M = np + pq - 1$.

Now, we are ready to prove our statement. For this purpose, let $w \in A$ be an arbitrary word. If $w = 1$, then the statement is obviously valid. Assume that $w \neq 1$. Depending on $\mathrm{alph}(w)$, we distinguish the following three cases.

Case 1: $\mathrm{alph}(w) \cap Y \neq \emptyset$ and $\mathrm{alph}(w) \cap Z \neq \emptyset$. Without loss of generality, we may assume that $\mathrm{alph}(w) \cap Y = \{a_1, \ldots, a_d\}$ and $\mathrm{alph}(w) \cap Z = \{b_1, \ldots, b_e\}$ for some positive integers $d \leq r$ and $e \leq s$. By our assumption, there are $x_1, \ldots, x_d \in \mathrm{alph}(w)^*$ such that $a_j^q x_j \in A$, $j = 1, \ldots, d$. Moreover, by (1), we may assume that, for all g, $g = 1, \ldots, d$, $|x_g|_{a_j} < k + p$, $j = 1, \ldots, d$ and $|x_g|_{b_i} < k + p$, $i = 1, \ldots, e$. Let $j \in \{1, \ldots, d\}$ be an arbitrary integer. Using (1) and the definition of q, we have that A contains the members of the following sequence:

$$a_j^{q-fp} x_j, \ a_j^{q-(f-1)p} x_j, \ a_j^{q-(f-2)p} x_j, \ a_j^{q-(f-3)p} x_j, \ldots.$$

Now, observe that, for any m_j, $m_j \in \{-fp, -fp+1, -fp+2, \ldots\}$, there are not necessarily distinct p words in the above sequence such that their shuffle product contains the word $a_j^{pq+m_jp} x_j^p$. Therefore,

$$a_j^{pq+m_jp} x_j^p \in A^{p\diamond}, \ m_j = -fp, -fp+1, -fp+2, -fp+3, \ldots.$$

Consider the above sequence of words for all j, $j = 1, \ldots, d$. Now, taking every shuffle product which contains one and only one member from each sequence, we obtain that

$$a_1^{pq+m_1p} a_2^{pq+m_2p} \cdots a_d^{pq+m_dp} x_1^p \cdots x_d^p \in A^{dp\diamond}$$

for all $(m_1, \ldots, m_d) \in \{-fp, -fp+1, -fp+2, -fp+3, \ldots\}^d$. Since $|x_g|_{a_j} < k+p$, $j = 1, \ldots, d$ and $|x_g|_{b_i} < k + p$, $i = 1, \ldots, e$, hold for all g, $g = 1, \ldots, d$,

$$a_1^{t_1p} a_2^{t_2p} \cdots a_d^{t_dp} b_1^{v_1p} \cdots b_e^{v_ep} \in \Psi^{-1}\Psi(x_1^p \cdots x_d^p)$$

where $t_j < (k + p)d$, $j = 1, \ldots, d$ and $v_i < (k + p)d$, $i = 1, \ldots, e$. Since $A^{dp\diamond}$ is commutative, $a_1^{pq+(m_1+t_1)p} a_2^{pq+(m_2+t_2)p} \cdots a_d^{pq+(m_d+t_d)p} b_1^{v_1p} \cdots b_e^{v_ep} \in A^{dp\diamond}$ for all $(m_1, \ldots, m_d) \in \{-fp, -fp + 1, -fp + 2, \ldots\}^d$. Then, by $t_j < (k + p)d$, $j = 1, \ldots, d$ and the definition of f, for arbitrary positive integers $\delta_1, \ldots, \delta_d$, there are $m_j \in \{-fp, -fp + 1, -fp + 2, -fp + 3, \ldots\}$, $j = 1, \ldots, d$, such that $pq + (m_j + t_j)p = \delta_j pq$, $j = 1, \ldots, d$. Consequently,

$$a_1^{\delta_1 pq} \cdots a_d^{\delta_d pq} b_1^{v_1p} \cdots b_e^{v_ep} \in A^{dp\diamond} \tag{2}$$

for all $(\delta_1, \ldots, \delta_d) \in \{1, 2, \ldots\}^d$.

Now, let m be an arbitrary positive integer. We shall prove that

$$w^{mpq\diamond} \subseteq A^{np\diamond\downarrow} \diamond \prod_{a\in Z}{}^{\diamond}(a^* \cap A^*).$$

By our assumption on w, $|w|_{a_j} = \alpha_j$, $j = 1, \ldots, d$, and $|w|_{b_i} = \beta_i$, $i = 1, \ldots, e$ for some positive integers $\alpha_1, \ldots, \alpha_d, \beta_1, \ldots, \beta_e$. Let $u = a_1^{\alpha_1} \cdots a_d^{\alpha_d} b_1^{\beta_1} \cdots b_e^{\beta_e}$. Then $\Psi^{-1}\Psi(w) = \Psi^{-1}\Psi(u)$. Let $u_0 = a_1^{\alpha_1 mpq} \cdots a_d^{\alpha_d mpq} b_1^{\beta_1 mpq} \cdots b_e^{\beta_e mpq}$. Then $w^{mpq\diamond} \subseteq \Psi^{-1}\Psi(u_0)$. By the definitions of f and q, we have $\beta_i mq \geq k + fp \geq (k + p)n \geq (k + p)d > v_i$, $i = 1, \ldots, d$. Therefore,

$$u_1 = a_1^{\alpha_1 mpq} \cdots a_d^{\alpha_d mpq} b_1^{v_1p} \cdots b_e^{v_ep} b_1^{\gamma_1p} \cdots b_e^{\gamma_ep} \in \Psi^{-1}\Psi(u_0)$$

where $\gamma_i = \beta_i mq - v_i > 0$, $i = 1, \ldots, e$. Now, observe that, by (2), we have $a_1^{\alpha_1 mpq} \cdots a_d^{\alpha_d mpq} b_1^{v_1p} \cdots b_e^{v_ep} \in A^{dp\diamond}$. On the other hand, by the definition of p, $b_1^{\gamma_1p} \cdots b_e^{\gamma_ep} \in \prod_{a\in Z}{}^{\diamond}(a^* \cap A^*)$. Therefore, $u_1 \in A^{dp\diamond} \diamond \prod^{\diamond}_{a\in Z}(a^* \cap A^*)$. By the commutativity of $A^{dp\diamond} \diamond \prod^{\diamond}_{a\in Z}(a^* \cap A^*)$, we obtain that $\Psi^{-1}\Psi(u_1) \subseteq A^{dp\diamond} \diamond \prod_{a\in Z}{}^{\diamond}(a^* \cap A^*)$. But $w^{mpq\diamond} \subseteq \Psi^{-1}\Psi(u_0) = \Psi^{-1}\Psi(u_1)$ and $d \leq n$ which imply the required inclusion, namely

$$w^{mpq\diamond} \subseteq A^{np\diamond\downarrow} \diamond \prod_{a\in Z}{}^{\diamond}(a^* \cap A^*) \tag{3}$$

is valid, for arbitrary positive integer m.

Since $w^\diamond = \cup\{w^{i\diamond} : i = 1, 2, \ldots\}$, it is enough to prove that $w^{i\diamond} \subseteq A^{M\diamond\downarrow} \diamond \prod^{\diamond}_{a\in Z}(a^* \cap A^*)$ for all i, $i = 1, 2, \ldots$. For this purpose, let i be an arbitrarily fixed positive integer. If $i < pq$, then

$$w^{i\diamond} \subseteq A^{i\diamond} \subseteq A^{(pq-1)\diamond\downarrow} \subseteq A^{M\diamond\downarrow} \diamond \prod_{a\in Z}{}^{\diamond}(a^* \cap A^*).$$

If $i \geq pq$, then there are integers $m > 0$ and $0 \leq \alpha < pq$ such that $i = mpq+\alpha$. On the other hand, $w^{i\diamond} = w^{mpq\diamond} \diamond w^{\alpha\diamond}$. Now, $w^{\alpha\diamond} \subseteq A^{(pq-1)\diamond\downarrow}$. Moreover, $w^{mpq\diamond} \subseteq A^{np\diamond\downarrow} \diamond \prod^{\diamond}_{a\in Z}(a^* \cap A^*)$ from (3). Therefore,

$$w^{mpq\diamond} \diamond w^{\alpha\diamond} \subseteq A^{np\diamond\downarrow} \diamond \prod_{a\in Z}{}^{\diamond}(a^* \cap A^*) \diamond A^{(pq-1)\diamond\downarrow} \subseteq A^{M\diamond\downarrow} \diamond \prod_{a\in Z}{}^{\diamond}(a^* \cap A^*)$$

which completes the proof of the considered case.

Case 2: $\mathrm{alph}(w) \cap Z = \emptyset$. Without loss of generality, we may assume that $\mathrm{alph}(w) \cap Y = \{a_1, \ldots, a_d\}$. Observe that $d > 1$. Indeed, in the opposite case, $a_1^t x_1 \in A$ for some $x_1 \in \mathrm{alph}(w)^* = a_1^*$ and $t > 0$ from our assumption and thus $a_1^{t+|x_1|} \in A$ which contradicts the fact that $a_1 \in Y$. Then, in a similar way as above, we obtain that

$$a_1^{\delta_1 pq} \cdots a_d^{\delta_d pq} \in A^{dp\diamond} \tag{4}$$

for all $(\delta_1, \ldots, \delta_d) \in \{1, 2, \ldots\}^d$. In this case, $w \in \Psi^{-1}\Psi(a_1^{\alpha_1} \cdots a_d^{\alpha_d})$ where $\alpha_j = |w|_{a_j} > 0$, $j = 1, \ldots, d$. Now, let m be a positive integer. Then $w^{mpq\diamond} \subseteq \Psi^{-1}\Psi(a_1^{\alpha_1 mpq} \cdots a_d^{\alpha_d mpq})$. On the other hand, $a_1^{\alpha_1 mpq} \cdots a_d^{\alpha_d mpq} \in A^{dp\diamond}$ from (4). Since $A^{dp\diamond}$ is commutative, we obtain that $w^{mpq\diamond} \subseteq A^{dp\diamond} \subseteq A^{np\diamond\downarrow}$ is valid for all positive integer m. Now, in the same way as above, it can be proved that $w^{i\diamond} \subseteq A^{M\diamond\downarrow}$ is valid for all i, $i = 1, 2, \ldots$ and thus $w^{\diamond} \subseteq A^{M\diamond\downarrow} \subseteq A^{M\diamond\downarrow} \diamond \prod_{a\in Z}{}^{\diamond}(a^* \cap A^*)$ which completes the proof of this case.

Case 3: $\mathrm{alph}(w) \cap Y = \emptyset$. Without loss of generality, we may assume that $\mathrm{alph}(w) \cap Z = \{b_1, \ldots, b_e\}$. Then $w \in \Psi^{-1}\Psi(b_1^{\beta_1} \cdots b_e^{\beta_e})$ where $\beta_i = |w|_{b_i} > 0$, $i = 1, \ldots, e$. Now, let m be an arbitrary positive integer. We show that

$$w^{mpq\diamond} \subseteq A^{M\diamond\downarrow} \diamond \prod_{a\in Z}{}^{\diamond}(a^* \cap A^*).$$

If $m = 1$, then $w^{pq\diamond} \subseteq A^{pq\diamond} \subseteq A^{M\diamond\downarrow} \diamond \prod^{\diamond}_{a\in Z}(a^* \cap A^*)$. Now, let $m > 1$. Then

$$w^{mpq\diamond} \subseteq \Psi^{-1}\Psi(b_1^{\beta_1 pq} \cdots b_e^{\beta_e pq}) \diamond \Psi^{-1}\Psi(b_1^{\beta_1(m-1)pq} \cdots b_e^{\beta_e(m-1)pq}).$$

Observe that $\Psi^{-1}\Psi(b_1^{\beta_1 pq} \cdots b_e^{\beta_e pq}) \subseteq A^{pq\diamond} \subseteq A^{M\diamond\downarrow}$. Furthermore, by the definition of p, we have $b_i^{\beta_i(m-1)pq} \in A^*$, $i = 1, \ldots, e$ and thus

$$\Psi^{-1}\Psi(b_1^{\beta_1(m-1)pq} \cdots b_e^{\beta_e(m-1)pq}) \subseteq \prod_{a\in Z}{}^{\diamond}(a^* \cap A^*).$$

Consequently, $w^{mpq\diamond} \subseteq A^{M\diamond\downarrow} \diamond \prod_{a\in Z}{}^{\diamond}(a^* \cap A^*)$ is valid for all positive integer m. Using this inclusion, one can prove, in the same way as above, that $w^{i\diamond} \subseteq A^{M\diamond\downarrow} \diamond \prod^{\diamond}_{a\in Z}(a^* \cap A^*)$ for all i, $i = 1, 2, \ldots$, which results in $w^{\diamond} \subseteq A^{M\diamond\downarrow} \diamond \prod_{a\in Z}{}^{\diamond}(a^* \cap A^*)$. This completes the proof of Lemma 4.

Theorem 5. *Let A be a commutative regular language. $A^{\diamond}$ is regular if and only if $1 \neq u \in A$, $a \in \mathrm{alph}(u)$ and $a^+ \cap A = \emptyset$ imply $a^t X^* \cap (A \cap \mathrm{alph}(u)^*) \neq \emptyset$ for all positive integer t.*

Proof. The necessity of the condition follows from Lemma 2. In order to prove the sufficiency, assume that the above condition is valid. We prove that $A^\diamond$ is regular. Let $X = \{a_1, \ldots, a_n\}$. Then, by Lemma 3, we obtain that $A^\diamond = A^{n\diamond\downarrow} \diamond F^\diamond$ where $F \subseteq A$ and F is a finite non-empty commutative language. It is easy to see that $F^\diamond = \prod^\diamond_{u \in F} u^\diamond$. By Lemma 4, there exists an $M > 0$ such that $u^\diamond \subseteq A^{M\diamond\downarrow} \diamond \prod^\diamond_{a\in Z}(a^* \cap A^*)$ for all $u \in F$. Consequently, $F^\diamond \subseteq \{1\} \cup A^{M|F|\diamond\downarrow} \diamond \prod^\diamond_{a\in Z}(a^* \cap A^*)$, and thus $A^\diamond \subseteq A^{(n+M|F|)\diamond\downarrow} \diamond \prod_{a\in Z}{}^\diamond(a^* \cap A^*)$. On the other hand, each member of the right-side is contained in $A^\diamond$, and consequently, $A^\diamond = A^{(n+M|F|)\diamond\downarrow} \diamond \prod_{a\in Z}{}^\diamond(a^* \cap A^*)$. Finally, observe that, by Lemma 1, we obtain that both $A^{(n+M|F|)\diamond\downarrow}$ and $\prod^\diamond_{a\in Z}(a^* \cap A^*)$ are regular and thus $A^\diamond$ is regular as well. This ends the proof of Theorem 5.

Using Lemma 2 and the same idea as in the above proof, we obtain the following statement which presents the fact that if $A^\diamond$ is regular, then it is equal to a shuffle product of a union of finite shuffle powers of A and a regular language. More precisely, we have the following statement.

Theorem 6. *Let A be a commutative regular language. Moreover, let $Z = \{a : a \in \mathrm{alph}(A)\ \&\ a^+ \cap A \neq \emptyset\}$. Then $A^\diamond$ is regular if and only if there exists a positive integer N such that $A^\diamond = A^{N\diamond\downarrow} \diamond \prod^\diamond_{a\in Z}(a^* \cap A^*)$.*

We note that it is decidable whether or not the condition of Theorem 5 is satisfied and thus it is decidable whether $A^\diamond$ is regular for a commutative regular language A. Using Theorem 6, we can provide another algorithm to decide whether $A^\diamond$ is regular for a commutative regular language A as follows: In Theorem 6, the value of N is given by A, e.g. $N = (n+1)p(k+p+1)(k+p)^p$. Using this N, we can check whether or not $A^{(N+1)\diamond\downarrow} \subseteq A^{N\diamond\downarrow} \diamond \prod^\diamond_{a\in Z}(a^* \cap A^*)$. If $A^{(N+1)\diamond\downarrow} \subseteq A^{N\diamond\downarrow} \diamond \prod^\diamond_{a\in Z}(a^* \cap A^*)$, then $A^\diamond$ is regular. Otherwise, $A^\diamond$ is not regular.

4 Characterization by additive decomposition

Lemma 7. *Let $u_1, \ldots, u_r \in X^+$ be given with* $\mathrm{alph}(u_1 \cdots u_r) = \{a_1, \ldots, a_s\}$ *and let $f_1, \ldots, f_s$ be arbitrary positive integers. Then the language R is regular where $R = \Psi^{-1}\Psi(u_1^+ \cdots u_r^+) \diamond (a_1^{f_1})^* \diamond \cdots \diamond (a_s^{f_s})^*$.*

Proof. We denote the product $f_1 \cdots f_s$ by M. Since $f_t > 0$, $t = 1, \ldots, s$, we have $M > 0$. Now, let $i \in \{1, \ldots, r\}$ be arbitrarily chosen. Then

$$\Psi^{-1}\Psi(u_i^+) = \Psi^{-1}\Psi(\cup\{u_i^k : k = 1, \ldots, M\}) \diamond \Psi^{-1}\Psi((u_i^M)^*) \subseteq$$
$$\Psi^{-1}\Psi(\cup\{u_i^k : k = 1, \ldots, M\}) \diamond (a_1^{f_1})^* \diamond \cdots \diamond (a_s^{f_s})^*.$$

Using these inclusions, we obtain that

$$R = \prod_{i=1}^{r}{}^\diamond \Psi^{-1}\Psi(u_i^+) \diamond \prod_{j=1}^{s}{}^\diamond (a_j^{f_j})^* \subseteq \prod_{i=1}^{r}{}^\diamond \Psi^{-1}\Psi(\cup\{u_i^k : k = 1, \ldots, M\}) \diamond \prod_{j=1}^{s}{}^\diamond (a_j^{f_j})^*.$$

We denote the language determined by the right-side of the above inclusion by L. Then $R \subseteq L$. On the other hand, it is easy to check that $R \supseteq L$ is also valid and consequently, $R = L$. Now, observe that each member of the right-side is regular and thus by Lemma 1, R is regular as well.

The following statement presents a decomposition procedure for commutative regular languages.

Lemma 8. *Let A be a commutative regular language with* $\mathrm{alph}(A) = \{a_1, \dots, a_n\}$. *For each $j \in \{1, \dots, n\}$, let k_j and p_j be positive integers such that $a_j^{k_j} \equiv a_j^{k_j+p_j}(P_A)$ where $k_j + p_j$ is the smallest positive integer with this property. Let $U = \{u : u \in A \;\&\; |u|_{a_j} < k_j + p_j \text{ for all } j,\ j = 1, \dots, n\} = \{u_1, \dots, u_m\}$ and*

$$f_{ij} = \begin{cases} 0 & \text{if } |u_i|_{a_j} < k_j, \\ p_j & \text{otherwise.} \end{cases}$$

Then A is a disjoint union of $A_1, \dots, A_m$ where $A_i = \Psi^{-1}\Psi(u_i) \diamond (a_1^{f_{i1}})^* \diamond \cdots \diamond (a_n^{f_{in}})^*$ for all i, $i = 1, \dots, m$.

Proof. First of all, we prove that the statement is correct. Since $A \neq \emptyset$ and $A \neq \{1\}$, we have $n \geq 1$. Let $j \in \{1, \dots, n\}$ be arbitrary. Since A is regular, there exist $k_j, p_j \geq 1$ such that $a_j^{k_j} \equiv a_j^{k_j+p_j}(P_A)$. Choosing the smallest positive integer k_j for which there exists a p_j with $a_j^{k_j} \equiv a_j^{k_j+p_j}(P_A)$ and moreover, choosing the smallest p_j with the above property, we obtain the required positive integers k_j, p_j. Let $\bar{u} = a_1^{d_1} \cdots a_n^{d_n}$ be an element of A with minimal length. If $\bar{u} \notin U$, then there is a $j \in \{1, \dots, n\}$ with $d_j \geq k_j + p_j$, and by $a_j^{k_j} \equiv a_j^{k_j+p_j}(P_A)$, we obtain that $a_1^{d_1} \cdots a_{j-1}^{d_{j-1}} a_j^{d_j - p_j} a_{j+1}^{d_{j+1}} \cdots a_n^{d_n} \in A$ which contradicts the minimality of the length of $\bar{u}$. Thus $\bar{u} \in U$ and therefore, U is a non-empty finite set. Then the matrix (f_{ij}) exists.

Now, we show that A can be represented in the required form. For the sake of simplicity, we denote $k_j + p_j - 1$ by q_j for all j, $j = 1, \dots, n$. Since A is commutative and $a_j^{k_j} \equiv a_j^{k_j+p_j}(P_A)$ for all j, $j = 1, \dots, n$, we have that $A_i = \Psi^{-1}\Psi(u_i) \diamond (a_1^{f_{i1}})^* \diamond \cdots \diamond (a_n^{f_{in}})^* \subseteq A$ is valid for all i, $i = 1, \dots, m$. Consequently, $\cup\{A_i : i = 1, \dots, m\} \subseteq A$. To prove the converse inclusion, observe that $\Psi^{-1}\Psi(U) \subseteq \cup\{A_i : i = 1, \dots, m\}$. Now, let $v \in A \setminus \Psi^{-1}\Psi(U)$ be an arbitrary word. Then we denote $|v|_{a_j}$ by d_j for all j, $j = 1, \dots, n$. By the definition of U, there exists at least one index j_1, $1 \leq j_1 \leq n$, such that $d_{j_1} > q_{j_1}$. By the commutativity of A, we have $a_1^{d_1} a_2^{d_2} \cdots a_{j_1}^{d_{j_1}} \cdots a_n^{d_n} \in \Psi^{-1}\Psi(v) \subseteq A$. Moreover, from $a_{j_1}^{k_{j_1}} \equiv a_{j_1}^{k_{j_1}+p_{j_1}}(P_A)$ it follows that $a_1^{d_1} a_2^{d_2} \cdots a_{j_1}^{d_{j_1} - s_1 p_{j_1}} \cdots a_n^{d_n} \in A$ with $k_{j_1} \leq d_{j_1} - s_1 p_{j_1} \leq q_{j_1}$ for some positive integer s_1. Let $v_1 = a_1^{d_1} a_2^{d_2} \cdots a_{j_1}^{d_{j_1} - s_1 p_{j_1}} \cdots a_n^{d_n}$. Then $v \in \Psi^{-1}\Psi(v_1) \diamond (a_{j_1}^{p_{j_1}})^*$ and $k_{j_1} \leq |v_1|_{a_{j_1}} \leq q_{j_1}$. If $v_1 \in A \setminus \Psi^{-1}\Psi(U)$, then we continue the same procedure. Finally, we obtain that $v \in \Psi^{-1}\Psi(v_r) \diamond (a_{j_1}^{p_{j_1}})^* \diamond \cdots \diamond (a_{j_r}^{p_{j_r}})^*$ for some integer r where $1 \leq r \leq n$, $v_r \in A$, $k_{j_t} \leq |v_r|_{a_{j_t}} \leq q_{j_t}$ for all t, $t = 1, \dots, r$, and $|v_r|_{a_j} \leq q_j$ for all the remaining indices. Therefore, $v_r = u_i \in U$ for some integer i, $1 \leq i \leq m$. On the other hand, by the definition of f_{ij}, we have $f_{ij_t} = p_{j_t}$ for all t, $t = 1, \ldots, r$. Consequently, $v \in \Psi^{-1}\Psi(u_i) \diamond (a_1^{f_{i1}})^* \diamond \cdots \diamond (a_n^{f_{in}})^* = A_i$, and thus $A = \cup\{A_i : 1 \leq i \leq m\}$. On the other hand, it is easy to see that $r \neq s \in \{1, \ldots, m\}$ implies $A_r \cap A_s = \emptyset$ which completes the proof of Lemma 8.

It is worth observing that the representation given above is unique. This unique representation is called the *additive decomposition* of A and the corresponding matrix (f_{ij}) is called the *decomposition matrix* of A. Now, we are ready to give a further characterization of those commutative regular languages whose closures are also regular. Since A° is regular if and only if A° is regular, it is enough to prove the following statement.

Theorem 9. *Let A be a commutative regular language having the additive decomposition with $U = \{u_1, \ldots, u_m\}$ and the decomposition matrix (f_{ij}). Then A° is regular if and only if, for all $1 \neq u_i \in U$ and $a_j \in$ alph(u_i), there exists $u_k \in U$ with* alph$(u_k) \subseteq$ alph(u_i) *such that $f_{kj} \neq 0$ or $u_k \in a_j^+$ is valid.*

Proof. To prove the necessity, assume that A° is regular and the conditions are not satisfied. Then there are $u_{i_0} \in U$ and $a_j \in$ alph(u_{i_0}) such that $f_{kj} = 0$ and $u_k \notin a_j^+$ for all $u_k \in U$ with alph$(u_k) \subseteq$ alph(u_{i_0}). Let

$$\Theta = \max\{\frac{|u_i|_{a_j}}{|u_i|} : 1 \leq i \leq m, u_i \neq 1\}.$$

Then $\Theta < 1$. Indeed, in the opposite case, $u_k \in a_j^+$ for some $u_k \in U$, but then alph$(u_k) \subseteq$ alph(u_{i_0}) and $u_k \in a_j^+$ which contradicts our assumption. Next we show that

$$\frac{|x|_{a_j}}{|x|} \leq \Theta \tag{5}$$

is valid for all $x \in A^\circ \setminus \{1\}$ with alph$(x) \subseteq$ alph(u_{i_0}). Since $A = \cup\{A_i : i = 1, \ldots, m\}$, we have $A^\circ = A_1^\circ \diamond \cdots \diamond A_m^\circ$. Then there are $y_i \in A_i^\circ$, $i = 1, \ldots, m$, such that $x \in y_1 \diamond \cdots \diamond y_m$. Since $x \neq 1$, without loss of generality, we may assume that only the first s members of the right-side differ from 1 for some $1 \leq s \leq m$, that is, $x \in y_1 \diamond \cdots \diamond y_s$ where $y_t \neq 1$, $t = 1, \ldots, s$. Now, let $t \in \{1, \ldots, s\}$ be arbitrary. By the definition of A_t, we have alph$(u_t) =$ alph(y_t). Moreover, alph$(y_t) \subseteq$ alph$(x) \subseteq$ alph(u_{i_0}) and thus alph$(u_t) \subseteq$ alph(u_{i_0}). Since $1 \neq y_t \in A_t^\circ$, we have $y_t \in \Psi^{-1}\Psi(u_t^p) \diamond (a_1^{f_{t1}})^* \diamond \cdots \diamond (a_n^{f_{tn}})^*$ for some $p > 0$ and $f_{tj} = 0$ from our assumption. Thus

$$\frac{|y_t|_{a_j}}{|y_t|} \leq \frac{p|u_t|_{a_j}}{p|u_t|} \leq \Theta.$$

Consequently, we have that $|y_t|_{a_j} \leq \Theta|y_t|$ for all t, $t = 1, \ldots, s$ and thus $\sum_{t=1}^{s} |y_t|_{a_j} \leq \Theta \sum_{t=1}^{s} |y_t|$. Since $|x|_{a_j} = \sum_{t=1}^{s} |y_t|_{a_j}$ and $|x| = \sum_{t=1}^{s} |y_t|$, the obtained relation implies the validity of (5) for the considered word x.

Since A° is regular and $a_j \in \mathrm{alph}(A)$, there are positive integers f, g such that $a_j^f \equiv a_j^{f+g}(P_{A^\circ})$. On the other hand, $\Psi^{-1}\Psi(u_{i_0}^+) \subseteq A_{i_0}^\circ \subseteq A^\circ$ and thus $\Psi^{-1}\Psi(u_{i_0}^f) \subseteq A^\circ$. Since $a_j \in \mathrm{alph}(u_{i_0})$, we have $a_j^f w \in \Psi^{-1}\Psi(u_{i_0}^f)$ for some word w where $\mathrm{alph}(w) \subseteq \mathrm{alph}(u_{i_0})$. Then, by $a_j^f \equiv a_j^{f+g}(P_{A^\circ})$, we obtain that $v_h = a_j^{f+hg} w \in A^\circ$ with $\mathrm{alph}(v_h) \subseteq \mathrm{alph}(u_{i_0})$ for all nonnegative integer h. Now, choosing a suitably large h, we get that $\frac{|v_h|_{a_j}}{|v_h|} > \Theta$ which contradicts (5).

In order to prove the sufficiency, assume that A satisfies the conditions of our theorem. We prove that A° is a regular language. As a first step, we show that, for every $\emptyset \neq I = \{i_1, \ldots, i_r\} \subseteq \{1, \ldots, m\}$, there exists a regular language R_I with $A_{i_1}^\diamond \diamond \cdots \diamond A_{i_r}^\diamond \subseteq R_I \subseteq A^\circ$.

We proceed by induction on $|I|$. Let $|I| = 1$. Without loss of generality, we may assume that $I = \{1\}$. If $u_1 = 1$ or $u_1 \in a_j^+$ for some $a_j \in \{a_1, \ldots, a_n\} = \mathrm{alph}(A)$, then the statement obviously holds by $R_I = \{1\}$ and $R_I = u_1^+ \diamond (a_j^{f_{1j}})^*$, respectively. In the opposite case, we may assume that $\mathrm{alph}(u_1) = \{a_1, \ldots, a_s\}$ with $s > 1$. We shall construct such new languages B_0, B_1, R_0, R_1 for which $A_1^\diamond = B_0 \cup B_1$, $B_0 \subseteq R_0 \subseteq A^\circ$, $B_1 \subseteq R_1 \subseteq A^\circ$ and R_0, R_1 are regular languages. Then the statement is satisfied by $R_I = R_0 \cup R_1$.

To construct the above sets, let $V = \{u_i : 1 \neq u_i \in U \;\&\; \mathrm{alph}(u_i) \subseteq \mathrm{alph}(u_1)\}$. Without loss of generality, we may assume that $V = \{u_1, \ldots, u_{1+t}\}$ for some integer $t \geq 0$. Let $R_0 = A_1^\diamond \diamond \cdots \diamond A_{1+t}^\diamond$. Then $R_0 \subseteq A^\circ$ and R_0

$$= \prod_{i=1}^{1+t}{}^{\diamond} \left(\Psi^{-1}\Psi(u_i) \diamond \prod_{j=1}^{n}{}^{\diamond} (a_j^{f_{ij}})^* \right)^{\diamond} = \left(\prod_{i=1}^{1+t}{}^{\diamond} \Psi^{-1}\Psi(u_i^+) \right) \diamond \left(\prod_{i=1}^{1+t}{}^{\diamond} \prod_{j=1}^{n}{}^{\diamond} (a_j^{f_{ij}})^* \right)$$

$$= \Psi^{-1}\Psi(u_1^+ \cdots u_{1+t}^+) \diamond \prod_{i=1}^{1+t}{}^{\diamond} \prod_{j=1}^{n}{}^{\diamond} (a_j^{f_{ij}})^*.$$

Let $j \in \{1, \ldots, n\}$ be arbitrary. By the definition of the decomposition matrix, $f_{ij} \in \{0, p_j\}$ for all i, $i = 1, \ldots, 1+t$. Moreover, if $j > s$, then $f_{ij} = 0$ for all i, $i = 1, \ldots, 1+t$. If $j \in \{1, \ldots, s\}$ and $f_{ij} = 0$, $i = 1, \cdots, 1+t$, then from our conditions it follows that there exists a $k \in \{1, \ldots, 1+t\}$ for which $u_k = a_j^{w_j}$ with some integer $w_j > 0$. We note that we may get different values w_j for different indices k. In this case, we can choose one of them arbitrarily. Now, we define the integers f_j, $j = 1, \ldots, s$ by

$$f_j = \begin{cases} w_j & \text{if } f_{ij} = 0,\ i = 1, \ldots, 1+t, \\ p_j & \text{otherwise.} \end{cases}$$

Then it is easy to see that

$$R_0 = \Psi^{-1}\Psi(u_1^+ \cdots u_{1+t}^+) \diamond (a_1^{f_1})^* \diamond \cdots \diamond (a_s^{f_s})^*. \tag{6}$$

Since $f_j > 0$, $j = 1, \ldots, s$, we have that R_0 is regular from Lemma 7.

In particular, if $t = 0$, then the statement is satisfied by $R_I = R_0$. Now, assume that $t > 0$.

Since $\text{alph}(u_2 \cdots u_{1+t}) \subseteq \text{alph}(u_1) = \{a_1, \ldots, a_s\}$, there are positive integers $d_1, \ldots, d_s$ such that $a_1^{d_1} \cdots a_s^{d_s} \in \Psi^{-1}\Psi(u_1)$. Let $|u_2 \cdots u_{1+t}|_{a_j} = c_j$ for all j, $j = 1, \ldots, s$ and let $N = 1 + \max\{c_j : 1 \leq j \leq s\}$. Then, by the positive integers $g_j = Nd_j - c_j$, $j = 1, \ldots, s$, we have

$$\Psi^{-1}\Psi(u_1^N) = \Psi^{-1}\Psi(u_2 \cdots u_{1+t}) \diamond a_1^{g_1} \diamond \cdots \diamond a_s^{g_s}.$$

Furthermore, let $M = f_1 \cdots f_s$. Then, by the above equality, we obtain that

$$\Psi^{-1}\Psi(u_1^{NM}) = \Psi^{-1}\Psi((u_2 \cdots u_{1+t})^M) \diamond a_1^{f_1 g_1'} \diamond \cdots \diamond a_s^{f_s g_s'} \tag{7}$$

where $g_j' = M g_j / f_j$, $j = 1, \ldots, s$. Now, we define the sets B_0, B_1, R_1 as follows:

$$B_0 = A_1^\diamond \diamond \Psi^{-1}\Psi((u_2 \cdots u_{1+t})^M) \diamond a_1^{f_1 g_1'} \diamond \cdots \diamond a_s^{f_s g_s'},$$

$$B_1 = (\cup\{\Psi^{-1}\Psi(u_1^k) : 1 \leq k \leq NM\}) \diamond (a_1^{f_{11}})^* \diamond \cdots \diamond (a_s^{f_{1s}})^*,$$

$$R_1 = B_1.$$

Then $B_1 \subseteq R_1 \subseteq A^\circ$ and R_1 is a regular language. By (7), the validity of $B_0 \cup B_1 \subseteq A_1^\diamond$ is obvious. To prove the converse inclusion, let $u \in A_1^\diamond$ be an arbitrary word. Then

$$u \in \Psi^{-1}\Psi(u_1^p) \diamond (a_1^{f_{11}})^* \diamond \cdots \diamond (a_s^{f_{1s}})^*$$

for some integer $p > 0$. If $p \leq NM$, then $u \in B_1$. If $p > NM$, then

$$u \in \Psi^{-1}\Psi(u_1^{p-NM}) \diamond (a_1^{f_{11}})^* \diamond \cdots \diamond (a_s^{f_{1s}})^* \diamond \Psi^{-1}\Psi(u^{NM})$$

and by (7), we obtain that $u \in B_0$. Consequently, $A_1^\diamond = B_0 \cup B_1$. Finally, we show that $B_0 \subseteq R_0$. Using (6), we can write R_0 in the following form:

$$R_0 = A_1^\diamond \diamond \Psi^{-1}\Psi(u_2^+ \cdots u_{1+t}^+) \diamond (a_1^{f_1})^* \diamond \cdots \diamond (a_s^{f_s})^*.$$

On the other hand, $\Psi^{-1}\Psi((u_2, \ldots, u_{1+t})^M) \subseteq \Psi^{-1}\Psi(u_2^+ \cdots u_{1+t}^+)$ and thus

$$B_0 = A_1^\diamond \diamond \Psi^{-1}\Psi((u_2 \cdots u_{1+t})^M) \diamond a_1^{f_1 g_1'} \diamond \cdots \diamond a_s^{f_s g_s'} \subseteq R_0.$$

Now, let $|I| > 1$ and assume that our statement is valid for any $\emptyset \neq J \subseteq \{1, \ldots, m\}$ with $|J| < |I|$. Without loss of generality, we may assume that $I = \{1, \ldots, r\}$ for some integer $r > 1$. If there exists a $j \in \{1, \ldots, n\}$ such that $u_i \in a_j^*$ for all i, $i = 1, \ldots, r$, then the statement obviously holds by $R_I = u_1^* \cdots u_r^* \diamond (a_j^{f_{1j}})^* \diamond \cdots \diamond (a_j^{f_{rj}})^*$. In the opposite case, we may assume that $\text{alph}(u_1 \cdots u_r) = \{a_1, \ldots, a_s\}$ for some integer $s > 1$. Now, we construct such languages $B_0', \ldots, B_r', R_0', \ldots, R_r'$ for which $A_1^\diamond \diamond \cdots \diamond A_r^\diamond = B_0' \cup \ldots \cup B_r'$, $B_i' \subseteq R_i' \subseteq A^\circ$ and R_i' is regular for all i, $i = 0, \ldots, r$. Then the statement is satisfied by $R_I = R_0' \cup \ldots \cup R_r'$.

For this purpose, let $V = \{u_i : 1 \neq u_i \in U \;\&\; \text{alph}(u_i) \subseteq \text{alph}(u_1 \cdots u_r)\}$. Without loss of generality, it can be assumed that $V = \{u_1, \ldots, u_{r+t}\}$ for some nonnegative integer t. Let $R_0' = A_1^\diamond \diamond \cdots \diamond A_{r+t}^\diamond$. Then $R_0' \subseteq A^\circ$ and in a similar way as above, R'_0 can be given in the form $R'_0 = \Psi^{-1}\Psi(u_1^+ \cdots u_{r+t}^+) \diamond (a_1^{f_1})^* \diamond \cdots \diamond (a_s^{f_s})^*$ where $f_j > 0$, $j = 1, \ldots, s$. Therefore, by Lemma 7, we obtain that R'_0 is a regular language. In particular, if $t = 0$, then the statement is valid by $R_I = R'_0$. Now, assume that $t > 0$. Since $\text{alph}(u_{r+1} \cdots u_{r+t}) \subseteq \text{alph}(u_1 \cdots u_r) = \{a_1, \ldots, a_s\}$, we can determine, in a similar way as above, positive integers g_j, $j = 1, \ldots, s$ such that

$$\Psi^{-1}\Psi((u_1 \cdots u_r)^N) = \Psi^{-1}\Psi(u_{r+1} \cdots u_{r+t}) \diamond a_1^{g_1} \diamond \cdots \diamond a_s^{g_s}.$$

Using this equality and the integer $M = f_1 \cdots f_s$, we obtain again that there are positive integers $g'_1, \ldots, g'_s$ such that

$$\Psi^{-1}\Psi((u_1 \cdots u_r)^{NM}) = \Psi^{-1}\Psi((u_{r+1} \cdots u_{r+t})^M) \diamond a_1^{f_1 g'_1} \diamond \cdots \diamond a_s^{f_s g'_s}. \quad (8)$$

Now, we define the following sets:

$$B'_0 = A_1^\diamond \diamond \cdots \diamond A_r^\diamond \diamond \Psi^{-1}\Psi((u_{r+1} \cdots u_{r+t})^M) \diamond a_1^{f_1 g'_1} \diamond \cdots \diamond a_s^{f_s g'_s},$$

$$Q_i = (\cup\{\Psi^{-1}\Psi(u_i^k) : 1 \leq k \leq NM\}) \diamond (a_1^{f_{i1}})^* \diamond \cdots \diamond (a_s^{f_{is}})^*, \ i = 1, \ldots, r,$$

$$B'_i = A_1^\diamond \diamond \cdots \diamond A_{i-1}^\diamond \diamond Q_i \diamond A_{i+1}^\diamond \diamond \cdots \diamond A_r^\diamond, \ i = 1, \ldots r,$$

$$R'_i = Q_i \diamond R_{\{1,\ldots,i-1,i+1,\ldots,r\}}, \ i = 1, \ldots, r.$$

The definitions are correct, since by our induction hypothesis, $R_{\{1,\ldots,i-1,i+1,\ldots,r\}}$ exists and $A_1^\diamond \diamond \cdots \diamond A_{i-1}^\diamond A_{i+1}^\diamond \diamond \cdots \diamond A_r^\diamond \subseteq R_{\{1,\ldots,i-1,i+1,\ldots,r\}}$ for all i, $i = 1, \ldots, r$. Now, we immediately obtain that $B'_i \subseteq R'_i \subseteq A^\diamond$ and R'_i is regular for all i, $i = 1, \ldots, r$. On the other hand, it can be proved in the same way as above that $B'_0 \subseteq R'_0$. Finally, by (8), the validity of $B'_0 \cup \ldots \cup B'_r \subseteq A_1^\diamond \diamond \cdots \diamond A_r^\diamond$ is obvious. Now, let $u \in A_1^\diamond \diamond \cdots \diamond A_r^\diamond$ be an arbitrary word. Then, for some nonnegative integers s_j, $j = 1, \ldots, s$, we have

$$u \in \Psi^{-1}\Psi(u_1^{s_1} \cdots u_r^{s_r}) \diamond \prod_{i=1}^{r}{}^\diamond \prod_{j=1}^{s}{}^\diamond (a_j^{f_{ij}})^*.$$

If $s_j \leq NM$ for some $j \in \{1, \ldots, r\}$, then $u \in B'_j$. If $s_j > NM$ for all j, $j = 1, \ldots, r$, then using (8), it is easy to check that $u \in B'_0$. Therefore,

$$A_1^\diamond \diamond \cdots \diamond A_r^\diamond = B'_0 \cup \ldots \cup B'_r.$$

Now, using the statement shown above, we can prove the sufficiency. It is obvious that $A^\diamond = \{1\} \cup (\cup\{\prod_{i \in I}{}^\diamond A_i^\diamond : \emptyset \neq I \subseteq \{1, \ldots, m\}\})$. But then we have $A^\diamond \subseteq \{1\} \cup (\cup\{R_I : \emptyset \neq I \subseteq \{1, \ldots, m\}\}) \subseteq A^\diamond$. Since every R_I is regular, $A^\diamond$ is regular as well, which completes the proof of Theorem 9.

References

1. T. Araki, N. Tokura, Decision problems for regular expressions with shuffle and shuffle closure operators, *Trans. Inst. Electronics Commun. Engrg.* Japan, **J64-D** (1981), 1069-1073.
2. T. Araki, N. Tokura, Flow languages equal recursively enumerable languages, *Acta Inform.* **15** (1981), 209-217.
3. B. Berard, Literal Shuffle, *Theor. Comp. Sci.* **51** (1987), 281-299.
4. S.L. Bloom, Z. Ésik, Nonfinite axiomatizability of shuffle inequalities. In: *Proceedings of TAPSOFT'95,* LNCS 915, Springer-Verlag, 1995, 318-333.
5. Z. Ésik, M. Bertol, Nonfinite Axiomatizability of the Equational Theory of Shuffle, *in: Proceedings of ICALP'95*, LNCS 944, Springer-Verlag, 1995, 27-38.
6. B. Imreh, M. Ito, A note on regular strongly shuffle-closed languages, *Acta Cybernetica* **11** (1994), 257-260.
7. M. Ito, G. Thierrin, S.S. Yu, Shuffle-closed Languages, *Publ. Math. Debrecen* **46/3-4** (1995),1-21.
8. M. Ito, G. Tanaka, Dense property of initial literal shuffles, *International Journal of Computer Mathematics* **34** (1990), 161-170.
9. K. Iwama, Universe problem for unrestricted flow languages, *Acta Inform.* **19** (1983), 85-96.
10. M. Jantzen, Extending regular operations with iterated shuffle, *Theoret. Comput.Sci.* **38** (1985), 223-247.
11. J. Jedrzejowicz, Nesting of Shuffle closure is important, *Inform. Process. Lett.* **25** (1987), 363-367.
12. J. Jedrzejowicz, Infinite hierarchy of shuffle expressions over a finite alphabet, *Inform. Process. Lett.* **36** (1990), 13-17.
13. A. Mateescu, On (left) partial shuffle, *Theoret. Comput. Sci.* **125** (1994), 361-374.
14. Gh. Paun, *Grammars for Economic Processes*, The Technical Publ. House, Bucaresti, 1980 (in Romanian).

Combinatorial Properties of Sequences in Groups and Semigroups*

A. V. Kelarev

Department of Mathematics
University of Tasmania
G.P.O. Box 252C, Hobart
Tasmania 7001, Australia
e-mail kelarev@hilbert.maths.utas.edu.au

1 Introduction

Combinatorics on words, or sequences, plays important roles in various areas. Many difficult problems were solved with the use of certain combinatorial properties of words (see, for example, [1]). There are several books and surveys devoted to various aspects of combinatorics on words (see [2, 3, 17, 24, 25]). We shall present new results on sequences of elements in groups and semigroups devoted to three combinatorial properties: repetitivity, growth, and permutational property. It is interesting that all these properties are interrelated in the problems we are going to consider.

2 Permutational property

Giving an answer to a question of Paul Erdös, B.H. Neumann proved the following theorem.

Theorem 1 **([26]).** *A group is center-by-finite if and only if every infinite sequence contains a pair of elements that commutes.*

The following generalization of this theorem has been obtained recently.

Theorem 2 **([7]).** *For any group G, the following conditions are equivalent:*

(i) for a positive integer n, every infinite subset of G contains n elements such that $x_1 \cdots x_n = x_{\sigma(1)} \cdots x_{\sigma(n)}$, where $\sigma \neq 1$ is a permutation.
(ii) G has a normal subgroup N of finite index, N is an FC-group, and the exponent of $N/Z(N)$ is finite.

A group G has the *permutation property* or is said to be *permutable,* if there exists a positive integer n such that the product of any n-tuples $x_1, \dots, x_n$ of elements of G can be rewritten as $x_1 \dots x_n = x_{\sigma(1)} \cdots x_{\sigma(n)}$ for a non-trivial

* This research has been supported by a grant of Australian Research Council.

permutation σ (see [27]). If the number n is known, then the group is said to be n-permutable. This concept is a natural generalization of commutativity. It turns out that various modifications of this property lead to interesting classes of groups.

A semigroup S is said to be *weakly permutational* if there exists an integer $n > 1$ such that for any $x_1 \cdots x_n \in S$ there exist two permutations $\sigma \neq \tau$ such that $x_{\sigma(1)} \dots x_{\sigma(n)} = x_{\tau(1)} \dots x_{\tau(n)}$.

We say that S is an *involutional* semigroup if there exists a positive integer n such that the product of any n-tuple $x_1, \dots, x_n$ of elements of G can be rewritten as $x_1 \dots x_n = x_{\sigma(1)} \dots x_{\sigma(n)}$ for an involution σ.

A semigroup S is *ω-permutable* if, for each infinite sequence $x_1, x_2, \dots$, there exists an integer n and a permutation $\sigma \neq 1$ such that $x_1 \dots x_n = x_{\sigma(1)} \dots x_{\sigma(n)}$.

Theorem 3 **([4, 6, 22]).** *For any group G, the following conditions are equivalent:*

(i) G is permutable;
(ii) G is weakly permutable;
(iii) G is involutional;
(iv) G is ω-permutable;
(v) G is finite-by-abelian-by-finite.

It is interesting that for semigroups not all properties (i) to (iv) are equivalent (see [13]).

Theorem 4 **([4]).** *Every 4-permutable group is solvable.*

Theorem 5 **([5]).** *For any group G, the following are equivalent:*

(i) for each infinite sequence $x_1, x_2, \dots$, there exists an integer n and a permutation $\sigma \neq 1$ such that $x_{\sigma(1)} \dots x_{\sigma(n)} = x_{\tau(1)} \dots x_{\tau(n)}$;
(ii) the largest normal FC-subgroup of G has finite index in G.

A semigroup S is f.p.f.-permutable if there exists an integer n such that every sequence of n elements can be rewritten as $x_1 \dots x_n = x_{\sigma(1)} \dots x_{\sigma(n)}$ and the permutation σ has no fixed points.

Theorem 6 **([14]).** *A finite semigroup is f.p.f-permutable if and only if its idempotents belong to the center.*

The following theorem shows that the class of f.p.f.-permutable groups coincides with the class considered by B.H. Neumann.

Theorem 7 **([14]).** *A group is f.p.f-permutable if and only if it is center-by-finite.*

The following theorem is motivated by the classical Burnside problem.

Theorem 8 ([6, 4, 14, 22]). *For a finitely generated periodic semigroup S, the following are equivalent:*

(i) *S is finite;*
(ii) *S is permutable;*
(iii) *S is weakly permutable;*
(iv) *S is f.p.f.-permutable;*
(v) *S is involutional.*

3 Repetitivity

The concept of a repetitive semigroup was introduced by Justin in 1971. Intuitively speaking, a group or semigroup is repetitive if it satisfies the analog of the famous Van der Waerden theorem on arithmetical progressions.

Let S be a group or a semigroup, and let $s = s_1, \ldots, s_m$ be a sequence of elements of S. A *k-factorization* of s is a sequence $t = t_1, \ldots, t_m$, where the t_j's are the values of k consecutive segments of s, i.e., $t_j = s_{i_j} s_{i_j+1} \cdots s_{i_{j+1}-1}$ for $j = 1, \ldots, k$ and some $1 \leq i_1 < i_2 < \ldots < i_{k+1} \leq m + 1$. We say that t is a *power* k-factorization if $t_1 = \ldots = t_k$. A semigroup S is said to be *repetitive* if and only if, for each finite subset X of S and every integer $k > 0$, there exists a positive integer $L = (S, X, k)$ such that every sequence $s_1, \ldots, s_L$ of elements of X has a power k-factorization.

The Van der Waerden theorem on arithmetical progressions is equivalent to saying that the infinite cyclic group is repetitive (see [8]). A whole chapter in Lothaire's book [25] is devoted to this property. Several papers dealt with the preservation of repetitivity by various constructions. Justin has described all repetitive nilpotent groups and commutative semigroups.

Theorem 9 ([9]). *A nilpotent group or commutative semigroup is repetitive if and only if it does not contain the free commutative semigroup of rank two.*

Complete description of repetitive Rees quotients of free monoids was obtained in [15]. The next theorem describes all FC-soluble and linear groups in the language of forbidden sections. Note that FC-groups play key roles in considerations of several other combinatorial properties of groups (see, for example, [5]), and contain soluble groups as a special case.

A group H is called a *section* of G if it is a homomorphic image of a subgroup of G. We shall call a semigroup S a section of G if it is a homomorphic image of a subsemigroup of G. We shall use $\mathcal{Z}$ and $\mathcal{N}$ to denote the additive group of all integers and the additive semigroup of all positive integers respectively. Recall that the free commutative group (semigroup) of rank two is isomorphic to $\mathcal{Z} \times \mathcal{Z}$ (respectively, $\mathcal{N} \times \mathcal{N}$). For a subset X of a semigroup S, put $X^{[n]} = \bigcup_{i=1}^{n} X^i$. As usual the wreath product of groups G and H is denoted by $G \wr H$. A subgroup generated in G by a set X is denoted by $\langle X \rangle$.

Theorem 10 ([21]). *Let G be an FC-soluble or linear group. Then the following are equivalent:*

(i) *G is repetitive;*
(ii) *G has linear growth;*
(iii) *every finitely generated subgroup of G is cyclic-by-finite;*
(iv) *G has no section isomorphic to the free commutative semigroup of rank two.*

Corollary 11 ([21]). *An algebraic linear semigroup is repetitive if and only if it does not contain a subsemigroup that can be homomorphically mapped onto the free commutative semigroup of rank two.*

The following example shows that the exact analog of Theorem 10 is not true for arbitrary groups.

Example 1. Let G be the infinite finitely generated Burnside group by Novikov and Adian (see [1]) satisfying the identity $x^n = 1$ for some n. Obviously, it has no section isomorphic to $\mathcal{N} \times \mathcal{N}$. However, it is not repetitive. Indeed, repetitivity and the identity $x^n = 1$ easily imply that G is locally finite, giving a contradiction. Thus, the analog of our main theorem for arbitrary groups is not true.

The groups described in Theorem 10 may be not cyclic-by-locally finite.

Example 2. Let $C = \langle c \rangle$ be the infinite cyclic group with a generator c, and let $C_i = \langle c_i \rangle$ be the cyclic group of order i with a generator $c_i; i = 2, 3, \ldots$. Denote by D the direct product of all these cyclic groups. We shall use the additive notation for the group operation in D. Let b_i denote the automorphism of D which maps c to $c + c_i$ and leaves all c_i's fixed. Put $B = \langle b_i; i = 2, 3, \ldots \rangle$ and consider the semidirect product $G = DB$. Every finitely generated subgroup of G is cyclic-by-finite, and so repetitive. However, G has no normal infinite cyclic subgroup. Thus, is a repetitive group which is not an extension of a cyclic group by a locally finite group.

In condition (iv) of Theorem 10 we cannot replace sections isomorphic to the free commutative semigroup of rank two by sections isomorphic to the free abelian group of rank two.

Example 3. For a prime p, consider the wreath product $W = \mathcal{Z} \wr \mathcal{Z}$. It has no section isomorphic to the free abelian group of rank two. It is easily seen that W is not repetitive.

The following modification of repetitivity is related to the Ramsey Theorem. A semigroup S is said to be *ultrarepetitive* if and only if for any $m > 1$ there exists n such that every sequence of elements $s_1, \ldots, s_n \in S$ has an m-power, i.e. for some $1 \le i_1 \le \cdots \le i_{m+1} \le n + 1$ the equalities $s_{i_1} \cdots s_{i_2 - 1} = \ldots = s_{i_m} \cdots s_{i_{m+1}-1}$ hold, cf. [10]. The difference with repetitivity is that the elements of the sequence do not have to form a finite set. Thus this concept is more restrictive, and all ultrarepetitive semigroups have been described.

Theorem 12 ([20]). *A semigroup S is ultrarepetitive if and only if there exists a finite ideal chain*

$$0 = S_0 \subseteq S_1 \subseteq \cdots \subseteq S_{2n-1} \subseteq S_{2n} = S^0$$

and a positive integer N such that, for any $m = 1, \ldots, n$, the factor S_{2m}/S_{2m-1} is a 0-direct union of a (possibly empty) set of completely 0-simple semigroups $D_i^{(m)}$, indexed by $i \in I_m$, such that $|D_i^{(m)}| < N$, and all other factors are nilpotent of index N, i.e. $S_{2m-1}^N \subseteq S_{2m-2}$.

Corollary 13 ([20]). *Every ultrarepetitive semigroup is locally finite and satisfies an identity $x^k = x^{k+m}$ for some positive integers k, m.*

Corollary 14 ([20]). *A commutative semigroup is ultrarepetitive if and only if there exist positive integers k, m, n such that every chain of idempotents of S has less than n elements, every subgroup of S consists of less than n elements, every nil factor of S is nilpotent of index n, and S satisfies the identity $x^k = x^{k+m}$.*

A sequence $s = (s_i)$ of elements s_i of a semigroup S is right (left; twosided) infinite if it has the form $s = s_1, s_2, \ldots,$ (respectively, $s = s \ldots, s_{-2}, s_{-1}; s = \ldots, s_{-2}, s_{-1}, s_0, s_1, s_2, \ldots,$). A finite subsequence $s_i, s_{i+1} \ldots s_{i+k}$ is called a *segment* of s. The *value* of the segment is the element $s_i s_{i+1} \ldots s_{i+k}$ of the semigroup S. We shall denote a segment and its value by the same letter when the meaning is clear from the context. An ω-factorization of a right (left; twosided) infinite sequence $s = (s_i)$ is a right (left; two-sided) infinite sequence $t = (t_k)$, where all t_k are the values of the consecutive segments of s, that is $t_k = s_{i_k} \ldots s_{i_{k+1}-1}$ for $k = 1, \ldots, m$ and some $i_1 < i_2 < \cdots$.

A sequence $s = (s_i)$ over S with at least two elements is said to be *Ramseyan* if and only if all its segments have the same value. Clearly, this is the case if and only if all elements are equal to the same idempotent e of S. We say that a semigroup S is *right (left; two-sided) infinitely Ramseyan* if every right (left; two-sided) infinite sequence of elements of S has a Ramseyan *omega*-factorization. A well-known corollary to the Ramsey theorem tells us that every finite semigroup is right and left infinitely Ramseyan.

Let S be a semigroup. By a *factor* of S we mean a Rees factor, i.e., the quotient semigroup I/J, where $I \supset J$ are ideals of S. Consider the relation $\leq$ defined on the set $E(S)$ of idempotents of S by $e \leq f \iff ef = fe = e$. Any subset of $E(S)$ which is linearly ordered by $\leq$ is called an *idempotent chain.* A semigroup with zero is said to be *left (right) T-nilpotent* if and only if every right (left) infinite sequence has a segment equal to zero. Clearly, every left T-nilpotent semigroup is nil. The following algebraic characterization of left T-nilpotent semigroups is well-known: A nil semigroup is left T-nilpotent if and only if all factors of S have nonzero left annihilators.

Recall that the *Cayley graph* of S has elements of S as vertices, and for $s, t, u \in S$ the graph has an arc from s to t labeled by u if and only if $su = t$. If G is a digraph and T a set of vertices of G, then by $G \setminus T$ we denote the subgraph obtained by deleting all vertices of T from G.

The following theorems describes all left, right and two-sided infinitely Ramseyan semigroups.

Theorem 15 ([19]). *For every semigroup S, the following conditions are equivalent:*

(i) S is right (left) infinitely Ramseyan;
(ii) every right (left) infinite sequence of elements of S has at least two disjoint segments with the same values;
(iii) S is periodic and satisfies the descending chain condition for principal right (left) ideals, all idempotent chains of S are finite, all nil factors of S contain nonzero left (right) annihilators, and all completely 0-simple factors of S are finite;
(iv) the Cayley graph G of S has a finite set T of vertices such that every connected component of $G \setminus T$ is finite.

Theorem 16 ([19]). *For every semigroup S, the following conditions are equivalent:*

(i) S is two-sided infinitely Ramseyan;
(ii) S is an extension of a rectangular band by a nil semigroup N such that every factor of N has nonzero right and left annihilators.

Theorem 16 shows that two-sided infinitely Ramseyan property is much more restrictive. In particular, not every finite semigroup is two-sided infinitely Ramseyan.

If in the definition of repetitivity we require, in addition, that all factors in the factorization be of the same length, then the semigroup is called uniformly repetitive. A semigroup is uniformly repetitive if and only if it is locally finite ([11]). Ramseyan factorizations were used in [18] to answer a question of I. Simon on factorization forests.

Theorem 12 has been used in the investigation of the following property. Let S be a semigroup. A class $\mathcal{K}$ of semigroups is said to be S-closed if and only if for any homomorphism $f : T \to S$ with all preimages $f^{-1}(e)$ belonging to $\mathcal{K}$, whenever $e = e^2 \in S$, it follows that $T \in \mathcal{K}$. This concept has been considered in the literature for various classes $\mathcal{K}$. Shevrin proved that the class of locally finite semigroups is S-closed for every band S. All semigroups S, such that the class of locally finite semigroups is S-closed, were described by Brown. It is known that many combinatorial classes are S-closed when S is finite (see, for example, [12]).

Theorem 17 ([20]). *For any semigroup S, the following are equivalent:*

(i) the class of all permutable semigroups is S-closed;
(ii) S is ultrarepetitive and contains a finite number of idempotents;
(iii) S^0 has a finite ideal chain $0 = S_0 \subseteq S_1 \subseteq \dots S_n = S^0$ such that each factor $S_m/S_{m-1}, 1 \le m \le n$ is finite or nilpotent.

4 Linear growth

If T is a finitely generated semigroup with a generating set V, and m is a positive integer, then the growth function $d_V(m)$ measures the cardinality of the set $V \bigcup V^2 \bigcup \cdots \bigcup V^m$. We write $d(m)$, if this is unambiguous. The Gelfand-Kirillov dimension of an arbitrary semigroup S, denoted by $GK(S)$, is equal to $\sup_V \overline{\lim} \log_n d_V(n)$, where V runs over all finite subsets of S, see [24]. We shall say that a semigroup S has *linear growth* if $GK(S)$ is finite. It has *polynomial growth* if $GK(S)$ is finite.

Problem 18 ([17]). Is it true that a group is repetitive if and only if it has a linear growth?

Justin proved, in particular, that this is so for every nilpotent or linear group and every group of polynomial growth (see [17]). In view of Gromov theorem, this gives a description of all repetitive groups of polynomial growth. However, the problem of characterizing repetitive groups remains open. Theorem 10 confirms the conjecture above for soluble groups. However, in general Problem 18 still remains open.

Since matrix semigroups play important roles in many areas and enjoy good properties similar to those of groups, it would be interesting to find out whether repetitivity is equivalent to linear growth in the case of matrix semigroups. We shall see that in general this is not true. However, if we impose a natural additional assumption (see Theorem 20) repetitivity becomes equivalent to linear growth.

Let us recall the necessary facts about the structure of arbitrary matrix semigroups. Suppose that S is a subsemigroup of the multiplicative monoid $M_n(K)$ of all $n \times n$ matrices over a field K. If G is a maximal subgroup of $M_n(K)$ intersecting S, then the subgroup of G generated by $S \bigcap G$ is called an *associated linear group* of S. If S has no free subsemigroups, then this is the quotient group $(S \bigcap G)(S \bigcap G)^{-1}$. Every matrix semigroup has a finite number of associated linear groups up to conjugacy. For $i = 0, 1, \ldots, n$, denote by M_i the set of all matrices of rank at most i in $M_n(K)$. Let S^0 be the semigroup S with zero 0 adjoined. Then S^0 has a finite ideal chain

$$\{0\} = S_0 \subseteq S_1 \subseteq \cdots \subseteq S_m = S^0$$

such that, for every $i = 1, \ldots, m$, the factor $S(i) = S_i/S_{i-1}$ of this chain either is nilpotent or is contained in a completely 0-simple subsemigroup $T(i)$ of a principal factor $M_j/M_{j-1}, j \in \{1, \ldots, n\}$ of the monoid $M_n(K)$. In the latter case, $S(i)$ intersects all nonzero $\mathcal{H}$-classes of $T(i)$ and every maximal subgroup H of $T(i)$ is generated by $H \bigcap S(i)$. (In particular, every two associated groups intersecting $S_i \setminus S_{i-1}$ are isomorphic.) The set $S_i/Si - 1$ (or, with a slight abuse of language, the semigroup $S(i)$) is called a *uniform component* of S.

Theorem 19 ([23]). *A matrix semigroup is repetitive if and only if all the associated linear groups are locally cyclic-by-finite.*

The following two examples show that there exist finitely generated repetitive matrix semigroups of subexponential growth and of polynomial but nonlinear growth.

Example 4. Let $\mathcal{Z}$ be the ring of integers, $M_3(\mathcal{Z})$ the ring of 3×3 matrices over $\mathcal{Z}$. Denote by S the subsemigroup generated in $M_3(\mathcal{Z})$ by two matrices

$$e = \begin{bmatrix} 1\,0\,0 \\ 0\,0\,0 \\ 0\,0\,0 \end{bmatrix} \quad \text{and} \quad g = \begin{bmatrix} 1\,0\,0 \\ 0\,0\,0 \\ 0\,0\,0 \end{bmatrix}.$$

Then S is repetitive and has subexponential growth.

On the other hand, there exist repetitive matrix semigroups with polynomial but nonlinear growth.

Example 5. Denote by S the subsemigroup generated in $M_2(\mathcal{Z})$ by the elements

$$a = \begin{bmatrix} \alpha & 0 \\ 0 & \alpha \end{bmatrix}, b = \begin{bmatrix} 0 & 0 \\ 0 & \beta \end{bmatrix} \quad \text{and} \quad e = \begin{bmatrix} 0 & 1 \\ 0 & 0 \end{bmatrix},$$

where $\alpha, \beta \in \mathcal{Z}$ are relatively prime. Then S is repetitive and $GK(S) = 2$.

The next two examples show that it is impossible to replace 'locally cyclic-by-finite' by 'cyclic-by-finite' in Theorem 19, even if the whole matrix semigroup is finitely generated.

Example 6. Let S be the subsemigroup generated in $M_3(\mathcal{Z})$ by

$$e = \begin{bmatrix} 1\,0\,0 \\ 0\,0\,0 \\ 0\,0\,1 \end{bmatrix} \quad \text{and} \quad g = \begin{bmatrix} 1 & 1 & 1 \\ 0 & 1/2 & 2 \\ 0 & 0 & 1 \end{bmatrix}.$$

Then S is repetitive. However, one of its associated groups is not cyclic-by-finite, and is not finitely generated.

A similar example can also be given in the case of positive characteristic.

Example 7. Let p be a prime, F_p a finite field of p elements, and let S be the subsemigroup generated in $M_3(F_p(x))$ by

$$e = \begin{bmatrix} 1\,0\,0 \\ 0\,0\,0 \\ 0\,0\,1 \end{bmatrix} \quad \text{and} \quad g = \begin{bmatrix} 1 & 1 & 0 \\ 0 & x & 1 \\ 0 & 0 & 1 \end{bmatrix}.$$

Then S is repetitive, but one of its associated groups G that is not not cyclic-by-finite, and is not finitely generated.

The following theorem describes all matrix semigroups with linear growth or, equivalently, of Gelfand-Kirillov dimension at most one.

Theorem 20 ([23]). *Let $S \subseteq M_n(K)$ be a matrix semigroup. Then the following conditions are equivalent:*

(1) $GK(S) \leq 1$;
(2) for every $x, y, z \in S$, there exist positive integers i, j, k, l such that $x^i y z^j = x^k y z^l$ and $(i, j) \neq (k, l)$.

If, in addition, S is finitely generated, then the groups associated to S are cyclic-by-finite.

Corollary 21 ([23]). *Every matrix semigroup with linear growth is repetitive.*

Corollary 22 ([23]). *Assume that $S \subseteq M_n(K)$ is a finitely generated matrix semigroup that is the union of its uniform components. Then the following conditions are equivalent:*

(i) $GK(S) \leq 1$;
(ii) S is repetitive and S is contained in the union of a finite number of $\mathcal{R}$- and $\mathcal{L}$-classes of $M_n(K)$;
(iii) the groups associated to S are cyclic-by-finite and S is contained in the union of a finite number of $\mathcal{R}$- and $\mathcal{L}$-classes of $M_n(K)$.

Acknowledgement

The editors gratefully acknowledge the contribution of Asat Arslanov in preparing the final form of this paper.

References

1. S. I. Adian, *The Burnside Problem and Identities in Groups*, Springer-Verlag, Berlin, 1978.
2. K. A. Baker, G. F. McNulty and W. Taylor, Growth problems for avoidable words, *Theoret. Comput. Sci.* 69 (1989), 319-345.
3. J. Berstel, Properties of Infinite Words : Recent Results, *STACS'89*, Springer-Verlag, Lect. Notes Comp. Sci. 349 (1989), 36-46.
4. R. D. Blyth, Rewriting products of group elements, I, II, *J. Algebra* 116 (1988), 506-521; 119 (1988), 246-259.
5. R. D. Blyth and A. H. Rhemtulla, Rewritable products in FC-by-finite groups, *Canad. J. Math.* 41 (1989), 369-384.
6. M. Curzio, P. Longobardi, M. Maj and D.J.S. Robinson, A permutational property of groups, *Arch. Math.* 44, 1985, 385-389.
7. M. Curzio, P. Longobardi, M. Maj and A. H. Rhemtulla, Groups with many rewritable products, *Proc. Amer. Math. Soc.* 115 (1992), 931-934.
8. J. Justin, Généralisation du théorème de van der Waerden sur les semigroupes répétitifs, *J. Combinatorial Theory. Ser. A.* 12 (1972), 357-367.
9. J. Justin, Characterization of the repetitive commutative semigroups, *J. Algebra* 21 (1972), 87-90.

10. J. Justin and G. Pirillo, Une condition de finitude pour les semigroupes finiment engendres, *C. R. Acad. Sci. Paris* 306, 1 (1988), 63-65.
11. J. Justin, G. Pirillo. and S. Varricchio, Unavoidable regularities and finiteness conditions for semigroups, *Proc. III Italian Conference "Theoretical Computer Science"*, Mantova, 2-4 November 1989, Edited by Bertoni-Bohm-Miglioli, World Scientific 1989.
12. J. Justin and G. Pirillo, Some remarks on the permutation property for semigroups, *Europ. J. Combinatorics* 11 (1990), 151-154.
13. J. Justin and G. Pirillo, Shirshov's Theorem and ω-permutability of semigroups, *Advances Math.* 87 (1991), 151-159.
14. J. Justin, B. Piochi and G. Pirillo, On fixed-point-free permutation properties in groups and semigroups, *Annali Matem. Pura Appl.* 39 (1991), 151-159.
15. J. Justin and G. Pirillo, Intracommutativity properties for groups and semigroups, *J. Algebra*, 153 (1992), 424-443.
16. J. Justin and G. Pirillo, Factorial languages and some combinatorial properties of semigroups, *Inter. J. Algebra Comput.* 3 (1993), 295-316.
17. J. Justin and G. Pirillo, On some questions and conjectures in combinatorial semigroup theory, *Southeastasian Bull. Math.* 18 (1994), 91-104.
18. J. Justin and G. Pirillo, On a question about factorization forests, *Theoret. Comput. Sci.* 141 (1995), 351-355.
19. J. Justin and A. V. Kelarev, Factorizations of infinite sequences in semigroups, *Annali Matem. Pura Applicata*, to appear in 1997.
20. A. V. Kelarev, Combinatorial properties and homomorphisms of semigroups, *Internat. J. Algebra Comput.* 4 (1994), 443-450.
21. A. V. Kelarev and P. V. Shumyatsky, Soluble and linear repetitive groups, *Bull. Austral. Math. Soc.* 52 (1995), 253-261.
22. A. V. Kelarev and G. Pirillo, A note on rewritable products in groups and semigroups, *Bollettino Unione Mat. Italiano* 9-A (1995), 1-4.
23. A. V. Kelarev and J. Okniński, Combinatorial properties for sequences of matrices, submitted.
24. G.R. Krause and T.H. Lenagan, *Growth of Algebras and Gelfand-Kirillov Dimension*, (Pitman Research Notes, 116, Boston, 1985).
25. M. Lothaire, *Combinatorics on Words*, Addison-Wesley, London, 1983.
26. B. H. Neumann, A problem of Paul Erdös on groups, *J. Austral. Math. Soc.* (1976), 467-472.
27. R. Restivo and C. Reutenauer, On the Burnside problem for semigroups, *J. Algebra* 89 (1984) 102-104.

Categoricity and Scott Families

Bakhadyr Khoussainov[1] and Richard Shore[2]

[1] The University of Auckland, Auckland, New Zealand
Cornell University, Ithaca, NY, USA
[2] Cornell University, Ithaca, NY, USA

1 Introduction

Effective model theory is an area of logic that analyzes the effective content of the typical notions and results of model theory and universal algebra. Typical notions in model theory and universal algebra are languages and structures, theories and models, models and their submodels, automorphisms and isomorphisms, embeddings and elementary embeddings. In this paper languages, structures, and models are assumed to be countable.

There are many ways to introduce considerations of effectiveness into the area of model theory or universal algebra. Here we will briefly explain considerations of effectiveness for theories and their models on the one hand, and for just structures on the other hand.

Let us begin by considering effectiveness for theories and their models. From the model theoretic point of view, given a first order theory, one is interested in finding models for the theory with specific algebraic or model-theoretic properties. In this sense theories are the basic objects in model theory. A natural way of introducing effectiveness is, therefore, to begin by considering decidable theories, that is, ones whose theorems form a decidable (computable or recursive) set.

Given a decidable complete theory, one can effectively carry out a Henkin type construction and build a model of the theory. This procedure guarantees that the satisfaction predicate for the model constructed is decidable. Thus we are led to the following definition.

Definition 1. A structure $\mathcal{A}$ is *decidable* if there is a computable enumeration a_i of the elements of $\mathcal{A}$ such that $Th(\mathcal{A}, \mathbf{a}_i)$ is decidable.

Just as model theory investigates the class of models of a given theory, effective model theory is concerned with decidable models of decidable theories. There have been a significant number of results about decidable models of decidable theories. These results typically discuss questions related to finding decidable prime, saturated, homogeneous models; omitting or realizing types by decidable models; the number of decidable models for decidable theories, etc. We refer the reader to [8] and [15] for survey articles in this area.

If we begin to introduce consideration of effectiveness just for the structures themselves, then we are essentially in the realm of general effective mathematics. Considerations of effectiveness for structures have been extensively developed since the early 60's beginning with Frölich and Shepherdson [5], Rabin [17], and Malcev [13]. However, it is worth to noting that even in the early 60's the idea of considering effectiveness in structures was not new. In the 30's Kleene and Church considered effectiveness in well-ordered sets and invented recursive ordinals. In the early 70's Nerode and his collaborators in the U. S., as well as Ershov and his colleagues in Novosibirsk, developed the powerful idea of combining model-theoretic and algebraic constructions with priority arguments from computability theory. This approach embodies the technical core of many results in effective model theory.

More recently, there have been many papers devoted to investigating effectiveness in structures. For example, Cenzer, Nerode and Remmel [1] have been developing the theory of p-time structures. Khoussainov and Nerode have begun the development of the theory of automatic structures [11]. These theories are based, respectively, on computations which can be performed in p-time and by finite automata. We will not discuss these topics but turn instead to structures in which the basic functions and relations can be computed by Turing machines.

Definition 2. A structure $\mathcal{A}$ for a language $\mathcal{L}$ is *computable* if its domain A is a computable subset of ω and its functions and relations are uniformly computable. A structure isomorphic to a computable structure is called *computably presentable*, and any such isomorphism is called a *computable presentation.*

The requirements of computability are significantly weaker than those for decidability. However, the definition captures what one normally means by an effective structure or presentation in mathematical discourse.

Identification of isomorphic structures is typical in model theory and universal algebra or, indeed, generally in classical mathematics. A typical model-theoretic or algebraic problem about isomorphisms can often be stated as follows: Find some invariants such that any two structures from the class are isomorphic if only if they have the same invariants.

Introducing effectiveness considerations into the area, we would like to understand the relationship between classical invariants and effective invariants; in particular, between isomorphism types and effective isomorphism types. Thus, while model theory identifies isomorphic structures, effective model theory is concerned with computable isomorphisms and finding characterizations for structures which have the same computable isomorphism type. A fundamental concept is therefore that of computable isomorphism type.

Definition 3. Two computable structures $\mathcal{A}$ and $\mathcal{B}$ are of the same *computable isomorphism type* if there is computable isomorphism taking $\mathcal{A}$ to $\mathcal{B}$. The *dimension* of a computable structure $\mathcal{A}$ is the number of its computable isomorphism types.

To what extent computable isomorphism types can differ from classical ones can be seen from the following result of Goncharov:

Theorem 4 [7]. *For each $n \leq \omega$ there is a computable structure with computable dimension n.*

There has been significant interest in understanding the nature of the structures of dimension 1. The basic model theoretic notion which motivated this interest is that of ω–categoricity. A theory T is ω–categorical if all countable models of T are isomorphic. A structure $\mathcal{A}$ is ω–categorical if its theory is ω–categorical. The analogous concept for effective model theory deals only with computable structures and isomorphisms:

Definition 5. A structure $\mathcal{A}$ is *computably categorical* if any two computable structures isomorphic to $\mathcal{A}$ are computably isomorphic.

The result of Nurtazin [16] is one of the first about the nature of computably categorical structures. His theorem characterizes structures whose decidable presentations form one computable isomorphism type.

Theorem 6 [16]. *For a structure $\mathcal{A}$ the following two conditions are equivalent:*

1. *Any two decidable presentations of $\mathcal{A}$ are computably isomorphic.*
2. *There exists a finite sequence $\bar{c}$ of constants from A such that $(\mathcal{A}, \bar{c})$ is the prime model of the theory $Th(\mathcal{A}, \bar{c})$ and the set of atoms of this theory is computable.*

In the late 70's Goncharov [6] and Remmel [18] independently gave an algebraic characterization for Boolean Algebras and Linear Orderings to be computably categorical.

Theorem 7 [6, 18].

1. *A Boolean Algebra is computably categorical if and only if it has finitely many atoms.*
2. *A linear ordering is computably categorical if and only if the number of pairs of adjacent elements is finite.*

2 Scott families

Interestingly, all the structures which have been shown to be computably categorical have one common property. They all have Scott families.

Definition 8. A *Scott family* for a structure $\mathcal{A}$ is a computable sequence

$$\phi_0(\bar{a}, x_1, \ldots, x_{n_0}), \phi_1(\bar{a}, x_1, \ldots, x_{n_1}), \ldots$$

of $\exists$-formulas satisfiable in $\mathcal{A}$, where $\bar{a}$ is a fixed tuple of elements from $\mathcal{A}$, such that every tuple in $\mathcal{A}$ satisfies one these formulas and any two tuples satisfying the same formula from the sequence can be sent to each other via an automorphism of $\mathcal{A}$.

The basic idea behind this definition is the following. If a computable structure $\mathcal{A}$ has a Scott family and $\mathcal{B}$ is a computable structure isomorphic to $\mathcal{A}$, then we can effectively carry out a back and forth argument to construct a computable isomorphism from $\mathcal{A}$ to $\mathcal{B}$.

Theorem 9. *If a structure $\mathcal{A}$ has a Scott family, then the structure is computably categorical. Moreover, for any n-tuple $(d_1, \ldots, d_n)$ the expanded structure $(\mathcal{A}, d_1, \ldots, d_n)$ also has a Scott Family.*

Proof. Let $\phi_0(\bar{a}, x_1, \ldots, x_{n_0}), \phi_1(\bar{a}, x_1, \ldots, x_{n_1}), \ldots$ be a Scott family for $\mathcal{A}$, where $\bar{a} = (a_0, \ldots, a_{m-1})$. Let $\mathcal{A}_1$ and $\mathcal{A}_2$ be computable presentations of $\mathcal{A}$. We define a mapping $f : A_1 \to A_2$ by stages. We can assume that for each $j \in \{0, \ldots, m-1\}$, a_j^i is the element in A_i corresponding to the constant a_j. At even stages we define images of elements from $\mathcal{A}_1$, at odd stages we define preimages of elements from $\mathcal{A}_2$.

Stage 0**:** Set $f_1 = \{(a_0^1, a_0^2), \ldots, (a_{m-1}^1, a_{m-1}^2)\}$.

Stage $2k$**:** We can suppose that the function f_{2k-1} has been defined. Assume that $f_{2k-1} = \{(a_0^1, a_0^2), \ldots, (a_{m-1}^1, a_{m-1}^2), (b_1, c_1), \ldots, (b_s, c_s)\}$ and that f_{2k-1} can be extended to an isomorphism of $\mathcal{A}_1$ to $\mathcal{A}_2$. Let b be the first number not in the domain of f_{2k-1}. Consider the tuple $(b_1, \ldots, b_s, b)$. Find an i such that $\phi_i(\bar{a}, b_1, \ldots, b_s, b)$ holds in $\mathcal{A}_1$. Hence $\exists x \phi_i(\bar{a}, c_1, \ldots, c_s, x)$ holds in $\mathcal{A}_2$. Find the first $c \in A_2$ for which $\phi_i(\bar{a}, c_1, \ldots, c_s, c)$ holds. Extend f_{2k-1} by letting $f_{2k} = f_{2k-1} \bigcup \{(b, c)\}$.

Stage $2k+1$**:** We define f_{2k+1} similarly so as to put the least element of A_2 not yet in the range of f_{2k} into that of f_{2k+1}.

Finally, let $f = \bigcup_{i \in \omega} f_i$. Thus, f is a computable isomorphism.

For the second part of the theorem, we slightly change the original Scott family. Namely, set $\psi_i = \phi_i(\bar{a}, x_1, \ldots, x_{n_i}) \& \exists y_1 \ldots \exists y_n (\&_j (d_j = y_j)$. Then, one can easily check that the sequence $\psi_0, \psi_1, \ldots$ is a Scott family for the expanded structure $(\mathcal{A}, d_1, \ldots, d_n)$. The theorem is proved.

Corollary 10. *If a structure $\mathcal{A}$ has a Scott family, then any expansion of $\mathcal{A}$ by finitely many constants is computably categorical.*

At this point we would like to make the following two observations about the effect of expanding computably categorical structures by finitely many constants. First, as we have mentioned, all the known examples of computably categorical structures have Scott families. Thus, it is natural to ask whether there exists a computably categorical structure without a Scott family. By Corollary 10, one possible way to build a such structure is to provide an example of a computably categorical structure whose expansion by finitely many constants is not computably categorical. Second, as we mentioned, the notion of ω–categoricity is a basic model-theoretic motivation in the investigation of computably categorical structures. It is an easy consequence of the Ryll-Nardzewski theorem that if a structure $\mathcal{A}$ is ω-categorical then so is $(\mathcal{A}, \bar{a})$, the structure expanded by finitely many constants. It is the analogous situation in effective model theory that we wish to consider.

Millar [14] proved that a small amount of decidability is enough to guarantee that categoricity is preserved under such expansions. Informally his theorem states that if a structure $\mathcal{A}$ is computably categorical and we can effectively solve systems of algebraic equations and inequations over this structure, then computable categoricity is preserved under expansions by a finite number of constants.

Theorem 11 [14]. *If a structure $\mathcal{A}$ is computably categorical and its existential theory is decidable, then the expansion of $\mathcal{A}$ by finitely many constants is also computably categorical.*

Without the assumption of the decidability of the existential diagram, the question (known as Ash-Goncharov problem) has been open:

> *Does there exist a computably categorical structure whose expansion by a finite number of constants is not computably categorical?*

An answer to this question has recently been found:

Theorem 12 [2]. *For each natural number n, there exists a computably categorical structure $\mathcal{A}$ such that, for every $a \in A$, the expanded structure $(\mathcal{A}, a)$ has dimension n.*

An immediate consequence of Corollary 10 is now the following result:

Corollary 13. *There exists a computably categorical structure without a Scott family.*

However, based on Corollary 10, one could suggest that the reason the structure constructed in Theorem 12 does not have a Scott family is that the structure has an expansion by a finite number of constants which is not computably categorical. We construct a counterexample to this suggestion in the next section.

3 Scott sequences for families of computably enumerable sets

Our basic result is the following theorem.

Theorem 14. *There exists a structure without a Scott family such that every expansion of the structure by a finite number of constants is computably categorical.*

The structure required to establish the theorem is constructed by coding certain (uniformly) computably enumerable families of sets of natural numbers.

Definition 15. A family S of sets of natural numbers has a *one-to-one computable enumeration* if there is a bijection $f : \omega \to S$ such that $\{(i, x) | x \in f(i)\}$ is computably enumerable. We then call f a *(computable) one-to-one enumeration* of S.

We wish to consider a preordering on the one-to-one computable enumerations of S that naturally induces an equivalence relation corresponding to computable isomorphism:

Definition 16. A computable enumeration f of S is *reducible to* g, $f \leq g$, if there is a computable Φ such that $f = g\Phi$. If $f \leq g$ and $g \leq f$, then we say that f and g are *equivalent.*

Note that if f is a one-to-one enumeration of S and $f = g\Phi$, then Φ is a permutation of ω and so $f \leq g$. Thus the equivalence classes of one-to-one enumerations are minimal elements in the induced partial ordering. These are the enumerations that we need to consider to define the family that supplies the structure required for Theorem 14. Informally, computable categoricity corresponds to there being a single such equivalence class and dimension corresponds to the number of such classes.

Definition 17. A computable sequence $D_0, D_1, \ldots$ of (canonical indices for) finite sets is a *Scott sequence* for a family S if the following properties hold:

1. For each D_i there exists exactly one $M_i \in S$ such that $D_i \subset M_i$.
2. The set $S \setminus \{M_0, M_1, \ldots\}$ is finite.

The reader can easily prove the following:

Theorem 18. *If S has a Scott sequence, then any two computable enumerations of S are equivalent.*

For any given family S, we want to construct a structure $\mathcal{A}_S$ such that $\mathcal{A}_S$ has a Scott family if and only if S has a Scott sequence. Thus, let S be a family of sets and let f be a one-to-one computable enumeration of S. We assume that each set in S has at least two elements. Based on f, we will construct a computable structure, indeed a graph, $\mathcal{A}_f = (\omega, P_f)$, where P_f is a computable binary predicate on ω.

Consider a uniformly effective, possibly finite, sequence $a_{i,0}, a_{i,1}, a_{i,2}, \ldots$ without repetitions such that, for each $i \in \omega$, $f(i) = \{a_{i,0}, a_{i,1}, a_{i,2}, \ldots\}$.

For each $i \in \omega$, we can consider a computable structure $\mathcal{G}_i^f = (G_i, P_i^f)$ defined as follows. G_i has an element d_i such that for each $a_{i,j}$ the predicate P_i^f defines a unique cycle $C_{i,j}$ of length $a_{i,j}$ for which $d_i \in C_{i,j}$. In addition, for all j, k the cycles $C_{i,k}$ and $C_{i,j}$ have only one element in common which is d_i. Thus, we see that for all $j \neq k$ we have $d_i \in C_{i,k}$, $d_i \in C_{i,j}$ and $(C_{i,j} \setminus \{d_i\}) \bigcap (C_{i,k} \setminus \{d_i\}) = \emptyset$. We call the element d_i a *cluster point.* Informally, the structure $\mathcal{G}_i^f$ codes the set $f(i)$. The structure (graph) $\mathcal{G}_i^f$ is computable and satisfies the following properties:

1. For every number t, t belongs to $f(i)$ if and only if there exist distinct elements $x_0, \ldots, x_t$ of the structure $\mathcal{G}_i^f$ such that the formula
$$P_i^f(x_0, x_1) \,\&\, \ldots P_i^f(x_{t-1}, x_t) \,\&\, P_i^f(x_t, x_0)$$
holds in the structure $\mathcal{G}_i^f$.

2. Any two cycles in $\mathcal{G}_i^f$ have only one element in common.

By the construction of $\mathcal{G}_i^f$ and the computability of f, we can conclude that there exists a computable sequence $\mathcal{A}_0^f = (A_0^f, P_0)$, $\mathcal{A}_1^f = (A_1^f, P_1)$, $\mathcal{A}_2^f = (A_2^f, P_2), \ldots$ of computable structures such that:

1.
2. For each i the structure $\mathcal{A}_i^f$ is isomorphic to the structure $\mathcal{G}_i^f$.
3. For each pair $i \neq j$, $A_i^f \bigcap A_j^f = \emptyset$ and $\omega = \bigcup_i A_i^f$.
4. The relation $P_f = \bigcup_i P_i$ is computable.

Consider the computable structure $\mathcal{A}_f$ defined in some canonical way so that $\mathcal{A}_f$ is isomorphic to (ω, P_f). Note that the set of all cluster points of $\mathcal{A}_f$ is recursive in every recursive presentation of $\mathcal{A}_f$. (We are here using the assumption that every set in the family S has at least two elements.) The following lemma describes the relationship between S and $\mathcal{A}_f$.

Lemma 19. *The structure $\mathcal{A}_f$ satisfies the following conditions.*

1. *If g is a one-to-one computable enumeration of S, then $\mathcal{A}_f$ is isomorphic to $\mathcal{A}_g$.*
2. *The structure $\mathcal{A}_f$ is rigid, that is it does not have any nontrivial automorphisms.*
3. *If g is a one-to-one computable enumeration of S, then $\mathcal{A}_f$ is computably isomorphic to $\mathcal{A}_g$ if and only if f and g are equivalent.*
4. *The dimension of the structure $\mathcal{A}_f$ is equal to the maximal number of nonequivalent one-to-one computable enumerations of S.*
5. *The structure $\mathcal{A}_f$ has a Scott family if and only if S has a Scott sequence.*

Proof. To prove 1, first, note that for any pair $i, j \in \omega$ the graphs $\mathcal{G}_i^f$ and $\mathcal{G}_j^g$ are isomorphic if and only if $f(i) = g(j)$. Hence, since f and g are one-to-one enumerations of S, we can conclude that $\mathcal{A}_f$ is isomorphic to $\mathcal{A}_g$.

Any automorphism α of $\mathcal{A}_f$ must be the identity by the construction of $\mathcal{A}_f$ and the fact that f is a one-to-one mapping. This proves 2.

Suppose that f and g are equivalent. There exists a recursive function Φ such that $f = g\Phi$. Hence the structure $\mathcal{G}_i^f$ is isomorphic to the structure $\mathcal{G}_{\Phi(i)}^g$. Hence $\mathcal{A}_f$ and $\mathcal{A}_g$ are computably isomorphic. Let $\mathcal{B}$ be a computable presentation of $\mathcal{A}_f$. Consider an effective sequence $e_0, e_1, e_2, \ldots$ without repetition of all cluster points in $\mathcal{B}$. We define a one-to-one computable enumeration $f_{\mathcal{B}}$ of S as follows:

$$f_{\mathcal{B}}(i) = \{n \mid e_i \text{ belongs to a circle of length } n\}.$$

It follows that $\mathcal{B}$ is computably isomorphic to $\mathcal{A}_g$ if and only if g is equivalent to $f_{\mathcal{B}}$. This proves 3.

Part 4 follows from the proof of Part 3.

We are left to prove the last part of the lemma. Suppose that S has a Scott sequence $D_0, D_1, D_2, \ldots$. Without lost of generality we can suppose that $D_i \subset$

$f(i)$. We have to prove that $\mathcal{A}_f$ has a Scott family. Take an $x \in \mathcal{A}_f$. Find a d_i which is connected to x via P_f. Suppose that the length of a path which connects x with d_i is n. Define the following formula: $\psi(x) =$ [there exists a path of length n which connects x with a cluster point y such that for each $m \in D_i$ the element y belongs to a cycle of length m]. Now for every s–tuple $(x_1, \ldots, x_s)$ let $\phi_{(x_1,\ldots,x_s)}$ be $\psi(x_1)\& \ldots \&\psi(x_s)$. It is not hard to check that the sequence $\{\phi_{(x_1,\ldots,x_s)}\}$ is a Scott family for $\mathcal{A}_f$.

Now suppose for simplicity that $\mathcal{A}_f$ has a Scott family

$$\phi_0(x_1, \ldots, x_{n_0}), \phi_1(x_1, \ldots, x_{n_1}), \ldots$$

without parameters. The proof below will show that we do not lose any generality by making this assumption. Let $d_0, d_1, d_2 \ldots$ be an effective sequence of all cluster points from $\mathcal{A}_f$. Let

$$\phi_{i_0}(x_0), \phi_{i_1}(x_1), \ldots$$

be an effective subsequence of the original sequence such that $\phi_{i_k}(d_k)$ holds for each $k \in \omega$. Since the formulas are all existential and the structure is computable, we can effectively find a finite substructure $\mathcal{B}_i$ of $\mathcal{A}_f$ such that $d_i \in B_i$ and $\phi_{i_k}(d_k)$ holds in $\mathcal{B}_i$. Define

$$D_i = \{n \mid d_i \text{ belongs to a circle of length } n \text{ in substructure } \mathcal{B}_i\}.$$

Since we have a Scott family for $\mathcal{A}_f$ and since the structure $\mathcal{A}_f$ is rigid, we can see that the sequence $D_0, D_1, \ldots$ is a Scott sequence for family S.

Corollary 20. *Any two one-to-one computable enumerations of S are equivalent if and only if $\mathcal{A}_f$ is computably categorical.*

Now, to prove Theorem 14 it suffices, by Lemma 19, to build a computably enumerable family S of sets without a Scott sequence any two computable one-to-one enumerations of which are equivalent.

Lemma 21. *There is a computably enumerable family S of sets with no Scott sequence any two computable one-to-one enumerations of which are equivalent.*

In order to build a such family S and its one-to-one enumeration f, we need to satisfy the following requirements:

$$D_e: \quad F_e \text{ is not a Scott sequence for } S$$
$$R_j: \quad g_j \equiv f \text{ or } g_j \text{ is not a one-to-one enumeration of } S$$

where $\{g_j\}_j$ is a computable sequence of all potential one-to-one enumerations of a family of sets, and $\{F_e\}_e$ is a computable sequence of all potential Scott sequences for S. These requirements are similar to the requirements for constructing a computable structure as needed for Theorem 12 (see [2]). In fact, to construct such a family S, we essentially use the ideas from the proof of Theorem 2.5. A detailed proof of this lemma and similar results will appear in [12].

References

1. D. Cenzer, J. Remmel, Polynomial Time vs Recursive Models, *Ann. Pure and Appl. Logic* **54**, 1991.
2. P. Cholak, S. Goncharov, B. Khoussainov and R. Shore, On Recursively Categorical Models, in preparation.
3. J. N. Crossley, ed., *Aspects of Effective Algebra, Proceedings of a Conference at Monash University, 1979.*
4. Yu. Ershov and S. S. Goncharov, eds., *Logic Notebook: Problems in Mathematical Logic*, Novosibirsk University, 1986.
5. A. Frölich and J. C. Shepherdson, Effective Procedures in Field Theory, *Phil. Trans. Roy. Soc. London ser. A* **248** (1956), 407-432.
6. V. D. Dzgoev and S. S. Goncharov, Autostability of Models, *Algebra i Logika* **19** (1980), 45-58.
7. S. S. Goncharov, The Problem of the Number Of Non-Self-Equivalent Constructivizations, *Algebra i Logika* **19** (1980), 621-539.
8. V. Harizanov, Pure Recursive Model Theory, preprint, to appear in *Handbook of Recursive Mathematics*, Y. Ershov, S. Goncharov, A. Nerode, J. Remmel, eds.
9. L. Harrington, Recursively Presentable Prime Models, *J. Symbolic Logic* **39** (1973), 305-309.
10. N. Khissamiev, On Strongly Constructive Models of Decidable Theories, Izvestia AN Kaz. SSR (1974), 83-84, 94.
11. B. Khoussainov and A. Nerode, Automata Representable Structures, in *Proceedings of Logic and Computational Complexity Conference, 1995*, Springer-Verlag Lecture Notes in Computer Science, to appear.
12. B. Khoussainov and R. Shore, Recursive Categoricity and Degree Spectra of Relations, in preparation.
13. A. I. Mal'cev, Constructive Algebras, *Uspekhi Matem. Nauk* **16** (1961), 3-60.
14. T. Millar, Recursive Categoricity and Persistence, *J. Symbolic Logic* **51** (1986), 430-434.
15. T. Millar, Abstract Recursive Model Theory, in *Handbook of Recursion Theory*, E. Griffor, ed., North-Holland, to appear.
16. A. Nurtazin, Strong and Weak Constructivizations and Computable Families, *Algebra i Logika* **13** (1974), 311-323,364.
17. M. O. Rabin, Computable Algebra, General Theory and Theory of Computable Fields, *Trans. Am. Math. Soc.* **95** (1960), 341-360.
18. J. B. Remmel, Recursive Isomorphisms Types of Recursive Boolean Algebras, *J. Symbolic Logic* **46** (1981), 572-593.
19. J. B. Remmel and J. N. Crossley, The Work of Anil Nerode: A Retrospective, in *Logical Methods*, J. N. Crossley, J. B. Remmel, R. A. Shore and M. E. Sweedler, eds., Birkhäuser, Boston, 1993, 1-91.

A Process Algebraic View of Linear Logic

Padmanabhan Krishnan
Jane Hopkins

Department of Computer Science
University of Canterbury, Private Bag 4800
Christchurch 1, New Zealand
E-Mail: {paddy,jane}@cosc.canterbury.ac.nz

Abstract. In this paper we present a process algebra where the combinators are borrowed from linear logic. The principal idea is to present a direct operational semantics for $\wp$. The approach considers two sets of actions (input and output) and a merge operation which describes the synchronisation of processes. We then present a few results satisfied by the bisimulation relation induced by the labelled transition system.

1 Introduction

Linear Logic [4] has been described as a resource conscious logic as it does not have the general rules of weakening and contraction. It has been used to study various aspects of computation including functional programming and concurrency. While most of the connectives have a direct operational meaning, there is no simple and direct semantics for $\wp$. Girard himself has referred to this connective as mysterious and its semantics has been mainly as a De Morgan dual to the connective $\otimes$.

In Section 2 a brief intuitive introduction to certain aspects of linear logic is provided. The reader is referred to [1] along with [4, 5] for the complete technical details. In this paper we focus on our approach to understanding linear logic.

Here we present a CCS [7] like process calculus where the principal focus is on a process semantics for $\wp$. We then show that other linear logic connectives can be interpreted in a process semantics fashion. As in CCS we have a set of names and co-names. But the synchronisation we consider is not that of a name and a co-name. Rather, in our framework two names can be merged to obtain a new name while two co-names can be merged to obtain a new co-name. The intuition is that two processes merge their names or co-names to obtain another name or co-name. If names represent input channels, the merging of two channels indicates that the two processes synchronised by agreeing to communication on a common channel. Engberg and Winskel in [3] present a petri-net based semantics for linear logic, but they do not consider the $\wp$ connective. An encoding of proofs using the π-calculus is presented in [2]. Kobayashi and Yonezawa in [6] also present proof construction in linear logic as an asynchronous concurrent calculus. They also address higher order issues such as process consumption. Our approach, more in the style of [3], is to have a direct process interpretation of the various connectives.

Section 3 describes the syntax and operational semantics of the process language, Section 4 develops the bisimulation relation and presents a few equivalences that hold. The final section (Section 5) presents a complete axiomatisation of the bisimulation relation.

2 Background

In this section we produce an intuitive explanation of some of the linear logic connectives.

Linear implication $\Gamma \multimap A$ is similar to implication in classical logic where the formula Γ is required to establish A. However, in producing A, Γ is consumed and is no longer available. Hence, formulas in linear logic can be thought of as resources which are produced and consumed unlike those of classical logic which may be duplicated or discarded.

Additive disjunction $A \oplus B$ means we have one of A or B, but cannot choose which. The choice is left up to the system. From $\Gamma \multimap A$ one can deduce $\Gamma \multimap (A \oplus B)$. If the user asks for $A \oplus B$, the system can decide which of A or B to satisfy. Hence having the resources to satisfy only A is sufficient to satisfy $A \oplus B$.

Additive conjunction $A \,\&\, B$ also means a choice between A and B but the difference here is that the choice is made by the user. From $\Gamma \multimap A$ one cannot deduce $\Gamma \multimap (A \,\&\, B)$ because if the user chooses B it is not clear that Γ is sufficient to produce B. However from $\Gamma \multimap A$ and $\Gamma \multimap B$, one can deduce $\Gamma \multimap (A \,\&\, B)$. Here the system lets the user choose from A and B, hence the system needs to have sufficient resources to satisfy both A and B. After having chosen A (or B) the user cannot ask for B (or A) because Γ can only be used once, satisfying only one request.

Multiplicative conjunction $A \otimes B$ means to have both A and B. As explained above, from $\Gamma \multimap A$ and $\Gamma \multimap B$, one cannot deduce $\Gamma \multimap (A \otimes B)$ because Γ cannot be reused as in classical logic. One can however deduce $\Gamma, \Gamma \multimap (A \otimes B)$.

The multiplicative disjunction connective $⅋$ does not have an elegant intuitive meaning as its semantics is mainly as a De Morgan dual to $\otimes$. The main motivation of our algebraic approach is to provide an explanation of $⅋$.

3 The process language

We assume two sets I and O which are in bijection. For an action μ in I we write $\overline{\mu}$ to identify the corresponding action in O. As a notational convenience, we use subscript i to indicate an element of I and subscript o to indicate an element of O. Rather than write the inverse symbol, we continue to write $\overline{\mu}$ for μ belonging to O with the exact semantics being clear from the context. We also assume functions $\cdot_I$ on I and $\cdot_O$ on O. These functions represent the merge of input or output channels to obtain input or output channels. We will require that both $\cdot_I$ and $\cdot_O$ are commutative and associative. Furthermore they will satisfy the following involutive and idempotent laws.

- For every μ_1, μ_2 in I, $\overline{\overline{\mu_1} \cdot_O \overline{\mu_2}} = \mu_1 \cdot_I \mu_2$
- For every μ_1, μ_2 in O, $\overline{\overline{\mu_1} \cdot_I \overline{\mu_2}} = \mu_1 \cdot_O \mu_2$.
- For every $\mu \in I$, $\mu \cdot_I \mu = \mu$.
- For every $\mu \in O$, $\mu \cdot_O \mu = \mu$.

We will also assume the existence of two special actions e_o (belonging to O) and e_i (belonging to I) such that

- $\overline{e_i} = e_o$.
- for every action $\mu_o \in O$, $e_o \cdot_o \mu_o$ is equal to μ_o.
- for every action $\mu_i \in I$, $e_i \cdot_i \mu_i$ is equal to μ_i.

The actions e_o and e_i act as identity elements for other channels. These can be interpreted to be dead channels or channels on which communication cannot occur. Hence merging them with any other channel does not result in a new channel. When there is no confusion we write $\cdot$ instead of $\cdot_I$ or $\cdot_O$.

We begin by presenting an operational semantics for the units for the various connectives. We then consider the multiplicative and additive fragments. In addition to these fragments, we also consider action prefix as a basic combinator.

The syntax for the units is presented below and the operational semantics is presented in Figure 1.

$$\mathrm{P} ::= \mathbf{1} \mid \bot \mid \mathbf{0} \mid \top$$

Unit $\mathbf{1}$	$\mathbf{1} \xrightarrow{e_i} \mathbf{1}$
Unit $\bot$	$\bot \xrightarrow{e_o} \bot$
Unit $\top$	$\top \xrightarrow{\mu_i} \top$ for every $\mu_i \in I$
Unit $\mathbf{0}$	$\mathbf{0} \xrightarrow{\mu_o} \mathbf{0}$ for every $\mu_o \in O$

Fig. 1. Units

The processes $\mathbf{1}$ and $\bot$ represent deadlocked processes which use the dead input and output channel respectively. The processes $\top$ and $\mathbf{0}$ are completely non-deterministic processes in that they can exhibit any input or output action. The choice of these rules will become clear once the semantics for the main connectives are presented.

The syntax for the multiplicative fragment along with action prefix is given below.

$$\mathrm{P} ::= \mu; \mathrm{P} \mid (\mathrm{P} \mathbin{⅋} \mathrm{Q}) \mid (\mathrm{P} \multimap \mathrm{Q}) \mid (\mathrm{P}^{\bot}) \mid (\mathrm{P} \otimes \mathrm{P})$$

where μ belongs to $(I \cup O)$.

Intuitively ";" denotes action prefix, ⅋ denotes synchronisation on all and only output actions; input actions result in asynchronous behaviour. The combinator $\bot$ denotes the relabelling of an input (or output) channel to its corresponding output (or input) channel. The combinator $\multimap$ represents the standard synchronisation (i.e., behaviour determined by combining an input action and an output action). The formal operational semantics based on a labelled transition system [8] (where the labels are elements of $I \cup O$) is presented in Figure 2.

Prefix	$\mu;\ \mathrm{P} \xrightarrow{\mu} \mathrm{P}$
Asynchronous	$\dfrac{\mathrm{P} \xrightarrow{\mu} \mathrm{P}'}{\mathrm{P} \parr \mathrm{Q} \xrightarrow{\mu} \mathrm{P}' \parr \mathrm{Q} \text{ and } \mathrm{Q} \parr \mathrm{P} \xrightarrow{\mu} \mathrm{Q} \parr \mathrm{P}'}\ \mu \in I$
Synchronous	$\dfrac{\mathrm{P} \xrightarrow{\mu} \mathrm{P}',\ \mathrm{Q} \xrightarrow{\mu'} \mathrm{Q}'}{\mathrm{P} \parr \mathrm{Q} \xrightarrow{\mu \cdot \mu'} \mathrm{P}' \parr \mathrm{Q}'}\ \mu \in O \text{ and } \mu' \in O$
Negation	$\dfrac{\mathrm{P} \xrightarrow{\mu} \mathrm{P}'}{\mathrm{P}^{\perp} \xrightarrow{\overline{\mu}} \mathrm{P}'^{\perp}}$
Unrestricted Implication$_1$	$\dfrac{\mathrm{P} \xrightarrow{\mu} \mathrm{P}'}{\mathrm{P} \multimap \mathrm{Q} \xrightarrow{\overline{\mu}} \mathrm{P}' \multimap \mathrm{Q}}\ \mu \in O$
Unrestricted Implication$_2$	$\dfrac{\mathrm{Q} \xrightarrow{\mu} \mathrm{Q}'}{\mathrm{P} \multimap \mathrm{Q} \xrightarrow{\mu} \mathrm{P} \multimap \mathrm{Q}'}\ \mu \in I$
Restricted Implication	$\dfrac{\mathrm{P} \xrightarrow{\mu} \mathrm{P}',\ \mathrm{Q} \xrightarrow{\mu'} \mathrm{Q}'}{\mathrm{P} \multimap \mathrm{Q} \xrightarrow{\overline{\mu} \cdot \mu'} \mathrm{P}' \multimap \mathrm{Q}'}\ \mu \in I \text{ and } \mu' \in O$
Asynchronous	$\dfrac{\mathrm{P} \xrightarrow{\mu} \mathrm{P}'}{\mathrm{P} \otimes \mathrm{Q} \xrightarrow{\mu} \mathrm{P}' \otimes \mathrm{Q} \text{ and } \mathrm{Q} \otimes \mathrm{P} \xrightarrow{\mu} \mathrm{Q} \otimes \mathrm{P}'}\ \mu \in O$
Synchronous	$\dfrac{\mathrm{P} \xrightarrow{\mu} \mathrm{P}',\ \mathrm{Q} \xrightarrow{\mu'} \mathrm{Q}'}{\mathrm{P} \otimes \mathrm{Q} \xrightarrow{\mu \cdot \mu'} \mathrm{P}' \otimes \mathrm{Q}'}\ \mu \in I \text{ and } \mu' \in I$

Fig. 2. Multiplicative Connectives

The additive connectives $\oplus$ and & are now considered. As they represent choice, the operational rules are similar to that of the '+' combinator in CCS. The syntax is described below and the operational semantics presented in Figure 3.

P ::= (P & P) | (P ⊕ P)

I-Choice	$\dfrac{P \xrightarrow{\mu} P'}{(P \oplus Q) \xrightarrow{\mu} P',\ (Q \oplus P) \xrightarrow{\mu} P'}\ \mu \in I$
O-Choice	$\dfrac{P \xrightarrow{\mu} P'}{(P \,\&\, Q) \xrightarrow{\mu} P',\ (Q \,\&\, P) \xrightarrow{\mu} P'}\ \mu \in O$
O-Agreement	$\dfrac{P \xrightarrow{\mu} P',\ Q \xrightarrow{\mu} Q'}{(P \oplus Q) \xrightarrow{\mu} (P' \oplus Q')}\ \mu \in O$
I-Agreement	$\dfrac{P \xrightarrow{\mu} P',\ Q \xrightarrow{\mu} Q'}{(P \,\&\, Q) \xrightarrow{\mu} (P' \,\&\, Q')}\ \mu \in I$

Fig. 3. Additive Connectives

Example 1. Consider the process $a_i; b_o;\ \mathbf{0} \parr d_o;\ \mathbf{1}$. It can exhibit the action a_i after which the actions b_o and d_o will synchronise to produce $b_o \cdot d_o$. This will be followed by a sequence of e_i's. $\mathbf{0}$ cannot exhibit any action as there is no output matching action to synchronise with any μ_o.

If instead of $\mathbf{1}$ the process $\bot$ is considered, the process can exhibit any output action after exhibiting $b_o \cdot d_o$. This is because $\bot$ produces a sequence of the identity output action.

Example 2. Consider the process $(a_i; b_o;\mathbf{1} \,\&\, a_i; c_o;\mathbf{1}) \otimes (b_i; \bot)$. Initially, the process will exhibit $a_i \cdot b_i$ and evolve to $(b_o;\mathbf{1} \,\&\, c_o;\mathbf{1}) \otimes (\bot)$. It can now exhibit an arbitrary number of e_o's. It can also exhibit exactly one of b_o and c_o and evolve to $(\mathbf{1} \otimes \bot)$ which will exhibit only e_o's.

Example 3. We can now justify our choice of semantics for the units. Consider the process $(P \parr \bot)$. As $\bot$ is the unit for $\parr$, $(P \parr \bot)$ should behave as P. As $\parr$ is asynchronous over input actions, $\bot$ should not exhibit any input action. Furthermore, for any output action μ_o from P there should be an output action from $\bot$ such that their merge yields μ_o. Also, this should hold for arbitrary P. Hence $\bot$ exhibits a unbounded sequence of e_o.

The justification for the other units follows a similar argument.

Given the above set of rules, it turns out the behaviour of all processes in the system is constrained. At any stage in the computation, a process can either exhibit a number of input actions or a number of output actions. It is not possible for a process to have the ability to exhibit either an input action or an output action. Hence the choice is restricted to either an action from I or an action from O. The following theorem formally describes the constraint.

Theorem 1. *If* $P \xrightarrow{\mu_1} P_1$ *and* $P \xrightarrow{\mu_2} P_2$*, then* $\mu_1 \in I$ *iff* $\mu_2 \in I$.

Proof. As usual the proof proceeds by structural induction. The base case is for the units which obviously hold. By simple induction it is easy to see that the property holds for μ;P. We now just present the proof for two connectives, namely, $\wp$ and &.

Let (P $\wp$ Q) have the ability to exhibit μ_1 and μ_2 such that $\mu_1 \in I$ and $\mu_2 \in O$. As synchronisation on O actions is necessary, μ_2 can be decomposed into two actions belonging to O. Hence both P and Q need to exhibit an O action. However, one of P or Q needs to exhibit μ_1. Without loss of generality one can assume that P $\xrightarrow{\mu_1}$ (as any action in I is asynchronous). Hence P can do both an I and an O action which violates the induction hypothesis.

The situation for & is similar. Let (P & Q) have the ability to exhibit μ_1 and μ_2 such that $\mu_1 \in I$ and $\mu_2 \in O$. As an I action does not resolve choice, both P and Q must be able to perform μ_1. But clearly one of P or Q must be able to perform μ_2 thus violating our induction hypothesis. □

The above theorem shows that the behaviour of any process can be split into two phases. It can either exhibit an I action or it can exhibit an O action. Thus, although we have non-determinism in the system, the non-determinism is always between the same type of objects.

4 Bisimulation

Bisimulation is one of the key relations which is used to identify behaviour in concurrent systems. The key idea is that at every stage of the computation, the identified processes have identical behaviour. The purpose of bisimulation is to check that a given implementation meets an abstract specification. The reader is referred to [7] for details of this approach.

In this section we show that our bisimulation relation can be related to equivalence of logic terms. For this section we will assume that the sets I and O are finite, otherwise we need to deal with unbounded non-determinism.

Definition 2. Bisimulation

We let $\sim$ be the largest equivalence relation satisfying the following properties. (P $\sim$ Q) iff

If (P $\xrightarrow{\mu}$ P′), there exists (Q $\xrightarrow{\mu}$ Q′) such that (P′ $\sim$ Q′).
If (Q $\xrightarrow{\mu}$ Q′), there exists (P $\xrightarrow{\mu}$ P′) such that (P′ $\sim$ Q′).

The next few lemmas are related to the phase semantics presented in [4]. The sets I and O are like phases in that they are commutative and associative monoids with a neutral element. The similarity ends there as the terms we are dealing with are processes over these sets rather than subsets of these sets. But the logical equivalence using the phase semantics is captured by processes being bisimilar.

As expected, negation is involutive.

Proposition 3. *$(P^{\perp})^{\perp} \sim P$.*

The following results indicate that it is possible to express $\parr$, $\multimap$ and $\otimes$ in terms of one another using negation.

Lemma 4. Basic Properties of Multiplicatives

1. $(\mathrm{P} \parr \mathrm{Q}) \sim \mathrm{P}^{\perp} \multimap \mathrm{Q}$
2. $\mathrm{P} \multimap \mathrm{Q} \sim \mathrm{P}^{\perp} \parr \mathrm{Q}$
3. $(\mathrm{P} \otimes \mathrm{Q}) \sim (\mathrm{P}^{\perp} \parr \mathrm{Q}^{\perp})^{\perp}$
4. $(\mathrm{P} \multimap \mathrm{Q}) \sim (\mathrm{P} \otimes \mathrm{Q}^{\perp})^{\perp}$

Proof.

1. Let $(\mathrm{P} \parr \mathrm{Q}) \xrightarrow{\mu_i} \mathrm{R}$. If $(\mathrm{P} \xrightarrow{\mu_i} \mathrm{P}')$, R is identical to $(\mathrm{P}' \parr \mathrm{Q})$. $\mathrm{P}^{\perp} \xrightarrow{\overline{\mu_i}}$, hence $(\mathrm{P}^{\perp} \multimap \mathrm{Q}) \xrightarrow{\overline{\overline{\mu_i}}} (\mathrm{P}'^{\perp} \parr \mathrm{Q})$.
   If $\mathrm{Q} \xrightarrow{\mu_i}$ the result is obvious.
   If $(\mathrm{P} \parr \mathrm{Q})$ exhibited an action belonging to O, P must exhibit some μ_o and Q must exhibit some μ_o'. That is, the output action is decomposed into two parts such that $(\mathrm{P} \parr \mathrm{Q}) \xrightarrow{\mu_o \cdot_o \mu_o'}$. $\mathrm{P}^{\perp} \xrightarrow{\overline{\mu_o}}$ hence $(\mathrm{P}^{\perp} \multimap \mathrm{Q}) \xrightarrow{\overline{\overline{\mu_o}} \cdot_O \mu_o'}$.
2. Similar to above argument.
3. If $\mathrm{P} \xrightarrow{\mu_o} \mathrm{P}'$, $(\mathrm{P} \otimes \mathrm{Q}) \xrightarrow{\mu_o} (\mathrm{P}' \otimes \mathrm{Q})$. $\mathrm{P}^{\perp} \xrightarrow{\overline{\mu_o}} \mathrm{P}'^{\perp}$, due to which $(\mathrm{P}^{\perp} \parr \mathrm{Q}^{\perp}) \xrightarrow{\overline{\mu_o}} (\mathrm{P}'^{\perp} \parr \mathrm{Q}^{\perp})$. Hence $(\mathrm{P}^{\perp} \parr \mathrm{Q}^{\perp})^{\perp} \xrightarrow{\mu_o} (\mathrm{P}'^{\perp} \parr \mathrm{Q}^{\perp})^{\perp}$.
   $\mathrm{Q} \xrightarrow{\mu_o}$ is verified analogously.
   If $\mathrm{P} \xrightarrow{\mu_i} \mathrm{P}'$, for $(\mathrm{P} \otimes \mathrm{Q})$ to exhibit an action will require Q to exhibit some μ_i' and hence $(\mathrm{P} \otimes \mathrm{Q})$ will exhibit $\mu_i \cdot_i \mu_i'$. Now, $\mathrm{P}^{\perp}$ and $\mathrm{Q}^{\perp}$ will exhibit $\overline{\mu_i}$ and $\overline{\mu_i'}$. Hence by the involutive properties imposed on actions $(\mathrm{P}^{\perp} \parr \mathrm{Q}^{\perp})^{\perp}$ will exhibit $\mu_i \cdot_i \mu_i'$.
4. Similar to above argument. It needs the property that $\overline{\mu_i} \cdot_O \mu_o$ is equal to $\mu_i \cdot_I \overline{\mu_o}$. This equality follows from the axioms assumed about the set of actions. □

The following lemma presents some of the unit laws for the multiplicative connectives.

Lemma 5. Multiplicative Unit Laws

1. $(\perp \parr \mathrm{P}) \sim \mathrm{P}$

2. $\mathbf{1} \otimes \mathrm{P} \sim \mathrm{P}$
3. $\mathbf{1} \multimap \mathrm{P} \sim \mathrm{P}$

Proof.

1. $(\mathrm{P} \xrightarrow{\mu_i} \mathrm{P}')$ implies $(\bot \parr \mathrm{P}) \xrightarrow{\mu_i} (\bot \parr \mathrm{P}')$. If $(\mathrm{P} \xrightarrow{\mu_o} \mathrm{P}')$,$(\bot \parr \mathrm{P}) \xrightarrow{\mu_o} (\bot \parr \mathrm{P}')$ as $(e_o \cdot_O \mu_o) = \mu_o$.
2. Similar to above argument.
3. $\mathbf{1} \multimap \mathrm{P} \sim (\mathbf{1} \otimes \mathrm{P}^{\bot})^{\bot}$. Hence it is bisimilar to $(\mathrm{P}^{\bot})^{\bot}$. □

The unit laws and the De Morgan duals for the additive connectives are presented below.

Lemma 6. De Morgan Laws

1. $(\mathrm{P} \oplus \mathrm{Q})^{\bot} \sim \mathrm{P}^{\bot} \,\&\, \mathrm{Q}^{\bot}$
2. $\top \,\&\, \mathrm{P} \sim \mathrm{P}$
3. $\mathbf{0} \oplus \mathrm{P} \sim \mathrm{P}$

Proof.

1. If $(\mathrm{P} \xrightarrow{\mu_i} \mathrm{P}')$, $(\mathrm{P} \oplus \mathrm{Q}) \xrightarrow{\mu_i} \mathrm{P}'$. Hence $(\mathrm{P} \oplus \mathrm{Q})^{\bot} \xrightarrow{\overline{\mu_i}} \mathrm{P}'^{\bot}$. Hence $(\mathrm{P}^{\bot} \xrightarrow{\overline{\mu_i}} \mathrm{P}'^{\bot})$ due to which $\mathrm{P}^{\bot} \,\&\, \mathrm{Q}^{\bot} \xrightarrow{\overline{\mu_i}} \mathrm{P}'^{\bot}$.
   If $(\mathrm{P} \oplus \mathrm{Q})$ has to exhibit μ_o, both P and Q have to exhibit μ_o. Hence $(\mathrm{P} \oplus \mathrm{Q})^{\bot}$ will exhibit $\overline{\mu_o}$. As both $\mathrm{P}^{\bot}$ and $\mathrm{Q}^{\bot}$ can exhibit $\overline{\mu_o}$ the result holds.
2. As $\top$ can exhibit only (and all) actions belonging to I, the choice between P and $\top$ is resolved only when (if) P can exhibit an O action.
3. The proof is similar to the above case. □

The following result indicates that $\mathbf{0}$ is an annihilator for $\otimes$ and $\top$ is an annihilator for $\parr$.

Lemma 7. Annihilator Laws

1. $\mathbf{0} \otimes \mathrm{P} \sim \mathbf{0}$
2. $\top \parr \mathrm{P} \sim \top$

Proof.

1. Any I action of P cannot be exhibited as $\mathbf{0}$ can exhibit no I action. As the composition of two O actions is an O action the result holds.
2. Follows by an analogous argument. □

The final set of results deal with the distributive laws.

Lemma 8. Distributive Laws

1. $\mathrm{P} \otimes (\mathrm{Q} \oplus \mathrm{R}) \sim (\mathrm{P} \otimes \mathrm{Q}) \oplus (\mathrm{P} \otimes \mathrm{R})$
2. $\mathrm{P} \parr (\mathrm{Q} \,\&\, \mathrm{R}) \sim (\mathrm{P} \parr \mathrm{Q}) \,\&\, (\mathrm{P} \parr \mathrm{R})$

Proof.

1. Consider the equivalence between process S_1 which is (P $\otimes$ (Q $\oplus$ R)) and process S_2 which is (P $\otimes$ Q) $\oplus$ (P $\otimes$ R).
   If S_1 has to exhibit $\mu \in O$, either P can exhibit μ or both Q and R can exhibit μ.
   If P $\xrightarrow{\mu}$ P′, it is clear that (P $\otimes$ Q) $\xrightarrow{\mu}$ (P′ $\otimes$ Q) and (P $\otimes$ R) $\xrightarrow{\mu}$ (P′ $\otimes$ R). Hence S_1 can evolve to (P′ $\otimes$ (Q $\oplus$ R)) while S_2 can evolve to (P′ $\otimes$ Q) $\oplus$ (P′ $\otimes$ R). These processes are equivalent under the induction hypothesis.
   If Q $\xrightarrow{\mu}$ Q′ and R $\xrightarrow{\mu}$ R′, the process S_1 will exhibit μ and evolve to (P $\otimes$ (Q′ $\oplus$ R′)) which process S_2 can match by evolving to (P $\otimes$ Q′) $\oplus$ (P $\otimes$ R′).
   If $S_1 \xrightarrow{\mu}$ where $\mu \in I$, both P and (Q $\oplus$ R) has to exhibit some I actions. This implies that either Q or R has to exhibit some μ' belonging to I. Clearly S_2 can match this such that the derivatives in the two cases are equivalent, i.e., of the form (P′ $\otimes$ Q′) or (P′ $\otimes$ R′).
2. Similar to above proof. □

In the phase semantics, half distributivity between $\otimes$ and & holds. That is P $\otimes$ (Q & R) is contained in (P $\otimes$ Q) & (P $\otimes$ R). While we do not directly have half distributivity, a similar result holds. If P $\xrightarrow{\mu_o}$ P′, the process P $\otimes$ (Q & R) will exhibit μ_o and evolve to P′ $\otimes$ (Q & R). However, an action belonging to O resolves choice and (P $\otimes$ Q) & (P $\otimes$ R) will evolve to either (P′ $\otimes$ Q) or (P′ $\otimes$ R). However if P $\xrightarrow{\mu_i}$ P′, both Q and R have to agree on some input action μ'_i. P $\otimes$ (Q & R) will evolve to some P′ $\otimes$ (Q′ & R′) while choice is not resolved and hence (P $\otimes$ Q) & (P $\otimes$ R) will become (P′ $\otimes$ Q′) & (P′ $\otimes$ R′). So distributivity is maintained by I actions and not by O actions.

Similarly, in the phase semantics (P $⅋$ Q) $\oplus$ (P $⅋$ R) is contained in P $⅋$ (Q $\oplus$ R). If P $\xrightarrow{\mu_i}$ P′ will resolve the choice. Furthermore, such a move is asynchronous with respect to $⅋$. Hence (P $⅋$ Q) $\oplus$ (P $⅋$ R) will evolve to either (P′ $⅋$ Q) or (P′ $⅋$ R) while P $⅋$ (Q $\oplus$ R) will evolve to P′ $⅋$ (Q $\oplus$ R). However, if P $\xrightarrow{\mu_o}$ P′, an output action (say μ'_o) is required from Q (as the $⅋$ is a merge on output actions). The semantics of $\oplus$ insists on μ'_o from R. Hence the choice is not resolved and all three process will evolve by exhibiting an output action. The output action exhibit by P $⅋$ (Q $\oplus$ R) and (P $⅋$ Q) $\oplus$ (P $⅋$ R) is $\mu_o \cdot_O \mu'_o$.

5 Complete axiomatisation

In this section we present a sound and complete axiomatisation of the bisimulation equivalence. As we are still within the domain of interleaving semantics, we show that any process is semantically equivalent to a process over the set of actions I and O, action prefix(;), &, $\oplus$ and the units. Following [7], we define a standard form to which all processes can be reduced. We then show that bisimilar processes have a similar standard form.

We assume the usual associativity and commutativity axioms for & and $\oplus$.

We use $\&_{j \in J} P^j$ to indicate $P^{j_1} \& P^{j_2} \& \ldots P^{j_n}$ and $\bigoplus_{j \in J} P^j$ to indicate $P^{j_1} \oplus P^{j_2} \oplus \ldots P^{j_n}$ where $J = \{j_1, j_2, \ldots, j_n\}$.

Definition 9. Standard Form

All the units are in standard form.

If for every k belonging to an index set K the process P^k is in standard form, we conclude that the process $\&_{k \in K}\ \mu_o^k; P^k$ is in standard form.

If for every k belonging to an index set K the process P^k is in standard form, we conclude that the process $\bigoplus_{k \in K} \mu_i^k; P^k$ is in standard form.

This is very similar to the standard form used in CCS. Instead of having a single form we are required to have two forms; one for each type of choice. As & resolves choice only over output actions the standard form involving & involves only the output action prefix. Similarly for $\otimes$ we consider only prefix over input actions.

We first present a few equations which deal with the units.

$$\mathbf{1} = e_i;\mathbf{1} \qquad \bot = e_o;\bot \qquad \mathbf{0} = \&_{\mu_o \in O}\ \mu_o;\mathbf{0} \quad \top = \bigoplus_{\mu_i \in I} \mu_i;\top$$

$$\begin{array}{lll} \mathbf{1} \parr \mathbf{1} = \mathbf{1} & \mathbf{1} \parr \top = \top & \mathbf{1} \parr \mathbf{0} = \mathbf{1} \\ \mathbf{0} \parr \top = \top & \mathbf{0} \parr \mathbf{0} = \mathbf{0} & \top \parr \top = \top \end{array}$$

$$\begin{array}{lll} \mathbf{0} \otimes \mathbf{0} = \mathbf{0} & \mathbf{0} \otimes \bot = \mathbf{0} & \mathbf{0} \otimes \top = \mathbf{0} \\ \bot \otimes \top = \bot & \bot \otimes \bot = \bot & \top \otimes \top = \top \end{array}$$

The above equations can be verified to be sound. The first set of equations only show that the units represent recursive but finite state processes.

We now present a few equations which resemble the expansion theorem for CCS which indicate that processes involving $\parr$ negation and $\otimes$ can be converted to a process using & and $\oplus$. There are four cases to consider; viz., P and Q standard forms over &, P and Q standard forms over $\oplus$, P a standard form over & and Q a standard form over $\oplus$ and finally P a standard form over $\oplus$ and Q a standard form over &. This contrasts with the single case for the expansion theorem in CCS; where there is only one form of choice.

Two of the four cases are covered by the following.

Let P be $\&_{k \in K}\ \mu_o^k; P^k$ and let Q be $\bigoplus_{j \in J} \mu_i^j; Q^j$. The following are sound equations for the above processes.

$$\mathrm{P}^{\perp} = \bigoplus_{k\in K} \mu_i^k; (P^k)^{\perp} \text{ where } (\overline{\mu_i^k} = \mu_o^k).$$

$$(\mathrm{P} \wp \mathrm{Q}) = \bigoplus_{j\in J} \mu_i^j;(\mathrm{P} \wp Q^j) \qquad (\mathrm{P} \otimes \mathrm{Q}) = \&_{k\in K}\, \mu_o^k; (P^k \otimes \mathrm{Q})$$

$$(\mathrm{P} \,\&\, \mathrm{Q}) = \mathrm{P} \qquad (\mathrm{P} \otimes \mathrm{Q}) = \mathrm{Q}$$

The third case, where both processes are choice over output actions is given below.

Let P be $\&_{k\in K}\, \mu_o^k; P^k$ and let Q be $\&_{j\in J} \mu_o'^j; Q^j$. The following are sound equations for the above processes.

$$(\mathrm{P} \wp \mathrm{Q}) = \&_{j\in J, k\in K} (\mu_o^k \cdot_O \mu_o'^j); (P^k \wp Q^j)$$

$$(\mathrm{P} \otimes \mathrm{Q}) = (\&_{k\in K}\, \mu_o^k; (P^k \otimes \mathrm{Q})) \,\&\, (\&_{j\in J}\, \mu_o'^j; (\mathrm{P} \otimes Q^j))$$

The final case, where both processes are choice over input actions is given below. Let P be $\bigoplus_{k\in K} \mu_i^k; P^k$ and let Q be $\bigoplus_{j\in J} \mu_i'^j; Q^j$. The following are sound equations for the above processes.

$$(\mathrm{P} \otimes \mathrm{Q}) = \bigoplus_{j\in J, k\in K} (\mu_i^k \cdot_I \mu_i'^j); (P^k \otimes Q^j)$$

$$(\mathrm{P} \wp \mathrm{Q}) = (\bigoplus_{k\in K} \mu_i^k; (P^k \wp \mathrm{Q})) \oplus (\bigoplus_{j\in J} \mu_i'^j; (\mathrm{P} \wp Q^j))$$

The soundness of the above equations follows from theorem 1 and the fact that input actions are asynchronous with respect to $\wp$ and that they resolve $\oplus$ choice while output actions are asynchronous with respect to $\otimes$ and that they resolve & choice.

There is only one outstanding matter before the completeness result can be stated. Consider the process $(a_o; P \oplus b_o; Q)$ where a_o and b_o are distinct. It is easy to check using the operational rules that the process can exhibit no action. Hence to obtain a good standard form we consider a new process (**Nil**) which cannot exhibit any action. **Nil** is like the **0** process in CCS. We will assume that **Nil** is also in standard form.

Now we consider the remaining class of equations.

$$\mu_o; P \oplus \mu_o; Q = \mu_o; (P \oplus Q) \qquad \mu_i; P \mathbin{\&} \mu_i; Q = \mu_i; (P \mathbin{\&} Q)$$

$$\mu_o; P \oplus \mu'_o; Q = \mathbf{Nil} \qquad \mu_i; P \mathbin{\&} \mu'_i; Q = \mathbf{Nil}$$

$$\mu_i; P \oplus \mu_o; Q = \mu_i; P \qquad \mu_i; P \mathbin{\&} \mu_o; Q = \mu_o; Q$$

$$\mu_o; P \parr \mathbf{Nil} = \mathbf{Nil} \qquad \mu_i; P \parr \mathbf{Nil} = \mu_i; (P \parr \mathbf{Nil})$$

$$\mathbf{Nil} \oplus \mu_o; P = \mathbf{Nil} \qquad \mathbf{Nil} \mathbin{\&} \mu_i; P = \mathbf{Nil}$$

$$\mathbf{Nil} \mid \mathbf{Nil} = \mathbf{Nil} \text{ (} \mid \text{ any binary combinator)}$$

Proposition 10. *Every process can be converted to a standard form by using the equations discussed earlier.*

Proof. The proof follows a case by case analysis for each combinator. □

We show a few examples to illustrate this process.

Example 4. Consider the process $a_o;\mathbf{1} \parr a_i; \bot$. This is equal to $a_i; (a_o;\mathbf{1} \parr \bot)$. This can then be transformed into $a_i; a_o; \mathbf{1}$.

Example 5. Consider the process $a_i; a_o;\mathbf{Nil} \parr b_i; a_o;\mathbf{Nil}$. The two input actions can be interleaved giving rise to $a_i; b_i; a_o;\mathbf{Nil} \oplus$ to $b_i; a_i; a_o;\mathbf{Nil}$. This follows from the expansion result presented as the final case and the fact that **Nil** combined with **Nil** results in **Nil**.

Proposition 11. *The axiomatisation is complete. That is, if (P $\sim$ Q) then (P = Q).*

Proof. It suffices to consider processes in standard form. Let P be $\&_{k \in K}\, \mu_o^k; P^k$. If (P $\sim$ Q), Q has to be in form $\&_{j \in J}\, \mu_o^j; Q^j$. Now it is easy to show that (P & Q) = Q = P. □

6 Conclusion and future work

In this paper we have presented a process algebra where the process combinators have been inspired by the connectives in linear logic. By considering a "merge" operation, we have been able to give a direct operational meaning to $\parr$. Although we have presented a sound and complete axiomatisation of the bisimulation equivalence, we still need a good notion of provability and validity in the process algebraic framework. Once that is available, issues such as soundness and completeness with respect to the original sequent calculus can be addressed. We have ignored the exponentials and a suitable definition for them needs to be found.

References

1. V. Alexiev. Applications of linear logic to computation: An overview. Technical Report TR93-18, University of Alberta, Edmonton, Alberta T6G 2H1, September 1994.
2. G. Bellin and P. Scott. On the π-calculus and linear logic. *Theoretical Computer Science*, 135(1):11–65, 1994.
3. U. H. Engberg and G. Winskel. Completeness Results for Linear Logic on Petri Nets. In *MFCS 93*, volume LNCS 711, pages 442–452. Springer Verlag, 1993.
4. J-Y. Girard. Linear logic. *Theoretical Computer Science*, 50(1):1–102, 1987.
5. J.-Y. Girard. Linear logic: Its syntax and semantics. In J.-Y. Girard, Y. Lafont, and L. Regnier, editors, *Advances in Linear Logic*, pages 1–42. Cambridge University Press, 1995. Proceedings of the Workshop on Linear Logic, Ithaca, New York, June 1993.
6. N. Kobayashi and A. Yonezawa. Asynchronous communication model based on linear logic. *Formal Aspects of Computing*, 7(2):113–149, 1995.
7. R. Milner. *Communication and Concurrency.* Prentice Hall International, 1989.
8. G. D. Plotkin. A Structural Approach to Operational Semantics. Technical Report DAIMI FN-19, Computer Science Department, Aarhus University, 1981.

Set Automata

Klaus-Jörn Lange and Klaus Reinhardt*

Wilhelm-Schickhard Institut für Informatik, Universität Tübingen
Sand 13, D-72076 Tübingen, Germany
e-mail: {lange,reinhard}@informatik.uni-tuebingen.de

Abstract. In order to characterize certain formal languages capturing some syntactic aspects of higher programming languages we introduce automata with the storage type *set*, called *set automata*. We show the corresponding language class to have an NP-complete membership problem and a decidable emptiness problem.

1 Introduction

One of the main issues of formal languages is to model the syntax of higher programming languages or parts of it. This leads for instance to grammars generating parenthesis expressions or representing recursive structures. A limit of these approaches has often been the existence of certain constraints which have to be fullfilled by the syntactical objects. A very typical example is the problem of declaring variables in programs written in languages like PASCAL. In its most simple form without concerning recursive or block structures it can be formulated as: *Each variable used in a program has to be declared exactly once!* Formally, this means that there exists a mapping from the set of all occurrences of variables being used in a program to the set of all declarations of variables in that program. This condition of being a mapping can be split into the two subconditions *totality on the domain* and *uniqueness on the range*, that is, first, that for each element of the domain (here the set of all occurences of variables) there is a corresponding element in the range (here the corresponding declaration), and second, that for no element in the domain there are two or more corresponding elements in the range (which here means that no variable is declared twice or more). These two conditions can be expressed in terms of the following two formal languages:

$$L_{td} = \left\{ x_1 c x_2 \cdots c x_n d y_1 c y_2 \cdots c y_m \;\middle|\; \begin{array}{l} m, n \geq 1, x_i, y_j \in \{a,b\}^*, \\ \bigwedge_{1 \leq i \leq n} \bigvee_{1 \leq j \leq m} x_i = y_j \end{array} \right\}$$

$$\text{and } L_{ur} = \left\{ y_1 c y_2 \cdots c y_m \;\middle|\; m \geq 1, y_i \in \{a,b\}^*, \bigwedge_{i \neq j} y_i \neq y_j \right\}.$$

The set L_{ur} is also known as the problem of *Element Distinctness*. On the one hand, being elements of AC^0 both languages have a very simple membership

* This research has been supported by the DFG Project La 618/3-1 KOMET.

problem. On the other hand, capturing them in terms of formal languages, that is describing by grammars or automata seems to be quite difficult. Fischer, addressing the problem of well-declaration, was able to construct an indexed grammar generating L_{td} ([2]). In fact, L_{td} is even an element of $ETOL$. But both these grammar types have in general an *NP*-complete membership problem.

It is the aim of this work to represent L_{ur} by an automaton type sharing typical properties of formal languages. In particular, we look for an automaton type generating a class of languages constructively closed under morphisms, inverse morphisms, and intersection with regular sets and having a decidable emptiness problem enabling the existence of pumping or copying results.

For this reason we introduce *set automata* which are automata with access to a datastructure of type set. A set automaton has a read-only one-way input tape, a finite set of states, and write-only one-way output tape, called the *query tape.* It can perform three different kind of steps: reading the next symbol from the input tape, writing another output symbol on the query tape, and *asking a question*, which means that the word currently written on the query tape (called the query) is compared with all queries asked previously in this computation. If it coincides with a previous query, the *answer* is yes, otherwise no. The answer is given to the automaton in terms of its finite states. Simultaneously, the query tape is erased.

It turns out that the class $\mathcal{SET}$ of all *set languages*, that is of all languages accepted by set automata, forms a trio, that is, is closed under morphisms, inverse morphisms, and intersection with regular sets and that these closure properties are constructive. As our first main result of this work we show the decidability of the emptiness problem for set automata. Our second main result proves the *NP*-completeness of the membership problem even if we restrict the automaton to accept only if all queries have been answered negatively.

We assume the reader to be acquainted with the basic notions and results of formal language theory as they are contained for instance in the books of Hopcroft and Ullman or by Salomaa ([3, 4]). Throughout of this paper we denote the empty word by λ and the cardinality of a set M by $|M|$.

For lack of space we can only give indications for most of our constructions and proofs.

2 Set automata

This section starts with the formal definitions of set automata and of set languages. After some results concerning formal language aspects it continues with questions of decidability and complexity.

2.1 Definition of the set-automaton

A set-automaton has a nondeterministic finite control, a one-way input-tape and a one-way output-tape, which we call the *query-tape.* There are three kinds of elementary operations: The reading of an input symbol (δ_r–transitions), the writing of a symbol to the query-tape (δ_w–transitions) and the posing of a query (δ_q–transitions). This posing of a query means that the following state depends on whether the word on the query-tape is in the set of words, which have been asked before. Simultaneaously the word is included in this set and the query-tape is cleared. Accordingly we have two special kinds of states: the *query-states* and the (positive and negative) *answer-states*. For technical reasons, we regard both the initial state and the final states as answer-states.

Definition 1. A *set-automaton* is a 8-tuple $A = (Z, \Sigma, \Gamma, \delta, Q, S, E, z_0)$, where

- Z is a finite set of *states*
- Σ is a finite *input-alphabet*,
- Γ is a finite *query-alphabet*,
- $Q \subseteq Z$ is a set of *query-states*
- $S \subseteq Z \setminus Q$ is a set of *answer-states*,
- $E \subseteq S$ is a set of *accepting states*,
- $z_0 \in S$ is an *initial state* and
- $\delta = \delta_r \cup \delta_b \cup \delta_q$ with

$$\delta_r \subset (Z \setminus Q) \times \Sigma \times (Z \setminus S),\ \delta_w \subset (Z \setminus Q) \times \Gamma \times (Z \setminus S),\ \delta_q \subset Q \times S \times S$$

is a finite set of *transitions*.

To describe calculations of a set-automaton, we have to define the notion of a *configuration*. A configuration consists of the actual state, the rest of the input-tape, the part of the query-tape written so far and the set of the priviously asked queries.

Definition 2. Let $A = (Z, \Sigma, \Gamma, \delta, Q, E, S, z_0)$ be a set-automaton.

1. A *configuration* of A is an element of $Z \times \Sigma^* \times \Gamma^* \times 2^{\Gamma^*}$.
2. A configuration (z', v', α', M') of A is the *successor* of a configuration (z, v, α, M) (written as $(z, v, \alpha, M) \vdash (z', v', \alpha', M')$) if and only if one of the following holds:
   Reading an input-symbol: There is a transition $(z, a, z') \in \delta_r$ and it holds $v = av'$, $\alpha = \alpha'$ and $M = M'$,
   Writing to the query-tape: There is a transition $(z, x, z') \in \delta_w$ and it holds $\alpha x = \alpha'$, $v = v'$ and $M = M'$, or
   Asking a question: There is a transition $(z, z_+, z_-) \in \delta_q$ and it holds either (positive answer) $\alpha \in M$, $M = M'$ and $z' = z_+$ or (negative answer) $\alpha \notin M$, $M' = M \cup \{\alpha\}$ and $z' = z_-$. In both cases we have $v = v'$ and $\alpha' = \lambda$.

   We use $\vdash^*$ for the transitive closure of the relation $\vdash$.
3. The *language accepted by* A is

$$L(A) = \left\{ w \in \Sigma^* \,\middle|\, \bigvee_{z \in E} \bigvee_{\alpha \in \Gamma^*} \bigvee_{M \subset \Gamma^*} (z_0, w, \lambda, \emptyset) \vdash^* (z, \lambda, \alpha, M) \right\}.$$

In the following we call a language accepted by a set automaton a *set language*. The class of all set languages is denoted by $\mathcal{SET}$.

2.2 Formal language properties

From the description of set-languages based on automata, we get directly some closure properties, which we state here without proof. We mention in passing that they are constructive in the sense that given descriptions of a set automaton A and of a morphism h or of a regular set R, the description of a set automaton accepting $h(L(A))$, $h^{-1}(L(A))$, or $L(A) \cap R$ is computable by a recursive function.

Theorem 3. *The family $\mathcal{SET}$ of set languages forms a TRIO, which means, they are closed under (possibly erasing) homomorphisms, inverse homomorphisms, and intersections with regular languages.*

In the following we present a generator playing the same rôle for $\mathcal{SET}$ that the Dyck sets have for the context-free languages. Just as the Dyck sets reflect the proper use of a push down store in terms of matching push- and pop-moves, we consider the set of all correct set computations with matching positive and negative queries to a set built by these queries.

Definition 4. The language of correct set computations is defined by:

$$L_{+,-} = \left\{ y_1\delta_1 c y_2 \delta_2 \cdots c y_m \delta_m \;\middle|\; \begin{array}{l} m \geq 1, y_i \in \{a,b\}^*, \delta_i \in \{+,-\}, \\ \bigwedge_{1 \leq i \leq m} (\delta_i = +) \Leftrightarrow \left(\bigvee_{j<i} y_j = y_i \right) \end{array} \right\}$$

Let us observe, that the original language L_{ur} as considered in the introduction corresponds to the subset of those words in $L_{+,-}$ where no $+$ appears, this means that no question is answered positive. By converting positive and negative answers we could view L_{ur} as conjunctive restriction of $L_{+,-}$, if we regard set-automata as oracle machines with set oracles.

Theorem 5. *$L_{+,-}$ is a trio-generator of the class of set-languages, this means for every set-language L there are two homomorphisms f and g and a regular language R such that $L = f(g^{-1}(L_{+,-}) \cap R)$.*

Proof. Obviously, $L_{+,-}$ is a set language. Thus, Theorem 3 implies the inclusion $\mathrm{TRIO}(L_{+,-}) \subseteq \mathcal{SET}$. Now, it suffices to show that every set-language L can be reduced to $L_{+,-}$ by a finite transducer. Such a finite transducer simulates the set-automaton accepting L by writing the symbols which it writes to the query tape to the output tape. If the set-automaton enters the query state, the finite transducer nondeterministically guesses the answer and either it writes a $+$ to the output tape and continues the simulation in the positive answer state or it writes a $-$ to the output tape and continues the simulation in the negative answer state. Obviously, the nondeterministic transducer can output an element of $L_{+,-}$ if and only if its input word belongs to L. □

It is easy to see that $\mathcal{SET}$ is closed under union and catenation, as well. Closure under union is provided by the machine's nondeterminism. The simple idea to prove closure under catenation is to use two disjoint query-alpabets when simulating the two underlying set automata. But this method does not work for the operation of Kleene's closure. We leave it as an open question whether $\mathcal{SET}$ is closed under Kleene's star which would make $\mathcal{SET}$ a full AFL. A possible counterexample is the language $(L_{+,-}\$)^*$, the marked Kleene's closure of $L_{+,-}$.

3 Decidability and complexity

Theorem 6. *The emptyness problem for set automata is decidable.*

Proof. The gist of the following construction is to separate in a computation of a set automaton the query transitions from the pure reading and writing steps. The latter, without access to the storage medium, can be described by rational transductions.

Let $A = (Z, \Sigma, \Gamma, \delta, Q, E, S, z_0)$ be a set-automaton. Let in the following m be the size of $S \times Q$, that is, $m := |S| \cdot |Q|$. Obviously we can regard A without the δ_q-transitions as a finite automaton with input and output, calculating a rational transduction $\tau_A \subset Z \times \Sigma^* \times \Gamma^* \times Z$ (see for example [1]). The relation τ_A consists of exactly those tuples (z, w, α, z') such that there is a path from z to z' of δ_r– and δ_w–transitions reading w and writing α. For every answer-state $s \in S$, and every query-state $q \in Q$, the set

$$R_{s,q} = \left\{ \alpha \in \Gamma^* \,\middle|\, \bigvee_{w \in \Sigma^*} (s, w, \alpha, q) \in \tau_A \right\}$$

is regular. Therefor for every subset $N \subseteq S \times Q$ the following set is regular as well:

$$R_N = \left\{ \alpha \in \Gamma^* \,\middle|\, \bigwedge_{(s,q) \in S \times Q} \alpha \in R_{s,q} \Leftrightarrow (s,q) \in N \right\}.$$

The 2^m (possibly empty) sets $R_N, N \subseteq S \times Q$, form a partition of Γ^*.

Let $F = \{N \subseteq S \times Q \,|\, R_N \text{ is finite}\}$ and $I := (S \times Q) \setminus F$. We now define the set of *global states*:

$$K := S \times \{0, \infty\}^{|I|} \times \prod_{N \in F} \{0, 1, \cdots, |R_N|\}.$$

Observe, that K is finite. We let the *global start state* be $(z_0, \overbrace{0, 0, \cdots, 0}^{m-times})$. A *global end state* is a global state $(s, i_1, \cdots, i_m)$ with $s \in E$. If the set-automaton M is in an answer state s after asking a question, then the global state not only contains the state s but also for every $N \subseteq S \times Q$ the number of previously asked words which are in R_N. For the case of an infinite R_N, we just store whether no $(i_N = 0)$ or at least one $(i_N = \infty)$ word of R_N was asked.

We now regard the elements of K as nodes of a directed graph having the following edges:

$$(s, i_1, i_2, \cdots, i_m) \longrightarrow (s', i'_1, i'_2, \cdots, i'_m)$$

if and only if there is a $(q, s_1, s_2) \in \delta_q$ with:

Case 1: 'Query with positive answer'

$$s' = s_1 \wedge \bigwedge_{N \subseteq S \times Q} (i'_N = i_N \wedge [(s,q) \in N \Rightarrow i_N \neq 0])$$

or

Case 2: 'Query with negative answer'

$$s' = s_2 \wedge \bigwedge_{N \subseteq S \times Q} ((s,q) \notin N \Rightarrow i'_N = i_N) \wedge$$

$$\bigwedge_{N \subseteq S \times Q} [((s,q) \in N \wedge N \in I) \Rightarrow i'_N = \infty] \wedge$$

$$\bigwedge_{N \subseteq S \times Q} [((s,q) \in N \wedge N \in F) \Rightarrow (i_N < |R_N| \wedge i'_N = i_N + 1)] .$$

The idea behind this construction is that we represent valid computations of A by paths in a graph. Sequences of reading and writing steps are captured by the sets $R_{s,q}$ and R_N. Query steps leading from s via q to s_1 or S_2 lead to an update of the counters. Positively answered queries, for instance, mean that all counters remain unchanged since nothing is inserted in the history of asked questions, but that for each set N containing (s,q) the counter for R_N must be nonzero, since there must exist a query leading from s to q which was posed previously. Hence, there is an accepting calculation of A if and only if there is a way in the graph of global states from the global start state to a global end state. The decidability follows from the finiteness of the set K. □

Because $w \in L(A)$ holds if and only if $L(A) \cap \{w\}$ is not empty, we get as a consequence of the effective closure of the set-languages under intersection with regular sets:

Corollary 7. *The word problem for set-languages is decidable.*

Theorem 8. *There exist NP-complete set languages. Hence the membership question for set languages is NP-hard.*

Proof. From the coding F of a boolean formula in conjunctive normal form we can compute (in logarithmic space) the word $V\$F$ where V is a list of all variables occuring in F. A set automaton A first reads the partial word V and guesses for every variable x_i its value by either posing the query x_i1 or x_i0, which, of course, are all answered negatively. Afterwards A reads the formula F and guesses for every clause the fulfilling literal. A proves the correctness by posing queries of the form x_i1 or x_i0, which now all have to be answered positive. In this case A accepts, otherwise A rejects. □

Theorem 9. *The word problem for set automata is NP-complete.*

Proof. Because of Theorem 8 it suffices to show that if the set automaton A accepts an input x, then it can do so making only $p(|x|)$ steps for a polynomial p. Then a nondeterministic polynomial timebounded Turing machine can simulate A.

We use the terminology and parameters of Theorem 6. Let $n = |x|$ be the length of the input x. Continuing the construction of Theorem 6 let $g := |K|$ be the number of global states. In the following, we distinguish in a computation of A between *special queries* which are generated by both reading and writing transitions and *normal queries* which are built using writing steps, only. During an accepting calculation of A on x, A can ask at most n special queries. For each positively answered special query asking the word α, we regard that query which asked this query word α the first time as a special query, as well. This means we have to consider at most $2n$ special queries. If the calculation of A on x reaches a global state twice without asking a special query inbetween, then this part of the calculation can be skipped and later positive queries related to negative queries in this part can be replaced by an earlier query in the same R_N. (Such a query must already exist in the set, otherwise the global state would have changed.) In this way we can reduce an accepting calculation of A on x to making at most $2gn = O(n)$ queries.

Let $(\beta_i)_{1 \leq i \leq m}$ for some $m = O(n)$ be the sequence of queries asked by A during an arbitrary, but fixed successful computation on input x. Now let $\Omega := \{\beta_i | 1 \leq i \leq n\}$ be the set of different query words. For each $\gamma \in \Omega$ let $index(\gamma)$ be the index of the first query asking for γ which inserted γ into Ω and had been answered negatively, that is, $index(\gamma) := \text{MIN}\{i | \beta_i = \gamma\}$. We now decompose each query β_i into pieces such that no input symbol has been read while writing a piece. Then we decompose each $\gamma \in \Omega$ into $\gamma = \gamma_0\gamma_1 \cdots \gamma_t$ by overlaying the decompositions of all queries β_i with $\beta_i = \gamma$. Since there are at most n input symbols to be read, we have $t = O(n)$.

For each $z, z' \in Z$ and for each $M \subseteq Z \times Z$ we consider the sets:

$$T_{z,z'} = \{\alpha \in \Gamma^* \,|(z, \lambda, \alpha, z') \in \tau_A\}$$

and

$$T_M = \bigcap_{(z,z') \in M} T_{z,z'}.$$

We partition $Z \times Z$ into the two parts *FIN*:$= \{M \subseteq Z \times Z | T_M \text{ is finite}\}$ and its complement *INF*:$= \{M \subseteq Z \times Z | T_M \text{ is infinite}\}$. We now let c_1 be the length of the longest word in some finite T_M, that is, $c_1 := \text{MAX}_{M \in FIN}\text{MAX}_{v \in T_M}|v|$. Further on, we let c_2 be the minimal integer such that the minimal period of the word length of each infinite T_M is a divisor of c_2, that is, there exists a positive integer c_3, minimal with respect to c_2, such that for all $M \in INF$ and for all $v \in T_M$ with $|v| > c_3$ there exist words $v', v'' \in T_M$ such that $|v'| = |v| - c_2$ and $|v''| = |v| + c_2$. We now let c be the maximum of c_1, c_2, and c_3.

The decomposition $\gamma = \gamma_0\gamma_1 \cdots \gamma_t$ induced by the decomposition of some queries $\beta_{i_1}, \cdots, \beta_{i_r}$ means for each γ_j the existence of pairs $(z_1, z_1'), \cdots, (z_r, z_r')$ such that γ_j has to be a member of $T_{z_i,z_i'}$ for each $1 \leq i \leq r$. For each γ_j let M_j be the set of these pairs of states. That is, the decomposition of γ asks for $\gamma_j \in T_{M_j}$ for each j. If now $M_j \in FIN$ for each $0 \leq j \leq t$ then we know $|\gamma| \leq (t+1) \cdot c_1 = O(n)$. Now let us assume, that $M_j \in INF$ for some $0 \leq j \leq t$. By the definition of c_2 we can replace each γ_j for which T_{M_j} is infinite by some γ_j' of length $|\gamma_j'| \leq c_3$. This would result in a query candidate γ' of length $|\gamma'| \leq (t+1) \cdot c = O(n)$. The problem is that this modified query could interfere with other modified or unmodified elements of Ω. To avoid this we will increase the length of γ' again such that we can be sure that all pumped queries have different length. So let $i := \mathit{index}(\gamma)$. We now simply increase the length of γ' in steps of length c_2 until we have $(n+j) \cdot c < |\gamma'| \leq (n+j+1) \cdot c$. This prevents any interference of the modified queries with each other while keeping $|\gamma'| \leq (n + \mathit{index}(\gamma) + 1) \cdot c = O(n)$.

Thus we know, that if there is an accepting computation, there is also one with $O(n)$ queries of length $O(n)$ which can be guessed in time bounded polynimial in n. □

Theorem 10. *The word problem for set languages accepted by set automata with only negative answers which is the trio generated by L_{ur} is NP-complete.*

Proof. Let F be the coding of a formula in KNF. By a log-reduction we can construct the word $x_1\$x_1\$\$x_1\$\$\$\ldots x_1\$^{k_1}x_2\$\ldots x_n\$^{k_n}F'$,where $k_i = \#_{x_i}(F)$ is the number of occurances of the variable x_i in F and F' is formula F modified in a way that the jth occurence of variable x_i is coded by $x_i\#^j$. While reading the first part of the input, a set automaton A guesses an assignment for every variable x_i and either puts all the strings $x_i1, \cdots, x_i1^{k_i}$ into the set if the assignment is false or all the strings $x_i0, \cdots, x_i0^{k_i}$ if the assignment is true. While reading F', for every clause A guesses the fulfilling literal and verifies the correctness by puting the query x_i1^j or x_i0^j, where j is given by the coding. So if and only if for example x_i is true, then it is possible to ask negatively for x_i1^j, which is still not in the set. The automaton A only accepts if all answers to queries are negative. □

Finally, we mention without proof that in the deterministic case we get a *P*-complete word problem:

Theorem 11. *The word problem for deterministic set automata is P-complete.*

4 Discussion

In designing the notion of a set automaton it was essential to restrict the model severely in order to end up with a decidable emptiness problem. On the other hand, this led to a rather 'uncomfortable' model. It is still open, whether $\mathcal{SET}$ is closed under Kleene's star. One possibility to gain a full AFL is to give set automata the possibility of emptying the sets of asked queries. In this way we come to the trio generated by the set $(L_{+,-}\$)^*$. Here the symbol \$ indicates the operation of emptying the set of old queries. This trio is obviously closed under the $*$-operation while still having a decidable emptiness problem.

On the other hand, with an *NP*-complete membership problem set automata are of a rather high complexity. If we enhance the model by allowing to put a query without puting the word into the set of previous questions, the complexity results remain the same. But if we also allow to delete the word on the query tape from the set, then the infinite sets R_N constructed in the proof of Theorem 6 behave like counters without zero test which means that the word and the emptyness problem become equivalent to the reachability problem in Petri nets.

Things change if we allow the set automata to put a query without emtying the query tape (and afterwards continue the query word). In this case the word and the emptyness problem become undecidable since it is possible to simulate automata with two counters by set automata of the extended type, even if there is only a single query symbol: The contents n_c of counter $c \in \{0,1\}$ is represented by the set containing the words $a^c, a^{c+4}, \dots a^{c+(o_c)4}$ and $a^{c+2}, a^{c+6}, \dots a^{c+(o_c+n_c)4+2}$. To perform a zero test of counter c, the set automaton asks all queries $a^c, a^{c+2}, \dots a^{c+(o_c+1)4+2}$, where exactly the last two must be answered negative. To perform a decrement of counter c, the set automaton asks all queries $a^c, a^{c+2}, \dots a^{c+(o_c+1)4+2}$, where exactly the query $a^{c+(o_c+1)4}$ must be answered negative. To perform a increment of counter c, the set automaton asks all queries $a^{c+2}, a^{c+6}, \dots a^{c+(o_c+n_c)4+6}$, where only the last query must be answered negative.

References

1. J. Berstel. *Transductions and Context-Free Languages.* Teubner Verlag, Stuttgart, 1979.
2. M.J. Fischer. Grammars with macro-like production. Ph.D. thesis, Harvard Univ., 1968.
3. J. Hopcroft and J. Ullman. *Introduction to Automata Theory, Language, and Computation.* Addison-Wesley, Reading Mass., 1979.
4. Arto Salomaa. *Formale Sprachen.* Springer Verlag, 1974.

A Condition for a Normal Semicycle to Separate an Orientable 3-Graph

Charles H. C. Little and Bruce van-Brunt

Massey University, Palmerston North, New Zealand

Abstract. In [1] it is proved that a b-cycle separates a 3-graph if and only if it induces a normal b-cycle whose sides are not linked by a semi-path. Here we present a new condition for a normal semicycle to separate an orientable 3-graph. It is an attempt to encapsulate the topological idea that if a simple closed curve C does not separate a surface into two regions, then any two points not on C may be joined by another curve which crosses C an even number of times.

1 Introduction

We denote the vertex set and edge set of a graph G by VG and EG respectively. If $S \subseteq EG$, we write $G[S]$ for the subgraph of G induced by S. Paths and circuits are regarded as sets of edges. If S is a path or a circuit, then we write $VS = VG[S]$.

A study of 3-graphs is undertaken in [1]. For the sake of completeness the relevant definitions from [1] are reproduced below, but the reader is referred to [1] for examples and motivation.

A *3-graph* is a cubic graph with a proper edge colouring in three colours, which we call *red, yellow* and *blue*. Circuits which use just two of the three colours are called *bigons*. A non-empty set C of edges of a 3-graph G is called a *b-cycle* if C is the union of disjoint circuits with at least one blue edge in each. A set $\mathcal{C}$ of b-cycles *induces* a b-cycle D if each blue edge of D is an element of $\bigcup \mathcal{C}$. (By $\bigcup \mathcal{C}$ we mean the union of the members of $\mathcal{C}$.) In the case where $\mathcal{C} = \{C\}$ for some b-cycle C, we also say that C induces D. The *boundary space* of G is the subspace of the cycle space spanned by the set of bigons. A b-cycle is said to *separate* if it induces a b-cycle which is a member of the boundary space. A b-cycle is *connected* if it is a circuit.

If C is a b-cycle, its *necklace* $N(C)$ is the set of red-yellow bigons it meets. The elements of its necklace are the *beads* of C. The *poles* of a bead B (with respect to C) are the vertices of B incident on a blue edge of C. If C is connected and each bead has just two poles, then C is a *semicycle*. A cycle C is *normal* if no blue edge of C has neighbours in C of distinct colours. The *conjugate* $\overline{C}$ of a semicycle C is defined as $C + (\bigcup N(C))$. (Throughout this paper, a sum of sets is taken to be their symmetric difference.) We define $I(C) = C \cap (\bigcup N(C))$ and $O(C) = \overline{C} \cap (\bigcup N(C))$, and refer to $I(C)$ and $O(C)$ as the *sides* of C.

Similar definitions may be made for paths. If P is a path, its *necklace* $N(P)$ is the set of red-yellow bigons it meets. Beads, poles, conjugates and normality are defined as for semicycles. In addition, a red-yellow bigon passing through an end of P is called a *terminal bead* of P. The beads in $N(P)$ are sometimes described as *internal*. A path is *elementary* if each bead (including the terminal beads) has at most two poles. An elementary path is a *semipath* if its terminal edges are blue.

A b-cycle C and a semipath P are said to *miss* if they are disjoint and no internal bead of P is a bead of C. If C and P miss, then P is said to *link* the sides of C if one end is in $VG[I(C)]$ and the other in $VG[O(C)]$. The following theorem is proved in [1, 2].

Theorem 1. *A necessary and sufficient condition for a b-cycle C to separate a 3-graph is for C to induce a normal b-cycle C' such that no semipath links the sides of C'.*

A 3-graph is *orientable* if it is bipartite. In this case, for any semicycle there is a normal semicycle with the same blue edges.

The definitions above are motivated by topological considerations. The connection with embeddings of graphs in surfaces is explained in detail in [1]. A 3-graph is a *gem* if every red-blue bigon is a square. Gems can be considered to model embeddings of graphs in surfaces. Under this interpretation, a semicycle in a gem corresponds to a path of length 1 or a circuit in the embedded graph. It is clear, at least intuitively, from the discussion in [1] that if a semicycle C in a gem does not separate, then in the embedded graph one may cut along the corresponding circuit (or path) without disconnecting the surface. The goal of this paper is to formulate a theorem to this effect in the general setting of orientable 3-graphs.

We shall assume given a connected orientable 3-graph G. We begin by considering two sets P_1 and P_2 of edges. The idea is to define a certain function $\mu(P_1, P_2)$ which takes the value 0 or 1. We then investigate the condition which is necessary to ensure that the arguments of μ commute. Specialising to the case where P_1 is a semicycle and P_2 an elementary path, we then modify μ so as to obtain another function κ which also takes the value 0 or 1. We then define P_1 and P_2 to cross if $\kappa(P_1, P_2) = 1$. Finally we show that if C is a normal semicycle which does not separate, then two edges which are not in C or a bead of C may be joined by a path which does not cross C.

2 The crossing index

If S is a set of edges in a 3-graph, we denote the collection of blue edges in S by $\beta(S)$. If T is a set of vertices, then we denote by ∂T the set of edges with just one end in T. Let $P_1, P_2 \subseteq EG$. We define

$$\mu(P_1, P_2) = |\partial V P_1 \cap \beta(P_2)| \ .$$

(In this paper all numerical equations are to be read as congruences modulo 2. Thus $\mu(P_1, P_2) \in \{0, 1\}$.)

Figure 1 gives a 3-graph in which two paths, P_1 and P_2, are defined by the edges labelled P_1 and P_2 respectively. In this example we have $\mu(P_1, P_2) = 1$ and $\mu(P_2, P_1) = 0$.

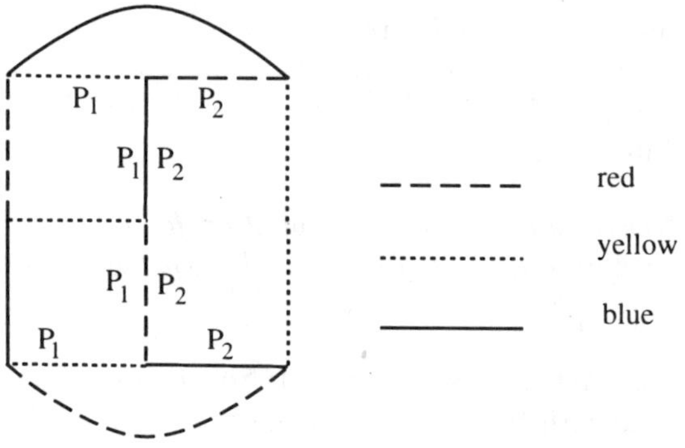

Fig. 1.

We give a condition for commutativity.

Lemma 2. *Suppose that* $|\partial(VP_1 \cap VP_2) \cap \beta(P_1 \cup P_2)|$ *is even. Then* $\mu(P_1, P_2) = \mu(P_2, P_1)$.

Proof. Choose $e \in \partial(VP_1 \cap VP_2) \cap \beta(P_1 \cup P_2)$. Then e is blue. Let e join vertices v and w, where $w \in VP_1 \cap VP_2$. Thus $v \notin VP_1 \cap VP_2$. Suppose without loss of generality that $v \notin VP_2$. Therefore $e \notin P_2$, and so $e \in P_1$. Moreover $e \in \partial VP_2$, so that $e \in \partial VP_2 \cap \beta(P_1)$. Hence $e \in (\partial VP_1 \cap \beta(P_2)) \cup (\partial VP_2 \cap \beta(P_1))$, regardless of whether $v \notin VP_1$ or $v \notin VP_2$.

Conversely, choose $e \in (\partial VP_1 \cap \beta(P_2)) \cup (\partial VP_2 \cap \beta(P_1))$. Suppose without loss of generality that $e \in \partial VP_1 \cap \beta(P_2)$. Then $e \in \beta(P_1 \cup P_2)$. Let e join vertices v and w, where $v \in VP_1$. Then $v \in VP_1 \cap VP_2$ and $w \in VP_2 - VP_1$. Hence $e \in \partial(VP_1 \cap VP_2)$. Thus $e \in \partial(VP_1 \cap VP_2) \cap \beta(P_1 \cup P_2)$ regardless of whether $e \in \partial VP_1 \cap \beta(P_2)$ or $e \in \partial VP_2 \cap \beta(P_1)$.

We conclude that

$$\begin{aligned}\partial(VP_1 \cap VP_2) \cap \beta(P_1 \cup P_2) &= (\partial VP_1 \cap \beta(P_2)) \cup (\partial VP_2 \cap \beta(P_1)) \\ &= (\partial VP_1 \cap \beta(P_2)) + (\partial VP_2 \cap \beta(P_1))\end{aligned}$$

since $\partial VP_1 \cap \beta(P_2) \cap \partial VP_2 \cap \beta(P_1) = \emptyset$. Therefore, by hypothesis,

$$\begin{aligned}0 &= |\partial(VP_1 \cap VP_2) \cap \beta(P_1 \cup P_2)| \\ &= |\partial VP_1 \cap \beta(P_2)| + |\partial VP_2 \cap \beta(P_1)| \\ &= \mu(P_1, P_2) + \mu(P_2, P_1)\ .\end{aligned}$$

□

If H_1 and H_2 are paths or circuits, then an H_1H_2-*arc* is a path of maximal length included in $H_1 \cap H_2$.

Lemma 3. *Let H_1, H_2 be normal cycles or normal paths in G. Suppose that no end of H_1 or H_2 is a vertex of the other. Then $\mu(H_1, H_2) = \mu(H_2, H_1)$.*

Proof. Each vertex of $VH_1 \cap VH_2$ belongs to an H_1H_2-arc. Let A be such an arc. By normality and the hypothesis concerning the ends of H_1 and H_2, we find that if $|A| > 1$ then neither terminal edge of A can be blue. Hence $|\partial VA \cap \beta(H_1 \cup H_2)|$ is even. Summation over all H_1H_2-arcs shows that $|\partial(VH_1 \cap VH_2) \cap \beta(H_1 \cup H_2)|$ is even, and the result now follows from Lemma 2. □

Let H_1, H_2 be elementary paths or semicycles in G. Suppose that no terminal bead of H_1 or H_2 is a bead of the other. Since G is orientable, we may let H_1^*, H_2^* be normal elementary paths or normal semicycles such that $\beta(H_1^*) = \beta(H_1)$ and $\beta(H_2^*) = \beta(H_2)$. Define $\kappa(H_1, H_2) = \mu(H_1^*, H_2^*)$.

Lemma 4. *The notation $\kappa(H_1, H_2)$ is well defined, and $\kappa(H_1, H_2) = \kappa(H_2, H_1)$.*

Proof. Since no terminal bead of H_1 or H_2 is a bead of the other, no end of H_1^* or H_2^* can be a vertex of the other. If $\overline{H_2^*}$ is the conjugate of H_2^*, then the same is true of H_1^* and $\overline{H_2^*}$. Thus $\mu(H_1^*, H_2^*) = \mu(H_2^*, H_1^*)$ and $\mu(H_1^*, \overline{H_2^*}) = \mu(\overline{H_2^*}, H_1^*)$ by Lemma 3. Similarly if $\overline{H_1^*}$ is the conjugate of H_1^* then $\mu(H_2^*, \overline{H_1^*}) = \mu(\overline{H_1^*}, H_2^*)$ and $\mu(\overline{H_2^*}, \overline{H_1^*}) = \mu(\overline{H_1^*}, \overline{H_2^*})$. Since $\beta(H_2^*) = \beta(\overline{H_2^*})$ and $\beta(H_1^*) = \beta(\overline{H_1^*})$ we have $\mu(H_1^*, H_2^*) = \mu(H_1^*, \overline{H_2^*}) = \mu(\overline{H_2^*}, H_1^*) = \mu(\overline{H_2^*}, \overline{H_1^*}) = \mu(\overline{H_1^*}, \overline{H_2^*}) = \mu(\overline{H_1^*}, H_2^*)$, so that $\kappa(H_1, H_2)$ is well defined and $\kappa(H_1, H_2) = \mu(H_1^*, H_2^*) = \mu(H_2^*, H_1^*) = \kappa(H_2, H_1)$. □

We call $\kappa(H_1, H_2)$ the *crossing index* of H_1 and H_2. We say that H_1 and H_2 *cross* if $\kappa(H_1, H_2) = 1$. In Fig. 2 we give an example of a semicycle C and elementary paths P and Q. The beads of C are not terminal beads of P or Q. Here P and C cross but Q and C do not.

Lemma 5. *Let $P_1, P_2, P_3 \subseteq EG$. Then*

$$\mu(P_1, P_2 + P_3) = \mu(P_1, P_2) + \mu(P_1, P_3) \ .$$

Proof. We have

$$\begin{aligned}
\mu(P_1, P_2 + P_3) &= |\partial VP_1 \cap \beta(P_2 + P_3)| \\
&= |\partial VP_1 \cap (\beta(P_2) + \beta(P_3))| \\
&= |(\partial VP_1 \cap \beta(P_2)) + (\partial VP_1 \cap \beta(P_3))| \\
&= |\partial VP_1 \cap \beta(P_2)| + |\partial VP_1 \cap \beta(P_3)| \\
&= \mu(P_1, P_2) + \mu(P_1, P_3) \ .
\end{aligned}$$

□

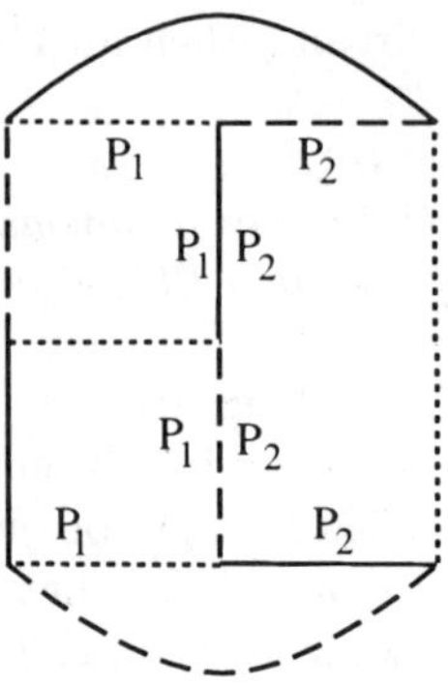

Fig. 2.

We now consider the case where P_1 is a semicycle C and P_2 is an elementary path P. Note first that if P_1 is a semicycle and P_2 and P_3 are disjoint paths such that $P_2 + P_3$ is an elementary path, then Lemma 5 shows that

$$\kappa(P_1, P_2 + P_3) = \kappa(P_1, P_2) + \kappa(P_1, P_3) \ .$$

Our next objective is to extend the definition of $\kappa(C, P)$ to deal with the case where a terminal bead of P is also a bead of C. The main purpose of the increased generality is to facilitate proofs. The idea behind the extended definition is to apply the previous definition to a modification of the 3-graph.

Suppose, as an example, that exactly one terminal edge e of P and exactly one end v of P belong to a bead of C. Let e join v to a vertex u. We may suppose e to be red, with no less generality. We present a construction which should motivate the definition of $\kappa(C, P)$ in this case. Let G' be the 3-graph formed from G in the following way. First, introduce vertices w, x, y, z not in VG. Next, delete e and add a red edge a joining u to w, a yellow edge b joining w to x, a red edge c joining x to v, a blue edge d joining w to y, a blue edge f joining x to z, a red edge g joining y to z and a yellow edge h joining the same two vertices. Now define $P' = (P - \{e\}) \cup \{a, d\}$ and $C' = (C - \{e\}) \cup \{a, b, c\}$. We observe that neither end of P' belongs to a bead of C', and so $\kappa(C', P')$ is defined. If C and P are normal, we now define $\kappa(C, P) = \kappa(C', P')$. It follows that

$$\kappa(C, P) = |\partial VC \cap \beta(P)| + \nu(C, P, v)$$

where $\nu(C, P, v) = 1$ if $e \in C$ and $\nu(C, P, v) = 0$ otherwise.

In general, let us suppose that P joins distinct vertices u and v. Considerations such as those of the previous paragraph will be used to motivate definitions for $\nu(C, P, u)$ and $\nu(C, P, v)$, and we then define

$$\kappa(C, P) = |\partial VC \cap \beta(P)| + \nu(C, P, u) + \nu(C, P, v)$$

if C and P are normal. Agreement with the original definition is secured by defining $\nu(C, P, w) = 0$ if w is an end of P not belonging to a bead of C. If w

does belong to a bead of C and so does the edge e of P incident on w, then the construction in the previous paragraph motivates us to define $\nu(C, P, w) = 1$ if $e \in C$, and $\nu(C, P, w) = 0$ otherwise. It remains to consider the case where w belongs to a bead of C but e is blue. If $P \neq \{e\}$, then we define $P' = P \cup \{f\}$, where f is the edge incident on w whose colour agrees with that of the edge g adjacent to e in $G[P]$. If P' is a path, then we may apply the reasoning in the previous paragraph. If $e \notin C$ then the conclusion is that $\nu(C, P, w) = 1$ if $w \in VC$ and $\nu(C, P, w) = 0$ otherwise, whereas if $e \in C$ then $\nu(C, P, w) = 1$ if $g \in C$ and $\nu(C, P, w) = 0$ otherwise. We define $\nu(C, P, w)$ in this way even if P' is not a path. If $P = \{e\}$ then we define $\nu(C, P, w) = 1$ if $w \in VC$ and $\nu(C, P, w) = 0$ otherwise. Thus if $P = \{e\}$ then $\kappa(C, P) = 0$ whether or not e is blue. We also define $\kappa(C, \emptyset) = 0$. If P is elementary and C is a semicycle but P and C are not necessarily normal, then we define $\kappa(C, P) = \kappa(C^*, P^*)$ where C^* is a normal semicycle such that $\beta(C^*) = \beta(C)$ and P^* is a normal elementary path such that $\beta(P^*) = \beta(P)$. Note that $\kappa(C, P)$ is well defined.

In order to investigate the additivity property for this more general definition of the crossing index, we need some new ideas. The following definition is an adaptation to 3-graphs of Tutte's concept of a bridge of a circuit in a graph. (See [3].) Let $\mathcal{C}$ be a set of b-cycles in G. Define $G' = G - VG[\bigcup N(\mathcal{C})]$, where $N(\mathcal{C})$ denotes the union of the necklaces of the members of $\mathcal{C}$. Let K be a component of G'. Then $EK \cup \partial VK$ is a *bridge* of $\mathcal{C}$ in G. If e is an edge in $EG - (\bigcup N(\mathcal{C}))$ joining two vertices of $VG[\bigcup N(\mathcal{C})]$, then $\{e\}$ is also reckoned as a bridge of $\mathcal{C}$ in G.

Let X be a bridge of $\mathcal{C}$ in G. An edge $e \in X$ is an *edge of attachment* of X if e is incident on a vertex of $VG[\bigcup N(\mathcal{C})]$. Any vertex of $VG[\bigcup N(\mathcal{C})]$ incident on an edge of attachment is called a *vertex of attachment* of X. It cannot be a vertex of attachment of any other bridge, since G is cubic.

We now specialise to the case where $\mathcal{C} = \{C\}$ for some semicycle C. Since $X \cap C = \emptyset$, no vertex of attachment can be a pole of a bead of C. The bridge X is *neutral* if there exist a vertex of attachment in VC and another not in VC. In the remaining case X is classified as *interior* or *exterior* according to whether or not its vertices of attachment are in VC. It follows from Theorem 1 that a normal semicycle C separates if and only if no bridge is neutral.

A path is said to be *external* to C if no edge belongs to C or a bead of C. A path is *internal* to C if each edge belongs to C or a bead of C.

We now investigate the additivity condition for the more general definition of the crossing index.

Lemma 6. *Let P and Q be disjoint paths such that $P+Q$ is an elementary path. Let e be the edge of P incident on a vertex of VQ. Suppose that $\{e\}$ is external to a semicycle C and that Q is internal to C. Then $\kappa(C, P+Q) = \kappa(C, P) + \kappa(C, Q)$.*

Proof. Since G is orientable, we may let C^* be a normal semicycle such that $\beta(C^*) = \beta(C)$. Let P^* and Q^* be normal paths such that $\beta(P^*) = \beta(P)$, $\beta(Q^*) = \beta(Q)$ and $P^* + Q^*$ is normal. It follows from the definitions that

$$\kappa(C, P+Q) = \kappa(C^*, P^* + Q^*)$$

$$= \kappa(C^*, P^*) + \nu(C^*, P^*, v) + \kappa(C^*, Q^*) + \nu(C^*, Q^*, v)$$
$$= \kappa(C, P) + \nu(C^*, P^*, v) + \kappa(C, Q) + \nu(C^*, Q^*, v)$$

where v is the unique vertex of $VP^* \cap VQ^*$. It is easily checked that $\nu(C^*, P^*, v) = \nu(C^*, Q^*, v)$, as required. □

The same argument establishes the following lemma.

Lemma 7. *Let P and Q be disjoint paths such that $P+Q$ is an elementary path. Suppose that Q is external to a semicycle C. Then $\kappa(C, P+Q) = \kappa(C, P) + \kappa(C, Q)$.* □

3 A condition for separation

Let P be an elementary path and C a semicycle. Suppose that $P = P_1 P_2 \ldots P_n$, where for each i the path P_i is internal or external to C, depending on the parity of i. Lemmas 6 and 7 show that

$$\kappa(C, P) = \sum_{i=1}^{n} \kappa(C, P_i) \ .$$

Motivated by this observation, we define $\kappa(C, P)$ by this formula, even if P is not elementary, provided only that P_i is elementary for each i.

Lemma 8. *Let e, f be edges in $EG - (C \cup (\bigcup N(C)))$, where C is a normal semicycle. Let X and Y be the bridges of C containing e and f respectively. Suppose that X and Y are either both interior or neutral or both exterior or neutral. Then there is a path, joining an end of e to an end of f, which does not cross C.*

Proof. If $X = Y$, then some subset of X is a path P joining an end of e to an end of f. It is clear in this case that P does not cross C. Suppose therefore that $X \neq Y$. Without loss of generality we can find a vertex w of attachment of X and a vertex z of attachment of Y such that $w \in VC$ and $z \in VC$. Moreover some subset of $X - \{e\}$ is an elementary path P_1 joining w to an end of e. Similarly some subset of $Y - \{f\}$ is an elementary path P_3 joining z to an end of f. Some subset of $G[C \cup (\bigcup N(C))]$ is an elementary path P_2 joining w to z. None of P_1, P_2, P_3 crosses C. The required path is $P_1 \cup P_2 \cup P_3$. □

Theorem 9. *Let G be a connected orientable 3-graph. Let e, f be edges in $G - VG[\beta(C)]$, where C is a normal semicycle which does not separate. Then there is a path, joining an end of e to an end of f, which does not cross C.*

Proof. Assume first that $\{e, f\} \subseteq EG - (C \cup (\bigcup N(C)))$. Let X and Z be the bridges containing e and f respectively. Thanks to Lemma 8 we may assume without less generality that X is interior and Z is exterior to C. Since C does not separate, there is also a neutral bridge Y. The argument is similar to the proof of Lemma 8. First some subset of $X - \{e\}$ is an elementary path P_1 joining a vertex w of attachment of X to an end of e. Similarly some subset of $Z - \{f\}$ is an elementary path P_5 joining a vertex z of attachment of Z to an end of f. Moreover Y has distinct vertices x and y of attachment such that just one of them is in VC. It also contains an elementary path P_3 joining x to y.

Let $H = G[C \cup (\bigcup N(C))]$. There is a path P_2 in H joining w to x which does not pass through y. We may assume P_2 to be chosen so that there is a path in H joining w to y and having no blue edge in common with P_2. Since x and y may be interchanged, we may also assume without less generality that z does not belong to an internal bead of P_2. In addition P_2 may be replaced by its conjugate, and so we may also assume that P_2 does not pass through z. Our choice of P_2 guarantees the existence of an elementary path P_4 in H, joining y to z, such that $VP_4 \cap (VP_1 \cup VP_2 \cup VP_3) = \{y\}$. None of P_1, P_3, P_5 crosses C, and P_2 crosses C if and only if P_4 crosses C. The required path is therefore $P_1 \cup P_2 \cup P_3 \cup P_4 \cup P_5$.

If $e \in \bigcup N(C)$, then we may take w to be an end of e, and put $P_1 = \emptyset$. Either the argument above or the proof of Lemma 8 is then applicable. The argument is similar if $f \in \bigcup N(C)$. □

Conversely we have the following theorem.

Theorem 10. *Let G be a connected orientable 3-graph separated by a normal semicycle C. Then either C or $\overline{C}$ is a bigon or there exist edges e and f such that every path joining an end of e to an end of f crosses C.*

Proof. Suppose that neither C nor $\overline{C}$ is a bigon. We may then choose edges $e \in I(C)$ and $f \in O(C)$ not adjacent to blue edges of C. Consider any path P joining an end of e to an end of f. We may write $P = P_1P_2 \ldots P_n$ where the segments $P_1, P_2, \ldots, P_n$ are alternately internal or external to C. Since C separates, there are no neutral bridges, and so for any i the path P_i crosses C if and only if exactly one end is incident on an edge of $I(C)$. Since P has this property, an odd number of these segments cross C, and therefore so does P. □

References

1. Bonnington, C.P., Little, C.H.C.: The Foundations of Topological Graph Theory. Springer-Verlag, New York, 1995
2. Bonnington, C.P., Little, C.H.C.: A combinatorial generalisation of the Jordan curve theorem. Austral. J. Combinatorics **9** (1994) 179–199
3. Tutte, W.T.: Graph Theory. Addison-Wesley, Reading, Mass., 1984

A New Proof Search Method for Linear Logic

Satoshi Matsuoka

Department of Electrical and Computer Engineering, Faculty of Engineering,
Nagoya Institute of Technology, Gokiso, Showa-Ku, 466, Japan.
matsuoka@juno.ics.nitech.ac.jp

Abstract. So far, in proof search for Linear Logic, two methods have been considered: one is the bottom up method, in works by Andreoli, Hodas and Miller, and the other is the top down method by Tammet. We propose a new proof search method for Linear Logic containing additive connectives as well as multiplicative connectives and it is possible to extend the method to exponentials. The method is based on a correctness condition for proof nets.

1 Introduction

So far, in proof search for Linear Logic, two methods have been considered: one is the bottom up method, see for example works by Andreoli [And92], Hodas and Miller [HM94], and the other is the top down method by Tammet [Tam94]. We propose a new proof search method for Linear Logic containing additive connectives as well as multiplicative connectives and it is possible to extend the method to exponentials. The method is based on a correctness condition for proof nets: when given a sequent $\vdash \Gamma$, construct a proof structure Θ for $\vdash \Gamma$ non-deterministically, and check whether Θ is a proof net by using the correctness condition. We use new proof nets [Gir95], in which boxes for additives are removed.

The proof search method is completely different from the bottom up and the top down methods. The approach is similar to model theoretic approach rather than the syntactical approach: constructing proof structures corresponds to model constructions and checking correctness corresponds to checking validity. In the method, non-deterministic guessing and correctness checking are completely divided.

2 Proof structures

In this section we review the proof structures of [Gir95].

Definition 1. A link L is an $n+m$-tuple of formulas with a type: $\dfrac{P_1, \ldots, Pn}{Q_1, \ldots, Q_m} L$
The type of a link is either ID, Cut, generalized axiom, $\otimes$, $\wp$, $\&$, $\oplus_1$, or $\oplus_2$. To each type, n, a number of its premises and m, a number of its conclusions are assigned($m, n \geq 0$, $m + n \neq 0$). The links with ID, Cut, generalized axiom, $\otimes, \wp, \&, \oplus_1$ and $\oplus_2$ as types have the following forms:

ID-links $\dfrac{}{A \quad A^{\perp}}$ Cut links $\dfrac{A \quad A^{\perp}}{\text{Cut}}$ generalized axiom-links $\dfrac{}{A_1 \cdots A_n}$

times $\dfrac{A \quad B}{A \otimes B}$ par $\dfrac{A \quad B}{A \wp B}$ with $\dfrac{A \quad B}{A \& B}$ plus $\oplus_1$ $\dfrac{A}{A \oplus B}$ $\oplus_2$ $\dfrac{B}{A \oplus B}$

Definition 2. To any &-link L with $A\&B$ as its conclusion, we associate an eigenweight p_L, which is a boolean variable. The intuitive meaning of p_L is the choice $\{l/r\}$ between the premises A and B: p_L stands for the selection "left", namely A, and $\neg p_L$ stands for the selection "right", namely B. We use $\epsilon.p_L$ to speak of p_L or $\neg p_L$.

Definition 3. A triple $\Theta = (V, E, w)$ is a proof structure if

- (V, E) is a pair such that V is a multiset of formulas and E is a multiset of links between formulas occurring in V.
- w is a function such that
  - For each formula A in V, a weight $w(A)$, that is, a non-zero element of the boolean algebra generated by the eigenweights $p_1, \ldots, p_n$ of the &-links of Θ;
  - For each link L in E, a weight $w(L)$, that is, a non-zero element of the boolean algebra generated by the eigenweights $p_1, \ldots, p_n$ of the &-links of Θ.

Moreover, the following conditions must be satisfied:

1. Each formula in V is the premise of at most one link and the conclusion of at least one link. The formulas which are not premises of some link are called the conclusions of Θ;
2. $w(A) = \sum w(L)$, the sum being taken over the set of links with conclusion A;
3. If A is a conclusion of Θ, then $w(A) = 1$;
4. If u is any element of the boolean algebra generated by the weights occurring in Θ, and L is a &-link, then $u \cdot \neg w(L)$ does not depend on p_L, that is, it belongs to the boolean algebra generated by the eigenweights distinct from p_L.
5. If u is any weight occurring in Θ, then u is a monomial $\epsilon_1.p_{L_1} \cdots \epsilon_n.p_{L_n}$ of eigenweights and negations of eigenweights;
6. If L is any non ID-link, with premises A and/or B then
   - if L is any of $\otimes, \wp$ and Cut, then $w(L) = w(A) = w(B)$;
   - if L is a $\oplus_1$-link, then $w(L) = w(A)$;
   - if L is a $\oplus_2$-link, then $w(L) = w(B)$;
   - if L is a &-link, then $w(A) = w(L) \cdot p_L$ and $w(B) = w(L) \cdot \neg p_L$ (hence $w(L) = w(A) + w(B)$).

7. For any $A \in V$, if the links whose conclusion is A are $L_1, \ldots, L_m$ then for each $1 \leq i, j \leq m$, whenever $i \neq j$, then $w(L_i) \neq w(L_j)$.

Definition 4. The size $\#(\Theta)$ of a proof structure Θ is the number of links in Θ.

Note that the number of formulas in a proof structure Θ is bounded by $2 \cdot \#(\Theta)$.

Definition 5. Let p_L be an eigenweight in a proof structure Θ; we say that the weight w (in Θ) depends on p_L if w contains p_L or the negation of p_L. A formula A of Θ is said to depend on p_L, if A is the conclusion of a link L' such that $w(L')$ depends on p_L.

Note that although $w(A)$ does not contain p_L, the formula A may depend on p_L.

3 Proof structure construction algorithm

Definition 6. We shall define pre-terminal forests of a sequent $\vdash A_1, \cdots, A_n$ as follows:

1. A sequence $A_1, \cdots, A_n$ is a pre-terminal forest with $A_1, \cdots, A_n$ as fronts.

2. Let @ be $\otimes, \wp, \oplus$, or &. If 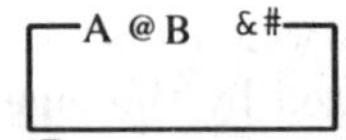

is a pre-terminal forest with $A@B, \Gamma$ as fronts, then 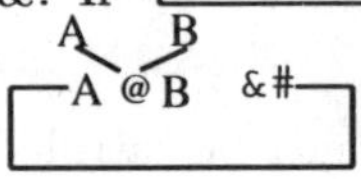

is a pre-terminal forest with A, B, Γ as fronts.

Definition 7. A pre-terminal forest is a *terminal forest* if its fronts are all atomic formulas or duals of atomic formulas, that is, literals.

We need the notion of the copies of the nodes in a terminal forest in order to present proof structure construction algorithm.

Definition 8 (copies in terminal forests). In a terminal forest Θ, a *copy* f of Θ is a function from non-leaf nodes in Θ to integers. According to f, the copy of each non-leaf node of Θ is defined. Any copy of each non-leaf node is represented as follows:

– Case for $\oplus$-link: $m + n - 1$ copy of

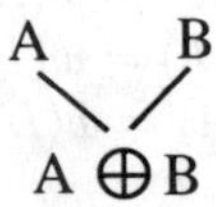

is

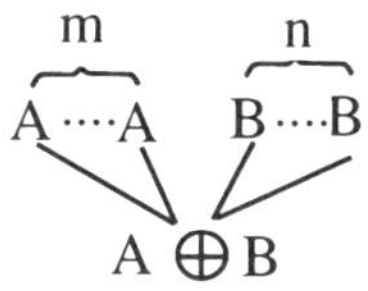

where $m \geq 0,\ n \geq 0$, and $m + n \neq 0$.

- Case for the other links: Let @ be any of $\otimes$, $\wp$, or &. $n - 1$ copy of

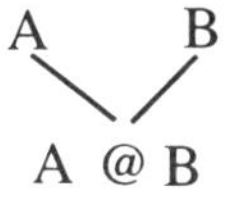

is

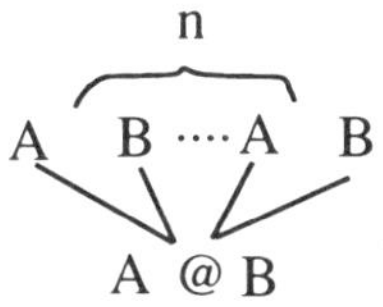

where $n \geq 1$.

According to a copy f, in each non-leaf node Θ, by duplicating the subtrees with the premises of the node as the roots, we can construct the unique forest Θ_f from Θ. We call Θ_f a duplicated forest of Θ by f.

Definition 9 (constrained forests). A duplicated forest Θ_f of Θ is a *constrained forest* if the sum of the number of the copies of all nodes in any path in Θ_f must be equal to or less than $\#\&(\Theta)$, the number of &-links in Θ.

Definition 10 (weights in constrained forests). To each of the nodes in a constrained forest Θ_f, a weight is assigned inductively from roots to leaves in the following way. Moreover, to Θ_f, equations are assigned:

- Base step: To any root in Θ_f weight 1 is assigned. Then, the set of equations is empty.
- Case for $\oplus$-nodes: If a $\oplus$-node with conclusion $A \oplus B$ in Θ_f has $n + m - 1$ copies and $A \oplus B$ has the weight w, then to the premises of the node, the weights are assigned as follows:

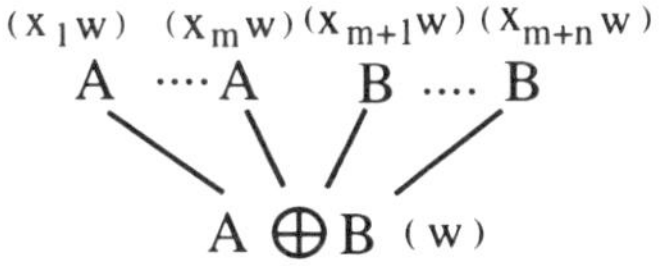

where $x_1, \ldots, x_{m+n}$ are new variables distinct from eigenweights for &-links. Then, the equation $x_1 \cdot w + \cdots + x_{m+n} \cdot w = w$ is added to the set of the equations for Θ_f.

- Case for &-nodes: If a &-node in Θ_f with the conclusion $A\&B$ in Θ_f has $n-1$ copies and $A\&B$ has the weight w, then to the premises of the link, the weights are assigned as follows:

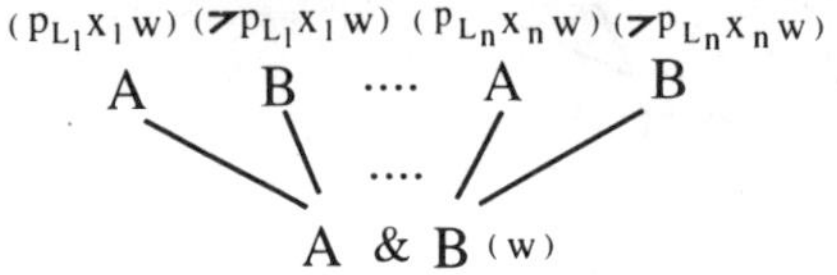

where p_{L_i} $(1 \leq i \leq n)$ is the eigenweights for &-link L_i with $A\&B$ as the conclusion and A and B as the premises, and $x_1, \ldots, x_n$ are the new variables distinct from eigenweights. Then, the equation $x_1 \cdot w + \cdots + x_n \cdot w = w$ is added to the set of the equations for Θ_f.
- Case for other nodes: Let @ be any of $\otimes$ or $\wp$. If an @-node is with the conclusion $A@B$ in Θ_f has $n-1$ copies and $A@B$ has the weight w, then to the premises of the link, the weights are assigned as follows:

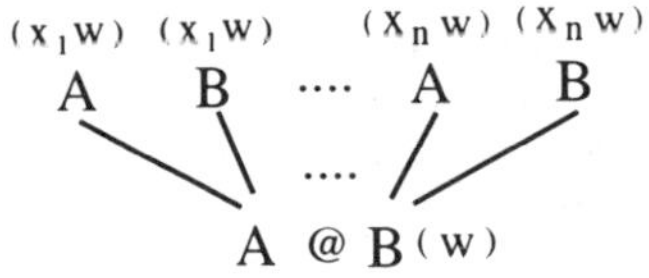

where $x_1, \ldots, x_n$ are the new variables distinct from eigenweights. Then, the equation $x_1 \cdot w + \cdots + x_n \cdot w = w$ is added to the set of the equations for Θ_f.

Definition 11 (ID-link connections). In a constrained forest Θ_f with the weights and the equations, a *ID-link connection* connects one literal α and many literals which are the duals of α by ID-links:

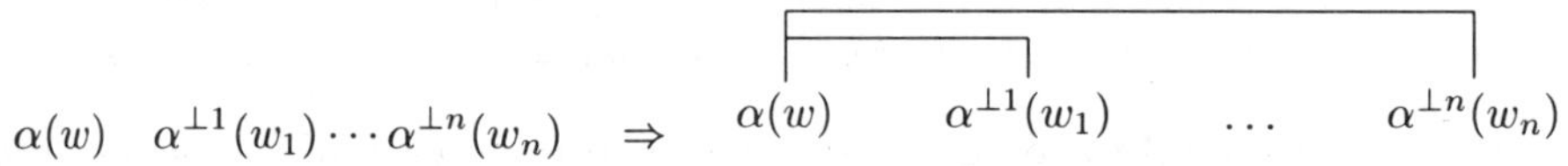

Then, the equation $w_1 + \cdots + w_n = w$ is added to the set of the equations for Θ_f, where n is equal to or less than $\#\&(\Theta) + 1$.

Definition 12 (proof structure construction algorithm). The following is a proof structure construction algorithm:

1. input: a sequent $\vdash s$;
2. construct the terminal forest Θ for s;
3. construct a constrained forest Θ_f from Θ non-deterministically;
4. compute the weight of each node in Θ_f and the equations Eq for Θ_f;
5. find the ID-link connections for Θ_f such that all the literals in Θ are the conclusions of ID-links. The resulting graph Θ'' is called *pseudo proof structure*.

   If such a pseudo proof structure cannot be constructed, then the algorithm terminates and outputs "reject". Let Eq' be the union of Eq and the equations for ID-link connections.

6. the equations Eq' are solved by equational first order unifications with the following axiom schemes and constraints:
   - the axioms of the boolean algebra;
   - In the solutions, variables do not occur: only eigenweights occur;
   - For variables distinct from eigenweights, monomial of eigenweights must be substituted;
   - All the weights of the nodes Θ'' are monomials.

   If a solution is not found, then the algorithm terminates and outputs "reject". If a solution is found, then the algorithm terminates and outputs the substitution θ and the graph $\Theta''\theta$ obtained by applying θ to Θ''.

Theorem 13 (Termination). *The algorithm above terminates.*

Proof. By the bounds of the number of the copies of all nodes in any path in any constrained terminal forest Θ_f and in ID-link connections. □

Theorem 14 (Soundness). *If the algorithm outputs Θ, then Θ is a proof structure.*

Theorem 15 (Completeness). *If Θ is a cut-free proof structure with conclusions Γ, then when given the terminal forest for Γ, the algorithm outputs Θ.*

Proof. It is obvious that any proof structure satisfies the constraints of the definitions above.

4 A p-time correctness condition for MALL proof nets

In this section, we shall define inductive empires and proof nets by ourselves and prove sequentialization theorem.
For a proof structure Θ and any formula A in Θ, we define the inductive empire $e_I(A)$ on Θ. In [Gir95], to each valuation ϕ for eigenweights and formula A in a proof structure Θ, the empire $e_\phi(A)$ is defined, while $e_I(A)$ does not depend on any valuation ϕ. In addition, $e_I(A)$ is defined on proof structures as well as on proof nets.
In the following, we only consider cut-free proof structures, which are sufficient for our purpose, proof search.

Definition 16. Let WList be a list of weights. A weight w is compatible with WList if for any u in WList, $u \cdot w \neq 0$.

Definition 17. Let Θ be a proof structure for MALL and let A occur in Θ. We inductively define the $e_I(A)$, inductive empire for A, in the following way. Let $\text{WList}_{e_I(A),X}$ be the list of the weights in the sequence of the formulas in the computation steps from A to X in the computation of $e_I(A)$. $\text{WList}_{e_I(A),X}$ and $e_I(A)$ is considered as a variable in imperative programming language and can be varied in the process of the computation.

(a) $e_I(A) := \{A\}$ (base step);
(b) let $\frac{X1 \quad X2}{Y}$ be a $\otimes$,$\wp$ or &-link in Θ: if $Y \in e_I(A)$ and, $w(X_1)$ and $w(X_2)$ is compatible with $\mathrm{WList}_{e_I(A),Y}$, then $e_I(A) := \{X_1, X_2\} \cup e_I(A)$ ($\uparrow_\otimes$,$\uparrow_\wp$ or $\uparrow_\&$-step);
(c) let $\frac{X}{Y}$ be a $\oplus$-link in Θ: if $Y \in e_I(A)$ and $w(X)$ is compatible with $\mathrm{WList}_{e_I(A),Y}$, then $e_I(A) := \{X\} \cup e_I(A)$ ($\uparrow_\oplus$-step);
(d) let $\overline{X \quad X^\perp}$ be an ID-link in Θ: if $X \in e_I(A)$ and $w(X^\perp)$ is compatible with $\mathrm{WList}_{e_I(A),X}$, then $e_I(A) := \{X^\perp\} \cup e_I(A)$ ($\rightarrow_{\mathrm{ID}}$-step);
(e) let $\overline{X_1 \cdots X_i \cdots X_n}$ be a generalized axiom-link in Θ: if $X_i \in e_I(A)$ for some $1 \leq i \leq n$ and $w(X_j)$ for any $1 \leq j \leq n$ is compatible with $\mathrm{WList}_{e_I(A),X_i}$, then $e_I(A) := \{X_1, \cdots, X_i, \cdots, X_n\} \cup e_I(A)$ ($\rightarrow_{\mathrm{ga}}$-step);
(f) let $\frac{X_1 \quad X_2}{X_1 \otimes X_2}$ be a $\otimes$-link in Θ: if $X_1 \neq A \neq X_2$ and $X_1 \in e_I(A)$ (resp. $X_2 \in e_I(A)$) then $e_I(A) := \{X_1 \otimes X_2, X_2\} \cup e_I(A)$ (resp. $e_I(A) := \{X_1 \otimes X_2, X_1\} \in e_I(A)$) ($\downarrow_\otimes$-step);
(g) let $\frac{X}{Y}$ be a $\oplus_1$-link or $\oplus_2$-link in Θ: if $X \neq A$ and $X \in e_I(A)$, then $e_I(A) := \{Y\} \cup e_I(A)$ ($\downarrow_\oplus$-step);
(h) let $\frac{X1 \quad X_2}{Y}$ be a $\wp$-link or &-link in Θ: if $X_1 \neq A \neq X_2$ and $\{X_1, X_2\} \subset e_I(A)$, then $e_I(A) = \{X_1 \wp X_2\} \cup e_I(A)$ (resp. $e_I(A) = \{X_1 \& X_2\} \cup e_I(A)$) ($\downarrow_\wp$-step or $\downarrow_\&$-step).

Remark. For any formula A in Θ, the empire $e_I(A)$ is a multiset of formulas. However, note that from $e_I(A)$, we can construct the multiset of links whose premises and conclusions are among $e_I(A)$.

Definition 18. A formula B in $e_I(A)$ is said to be a door of $e_I(A)$ if

- either it is the premise of a link L whose conclusion does not belong to $e_I(A)$;
- or it is a conclusion of Θ.

Note that in proof structures, A is not necessarily a door of $e_I(A)$.

The following propositions immediately follow from the definition of inductive empires.

Proposition 19. *Let C be a door of $e_I(A)$ which is not a conclusion of Θ and is different from A. Then, C is the premise of a $\wp$-link or &-link.*

Proposition 20. *For any proof structure Θ and any formula A in Θ, $e_I(A)$ is computed in p-time w.r.t $\#(\Theta)$.*

Proof. Starting from $A \in e_I(A)$ and traversing the links in Θ by following the inductive definition of $e_I(A)$, $e_I(A)$ is computed. Then, the links in Θ are visited at most once. □

Definition 21 (Proof nets). A MALL proof structure Θ is a proof net if Θ satisfies the following conditions:

1. For any link $\frac{A \quad B}{A \otimes B}$, we have $e_I(A) \cap e_I(B) = \emptyset$ (empire-acyclicity);
2. For any link $\frac{A \quad B}{A \wp B}$, we have $e_I(A) = e_I(B)$ (empire-connectedness);

3. Let Γ be the multiset of conclusions of Θ: if A is in Γ, then $e_I(A) = \Theta$ (empire-connectedness);
4. For any &-link L $\frac{A \quad B}{A\&B}$, the multiset of the doors Γ_A of $e_I(A)$,Γ_B of $e_I(B)$ and $\Gamma_{A\&B}$ of $e_I(A\&B)$ are (multi) set-theoretically equal except for A, B, and $A\&B$ (empire-synchronization);
   In addition, if X is in $\Gamma_{A\&B}$, then $e_I(A\&B) \subseteq e_I(X)$ (empire-connectedness);
5. For any &-link L $\frac{A \quad B}{A\&B}$, formulas that depend on p_L are contained in $e_I(A\&B)$. Moreover, all the formulas with weights that contain positive p_L are contained in $e_I(A)$ and are not contained in $e_I(B)$, and all the formulas with weights that contain $\neg p_L$ are contained in $e_I(B)$ and are not contained in $e_I(A)$.

The definition of proof nets differs from that of proof nets in [Gir95]. However, we can prove sequentialization theorem for proof nets defined in this paper.

The following lemma is easy to prove.

Lemma 22. *Let Θ be a proof structure. In Θ each condition of the Definition 21 can be checked in p-time.*

On the other hand, in proof structures (in particular, in proof nets), transitivity property "if $B \in e_I(A)$ and $C \in e_I(B)$, then $C \in e_I(A)$" does not hold. However, the following lemma holds.

Lemma 23. *Let A and B be distinct formulas in a proof net Θ and assume that $A \notin e_I(B)$; then*

$$\text{if } B \in e_I(A), \text{ then } e_I(B) \subset e_I(A).$$

Proof. Assume $e_I(B) \not\subset e_I(A)$. Since $B \in e_I(A)$, in the computation of $e_I(B)$, there is the computation step by which $e_I(B)$ exits from $e_I(A)$ for the first time. This step must be $\downarrow$-step. Moreover, in this $\downarrow$-step, one of the premises of the link must be A (if both premises of the link were in $e_I(A)$ and were not A, then the conclusion would be in $e_I(A)$). However, this contradicts the assumption $A \notin e_I(B)$. □

The above lemma corresponds to the first part of Lemma 5 in [Gir95]. However, with respect to inductive empire e_I, the second part of Lemma 5 in [Gir95], i.e, "if $B \notin e_I(A)$ and $A \notin e_I(B)$, then $e_I(A) \cap e_I(B) = \emptyset$" does not hold. For example, in a proof net with &-link $\frac{A \quad B}{A\&B}$, consider $e_I(A)$ and $e_I(B)$. However, as stated below, in a restricted formula, the second part of Lemma 5 in [Gir95] is proved.

Definition 24. In a proof structure Θ, a formula $A\&B$ in Θ is *maximal &-formula* if it has weight 1 and $e_I(A\&B)$ is maximal with respect to inclusion among the inductive empires of other &-formulas.

Proposition 25. *In a proof net Θ, if &-link exists in Θ, then maximal &-formula exists in Θ.*

Lemma 26. *Let Θ be a proof net, A a maximal &-formula in Θ and B any formula in Θ. Then*

$$\text{if } B \notin e_I(A) \text{ and } A \notin e_I(B), \text{ then } e_I(A) \cap e_I(B) = \emptyset.$$

Proof. We show that whenever $B \notin e_I(A)$ and $e_I(A) \cap e_I(B) \neq \emptyset$, then $A \in e_I(B)$. Assume that $B \notin e_I(A)$ and $e_I(A) \cap e_I(B) \neq \emptyset$. We consider the process of the computation of $e_I(B)$. In order that the computation of $e_I(B)$ reaches the region of $e_I(A)$, the step of the crossing must be $\uparrow$-step. By proposition 19, the formula C in $e_I(A)$ which the computation of $e_I(B)$ reaches for the first time must be A or a premise of a $\wp$-link or a premise of a &-link. However, since A is a maximal &-formula, the doors of $e_I(A)$ (containing A) can not be a premise of &-link. Hence the door of $e_I(A)$ which the computation of $e_I(B)$ crosses for the first time, is A or a premise of a $\wp$-link. If the formula is A, then we are done. Otherwise, let the formula be C. C is the premise of a $\wp$-link $L : \frac{C \quad D}{C \wp D}$, and the other premise D is not in $e_I(A)$. Since the computation of $e_I(B)$ crosses C by $\uparrow$-step, $C \wp D$ must be in $e_I(B)$. Moreover, since D is in $e_I(B)$ and $B \notin e_I(D)$ (otherwise, there is a $\otimes$-formula $E \otimes F$ that is a conclusion of Θ and is in the computation from B to $C \wp D$ in $e_I(B)$ such that $B \in e_I(E)$ and $B \in e_I(F)$. That contradicts Definition 21 (a)), by Lemma 23, $e_I(D) \subset e_I(B)$. By the empire-connectedness (b) of the Definition 21, $C \in e_I(D)$ and $D \in e_I(C)$. Then the computation of $e_I(D)$ must reach the region of $e_I(A)$, the step of the crossing must be $\uparrow$-step, and the crossed formula in $e_I(A)$ must be a $\wp$-formula different from C. Then, since $e_I(D) \subset e_I(B)$, the computation of $e_I(B)$ also crosses the formula. If the formula is A, then we are done. Otherwise, repeating the procedure, finally we can prove $A \in e_I(B)$. □

Lemma 27. *Let Θ be a proof net which does not contain &-link, A and B be distinct formulas in Θ. Then*

$$\text{if } B \notin e_I(A) \text{ and } A \notin e_I(B), \text{ then } e_I(A) \cap e_I(B) = \emptyset.$$

Proof. We can prove the lemma by the same method as the proof of Lemma 26. □

Lemma 28. *Let Θ be a proof net with $A\&B$ (which is the conclusion of link L) as one of conclusions of Θ. Let $\Theta_{\{p_L=1\}}$ (resp. $\Theta_{\{p_L=0\}}$) be the proof structure obtained by substituting 1 (resp. 0) for p_L in Θ and removing $A\&B$ and L. Then $\Theta_{\{p_L=1\}} = e_I(A)$ (resp. $\Theta_{\{p_L=0\}} = e_I(B)$).*

Proof. First, note that $X \in \Theta_{\{p_L=1\}} \Leftrightarrow w(X)$ does not contain $\neg p_L$ for $X \in \Theta$. Since any formula in $e_I(A)$ does not contain $\neg p_L$ (by definition of proof nets), $e_I(A) \subseteq \Theta_{\{p_L=1\}}$.

Conversely, assume that $X \in \Theta_{\{p_L=1\}}$. If $X \in \Theta_{\{p_L=1\}}$, then $w(X)$ in Θ does not contain $\neg p_L$. Assume that $X \notin e_I(A)$. By $\Theta = e_I(A\&B) = e_I(A) \cup e_I(B) \cup \{A\&B\}$, $X \in e_I(B)$. Then there exists a sequence s of formulas beginning at B and ending at X, tracing the steps of the computation of $e_I(B)$, such that there is a subsequence $[C, D]$ of s such that C is a premise of a link L, D the conclusion of L, $w(C)$ contains $\neg p_L$, $w(D)$ does not contain $\neg p_L$, and the subsequence of s following D immediately does not contain $\neg p_L$. Then by the definition of proof structures, there is a link L' and a formula C' such that D is the conclusion of L' and C' is a premise of L' and $w(C')$ contains p_L. Thus, $C' \in e_I(A)$. Since L and L' are links with same type, $D \in e_I(A)$. Since the subsequence of s following D does not contain $\neg p_L$, $X \in e_I(A)$. Contradiction.

With respect to the case for $\Theta_{\{p_L=0\}}$, the same method applies. □

Lemma 29. *Suppose that the assumptions are the same as the previous lemma. In $\Theta_{\{p_L=1\}}(= e_I(A))$ (resp. $\Theta_{\{p_L=0\}}(= e_I(B))$), each empire-connectedness condition holds.*

Proof. Assume that in $\frac{C \quad D}{C \wp D}$ in $\Theta_{\{p_L=1\}}$, the empire-connectedness condition does not hold, that is, $e_I(C) \neq e_I(D)$ in $\Theta_{\{p_L=1\}}$. If $w(C)$ and $w(D)$ contain p_L, then since $e_I(C) = e_I(D)$ in Θ and all the formulas of $e_I(C)(= e_I(D))$ in Θ do not contain $\neg p_L$, $e_I(C) = e_I(D)$ in $\Theta_{\{p_L=1\}}$. Hence we can assume $w(C)$ and $w(D)$ do not contain p_L (of course, do not contain $\neg p_L$). We can divide the proof into two subcases:

1. Case that $D \in e_I(C)$ and $C \in e_I(D)$ in $\Theta_{\{p_L=1\}}$:
Without loss of generality, we can assume that $X \in e_I(C)$ and $X \notin e_I(D)$ for some formula X in $\Theta_{\{p_L=1\}}$. Then in any sequence s of $e_I(D)$ in Θ from D to X, there is a link L' such that the conclusion Y of L' does not contain $\neg p_L$, the subsequence following Y does not contain $\neg p_L$ and the premise Y' of L' in s contains $\neg p_L$. Obviously, Y is reached via any $\downarrow$-step in $e_I(D)$. Then there is a formula Y'' that is the conclusion of a &-link L'' where L'' has Y as the conclusion and is the same type as L' that contains p_L. Then, since in at least one of any such sequences s, $Y \in e_I(C)$ in $\Theta_{\{p_L=1\}}$, the computation of $e_I(C)$ in $\Theta_{\{p_L=1\}}$ reach Y from C via any $\uparrow$-step or a premise Y'' of L'', which has Y as the conclusion and that contains p_L. From $Y'' \in e_I(C)$, we yield $Y'' \in e_I(D)$ in Θ. Since $Y'' \in e_I(D)$ in Θ and $w(Y'')$ contains p_L, $Y'' \in e_I(D)$ in $\Theta_{\{p_L=1\}}$. Hence by $\downarrow$-step, $Y \in e_I(D)$ in $\Theta_{\{p_L=1\}}$. Then since there is a computation from D to Y'' in $e_I(D)$ and there is a sequence from Y to X that does not contain both p_L and $\neg p_L$ in the computation of $e_I(D)$. Therefore $X \in e_I(D)$. Contradiction.
2. Otherwise:
Without loss of generality, we can assume $D \notin e_I(C)$ in $\Theta_{\{p_L=1\}}$. Since in Θ, $e_I(C) = e_I(D)$, any sequence s of formulas in the computation from C to D of $e_I(C)$ in Θ may contain $\neg p_L$, but does not contain p_L. In the any sequence s, there is a link L' such that the conclusion Y of L' does not contain $\neg p_L$, the subsequence following Y does not contain $\neg p_L$ and the premise of L' in s contains $\neg p_L$. Obviously, Y is reached via any $\downarrow$-step in $e_I(C)$. Then, in at least one of any such sequences s, in the computation of $e_I(D)$ in Θ, Y is reached from D via any $\uparrow$-step since the computation of $e_I(C)$ in $\Theta_{\{p_L=1\}}$ does not reach D via $\uparrow$-step. Therefore, all the premises

of the links that has Y as the conclusion are in $e_I(D)$ and then one of the premises contains p_L. Let us call the formula X. Since $e_I(C) = e_I(D)$ in Θ, $X \in e_I(C)$ in Θ. Moreover since $w(X)$ contains p_L, $X \in e_I(C)$ in $\Theta_{\{p_L=1\}}$. Therefore since there is a sequence from C to D via X in the computation of $e_I(C)$ in $\Theta_{\{p_L=1\}}$, $D \in e_I(C)$ in $\Theta_{\{p_L=1\}}$. Contradiction.

With respect to the case for $\Theta_{\{p_L=0\}}$, the same method applies. □

In the next theorem, as to the definitions of the removals of links in proof structures and the sequentializability of proof structures, see [Gir95].

Theorem 30. *Θ is a proof net iff it is sequentializable.*

Proof. The proof is the same as that in [Gir95] except that the definition of proof nets is different. If-part is shown by induction on the steps of the removal of Θ. Only-if-part is shown by induction on the number n of &-links in Θ. In some of the induction steps, the lemmas proved above are used. □

Corollary 31. *When given a proof structure Θ, we can decide whether if Θ is a proof net in polynomial time.*

5 Proof net construction algorithm

Definition 32 (proof net construction algorithm). We present a proof net construction algorithm as follows:

1. input: a sequent s;
2. by proof structure construction algorithm, construct a proof structure Θ for s non-deterministically.
   if the algorithm outputs "reject", then proof net construction algorithm outputs "reject";
3. to Θ, apply the conditions of Definition 21 in Section 4.
   if Θ is a proof net then the algorithm outputs "success" and Θ and not a proof net then outputs "reject".

Theorem 33 (Termination, Soundness, and Completeness). *The above algorithm terminates. Moreover if the algorithm outputs Θ, then Θ represents a proof of MALL, and if a sequent $\vdash s$ is provable, then the algorithm above outputs some of proof nets for s.*

Proof. The termination of the algorithm is shown by the terminations of proof structure construction algorithm and the algorithm for correctness condition for proof nets. The soundness and the completeness are shown by completeness for proof structure construction algorithm and sequentialization theorem. □

6 Related works

Before Linear Logic was born, a proof search method similar to our approach had been studied [KW84], while the logic was affine logic (in current terminology), the fragment was restricted to multiplicative fragment, and they did not consider the issues of complexity. With respect to the multiplicative fragment of Linear Logic, there is a work of proof search based on proof nets such as [GP91], while the fragment was restricted to multiplicative fragment, they did not use a correctness condition and, they did not consider the issues of complexity.

As to model theoretical method of proof search for Linear Logic, there is a work of [Laf95]. However though the work has some theoretical interests, the proof search method is a syntactical method, not semantical method, since the model used for proof search is defined by syntax.

7 Concluding remarks —— an extension to the exponential connectives

In this section, we shall present the sketch of the extension of the proof search algorithm for MALL to the standard system with the exponential connectives of Linear Logic.
First, in addition to the links in Section 2, the following links are added:

$$\text{d-links } \frac{A}{?A} \qquad \text{w-links } ?A \qquad \text{?-links } \frac{?A \cdots ?A}{?A}$$

where the number m of premises in ?-links is greater than 0.

In the constructions of terminal forests in the proof structure construction algorithm, the number m of the premises of any ?-subformula in a given sequent is chosen non-deterministically, where $m \geq 0$. If $m = 0$, then the premise of the ?-link is the conclusion of a w-link, if $m = 1$, then the premise of the ?-link is the conclusion of a d-link, and if $m > 1$, the that means in the premise of the ?-link, contraction rules apply $m - 1$ times. Moreover, in the construction the following link is needed: !-links $\frac{A}{!A}$.
The notion of copy in Definition 8 of Section 3 is extended. The copy operations do not apply to d-links and w-links, but apply to ?-links. The constraints for the number of copy do not change. The weights of the premises and that of the conclusion of the links added here are the same. Hence, the proof structure algorithm are easily extended to the case containing the exponential connectives. The correctness condition for a given proof structure Θ is the same as the MALL case except for finding !-boxes:

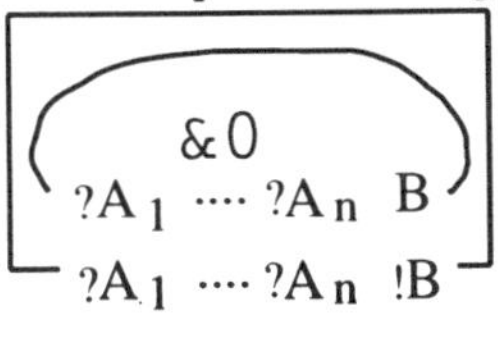

Non-deterministically, to any !-formula A in Θ, a finite set of ?-formulas in Θ (which have the same weight as that of A) is assigned. According to the assignment, some boxes are introduced in Θ. From inner boxes to outer ones in Θ, according to the correctness condition, any box in Θ are tested whether if it is a proof net. Then d-links behave like plus links, w-links like generalized axiom links with one conclusion, and ?-links like $\wp$-links with unspecified number of the premises.

References

[And92] J.-M. Andreoli. Logic programming with focusing proofs in linear logic. *Journal of Logic and Computation*, 2(3), 1992.

[Bel94] G. Bellin. On Proof Nets for Multiplicative and Additive Linear Logic: Correctness Conditions and Intuitionistic Translations. Available by ftp anonymous on `theory.doc.ic.ac.uk`, in `papers/Bellin`, 1994.

[GP91] D. Galmiche and G. Perrier. A procedure for automatic proof nets construction. LNAI 624, 1992.

[Gir87] J.-Y. Girard. Linear logic. *Theoretical Computer Science*, 50:1–102, 1987.

[Gir95] J.-Y. Girard. Proof-nets: the parallel syntax for proof-theory. In Ursini and Agliano, editors, *Logic and Algebra, New York, Marcel Dekker*, 1995.

[HM94] J.S. Hodas and D. Miller. Logic Programming in a Fragment of Intuitionistic Linear Logic. *Information and Computation*, 110(2):327-365, 1994.

[KW84] J. Ketonen and R. Weyhrauch. A decidable fragment of predicate calculus. *Theoretical Computer Science*, 32:297–307, 1984.

[Laf95] Y. Lafont. The finite model theory for various fragments linear logic. Available by ftp anonymous on `lmd.univ-mrs.fr`, in `pub/lafont`, 1995.

[Tam94] T. Tammet. Proof search strategies in linear logic. *Journal of Automated Reasoning*, 12:273-304, 1994.

Computability Structures on Metric Spaces

Takakazu Mori, Yoshiki Tsujii and Mariko Yasugi

Faculty of Science, Kyoto Sangyo University, Kita-ku, Kyoto 603, Japan

Abstract. The *computability structure* on a metric space and some related properties are considered. A *computability structure* on a metric space $\langle X, d\rangle$ is a set of sequences from X which satisfies three axioms; the computability of metrics (M1), the closedness under recursive re-enumeration (M2) and the closedness under effective convergence (M3). *Effective completion*, *effectively dense r.e. open sets* and *computable functions* are studied.

1 Introduction

In [P-R], Pour-El and Richards define a *computability structure* on a Banach space, and prove that, for a linear operator, *boundedness* and *preservation of the computability* are equivalent, presuming that certain conditions are satisfied.

A *computability structure on a Banach space* is a set of sequences, which reflects the original space in certain sense. A computability structure is closed with respect to effective linear forms and effective convergence, and the norms of computable sequence form a computable sequence of reals.

Washihara in [W] extended this idea to Fréchet spaces. Washihara and Yasugi in [W-Y] consider the computability problems of *metrizations* of Fréchet spaces. Some metrics induced from seminorms yield computable reals from a computable sequence. Whether it is the case with every metrization is yet an open problem.

The observation of these preceding works triggered us to investigate the computability structure on a metric space. We report here some basic results on this theme which have been obtained so far. We are planning to give an extended treatment in a forthcoming work.

The *computability structure of a metric space* is defined to be a set $\mathcal{S}$ of sequences from the space which satisfies the following axioms (§3):

Axiom M1 (Metrics). *If $\{x_m\}, \{y_n\} \in \mathcal{S}$, then $\{d(x_m, y_n)\}_{m,n}$ forms a double sequence of computable reals.*

Axiom M2 (Re-enumerations). *If $\{x_n\} \in \mathcal{S}$, then $\{x_{\Phi(n)}\} \in \mathcal{S}$ for any recursive function Φ.*

Axiom M3 (Limits). *If $\{x_{mn}\} \in \mathcal{S}$, $\{x_m\} \subset X$ and $\{x_{mn}\}$ converges to $\{x_m\}$ as $n \longrightarrow \infty$, effectively in m and n, then $\{x_m\} \in \mathcal{S}$.*

The notion of an *effective separating set* is also defined. Several examples of computability structures are given in §4.

Effective completion of a metric space with a computability structure and with an effective separating set is presented in §5. Here it is required that *effectively Cauchy* sequences have limits within the computability structure (Axiom M3′). Let X be a metric space with a computability structure $\mathcal{S}$, which is not complete, and let $\bar{X}$ be its completion. We extend $\mathcal{S}$ to $\bar{\mathcal{S}}$ so that the metric space $\bar{X}$ with $\bar{\mathcal{S}}$ is effectively complete, i.e., $\bar{\mathcal{S}}$ satisfies Axiom M3′ (Theorem 12).

This fact indicates that the notion of computable sequences of reals in the sense of Pour-El and Richards is natural from the viewpoint of the effective completion of the spaces of rationals.

The notion of *effectively dense open sets* is defined, and *the effective Baire's theorem* is proved in §6. In §7 *effective total boundedness, effective compactness, computable maps* and *co-r.e. closed sets* are defined, and the effectivization of some fundamental results on metric spaces, such as *the effective Urysohn's lemma*, are proved. In particular, Theorem 24 states that any continuous function on an effective compact space is approximated by computable functions with respect to the maximum norm.

2 Preliminaries

We give some preliminary definitions and their consequences here. Some of the terminology is quoted from Chapter 0 of [P-R].

Definition 1.

1. A sequence of rationals, say $\{r_k\}$, is said to be computable if there exist three recursive functions α, β and σ such that

$$r_k = (-1)^{\sigma(k)} \frac{\alpha(k)}{\beta(k)},$$

where $\beta(k)$ is assumed to take positive values for all k.
2. A double sequence $\{r_{kn}\}$ of rational numbers converges effectively to a sequence of real numbers $\{x_n\}$ if there exists a recursive function α such that, for all p and all $k \geq \alpha(p,n)$, $|r_{kn} - x_n| \leq 1/2^p$.
3. A sequence $\{x_n\}$ of reals is said to be computable if there exists a computable double sequence of rationals which converges effectively to $\{x_n\}$.

Definition 2. Let $\{r_k\}$ be a computable sequence of rationals. $\{r_k\}$ is said to be effectively Cauchy if there exists a recursive function α such that, for all p and all $m, n \geq \alpha(p)$, $|r_m - r_n| \leq 1/2^p$.

Proposition 3. *For a computable sequence of rationals $\{r_k\}$, the two conditions below are equivalent.*

1. *$\{r_k\}$ effectively converges to a (computable) real.*
2. *$\{r_k\}$ is effectively Cauchy.*

Proof. 1 implies 2: Suppose $\{r_k\}$ effectively converges to a real number a by a recursive function α. (It follows that a is computable.)

For every p, and for all $m, n \geq \alpha(p)$,

$$|r_n - r_m| \leq |a - r_n| + |a - r_m| \leq \frac{1}{2^p} + \frac{1}{2^p} = \frac{1}{2^{p-1}}.$$

If we put $\beta(p) = \alpha(p+1)$, then $\{r_k\}$ is effectively Cauchy by β.

2 implies 1: Suppose $\{r_k\}$ is effectively Cauchy by a recursive function β. Then there is a real number a to which $\{r_k\}$ classically converges. So it holds

$$|r_n - a| \leq \limsup_{m \to \infty} |r_n - r_m|.$$

On the other hand, since $|r_n - r_m| \leq 1/2^p$ for every p and all $n, m \geq \beta(p)$,

$$|r_n - a| \leq \limsup_{m \to \infty} |r_n - r_m| \leq \frac{1}{2^p}.$$

Putting $\alpha(p) = \beta(p)$, we obtain that $\{r_k\}$ effectively converges to a real number a by α. □

Definition 4. Let $\langle X, d \rangle$ be a metric space, let $\{x_n\}$ be a sequence from X and let $\{x_{nk}\}$ be a double sequence from X. $\{x_{nk}\}$ is said to converge to $\{x_n\}$ as $k \longrightarrow \infty$, effectively in n and k, if there exists a recursive function β such that, for all p and all $k \geq \beta(p, n)$, $d(x_{nk}, x_n) \leq 1/2^p$.

$\{x_n\}$ is said to be effectively Cauchy if there exists a recursive function α such that, for all p and all $m, n \geq \alpha(p)$, $d(x_m, x_n) \leq 1/2^p$.

3 Computability structure on a metric space

We define the notion of a computability structure on a metric space, and list some basic properties which are immediate consequences of the definitions.

We denote a metric space by $\langle X, d \rangle$. $\mathcal{S}$ will denote a set of sequences from X.

Remark. A double sequence $\{x_{nk}\}$ can be regarded as a sequence in terms of standard recursive pairing functions. The same holds with n-fold sequences. We will thus regard $\mathcal{S}$ as a set of sequences from X with arbitrary numbers of indices.

We define a computability structure on X and related notions in a manner similar to Section 1 of Chapter 2 in [P-R].

Definition 5 (Computability Structure). $\mathcal{S}$ will be called a computability structure on $\langle X, d \rangle$ if it satisfies the three axioms below. A sequence in $\mathcal{S}$ is called a computable sequence (relative to $\mathcal{S}$).

Axiom M1 (Metrics). *If $\{x_m\}, \{y_n\} \in \mathcal{S}$, then $\{d(x_m, y_n)\}_{m,n}$ forms a double sequence of computable reals.*

Axiom M2 (Re-enumerations). *If $\{x_n\} \in \mathcal{S}$, then $\{x_{\Phi(n)}\} \in \mathcal{S}$ for any recursive function Φ.*

Axiom M3 (Limits). *If $\{x_{mn}\} \in \mathcal{S}$, $\{x_m\} \subset X$ and $\{x_{mn}\}$ converges to $\{x_m\}$ as $n \longrightarrow \infty$, effectively in m and n, then $\{x_m\} \in \mathcal{S}$. (See Definition 4 for effective convergence.)*

Corollary 6. *If $\{x_n\}, \{y_n\} \in \mathcal{S}$, then $\{x_0, y_0, x_1, y_1, x_2, y_2, \ldots\} \in \mathcal{S}$.*

Remark. When $\{x, x, x, \ldots\}$ is computable, we say x is a computable element of X(relative to $\mathcal{S}$).

Subsequently, $\mathcal{S}$ will denote a computability structure, and the triplet $\langle X, d, \mathcal{S} \rangle$ will denote a metric space with a computability structure $\mathcal{S}$. The propositions below are metric space versions of the corresponding results in [P-R] on a Banach space.

Definition 7 (Effective Separability). $\langle X, d, \mathcal{S} \rangle$ is said to be effectively separable (with respect to $\mathcal{S}$) if there exists a sequence $\{e_n\}$ in $\mathcal{S}$ which is dense in X.

In this case, $\{e_n\}$ is called an effective separating set for $\langle X, d, \mathcal{S} \rangle$.

Proposition 8 (Effective Density Lemma, [P-R, Chapter 2, Section 5]). *Let $\{e_n\}$ be an effective separating set, and let $\{x_n\}$ be a sequence from X, then the following are equivalent.*

1. *$\{x_n\} \in \mathcal{S}$.*
2. *There is a recursive function α such that $\{e_{\alpha(n,k)}\}$ converges to $\{x_n\}$ as $k \longrightarrow \infty$, effectively in n and k.*

Proof. Axioms M2 and M3 yield that 2 implies 1.

Assume 1. Then $\{d(e_i, x_n)\}$ is a computable double sequence of reals. By the denseness of $\{e_i\}$, it holds that, for every n and every k, there is an i satisfying

$$d(e_i, x_n) < \frac{1}{2^k}. \tag{1}$$

There is an algorithm to obtain such an i recursively in n and k. (By Proposition 0 in Chapter 0 of [P-R], $a < b$ can be effectively decided for computable a and b when the inequality is true. By running the evaluation of the inequality in 1 and changing i simultaneously, one eventually hit a desired i.) This algorithm is defined uniformly in n and k.

Write $\alpha(n, k)$ for such i. Then α is recursive, and has the desired property. □

Proposition 9 (Stability Lemma, [P-R, Chapter 5, Section 5]). *Suppose there are two computability structures, $\mathcal{S}_1$ and $\mathcal{S}_2$, on $\langle X, d \rangle$, and suppose $\{e_n\}$ is an effective separating set for $\langle X, d, \mathcal{S}_1 \rangle$ and $\langle X, d, \mathcal{S}_2 \rangle$. Then $\mathcal{S}_1 = \mathcal{S}_2$.*

Proof. Suppose $\{e_n\} \in \mathcal{S}$ for a computability structure $\mathcal{S}$. $\mathcal{S}$ is uniquely determined by the condition 2 in Proposition 8. So follows the proposition. □

4 Examples

We give some examples of metric spaces with computability structures and effective separating sets.

Example 1 (Rational Numbers). Consider $\langle \mathbf{Q}, d \rangle$, where $\mathbf{Q}$ is the set of all rationals and $d(x, y) = |x - y|$. Let $\mathcal{S}$ be all computable sequences of rationals.

In this case, every element of $\mathbf{Q}$ is computable. Let $\{e_n\}$ be an effective enumeration of $\mathbf{Q}$. Then $\{e_n\}$ is an effective separating set for $\langle \mathbf{Q}, d, \mathcal{S} \rangle$.

Example 2 (Real Numbers). Consider $\langle \mathbf{R}, d \rangle$, where $\mathbf{R}$ is the set of all real numbers and $d(x, y) = |x - y|$. Let $\mathcal{S}$ be the set of all computable sequences of reals.

$\{e_n\}$ in Example 1 is an effective separating set for this metric space.

Example 3 (Product Space). Let $\langle X_j, d_j, \mathcal{S}_j \rangle$, $j = 1, 2, \ldots, m$, be metric spaces with computability structures.

Put $X \equiv X_1 \times X_2 \times \cdots \times X_m$ and

$$d(\boldsymbol{x}, \boldsymbol{y}) = \sqrt{\sum_{j=1}^{m} d_j(x_j, y_j)^2}$$

where $\boldsymbol{x} = (x_1, x_2, \ldots, x_m)$ and $\boldsymbol{y} = (y_1, y_2, \ldots, y_m)$.

Also put

$$\mathcal{S} = \left\{ \{(x_1^{(i)}, x_2^{(i)}, \ldots, x_m^{(i)})\}_i \mid \{x_j^{(i)}\}_i \in \mathcal{S}_j \text{ for } j = 1, 2, \ldots, m \right\}$$

Suppose $\{e_{jn}\}_n$ is an effective separating set for $\langle X_j, d_j, \mathcal{S}_j \rangle$, $j = 1, 2, \ldots, m$. Then $\{(e_{1n_1}, e_{2n_2}, \ldots, e_{mn_m})\}_{n_1, n_2, \ldots, n_m}^{\infty}$ is an effective separating set for $\langle X, d, \mathcal{S} \rangle$.

Example 4 (Banach space, [P-R, Chapter 2]). Let $B \equiv \langle X, || \cdot || \rangle$ be a Banach space, $\mathcal{S}_B$ be a computability structure on B and $\{f_m\}$ be an effective generating set relative to $\mathcal{S}_B$. We obtain a metric space $\langle X, d \rangle$ by $d(x, y) = ||x - y||$. Then $\mathcal{S} \equiv \mathcal{S}_B$ is a computability strucrure for $\langle X, d \rangle$. Let $\{e_n\}$ be an effective enumeration of all the rational combinations of $\{f_m\}$. Then $\{e_n\}$ is an effective separating set for $\langle X, d, \mathcal{S} \rangle$ relative to $\mathcal{S}$.

Example 5 (Fréchet space). Let $\langle X, \{p_k\} \rangle$ be a Fréchet space wth seminorms $\{p_k\}$. These seminorms induce several metrics.

1. Metrics d_1 and d_2.

$$d_1(x, y) = \sum_{k=1}^{\infty} \frac{p_k(x - y)}{2^k(1 + p_k(x - y))}$$

$$d_2(x, y) = \max_k \frac{p_k(x - y)}{2^k(1 + p_k(x - y))}$$

In (2) of **1** in [W], the notion of a computability structure $\mathcal{S}$ on $\langle X, \{p_k\}\rangle$ is defined. In 3 of [W-Y], it is claimed that, for computable sequences $\{x_n\}$ and $\{y_n\}$ in $\mathcal{S}$, $\{d_i(x_m, y_n)\}_{m,n}$ is a computable double sequence of reals, for $i = 1, 2$.
We claim that $\langle X, d_i, \mathcal{S}\rangle$ is a metric space with the computability structure $\mathcal{S}$. If $\{f_n\}$ is an effective generating set of $\langle X, \{p_n\}, \mathcal{S}\rangle$ (Definition 2 in **1** of [W]), then an effective enumeration of all the rational combinations of $\{f_n\}$ is an effective separating set for $\langle X, d_i, \mathcal{S}\rangle$.

2. Metric d_3.
Put $U_n = \{x \in X \mid p_n(x) < 1/2^n\}$. Let r be a positive rational which can be expanded as $r = 1/2^{i_1} + 1/2^{i_2} + \cdots + 1/2^{i_\ell}$, where $i_1, i_2, \ldots, i_\ell$ are integers satisfying $1 \le i_1 < i_2 < \cdots < i_\ell$. Write $U_{i_1} + U_{i_2} + \cdots + U_{i_\ell}$ as $A(r)$.
Now define a function by

$$\varphi(x) = \begin{cases} \inf\{r \mid x \in A(r)\} & \text{if } x \in A(r) \text{ for some } r \\ 1 & \text{otherwise.} \end{cases}$$

$d_3(x, y) = \varphi(x - y)$ is a metric on X defined in [R].
If the condition

$$V(i_1, r) \subset A(r) \qquad \text{where } V(i_1, r) = \{x \in X \mid p_{i_1}(x) < r\} \tag{2}$$

holds, then we can show that $\{\varphi(x_n)\}$ is a computable sequence of reals when $\{x_n\}$ is a computable sequence from X (Theorem 1 in **4** of [W-Y]). Condition (2) holds for $X = \mathbf{R}^\infty$ (Proposition 4.1 in [W-Y]) and $X = C(\mathbf{R})$ (Proposition 4.3 in [W-Y]), but not for $X = C^\infty[0, 1]$.

Example 6 (Infinite product of $\{0, 1\}$*).* Let X_0 be the set $\{0, 1\}$ and let d_0 be a metric on X_0 defined by

$$d_0(x, y) = \begin{cases} 1 \text{ if } x \neq y \\ 0 \text{ if } x = y. \end{cases}$$

Now we define X, d and $\mathcal{S}$ as follows. Put $X = X_0^{\mathbf{N}}$ where $\mathbf{N} = \{1, 2, 3, \ldots\}$.

$$d(\boldsymbol{x}, \boldsymbol{y}) = \sum_{k=1}^{\infty} \frac{d_0(x_k, y_k)}{2^k} \quad \text{for } \boldsymbol{x} = (x_k) \text{ and } \boldsymbol{y} = (y_k) \text{ from } X$$

$$\mathcal{S} = \left\{ \{(x_k^{(n)})_k\}_n \mid \{x_k^{(n)}\}_{k,n} \text{ is a recursive double sequence of } 0\text{'s and } 1\text{'s} \right\}$$

Let E be the set of the elements of X which have finitely many 1's. There is an enumeration of E, say $\{(e_k^{(n)})_k\}_n$, which belongs to $\mathcal{S}$. This serves as an effective separating set for $\langle X, d, \mathcal{S}\rangle$.

5 Complete metric space with computability structure

In this section, we deal with complete metric spaces. For a complete metric space with computability structure, say $\langle X, d, \mathcal{S} \rangle$, it is natural to require the following Completeness Axiom M3$'$ instead of Axiom M3.

Axiom M3$'$ (Completeness). *Let $\{x_{mn}\}$ be a computable sequence of effectively Cauchy sequences; $d(x_{mk}, x_{m\ell})$ converges to 0 as $k, \ell \longrightarrow \infty$, effectively in k, ℓ and m. Then there exists $\{x_m\} \in \mathcal{S}$ such that $\{x_{mk}\}$ converges to $\{x_m\}$ as $k \longrightarrow \infty$, effectively in k and m.*

Definition 10. Let $\langle X, d \rangle$ be a complete metric space. The triplet $\langle X, d, \mathcal{S} \rangle$ is called an effectively complete metric space with a computability structure $\mathcal{S}$ if $\mathcal{S}$ satisfies Axioms M1, M2 and M3$'$.

Let $\langle X, d, \mathcal{S} \rangle$ be a metric space with a computability structure $\mathcal{S}$. Suppose that $\langle X, d \rangle$ is not complete and let $\langle \bar{X}, \bar{d} \rangle$ be the completion of $\langle X, d \rangle$. We will define the completion $\bar{\mathcal{S}}$ of $\mathcal{S}$ such that $\langle \bar{X}, \bar{d}, \bar{\mathcal{S}} \rangle$ is an effectively complete metric space $\langle \bar{X}, d \rangle$ with the computability structure $\bar{\mathcal{S}}$.

Definition 11. A sequence $\{\boldsymbol{x}_n\} \subset \bar{X}$ is called an effective sequence if there exists a double sequence $\{x_{nk}\} \in \mathcal{S}$ such that

1. $\{\{x_{nk}\}_k\}_n$ is a computable sequence of effectively Cauchy sequences,
2. $\{x_{nk}\}_k$ is a representation of $\boldsymbol{x}_n$ for any $n = 1, 2, \ldots$.

Theorem 12. *Put $\bar{\mathcal{S}} = \{\{\boldsymbol{x}_n\} \mid \{\boldsymbol{x}_n\}$ is an effective sequence$\}$. Then $\langle \bar{X}, \bar{d}, \bar{\mathcal{S}} \rangle$ is an effectively complete metric space with computability structure $\bar{\mathcal{S}}$.*

Proof.

1. (Axiom M1) Let $\{\boldsymbol{x}_m\}, \{\boldsymbol{y}\}_n \in \bar{\mathcal{S}}$, and their representations be respectively $\{x_{mk}\} \in \mathcal{S}$ with a recursive function $\alpha(p, m)$ and $\{y_{n\ell}\} \in \mathcal{S}$ with $\beta(p, n)$. Note that $\{d(x_{mk}, y_{nk})\}_{m,n,k}$ is a computable sequence of reals, and that if $k, \ell \geq \max\{\alpha(p, m), \beta(p, n)\}$,
$$\begin{aligned} &|d(x_{mk}, y_{nk}) - d(x_{m\ell}, y_{n\ell})| \\ &\leq |d(x_{mk}, y_{nk}) - d(x_{m\ell}, y_{nk})| + |d(x_{m\ell}, y_{nk}) - d(x_{m\ell}, y_{n\ell})| \\ &\leq d(x_{mk}, x_{m\ell}) + d(y_{nk}, y_{n\ell}) \leq 2\,\frac{1}{2^p}. \end{aligned}$$
Since $\bar{d}(\boldsymbol{x}_m, \boldsymbol{y}_n) = \lim_{i \to \infty} d(x_{mi}, y_{ni})$ and the convergence is effective, it holds that $\{\bar{d}(\boldsymbol{x}_m, \boldsymbol{y}_n)\}_{m,n}$ is a computable double sequence of reals.
2. (Axiom M2) Let $\{\boldsymbol{x}_n\} \in \bar{\mathcal{S}}$ and $\{x_{nk}\}$ be their representation with a recursive function $\alpha(p, n)$. Then $\{x_{\Phi(n),k}\}$ is a representation of $\{\boldsymbol{x}_{\Phi(n)}\}$ with a recursive function $\alpha(p, \Phi(n))$.

3. (Axiom M3′) Suppose that $\{x_{mn}\} \in \bar{\mathcal{S}}$, and that there is a recursive function $\beta(p,m)$ such that $\bar{d}(x_{mk}, x_{m\ell}) \le 1/2^p$ if $k,\ell \ge \beta(p,m)$, that is, $\{x_{mn}\}_n$ is effectively Cauchy with respect to n.
Let $\{x_{mnk}\}$ $(\in \mathcal{S})$ be a representation of $\{x_{mn}\}$ with the recursive function $\alpha(p,m,n)$. Put $y_{m,i} = x_{m,i,\alpha(i,m,i)}$ for all m and i. We show that $\{y_{m,i}\}_{m,i}$ defines an effective sequence $\{y_m\}_m \in \bar{\mathcal{S}}$ and that $\{x_{mn}\}$ effectively converges to $\{y_m\}$.
Note that for any ℓ,

$$\begin{aligned} &d(x_{m,i,\alpha(i,m,i)}, x_{m,j,\alpha(j,m,j)}) \\ &\le d(x_{m,i,\alpha(i,m,i)}, x_{mi\ell}) + d(x_{mi\ell}, x_{mj\ell}) + d(x_{mj\ell}, x_{m,j,\alpha(j,m,j)}). \end{aligned}$$

Since $\{x_{m,i,\ell}\}_\ell$ is a Cauchy sequence for any m and i, it follows that for all $i, j \ge \max\{\beta(p,m), p\}$,

$$\begin{aligned} &d(y_{mi}, y_{mj}) \\ &= d(x_{m,i,\alpha(i,m,i)}, x_{m,j,\alpha(j,m,j)}) \\ &\le \lim_{\ell\to\infty} d(x_{m,i,\alpha(i,m,i)}, x_{mi\ell}) + \bar{d}(x_{mi}, x_{mj}) + \lim_{\ell\to\infty} d(x_{mj\ell}, x_{m,j,\alpha(j,m,j)}) \\ &\le 1/2^i + 1/2^p + 1/2^j \\ &\le 3/2^p. \end{aligned}$$

This shows that $\{y_m\}_m \in \bar{\mathcal{S}}$.
By the definition of $\{y_m\}_m$, it is easy to see that $\{x_{mk}\}$ converges to $\{y_m\}$ as $k \longrightarrow \infty$, effectively in k and m. □

Remark. Let $\langle \mathbf{Q}, d, \mathcal{S} \rangle$ be the metric space of all rationals with the computability structure $\mathcal{S}$ defined in Example 1 (§4). We obtain its completion $\langle \mathbf{R}, \bar{d}, \bar{\mathcal{S}} \rangle$ according to the construction described above. Then $\bar{\mathcal{S}}$ consists of all computable sequences of reals in the sense of Chapter 0 of [P-R]. That is, $\langle \mathbf{R}, \bar{d}, \bar{\mathcal{S}} \rangle$ coincides with the space in Example 2.

This fact suggests us that the notion of computable sequences of reals in the sense of Pour-El and Richards is natural from the viewpoint of the effective completion of **Q**.

Let $\langle X, d, \mathcal{S} \rangle$ be a metric space with a computability structure and with an effective separating set $\{e_i\}$. We define a computability structure $\bar{\mathcal{S}}(\{e_i\})$ derived from $\{e_i\}$ as follows:

$$\bar{\mathcal{S}}(\{e_i\}) := \left\{ \{e_{\Phi(n,j)}\}_{n,j} \;\middle|\; \begin{array}{l} \text{(i) } \Phi(n,j) \text{ is a recursive function} \\ \text{(ii) } \{e_{\alpha(n,j)}\}_{n,j} \text{ is a computable sequence of} \\ \text{effectively Cauchy sequences} \end{array} \right\}.$$

Note that $\{e_{\Phi(n,j)}\}_{n,j}$ is an effective sequence in $\bar{X}$.

Theorem 13. *$\bar{\mathcal{S}}(\{e_i\}) = \bar{\mathcal{S}}$. That is, $\bar{\mathcal{S}}$ is constructed from any effective separating set (relative to $\mathcal{S}$) by the above procedure.*

This result follows from the Effective Density Lemma (Proposition 8).

6 Effective Baire's Theorem

In this section, we prove an effective version of Baire's Theorem, assuming that $\langle X, d, \mathcal{S} \rangle$ is a complete metric space with a computability structure.

Definition 14. A sequence of open sets $\{U_k\}$ is said to be effectively dense if there exist an effective separating set $\{e_j\}$, an effective enumeration of all positive rationals $\{s_\ell\}$ and recursive functions β_1 and β_2 such that

$$U_k \cap \mathbf{B}(e_j, s_\ell) \supset \mathbf{B}\left(e_{\beta_1(k,j,\ell)}, s_{\beta_2(k,j,\ell)}\right) \quad \text{for all} \quad k, j, \ell,$$

where $\mathbf{B}(e, s) = \{x \mid d(x, e) < s\}$.

Proposition 15 (Effective Baire's theorem). *Suppose $\{U_k\}$ is a sequence of effectively dense open sets. Then there exists a sequence $\{x_q\}$ in $\mathcal{S}$ such that*

$$x_q \in \left(\bigcap_k U_k\right) \cap \mathbf{B}(e_{q_1}, s_{q_2})$$

for all (q_1, q_2), where $q = j(q_1, q_2)$ for a standard pairing function j.

7 Computable functions on an effectively compact space

We will denote a metric space with a computability structure and an effective separating set $\{e_n\}$ by $\langle X, d, \mathcal{S}, \{e_n\} \rangle$.

Definition 16. $\langle X, d, \mathcal{S} \rangle$ is said to be effectively totally bounded if there exists recursive function α such that $X = \bigcup_{n=1}^{\alpha(p)} \mathbf{B}(e_n, 1/2^p)$ for all p.

If $\langle X, d, \mathcal{S} \rangle$ is effectively totally bounded and effectively complete, then we say that $\langle X, d, \mathcal{S} \rangle$ is effectively compact.

Assumption. *In the discussion that follows, $\mathcal{X} = \langle X, d, \mathcal{S}, \{e_n\} \rangle$ is assumed to be effectively compact and $\mathcal{Y} = \langle Y, d', \mathcal{S}', \{e'_n\} \rangle$ is assumed to be effectively complete.*

Definition 17. A map $f : \mathcal{X} \longrightarrow \mathcal{Y}$ is said to be computable if it satisfies the following conditions.

1. f maps $\mathcal{S}$ into $\mathcal{S}'$.
2. f is effectively uniformly continuous on X, that is, there exists a recursive function α such that, for all p and for all $x, y \in X$, $d(x, y) \leq 1/\alpha(p)$ implies $d'(f(x), f(y)) \leq 1/2^p$.

Remark. A sequence $\{f_n\}$ of maps from $\mathcal{X}$ to $\mathcal{Y}$ is said to be computable if

1. a double sequence $\{f_n(x_m)\}_{n,m} \in \mathcal{S}'$ for any $\{x_m\} \in \mathcal{S}$, and
2. $\{f_n\}$ are effectively uniformly continuous on X, that is, there exists a recursive function $\alpha(p,n)$ such that, for all p,n and all $x,y \in X$,

$$d(x,y) \leq \frac{1}{\alpha(p,n)} \quad \text{implies} \quad d'(f(x),f(y)) \leq \frac{1}{2^p}.$$

From the definition, we immediately obtain the next corollary.

Corollary 18.

1. *The identity on X is computable.*
2. *For every $y \in Y$, if $\langle y,y,\ldots\rangle \in \mathcal{S}'$, then the constant function $C_y(x) \equiv y$ is computable.*

We will restrict ourselves to real-valued functions in the rest of this article.

Proposition 19 (Maximum values). *Let $f_n : X \longrightarrow \mathbf{R}$ be a computable sequence of functions. Then the maximum values $s_n = \max\{f_n(x) : x \in X\}$ form a computable sequence of reals.*

Proposition 20 (Closure under effective uniform convergence). *Let $f_n : X \longrightarrow \mathbf{R}$ be a computable sequence of functions such that $f_n \longrightarrow f$ as $n \longrightarrow \infty$, uniformly in x, effectively in n; that is, there exists a recursive function α such that, for all n and p,*

$$n \geq \alpha(p) \quad \textit{implies} \quad |f_n(x) - f(x)| \leq \frac{1}{2^p}.$$

Then f is a computable function.

Definition 21. An open set $U \subset X$ is said to be recursively enumerable (r.e.) if there exist a sequence $\{a_n\}$ in $\mathcal{S}$ and a computable sequence of positive rationals r_n such that $U = \bigcup_n \mathbf{B}(a_n, r_n)$. If $X - V$ is r.e., we call V co-r.e.

Remark. $X - U$ is not necessarily r.e. even if U is r.e.

If a point $x \in X$ is computable, then the set $\{x\}$ is co-r.e.

Lemma 22. *Let $U \subset X$ be a r.e. open set with respect to $(\{a_n\},\{r_n\})$. Then there exists a computable, real-valued function $h_U : X \longrightarrow [0,1]$ such that $x \in U$ if and only if $h_U(x) > 0$.*

Proof. Put

$$\varphi(n,x) = \frac{\max(0, r_n - d(a_n,x))}{r_n 2^{n+1}}.$$

$\{\varphi(n,x)\}_n$ is a computable sequence of functions, and it holds that

$$0 \leq \varphi(n,x) \leq \frac{1}{2^{n+1}} < \frac{1}{2^n}.$$

Note that $x \in U$ if and only if there is an n such that $\varphi(n,x) > 0$.

Put $\psi(n,x) = \sum_{m=1}^{n} \varphi(m,x)$. Then $\{\psi(n,x)\}$ is a computable sequence of real-valued functions. Define

$$h_U(x) = \lim_{n\to\infty} \psi(n,x).$$

It is obvious that $0 \leq h_U(x) \leq 1$ and that $x \in U$ if and only if $h_U(x) > 0$. Since the convergence of $\psi(n,x)$ to $h_U(x)$ is uniformly in x and effective in n, h_U is computable by Proposition 20. □

Using this lemma, we obtain the following proposition.

Proposition 23 (Effective Urysohn's Lemma). *Let $C_0, C_1 \subset X$ be closed co-r.e. sets which are mutually disjoint. Then there exists a computable function $g : X \longrightarrow [0,1]$ such that $x \in C_i$ if and only if $g(x) = i$ $(i = 0,1)$.*

Proof. Put $U_i = X - C_i$ $(i = 0,1)$ and let h_i be the h_{U_i} in Lemma 22. Define g by

$$g(x) = \frac{h_0(x)}{h_0(x) + h_1(x)}.$$

This function g satisfies the desired property. □

This proposition means that the family of computable functions separates two computable points in X. This in turn implies that computable functions are dense in the space of continuous functions on X.

Theorem 24 (Approximation). *Let $f : X \longrightarrow \mathbf{R}$ be a continuous function. For any $\epsilon > 0$, there exists a computable function $g : X \longrightarrow \mathbf{R}$ such that*

$$\max\{|f(x) - g(x)| : x \in X\} < \epsilon.$$

Since the effective compactness is defined in terms of the effective total boundedness and the effective completeness, we have proved the theorem by constructing the function g in an explicit way. The proof is rather long, and hence will be omitted. It will be included in a later work.

Remark. The theorem immediately above cannot be effectivized in ϵ, nor can it be extended to a sequence of functions.

Acknowledgements. The article has been revised according to the referee's comments. Our interest in computability structure in analysis was ignited by the communications with Marian B. Pour-El, especially by the discussions with her at Kyoto Sangyo University. We express our gratitude to her.

References

[P-R] M. Pour-El and J. Richards: Computability in Analysis and Physics. Springer-Verlag, 1989.

[W] M. Washihara: Computability and Fréchet spaces. Mathematica Japonica **42** (1995) 1–13.

[W-Y] M. Washihara and M. Yasugi: Computability and Metrics in a Fréchet space. Mathematica Japonica **43** (1996) 431–443.

[R] W. Rudin: Functional Analysis. McGraw-Hill Inc. , 1973.

EDT0L Structural Equivalence is Decidable*

Kai Salomaa and Sheng Yu

Department of Computer Science
University of Western Ontario
London, Ontario, Canada N6A 5B7

Abstract. We reduce the structural equivalence problem for EDT0L grammars to deciding emptiness of the tree language recognized by a tree pushdown automaton. This implies that structural equivalence of EDT0L grammars can be decided effectively. The result contrasts the known undecidability result for ET0L structural equivalence.

1 Introduction

Context-free grammars G_1 and G_2 are said to be structurally equivalent if corresponding to each syntax tree of G_1 producing a terminal word, the grammar G_2 has a syntax tree with the same structure, and vice versa. The structure of a syntax tree t is the leaf-labeled tree obtained from t by removing the nonterminals labeling the internal nodes. The importance of the notion of structural equivalence for context-free grammars is due to the fact that it can be decided effectively [11, 13, 17] whereas language equivalence is undecidable.

Structural equivalence remains decidable also for parallel context-free (E0L) grammars [14, 15, 16, 24]. Surprisingly it was shown in [23] that when the parallel derivations are controlled by a finite set of tables (in an ET0L grammar), structural equivalence is undecidable. We cannot even decide whether an E0L grammar and an ET0L grammar having two tables are structurally equivalent.

Here we show that structural equivalence becomes decidable if the tables of the grammars are restricted to be homomorphisms, that is, we have EDT0L grammars. Our proof uses automata theoretic methods but differs considerably from the automata theoretic decidability proofs for E0L structural equivalence [24] and ET0L strong structural equivalence [9]. The E0L structure trees, as well as ET0L structure trees augmented with the information about the control-sequence used, can be recognized deterministically bottom-up using an automaton model for which equivalence is decidable. This appears not possible for EDT0L structure trees since the arbitrary choice of the sequence of tables makes the derivation inherently nondeterministic.

We reduce EDT0L structural equivalence to the emptiness problem for the tree pushdown automata of Guessarian [7]. These automata recognize exactly the context-free tree languages and emptiness can be decided algorithmically.

* This work has been supported by the Academy of Finland Grant 14018 and the Natural Sciences and Engineering Research Council of Canada grant OGP0041630.

We show that, given EDT0L grammars G_1 and G_2, a tree pushdown automaton can verify whether a nondeterministically guessed structure tree of G_1 does not correspond to any valid derivation of G_2. The automaton pushes into the stack a representation of the structure tree of G_1 and the computations along different paths of the input verify that any control-sequence of G_2 does not have the same structure. The automaton receives as input a balanced tree t such that the paths corresponding to different leaves of t give all possible control-sequences of G_2 of a given length.

The decidability proof relies strongly on nondeterminism and an actual algorithm following the proof requires multiple exponential time. It is seen easily that EDT0L structural equivalence is PSPACE-hard, so one cannot expect to find a very efficient algorithm. It has been shown in [22] that E0L structural equivalence is hard for deterministic exponential time. For the EDT0L case we have not obtained an exponential time lower bound.

2 Preliminaries

We assume that the reader is familiar with the basics of formal language theory and with parallel context-free type grammars in particular, see for example [20, 28].

The cardinality of a finite set A is denoted by $\#A$ and the power set of A is $\wp(A)$. Sometimes we identify a singleton set $\{a\}$ with a. The sets of positive and nonnegative integers are, respectively, $\mathbb{N}$ and $\mathbb{N}_0$.

The set of (nonempty) finite words over A is A^* (A^+) and λ denotes the empty word. The length of $w \in A^*$ is $|w|$. Let A_i, $i = 1, 2$, be finite sets. We define the mappings $\Pi_i^{A_1 \times A_2} : (A_1 \times A_2)^* \longrightarrow A_i^*$, $i = 1, 2$, by setting for $w = (a_1, b_1) \cdots (a_m, b_m)$, $m \geq 0$, $a_1, \ldots, a_m \in A_1$, $b_1, \ldots, b_m \in A_2$,

$$\Pi_1^{A_1 \times A_2}(w) = a_1 \cdots a_m, \quad \Pi_2^{A_1 \times A_2}(w) = b_1 \cdots b_m.$$

A *tree domain* D [7] is a nonempty finite subset of $\mathbb{N}^*$ that satisfies the two conditions below:

1. If $u \in D$, then every prefix of u is in D.
2. For every $u \in D$ there exists $\mathrm{rank}_D(u) \in \mathbb{N}_0$ such that $ui \in D$ for $i = 1, \ldots, \mathrm{rank}_D(u)$ and $ui \notin D$ for $i > \mathrm{rank}_D(u)$. (If $\mathrm{rank}_D(u) = 0$, the node u has no successors.)

An *A-labeled tree* is a mapping $t : \mathrm{dom}(t) \longrightarrow A$, where $\mathrm{dom}(t)$ is a tree domain. A node $u \in \mathrm{dom}(t)$ is said to be labeled by $t(u) \in A$. A node v is a *successor* (respectively, an immediate successor) of $u \in \mathrm{dom}(t)$ if $v = ux$, $x \in \mathbb{N}^*$ (respectively, $x \in \mathbb{N}$.) We assume that notions such as the height, the root, a leaf, an internal node, and a subtree of a tree t are known. The height of t is denoted $\mathrm{hg}(t)$. We use the convention that the height of a one-node tree is zero. The subtree of t at node u is t/u.

By the *level* of a node $u \in \mathrm{dom}(t)$ we mean the distance of u from the root, i.e., $|u|$. Clearly, the maximal level of a node of t is $\mathrm{hg}(t)$. The tree t is said to be *balanced* if all leaf nodes have the same level. The set of level k subtrees of t, $\text{sub}_k(t)$, $0 \le k \le \text{hg}(t)$, is defined as

$$\text{sub}_k(t) = \{t/u \mid |u| = k\}.$$

Note that $\text{sub}_k(t)$ is a set of trees, i.e., it contains only one copy of each tree t' that occurs as a subtree defined by a level k node of t. Also $\text{sub}_0(t) = \{t\}$.

In our decidability proof we use the *top-down tree pushdown automaton* model of [7]. Below we give a brief informal description of this model which will be sufficient for our purposes. The interested reader can find the formal algebraic definition in [7]. A tree pushdown automaton $\mathbf{A}$ is an extension of a finite tree automaton where each copy of the finite-state control has access to an auxiliary pushdown store. The automaton begins the computation at the root of the input tree, in a given initial state q_0 and with an initial symbol in the pushdown store. When being in a state q at a node u labeled by b in the input tree and having Z as the topmost stack symbol, depending on the tuple (q, b, Z), $\mathbf{A}$ can either (i) change the internal state and pop Z from the stack, (ii) change the state and push some symbol to the stack, or (iii) go to the m immediate successor nodes of u in some states $q_1, \ldots, q_m$ sending to each of the m nodes a copy of the pushdown stack (u is assumed to have rank m). The automaton accepts an input tree t if it reaches all leaves of t in an accepting state. The tree language recognized by $\mathbf{A}$ is denoted $T(\mathbf{A})$.

In general, the automata of [7] can have trees also as stacks. The automata described above are the so called restricted tree pushdown automata of [7]. Both the restricted and general tree pushdown automata recognize exactly the family of context-free tree languages [3, 4, 5, 19]. Given a context-free tree language T (in terms of a tree pushdown automaton or a context-free tree grammar), we can effectively construct an indexed grammar that generates the yield of T. Since emptiness is decidable for indexed grammars [1, 2], we have the following result.

Proposition 1. *We can decide effectively whether the tree language recognized by a given tree pushdown automaton is empty.*

3 Structural equivalence

We recall and invent some notations concerning parallel context-free grammars ([20]) and the structure trees of their derivations.

An *ET0L grammar* G is a tuple

$$G = (V, \Sigma, s_0, H), \tag{1}$$

where V is a finite alphabet of nonterminals, Σ is a finite alphabet of terminals, $s_0 \in V$ is the initial nonterminal, and H is a finite set of *tables*. A table $h \in H$ is a finite set of rewrite rules $a \to w$, where $a \in V$ and $w \in (V \cup \Sigma)^*$. (We do not allow rewriting of terminals.) The grammar G is an EDT0L (deterministic ET0L) grammar if every table $h \in H$ contains exactly one rule with left side a, for each nonterminal $a \in V$. Thus, h is a morphism $V^* \longrightarrow (V \cup \Sigma)^*$ and $h(a)$ denotes the right side of the rule of h having a as the left side. The grammar G is an E0L grammar if H contains only one table. The grammar is said to be *propagating* if the right side of any production is not the empty word, i.e., for all $h \in H$ and $a \in V$: $(a \longrightarrow \lambda) \notin h$.

In this paper we will be dealing mainly with the structure trees of EDT0L derivations. We give the definitions for arbitrary ET0L grammars since they are not essentially simpler in the deterministic case. In the remainder of this section, G is always an ET0L grammar as in (1).

Let F_G denote the set of $(V \cup \Sigma \cup \hat{\lambda})$-trees. Here $\hat{\lambda}$ is a new symbol that will be used to label nodes corresponding to the empty word. We define the parallel derivation relation of G, $\rightarrow_G^{\text{par}} \subseteq F_G \times F_G$, as the union of relations $\rightarrow_{G,h}^{\text{par}} \subseteq F_G \times F_G$, $h \in H$. Let $t_1, t_2 \in F_G$ and let $h \in H$. Then $t_1 \rightarrow_{G,h}^{\text{par}} t_2$ if and only if t_2 is obtained from t_1 as follows. Assume that t_1 has m leaves $u_1, \ldots, u_m$ where $t_1(u_i) \in V \cup \{\hat{\lambda}\}$, $i = 1, \ldots, m$. (Note that if some leaf of t_1 is labeled by a terminal, the derivation cannot be continued from t_1.) For each $i = 1, \ldots, m$ such that $t_1(u_i) \neq \hat{\lambda}$ choose a rule

$$t_1(u_i) \rightarrow a_1^i \cdots a_{k_i}^i \in h, \tag{2}$$

$a_j^i \in V \cup \Sigma$, $j = 1, \ldots, k_i$, $k_i \geq 0$. If $t_1(u_i) \neq \hat{\lambda}$ and $k_i \geq 1$, then in t_2 the node u_i has k_i successors labeled respectively by the symbols $a_1^i, \ldots, a_{k_i}^i$. If $t_1(u_i) \neq \hat{\lambda}$ and $k_i = 0$, then in t_2 the node u_i has one successor labeled by the symbol $\hat{\lambda}$. If $t_1(u_i) = \hat{\lambda}$, then u_i is a leaf of t_2.

We denote by t_0 the singleton tree with the only node labeled by the initial nonterminal s_0. The set of *syntax trees* $S(G)$ of G is defined by

$$S(G) = \{t \in F_G \mid t_0 \, (\rightarrow_G^{\text{par}})^* \, t\}.$$

A syntax tree $t \in S(G)$ is *terminal* if all leaves of t are labeled by elements of $\Sigma \cup \hat{\lambda}$. The set of terminal syntax trees of G is denoted $TS(G)$.

If above G is an EDT0L grammar, then the rule (2) is determined uniquely by the table h. We call words over the alphabet H *control-sequences*. Given an EDT0L grammar G and a control-sequence $\omega = h_1 \cdots h_n$, $n \geq 1$, $h_i \in H$, $i = 1, \ldots, n$, we denote by $G(\omega)$ the syntax tree obtained from the initial nonterminal by applying the sequence of tables specified by ω. Thus $G(\omega)$ is the unique syntax tree t such that

$$t_0 \, (\rightarrow_{G,h_1}^{\text{par}} \circ \ldots \circ \rightarrow_{G,h_n}^{\text{par}}) \, t, \tag{3}$$

if a tree t as in (3) exists, and otherwise $G(\omega)$ is undefined.

If G is propagating, then in every syntax tree t of G all paths from the root to a leaf have the same length, i.e., t is balanced. For non-propagating grammars all paths from the root to a leaf labeled by an element of $V \cup \Sigma$ have the same length. Note that our definition does not allow the rewriting of terminal symbols, i.e., we consider only synchronized grammars [20]. This is not a restriction since an arbitrary E(D)T0L grammar can be easily transformed into a synchronized

E(D)T0L grammar in such a way that the transformation preserves structural equivalence of grammars [14].

For $t \in F_G$ denote by w_t ($\in (V \cup \Sigma \cup \{\hat{\lambda}\})^+$) the word obtained by catenating, in the natural left-to-right order of the leaves, the labels of the leaves of t. The *yield* of a syntax tree t is defined as

$$\text{yield}(t) = e_\lambda(w_t),$$

where $e_\lambda : (V \cup \Sigma \cup \{\hat{\lambda}\})^* \longrightarrow (V \cup \Sigma)^*$ is the morphism defined by the conditions $e_\lambda(a) = a$ for $a \in V \cup \Sigma$, and $e_\lambda(\hat{\lambda}) = \lambda$.

For a syntax tree t, yield(t) is the sentential form generated by G in the derivation corresponding to t. The language $L(G)$ generated by G consists of the terminal words generated by G, i.e.,

$$L(G) = \{\text{yield}(t) \mid t \in \text{TS}(G)\}.$$

Clearly the above definition is equivalent to the standard definition of the language generated by an E(D)T0L grammar [20].

The *structure of a syntax tree* $t \in \text{S}(G)$, $\mathbf{str}(t)$, is the tree obtained from t by relabeling each internal node with σ, where σ is a new symbol not in $V \cup \Sigma$. We denote

$$\text{STS}(G) = \{\mathbf{str}(t) \mid t \in \text{TS}(G)\}.$$

Elements of STS(G) are called *(terminal) structure trees* of G. When G is known, or can be arbitrary, we sometimes speak simply about (Σ-)structure trees since the leaves are labeled by elements of Σ. Note that the rules of G determine the maximal number of immediate successors of any node of a structure tree of G. Thus, strictly speaking, the alphabet Σ does not alone determine the set of Σ-structure trees.

Grammars G_1 and G_2 are said to be *language equivalent* if $L(G_1) = L(G_2)$. It is well known that language equivalence is undecidable already for context-free grammars. Here we will consider the following two more restricted notions of equivalence. Let G_1 and G_2 be ET0L grammars. The grammars G_1 and G_2 are

- *structurally equivalent* if $\text{STS}(G_1) = \text{STS}(G_2)$, and
- *syntax equivalent* if $\text{TS}(G_1)$ and $\text{TS}(G_2)$ are equal modulo a renaming of the nonterminals.

Note that syntax equivalence implies structural equivalence, and structurally equivalent grammars in turn are always language equivalent. Both syntax and structural equivalence are decidable for context-free and E0L grammars [6, 11, 13, 14, 17, 24]. Syntax equivalence remains decidable also for ET0L grammars but ET0L structural equivalence is undecidable [23]. Here we will consider this question for the restricted version of EDT0L grammars.

4 The main result

We show that EDT0L structural equivalence can be decided effectively. First we introduce some notations concerning the width of trees. Intuitively, a width M structure tree has at most M distinct subtrees at any level.

Definition 2. A set of Σ-structure trees $\{t_1, \ldots, t_m\}$ is said to have *subtree-width* M ($\in \mathbb{N}$) if, for all $0 \leq k \leq \max_{1 \leq j \leq m} \text{hg}(t_j)$, the trees $t_1, \ldots, t_m$ have at most M distinct subtrees at level k, i.e., $\# \bigcup_{j=1}^{m} \text{sub}_k(t_j) \leq M$.

In the above definition, note that $\text{sub}_k(t_j)$ is a set of $(\Sigma \cup \{\sigma\})$-labeled trees, i.e., its elements do not consist of occurrences of subtrees in t_j. Note also that the subtree-width M does not need to be the minimal number of distinct subtrees at any given level.

We code the structure of an arbitrary derivation of an EDT0L grammar G_1 as a unary string. Then using the unary encoding as a stack, a tree pushdown automaton can verify that *any* control-sequence of another grammar G_2 does not generate the same structure tree. The construction relies essentially on the fact that the failure of an EDT0L derivation with respect to a given control-sequence can be checked by following only one path of the tree.

Lemma 3. *Let $G = (V, \Sigma, s_0, H)$ be an EDT0L grammar. Then there exists an integer $M \in \mathbb{N}$ such that, for every control-sequence $\omega \in H^*$, the structure tree* $\mathbf{str}(G(\omega))$ *has width M.*

Proof. Since the tables of H are homomorphisms, it follows that always when nodes $u_1, u_2 \in \text{dom}(G(\omega))$ have the same length and are labeled by the same nonterminal, then $G(\omega)/u_1 = G(\omega)/u_2$. By choosing $M = |V \cup \Sigma| + 1$ we see that $\mathbf{str}(G(\omega))$ has width M. □

Structure trees having constant subtree-width can be coded as strings where the ith symbol from the left codes the information which of the level i subtrees are direct descendants of which of the level $i-1$ nodes. The number of distinct level i subtrees is bounded by a constant and, furthermore, we need to consider only a constant number of level $i-1$ nodes that correspond to pairwise different subtrees.

Below we define the above described coding of structure trees having subtree-width M and prove a regularity property of such codings (in Lemma 7) only for propagating EDT0L grammars. The restriction to propagating grammars is done just to avoid unnecessarily complicated notations and afterwards we explain how the result can be straightforwardly extended for grammars having erasing productions.

Definition 4. Let $G = (V, \Sigma, s_0, H)$ be a propagating EDT0L grammar and denote $r = \max\{|h(a)| \mid h \in H, a \in V\}$. For each $M \in \mathbb{N}$ we define the set $\Omega(M)$ to consist of tuples

$$(m_1, m_2, \chi), \tag{4}$$

where $1 \leq m_1, m_2 \leq M$ and χ is a mapping of $\{1, \ldots, m_1\}$ into $\bigcup_{i=1}^{r}\{1, \ldots, m_2\}^i$ such that

$$\text{every element of } \{1, \ldots, m_2\} \text{ occurs in some tuple } \chi(i),\ 1 \leq i \leq m_1. \tag{5}$$

The set of final symbols $\Omega_f(M)$ is defined to consist of tuples

$$(m_1, \Sigma, \chi), \tag{6}$$

where $1 \leq m_1 \leq M$ and χ is a mapping of $\{1, \ldots, m_1\}$ into $\bigcup_{i=1}^{r} \Sigma^i$. A sequence of $\Omega(M)^*\Omega_f(M)$,

$$(m_{1,1}, m_{1,2}, \chi_1)(m_{2,1}, m_{2,2}, \chi_2) \cdots (m_{n-1,1}, m_{n-1,2}, \chi_{n-1})(m_{n,1}, \Sigma, \chi_n) \tag{7}$$

is said to be *well-formed* if $m_{i+1,1} = m_{i,2}$ for all $i = 1, \ldots n-1$.

Consider an r-tuple of structure trees $(t_1, \ldots, t_r)$, $r \geq 1$. Then $\sigma(t_1, \ldots, t_r)$ denotes the structure tree where the level one subtrees, from left to right, are $t_1, \ldots, t_r$. (This is just the standard algebraic notation for trees where we allow the symbol σ to have variable arity.)

A well-formed sequence W as in (7) is defined inductively to *represent* an m_1-tuple of Σ-structure trees $\Xi(W)$ as follows. First a final symbol (6) represents the m_1-tuple of structure trees

$$\sigma(\chi(1)), \ldots, \sigma(\chi(m_1)).$$

For the inductive definition, denote by W' the suffix of W obtained by deleting the first symbol $(m_{1,1}, m_{1,2}, \chi_1)$ and assume that

$$\Xi(W') = (t_1, \ldots, t_{m_{2,1}}).$$

Note that $m_{1,2} = m_{2,1}$ since W is well-formed. Denote

$$\chi_1(j) = (a_{j,1}, \ldots, a_{j,r(j)}) \in \{1, \ldots, m_{1,2}\}^{r(j)},$$

$j = 1, \ldots m_{1,1}$. Then

$$\Xi(W) = (\sigma(t_{a_{1,1}}, \ldots, t_{a_{1,r(1)}}), \ldots, \sigma(t_{a_{m_{1,1},1}}, \ldots, t_{a_{m_{1,1},r(m_{1,1})}})).$$

The following lemma can be proved using induction on the maximal height of the trees $t_1, \ldots, t_m$.

Lemma 5. *Let $M \in \mathbb{N}$ be fixed. Let $\{t_1, \ldots, t_m\}$ be a set of Σ-structure trees having subtree-width M.*

Then there exists a well-formed sequence $W \in \Omega(M)^*\Omega_f(M)$ as in (7) such that $\Xi(W) = (t_1, \ldots, t_m)$. Furthermore, $m_{i,1}$ is the cardinality of the set of level $i-1$ subtree representatives in $\{t_1, \ldots, t_m\}$, $i = 1, \ldots, n$. □

A word $W \in \Omega(M)^*\Omega_f(M)$ is said to be *simple* if the first symbol of W is of the form $(1, m, \chi)$, $m \leq M$. If W is simple and well-formed, then $\Xi(W)$ is a one-tuple of Σ-structure trees (t) and, in practice, we identify $\Xi(W)$ with t.

As an immediate consequence of Lemma 5 we have:

Corollary 6. *For every Σ-structure tree t having subtree-width M there exists a simple well-formed sequence $W \in \Omega(M)^* \Omega_f(M)$ such that $\Xi(W) = t$.*

The following lemma states that given a simple well-formed sequence $W \in \Omega(M)^* \Omega_f(M)$ and a control-sequence ω of an EDT0L grammar G, a finite automaton can determine whether or not $\Xi(W) = \mathbf{str}(G(\omega))$.

Lemma 7. *Let $G = (V, \Sigma, s_0, H)$ be a propagating EDT0L grammar and $M \in \mathbb{N}$. Let the alphabets $\Omega(M)$ and $\Omega_f(M)$ be as in Definition 4. We denote by L the set of words $\mu \in [(\Omega(M) \cup \Omega_f(M)) \times H]^+$ such that*

1. *$\Pi_1(\mu) \in \Omega(M)^* \Omega_f(M)$ is simple and well-formed,*
2. *$\Xi(\Pi_1(\mu)) \neq \mathbf{str}(G(\Pi_2(\mu)))$.*

Here we denote the mappings $\Pi_i^{(\Omega(M) \cup \Omega_f(M)) \times H}$ simply by Π_i, $i = 1, 2$.
We claim that L is a regular language.

Proof. Condition 1 can be easily verified by a finite automaton. Hence it is sufficient to show that given $\mu \in [(\Omega(M) \cup \Omega_f(M)) \times H]^+$ satisfying condition 1, a finite automaton $\mathcal{A}$ can verify whether or not condition 2 holds. (Note that if condition 1 does not hold, then $\Xi(\Pi_1(\mu))$ is not necessarily defined or may denote an ordered tuple of trees.)

The set of states of $\mathcal{A}$ is

$$\bigcup_{1 \leq i \leq M} [\wp(V) - \emptyset]^i \cup \{\mathbf{acc}, \mathbf{rej}\},$$

and the initial state is $\{s_0\}$. Assume that $\mathcal{A}$ is in a state $(U_1, \ldots, U_m)$, $\emptyset \neq U_i \subseteq V$, $1 \leq i \leq m \leq M$, and reading an input symbol

$$[(m_1, m_2, \chi), h].$$

If $m \neq m_1$, then $\mathcal{A}$ goes to the rejecting state **rej**. From our construction it follows that this is possible only when condition 1 does not hold. Assume then that $m = m_1$. If for some $x \in U_i$, $1 \leq i \leq m$,

$$\chi(i) \notin \{1, \ldots, m_2\}^{|h(x)|}, \tag{8}$$

then $\mathcal{A}$ goes to the accepting state **acc**. Also, $\mathcal{A}$ goes to the accepting state **acc** if for some $x \in U_i$, $1 \leq i \leq m$, $h(x) \notin V^+$. After this $\mathcal{A}$ reads the rest of the input verifying only that condition 1 holds.

The remaining possibility is that for all $x \in U_i$, $1 \leq i \leq m$, $\chi(i)$ is of the form $(s_1, \ldots, s_{|h(x)|})$, $s_j \in \{1, \ldots, m_2\}$, $j = 1, \ldots, |h(x)|$, and $h(x) \in V^+$. In this case after reading the symbol $[(m_1, m_2, \chi), h]$, $\mathcal{A}$ goes to the state $(Z_1, \ldots, Z_{m_2})$ where the sets $Z_j \subseteq V$, $j = 1, \ldots, m_2$, are constructed as follows. For each $i = 1, \ldots, m$ and each $x \in U_i$ do the following. If

$$h(x) = a_1 \cdots a_n \text{ and } \chi(i) = (s_1, \ldots, s_n), \ (s_j \in \{1, \ldots, m_2\}),$$

then add the element $a_j \in V$ to the set Z_{s_j}, $j = 1, \ldots, n$. Note that each of the sets $Z_1, \ldots, Z_{m_2}$ will be nonempty because χ satisfies the condition (5).

Intuitively, U_i consists of all elements of V that the derivation of G (following the morphisms $h' \in H$ read so far in the second components) reaches at nodes corresponding to the ith subtree representative at this, say the kth, level of $\Xi(\Pi_1(\mu))$. Thus if condition (8) holds, the derivation reaches some level k node u of $\Xi(\Pi_1(\mu))$ with a symbol $a \in V$ such that the number of immediate successors of u is not equal to $|h(a)|$, or $h(a)$ contains a terminal symbol. Thus a derivation using next the table h cannot have the structure $\Xi(\Pi_1(\mu))$ and condition 2 holds. The case where (8) does not hold corresponds to the situation where the parallel derivation step determined by h at level k does not immediately violate the structure of the tree. Then the sets $Z_1, \ldots, Z_{m_2}$ are constructed to consist, respectively, of all nonterminals that will appear in the m_2 subtree representatives at the following level.

It remains to define the operation of $\mathcal{A}$ when it reaches a final symbol $[(m_1, \Sigma, \chi), h] \in [\Omega_f(M) \times H]$ in a state $(U_1, \ldots, U_m)$. Following the above idea this is done so that $\mathcal{A}$ accepts exactly then when the derivation step h produces for some leaf representative $a \in \Sigma$ the wrong terminal symbol (or a nonterminal).

The possibility $m \neq m_1$ corresponds again to a situation where condition 1 does not hold. We need to consider only the possibility $m = m_1$. If for all $i \in \{1, \ldots, m\}$ and for all $x \in U_i$, $\chi(i) = h(x)$, then the derivation step h produces the correct terminal symbols at all the leaves. This means that condition 2 does not hold and $\mathcal{A}$ rejects. On the other hand, $\mathcal{A}$ enters the accepting state if for some $x \in U_i$, $\chi(i) \neq h(x)$. □

Above Lemma 7 was formulated and proved only for propagating grammars. However, exactly the same proof works also for general EDT0L grammars: the possibility of having erasing productions just adds at most one more subtree representative to each level of the structure tree. We can modify Definition 4 so that in the symbols (m_1, m_2, χ) (as in (4)) χ is a partial function where $\chi(i)$ is undefined if i represents a node having $\hat{\lambda}$ as immediate successor (or in such cases $\chi(i)$ is defined to be a new symbol not belonging to $\{1, \ldots, m_2\}$.) Also, in the proof of Lemma 7, the condition (8) has to be divided into cases depending on whether $\chi(i)$ is defined or not. In the following, we will use the extension of Lemma 7 where the grammar is not necessarily propagating.

Now we can show that a tree pushdown automaton can verify whether all possible derivations of a given EDT0L grammar fail on a given structure tree having constant width.

Lemma 8. *Let $G_i = (V_i, \Sigma, s_{0,i}, H_i)$, $i = 1, 2$, be EDT0L grammars. Let M be the constant of Lemma 3 corresponding to the grammar G_1. Then we can effectively construct a tree pushdown automaton* **A** *such that $T(\mathbf{A}) \neq \emptyset$ if and only if there exists*

$$t \in STS(G_1) - STS(G_2) \tag{9}$$

such that t has width M.

Proof. Denote $\#H_2 = k$. The tree pushdown automaton **A** receives as inputs trees where each internal node has exactly k immediate successors. The internal nodes are labeled by a symbol a_0. The set of stack symbols is $(\Omega(M) \cup \Omega_f(M)) \times H_1$. At the beginning of the computation, **A** nondeterministically pushes into the stack a word $(\alpha_1, h_1) \cdots (\alpha_n, h_n)(\beta, h_{n+1})$, $\alpha_1, \ldots, \alpha_n \in \Omega(M)$, $\beta \in \Omega_f(M)$, $h_1, \ldots, h_{n+1} \in H_1$. (The top of the stack is to the left.)

Intuitively, the stack contents is guessed so that $W = \alpha_1 \ldots \alpha_n \beta$ is simple and well-formed and if $\omega = h_1 \cdots h_{n+1}$, then

$$\Xi(W) = \mathbf{str}(G_1(\omega)) \tag{10}$$

and $\Xi(W) \in \mathrm{STS}(G_1) - \mathrm{STS}(G_2)$ has width M, assuming that a tree as in (9) exists. This is always possible by Corollary 6.

When reading an input symbol, **A** always pops the topmost stack symbol and the remaining stack contents is forwarded to the successor nodes. After the initial nondeterministic guesses **A** does not push any more symbols into the stack. It is clear that **A** can check that the input tree is a balanced k-ary tree of height n (where each internal node is labeled by a_0.)

The states of **A** consist of two components that operate in parallel. The first component verifies that the condition (10) holds, by the result of Lemma 7 extended to nonpropagating grammars this can be done using only a finite-state memory. The first component treats the initial stack contents as input and operates identically on all paths of the input tree.

The second component of **A** ignores the second components $h_i \in H_1$ of the stack symbols. On a path to a leaf $i_1 i_2 \cdots i_{n+1}$ of the input, $1 \le i_j \le k$, $j = 1, \ldots, n+1$, the second component of **A** verifies that

$$\Xi(W) \neq \mathbf{str}(G_2(i_1 i_2 \cdots i_{n+1})).$$

By Lemma 7, this is again possible using the finite-state control of **A**.

Each branch of the computation has to consume the entire stack so an accepted input tree is necessarily balanced. Hence, on different paths of the input **A** verifies that $\Xi(W)$ is not the structure of a syntax tree $G_2(\omega')$ for any control-sequence ω' of length $n+1$, i.e., **A** verifies that $\Xi(W) \notin \mathrm{STS}(G_2)$. Thus **A** accepts some input if and only if there exists a tree t such that (9) holds and t has subtree-width M. □

Note that in the proof of Lemma 8 it is essential that the guessed instance of $t \in \mathrm{STS}(G_1)$ in the pushdown stack has a unary encoding, although the general tree pushdown automaton model of [7] allows, in fact, trees also in the stack. It can be shown that a tree pushdown automaton cannot nondeterministically push a balanced tree of arbitrary height into the stack [21], so one could not directly store an arbitrary $t \in \mathrm{STS}(G_1)$ in the tree stack at the beginning of the computation. Furthermore, simulating the derivations of G_2 given by different control-sequences directly on the tree t would have the problem that the paths leading to "failure of the derivation of G_2 in t" may branch out much before than the corresponding control-sequences branch out in the input tree. More specifically, the tree pushdown automaton **A** has to find, for each control-sequence

ω of G_2, at least one path ν_ω in t that leads to failure. The control-sequences correspond to different paths in the input and it is, in general, possible that two control-sequences ω_1 and ω_2 have a very long common prefix whereas the corresponding paths ν_{ω_1} and ν_{ω_2} in t branch out already at the root of t. Situations like this do not cause problems when **A** has a unary encoding $\Xi(W)$ of t in the pushdown stack. Then **A** can on all paths of the input simulate all distinct subderivations of G_2 within t.

Combining Lemmas 3 and 8 and Proposition 1 we have proved the following result. Note that the constant M in Lemma 3 is independent of the chosen control-sequence.

Theorem 9. *For given EDT0L grammars G_1 and G_2 we can decide effectively whether or not*

$$STS(G_1) = STS(G_2).$$

□

Exactly as in the proof of Lemma 8, for given EDT0L grammars G_1 and G_2 the nonemptiness of $\mathrm{TS}(G_1) - \mathrm{TS}(G_2)$ can be reduced to deciding whether a tree pushdown automaton recognizes a nonempty tree language. Since the number of nonterminals of G_1 and G_2 is finite, we have a new proof for the decidability of syntax equivalence. The result follows also from [23].

Theorem 10. *Syntax equivalence is decidable for EDT0L grammars.* □

5 Discussion and open problems

The proof of Theorem 9 gives only a multiple exponential time algorithm for EDT0L structural equivalence. We do not know what is the exact complexity of the problem. The deterministic exponential time hardness result obtained in [22] for E0L structural equivalence cannot, at least not directly, be used to prove a similar lower bound for the complexity of EDT0L structural equivalence. On the other hand, we cannot expect to obtain a polynomial time algorithm since it is known that already the structural equivalence problem for linear grammars is PSPACE-complete [8], and structural equivalence for linear grammars is easily logspace reducible to EDT0L structural equivalence. Note that when the sentential forms have only one nonterminal (or any constant number of occurrences of nonterminals), an EDT0L grammar can simulate a context-free derivation simply by having a different table for each rule.

Intuitively, the decidability proof of the previous section relies on the following two properties of EDT0L derivations:

1. the current nonterminal and the remaining control-sequence always uniquely determine a subderivation,
2. all control-sequences generate the structure tree one level at a time.

These properties enabled us to produce a unary encoding W of a structure tree such that the failure of *all possible* control-sequences to produce this structure can be verified by a finite automaton reading W and the control-sequence in parallel.

The necessity of both conditions 1 and 2 can be illustrated by considering the *Indian parallel* (IP) grammars. An IP grammar is a context-free grammar with the derivation relation defined so that at each derivation step one rewrites all occurrences of one (nondeterministically chosen) nonterminal b in the given sentential form using the same rule having left side b. The other nonterminals are not rewritten. For the formal definition the reader may consult [2, 18, 25, 26]. It is well known that languages generated by IP grammars are strictly included in the EDT0L languages.

When the sequence of rules used in a derivation of an IP grammar is viewed as a control-sequence, the IP grammars clearly have the above property 1. However, different sequences of rules can generate distinct parts of a given structure tree in completely different order, and any analogy of condition 2 does not seem to hold for IP grammars. Thus in spite of the fact that EDT0L grammars are strictly more powerful than IP grammars in terms of the family of generated languages, it appears not possible to use the proof method of the previous section to decide the structural equivalence problem for IP grammars. We conjecture that IP structural equivalence is decidable. For the E0LIP grammars of [10] structural equivalence can be shown to be decidable exactly as in the proof of Theorem 9. (E0LIP grammars combine the Indian parallel and E0L rewriting mechanisms: at each derivation step all occurrences of every nonterminal are rewritten using the same rule.)

Russian parallel (RP) grammars [2, 12, 27] extend IP grammars by allowing also (sequential) context-free derivation steps. The RP structural equivalence problem remains open.

References

1. Aho, A.V.: Indexed grammars: An extension of the context-free case. J. Assoc. Comput. Mach. **16** (1969) 383–406.
2. Dassow, J., Păun, G.: *Regulated rewriting in formal language theory.* EATCS Monographs in Theoretical Computer Science, Vol. 18, Springer-Verlag, 1989.
3. Engelfriet, J., Schmidt, E.M.: IO and OI. J. Comput. System Sci. **15** (1977) 328–353; **16** (1978) 67–99.
4. Fischer, M.J.: Grammars with macro-like productions. Proc. of 9th Ann. IEEE Symp. on Switching and Automata Theory (1968) 131–142.
5. Gécseg, F., Steinby, M.: Tree languages. In "Handbook of Formal Languages", G. Rozenberg and A. Salomaa (eds.), Springer-Verlag, 1996, (to appear).
6. Ginsburg, S., Harrison, M.: Bracketed context-free languages. J. Comput. System Sci. **1** (1967) 1–23.
7. Guessarian, I.: Pushdown tree automata. Math. Systems Theory **16** (1983) 237–263.

8. Hunt III, H.B., Rosenkrantz, D.J., Szymanski, T.G.: On the equivalence, containment, and covering problems for the regular and context-free languages. J. Comput. System Sci. **12** (1976) 222–268.
9. Istrate, G.: The strong equivalence of ET0L grammars. In "Developments in Language Theory", G. Rozenberg and A. Salomaa (eds.), World Scientific, 1994, pp. 81–89.
10. Kleijn, H.C.M., Rozenberg, G.: A study in parallel rewriting systems. Inform. Control **44** (1980) 134–163.
11. Knuth, D.E.: A characterization of parenthesis languages. Inform. Control **11** (1967) 269–289.
12. Levitina, M.K.: On some grammars with global productions (in Russian). Akad. Nauk. SSSR NTI, Ser. 2, No. 3 (1972) 32–36.
13. McNaughton, R.: Parenthesis grammars. J. Assoc. Comput. Mach. **14** (1967) 490–500.
14. Niemi, V.: A normal form for structurally equivalent E0L grammars. In "Lindenmayer Systems: Impacts on Theoretical Computer Science, Computer Graphics, and Developmental Biology", G. Rozenberg and A. Salomaa (eds.), Springer-Verlag, 1992, pp. 133–148.
15. Ottmann, Th., Wood, D.: Defining families of trees with E0L grammars. Discrete Applied Math. **32** (1991) 195–209.
16. Ottmann, Th., Wood, D.: Simplifications of E0L grammars. In "Lindenmayer Systems: Impacts on Theoretical Computer Science, Computer Graphics, and Developmental Biology", G. Rozenberg and A. Salomaa (eds.), Springer-Verlag, 1992, pp. 149–166.
17. Paull, M., Unger, S.: Structural equivalence of context-free grammars. J. Comput. System Sci. **2** (1968) 427–463.
18. Reichel, B.: Some classifications of Indian parallel languages. J. Inform. Process. Cybern. EIK **26** (1990) 85–99.
19. Rounds, W.C.: Mappings and grammars on trees. Math. Systems Theory **4** (1970) 257–287.
20. Rozenberg, G., Salomaa, A.: *The Mathematical Theory of L Systems.* Academic Press, New York, 1980.
21. Salomaa, K.: Deterministic tree pushdown automata and monadic tree rewriting systems. J. Comput. System Sci. **37** (1988) 367–394.
22. Salomaa, K., Wood, D., Yu, S.: Complexity of E0L structural equivalence. RAIRO Theoretical Informatics **29** (1995) 471–485.
23. Salomaa, K., Wood, D., Yu, S.: Structural equivalence and ET0L grammars. Theoret. Comput. Sci. **164** (1996) 123–140.
24. Salomaa, K., Yu, S.: Decidability of structural equivalence of E0L grammars. Theoret. Comput. Sci. **82** (1991) 131–139.
25. Siromoney, R., Krithivasan, K.: Parallel context-free languages. Inform. Control **24** (1974) 155–162.
26. Skyum, S.: Parallel context-free languages. Inform. Control **26** (1974) 280–285.
27. Skyum, S.: Decomposition theorems for various kinds of languages parallel in nature. SIAM J. Comput. **5** (1976) 284–296.
28. Wood, D.: *Theory of Computation.* John Wiley & Sons, New York, NY, 1987.

k–Regular Matroids

Charles Semple

Department of Mathematics, Victoria University of Wellington,
PO Box 600, Wellington, New Zealand

Abstract. The class of matroids representable over all fields is the class of regular matroids. The class of matroids representable over all fields except perhaps $GF(2)$ is the class of near-regular matroids. This paper considers a generalisation of these classes to the so called k–regular matroids. The main result of the paper determines the automorphisms of the algebraic structure associated with the class of k–regular matroids. This result is the first step in establishing a unique representation property for k–regular matroids.

1 Introduction

It follows from a result of Tutte [7] that a matroid is representable over all fields if and only if it can be represented by a totally unimodular matrix, that is, by a matrix over the rationals with the property that all non-zero subdeterminants are in $\{1, -1\}$. This is the class of regular matroids. In [10] Whittle gives an analogous matrix characterisation for the class of matroids representable over all fields except perhaps $GF(2)$. Let $\mathbf{Q}(\alpha)$ denote the field obtained by extending the rationals by the transcendental α. A matrix over $\mathbf{Q}(\alpha)$ is *near-unimodular* if all non-zero subdeterminants are in $\{\pm\alpha^i(\alpha-1)^j : i, j \in \mathbb{Z}\}$. A *near-regular* matroid is one that can be represented by a near-unimodular matrix. The class of matroids representable over all fields except perhaps $GF(2)$ is the class of near-regular matroids [10, Theorem 1.4]. These classes invite generalisation. This paper considers a generalisation to the so called "k–regular" matroids.

One reason why strong results for regular and near-regular matroids exist is that each of these classes has a unique representation property. Regular matroids are uniquely representable over any field. In particular, all totally unimodular representations of a regular matroid are equivalent. For near-regular matroids we have a weaker, but just as crucial, unique representation property [9, Theorem 5.11]. All strong results in matroid representation theory use some notion of unique representation in an essential way [2, 6, 7, 8, 9, 10]. With this in mind one would want immediate generalisations of regular and near-regular matroids to have a unique representation property.

In [4] an algebraic structure called a partial field is associated with classes of matroids which are obtained, like the classes of regular and near-regular matroids, by restricting the values of all non-zero subdeterminants in a particular way. The classes of regular and near-regular matroids can be interpreted as classes of matroids representable over a partial field. The class of k–regular matroids can also be interpreted in this way. The theory of matroid representation over partial fields is similar to that for fields. In particular there is a well-defined notion of an automorphism of a partial field and equivalence of representations over partial fields similar to that for fields. Automorphisms of a partial field **P** play the same role in determining the equivalence of representations over **P** as automorphisms of a field **F** play in the equivalence of representations over **F**. This paper, as the first step towards finding a unique representation property for the class of k–regular matroids, establishes the automorphisms of the natural partial field whose associated class of matroids is the class of k–regular matroids. This is stated as Theorem 7, which is the main result of this paper.

Note that the partial field that we will associate with the class of k–regular matroids can be embedded in the field $\mathbf{Q}(\alpha_1, \alpha_2, \ldots, \alpha_k)$, the field obtained by extending the rationals by the algebraically independent transcendentals $\alpha_1, \alpha_2, \ldots, \alpha_k$. It appears that if $k > 2$, then the automorphisms of such fields are unknown (see [1, Section 5.2]). The fact that we have determined the automorphisms of a partial field that can be embedded in $\mathbf{Q}(\alpha_1, \alpha_2, \ldots, \alpha_k)$ may reward studying partial fields for reasons other than the desire to solve problems in matroid representation theory.

The paper is organised as follows. Section 2 contains a general discussion of partial fields and matroid representation over partial fields. Section 3 defines k–regular matroids and presents two results. The first result shows that if a matroid is k–regular, then it is representable over all fields whose cardinality is at least $k+2$. The other result is needed as a lemma for Theorem 7, which is proved in Section 4.

2 Preliminaries

Familiarity is assumed with the elements of matroid theory, see for example [3]. In particular we assume familiarity with the theory of matroid representations. Essentially, it is assumed the reader is familiar with the substance of [3, Chapter 6].

Partial fields are studied in [4]. Essentially a partial field **P** is a structure that has all the properties of a field except that addition may only be a partial operation, that is, there may exist elements $a, b \in \mathbf{P}$ such that $a+b$ is not defined. The following special case of [4, Proposition 2.2] gives one way to obtain a partial field.

Theorem 1. *Let* **F** *be a field, and let* G *be a multiplicative subgroup of* $\mathbf{F}^*$ *with the property that* $-a \in G$ *for all* $a \in G$. *Then* $G \cup \{0\}$ *with the induced operations from* **F** *is a partial field.*

The partial field obtained via Theorem 1 is denoted $(G, \mathbf{F})$. All the partial fields referred to in this paper can be obtained in this way.

Interest in partial fields is due to the fact that classes of matroids can be associated with them. An $m \times n$ matrix A over a partial field $\mathbf{P} = (G, \mathbf{F})$ is a **P**–*matrix* if $\det(A') \in G \cup \{0\}$ for every square submatrix A' of A. If A is a **P**–matrix, then the matroid obtained in the usual way from A is denoted $M[A]$.

A matroid M is *representable* over $\mathbf{P}$ or is $\mathbf{P}$–*representable* if it is equal to $M[A]$ for some $\mathbf{P}$–matrix A; in this case A is said to be a *representation* of M.

In the language of partial fields, the matroids representable over the partial fields $(\{-1,1\},\mathbf{Q})$ and $(\{\pm\alpha^i(\alpha-1)^j : i,j \in \mathbb{Z}\},\mathbf{Q}(\alpha))$ are the classes of regular and near-regular matroids respectively. These partial fields are labelled **Reg** and **NR** respectively. Before going any further we make the following observations. We first note that the choice of $\mathbf{Q}$ and $\mathbf{Q}(\alpha)$ in defining these partial fields is not unique. In fact $\mathbf{Q}$ and $\mathbf{Q}(\alpha)$ could be replaced by $\mathbf{F}$ and $\mathbf{F}(\alpha)$, respectively, where $\mathbf{F}$ is any field whose characteristic is not 2 or 3. The point is that we require $1+1$ and $-1-1$ to be not defined in both partial fields. Secondly, in general, partial fields need not arise from fields. However if a partial field can be embedded in some field, as the ones discussed in this paper can, then we can regard the elements of the partial field as elements of the embedding field.

Let $\mathbf{P_1}$ and $\mathbf{P_2}$ be partial fields. A function $\varphi : \mathbf{P_1} \to \mathbf{P_2}$ is a *homomorphism* if, for all $a,b \in \mathbf{P_1}$, $\varphi(ab) = \varphi(a)\varphi(b)$, and whenever $a+b$ is defined, then $\varphi(a)+\varphi(b)$ is defined, and $\varphi(a+b) = \varphi(a)+\varphi(b)$.

Theorem 2. ([4, Corollary 5.3]) *Let $\mathbf{P_1}$ and $\mathbf{P_2}$ be partial fields. If there exists a non-trivial homomorphism $\varphi : \mathbf{P_1} \to \mathbf{P_2}$, then every matroid representable over $\mathbf{P_1}$ is also representable over $\mathbf{P_2}$.*

The homomorphism $\varphi : \mathbf{P_1} \to \mathbf{P_2}$ is an *isomorphism* if it is a bijection and has the property that $a+b$ is defined if and only if $\varphi(a)+\varphi(b)$ is defined. By extending the argument in the proof of [5, Proposition 2.4.4], we can simplify the task of showing that a function is an isomorphism.

Theorem 3. *Let $\mathbf{P_1}$ and $\mathbf{P_2}$ be partial fields and let $\varphi : \mathbf{P_1} \to \mathbf{P_2}$ be a function. Then φ is an isomorphism if and only if φ satisfies the following conditions:*

(i) *φ is a bijection.*
(ii) *For all $x,y \in \mathbf{P_1}$, $\varphi(xy) = \varphi(x)\varphi(y)$.*
(iii) *For all $z \in \mathbf{P_1}$, $z-1$ is defined if and only if $\varphi(z)-1$ is defined and $\varphi(z-1) = \varphi(z)-1$.*

An *automorphism* of a partial field $\mathbf{P}$ is an isomorphism $\varphi : \mathbf{P} \to \mathbf{P}$. From a matroid-theoretic point of view the main interest in automorphisms is the role they play in determining whether representations of a matroid are equivalent. As for fields two matrix representations of a matroid M over a partial field $\mathbf{P}$ are *equivalent* if one can be obtained from the other by a sequence of the following operations: interchanging two rows; interchanging two columns (together with labels); pivoting on a non-zero element; multiplying a row or column by a non-zero member of $\mathbf{P}$; and replacing each entry of the matrix by its image under some automorphism of $\mathbf{P}$. A matroid is *uniquely representable* over $\mathbf{P}$ if all representations of M over $\mathbf{P}$ are equivalent.

3 k–regular matroids

Let $\mathbf{Q}(\alpha_1, \alpha_2, \ldots, \alpha_k)$ denote the field obtained by extending the rationals by the algebraically independent transcendentals $\alpha_1, \alpha_2, \ldots, \alpha_k$. Let $\mathcal{A}_k$ denote the set

$$\{\pm \prod_{i=1}^{k} \alpha_i^{l_i} \prod_{i=1}^{k} (\alpha_i - 1)^{m_i} \prod_{i,j \subseteq \{1,2,\ldots,k\}, i \neq j} (\alpha_i - \alpha_j)^{n_{i,j}} : l_i, m_i, n_{i,j} \in \mathbb{Z}\}.$$

Evidently $\mathcal{A}_k$ is a subgroup of the multiplicative group of $\mathbf{Q}(\alpha_1, \alpha_2, \ldots, \alpha_k)$. Since $-a \in \mathcal{A}_k$ for all $a \in \mathcal{A}_k$, it follows by Theorem 1 that $\mathcal{A}_k \cup \{0\}$ is a partial field. Set $\mathbf{R_k} = (\mathcal{A}_k, \mathbf{Q}(\alpha_1, \alpha_2, \ldots, \alpha_k))$. A k–regular matroid is one that can be represented by an $\mathbf{R_k}$–matrix. When $k = 0$ we have the partial field **Reg** which carries the class of regular matroids. When $k = 1$ we have the partial field **NR** which carries the class of near-regular matroids. Some properties of 2–regular matroids are established in [5].

The rest of this section presents two results. It was noted in the introduction that the class of regular matroids is the class of matroids representable over all fields and the class of near-regular matroids is the class of matroids representable over all fields except possibly $GF(2)$. We next show that k–regular matroids are representable over all fields whose cardinality is at least $k+2$. Before doing this, however, we note that the converse of this is not true. For example it will be shown at the end of Section 4 that $U_{3,6}$, which is representable over a field $\mathbf{F}$ if and only if $|\mathbf{F}| \geq 4$ [3, p. 504], is not 2–regular.

Proposition 4. *Let M be a k–regular matroid and $\mathbf{F}$ be a field such that $|\mathbf{F}| \geq k+2$. Then M is representable over $\mathbf{F}$.*

Proof. Since $|\mathbf{F}| \geq k+2$, we can choose k distinct elements $a_1, a_2, \ldots, a_k$ from $\mathbf{F} - \{0, 1\}$. Consider the function $\varphi : \mathbf{R_k} \to \mathbf{F}$ defined by $\varphi(0) = 0$ and

$$\varphi(\pm \prod_{i=1}^{k} \alpha_i^{l_i} \prod_{i=1}^{k} (\alpha_i - 1)^{m_i} \prod_{i,j \subseteq \{1,2,\ldots,k\}, i \neq j} (\alpha_i - \alpha_j)^{n_{i,j}})$$

$$= \pm \prod_{i=1}^{k} a_i^{l_i} \prod_{i=1}^{k} (a_i - 1)^{m_i} \prod_{i,j \subseteq \{1,2,\ldots,k\}, i \neq j} (a_i - a_j)^{n_{i,j}},$$

where $\varphi(\alpha_1) = a_1, \varphi(\alpha_2) = a_2, \ldots, \varphi(\alpha_k) = a_k$. It is easily seen that φ is a homomorphism and so, by Theorem 2, the proposition is proved.

The other result of this section is needed as a lemma for Theorem 7, but it has independent interest so we call it a theorem. We first note that, for all $x, y \in \mathbf{P}^*$, $x + y$ is defined if and only if $-y(-xy^{-1} - 1)$ is defined, and the latter expression is defined if and only if $-xy^{-1} - 1$ is defined. It follows that to know whether the sum of a pair of elements in $\mathbf{P}$ is defined it suffices to know those elements z of $\mathbf{P}$ for which $z - 1 \in \mathbf{P}$. An element z of a partial field $\mathbf{P}$ is *fundamental* if $z-1$ is defined. Note that 0 and 1 are fundamental in all partial fields.

We now determine the fundamental elements of $\mathbf{R_k}$. The following observation is used in the proof of this characterisation. If z is an element of $\mathbf{R_k}$, then z is the quotient of two polynomials in $\mathbf{Q}[\alpha_1, \alpha_2, \ldots, \alpha_k]$. Moreover, as elements of $\mathbf{Q}[\alpha_1, \alpha_2, \ldots, \alpha_k]$, these polynomials have factors of the form $a-b$, where a and b are distinct elements of $\{0, 1, \alpha_1, \alpha_2, \ldots, \alpha_k\}$. Therefore we can regard an element of $\mathbf{R_k}$ as a quotient of two polynomials in $\mathbf{Q}[\alpha_1, \alpha_2, \ldots, \alpha_k]$. In the proof of Theorem 5 we regard all elements of $\mathbf{R_k}$ in this way. Furthermore to simplify the proof of Theorem 5 we make the following definitions. Let p be a polynomial in $\mathbf{R_k}$. By an abuse of language we say that $a-b$ is a *factor* of p if $a-b$ is a linear factor of p in the usual sense or $\{a, b\} = \{0, 1\}$. In the former case $a-b$ is defined to be a *normal* factor of p.

Theorem 5. *Let z be an element of $\mathbf{R_k}$ such that $z \notin \{0, 1\}$. Then z is a fundamental element of $\mathbf{R_k}$ if and only if z can be written in one of the following forms:*

(i)

$$\frac{a-b}{c-b}$$

where a, b, and c are distinct elements of $\{0, 1, \alpha_1, \alpha_2, \ldots, \alpha_k\}$.

(ii)

$$\frac{(a-b)(c-d)}{(c-b)(a-d)}$$

where a, b, c, and d are distinct elements of $\{0, 1, \alpha_1, \alpha_2, \ldots, \alpha_k\}$.

Proof. From the remarks preceding the statement of the proposition, we can regard z as a quotient of two polynomials p_1 and p_2 of $\mathbf{R_k}$. Without loss of generality we may assume that p_1 and p_2 are relatively prime polynomials. It follows that z is a fundamental element of $\mathbf{R_k}$ if and only if there is a polynomial p_3 of $\mathbf{R_k}$ such that $p_1 - p_2 = p_3$. Now $z \notin \{0, 1\}$, so by rearranging if necessary, we may assume that $p_1 \notin \{1, -1\}$. The proof finds all pairs of polynomials p_1 and p_2 in $\mathbf{R_k}$ with the property that $p_1 - p_2$ is also a polynomial in $\mathbf{R_k}$. In doing this we immediately establish all the fundamental elements of $\mathbf{R_k}$.

First we show that p_1, p_2, and p_3 are relatively prime. If p_1 and p_3 are not relatively prime, then they have a common normal factor q. Since $p_2 = p_1 - p_3$, q is also a normal factor of p_2, contradicting the fact that p_1 and p_2 are relatively prime. Similarly p_2 and p_3 are relatively prime. In the proof we repeatedly use this fact.

Since $p_1 \notin \{1, -1\}$, it has a normal factor $a-b$ where a and b are distinct elements of $\{0, 1, \alpha_1, \alpha_2, \ldots, \alpha_k\}$. Without loss of generality assume that $a = \alpha_i$ for some $i \in \{1, 2, \ldots, k\}$. Let $p(\alpha_i = b)$ denote the polynomial obtained by substituting b for α_i in p. Then $p_1(\alpha_i = b) = 0$ and so $-p_2(\alpha_i = b) = p_3(\alpha_i = b)$. Since p_1, p_2, and p_3 are relatively prime, it follows that there is an element c in $\{0, 1, \alpha_1, \alpha_2, \ldots, \alpha_k\} - \{a, b\}$ such that either $c-b$ or $a-c$ is a factor of p_2.

If $c-b$ is a factor of p_2, then $a-c$ is a factor of p_3. If $a-c$ is a factor of p_2, then $c-b$ is a factor of p_3. The rest of the proof is a case analysis based on the factors of p_2.

Sublemma 1. *If p_2 has at most one normal factor, then one of the following holds: $p_1=a-b$ and $p_2\in\{c-b,a-c\}$; $p_1=b-a$ and $p_2\in\{b-c,c-a\}$; $p_1=(a-b)(c-d)$ and $p_2=(c-b)(a-d)$; or $p_1=(b-a)(c-d)$ and $p_2=(b-c)(a-d)$.*

Proof. Assume that p_2 has no normal factor. Then $p_2\in\{1,-1\}$. Since $a\in\{\alpha_1,\alpha_2,\ldots,\alpha_k\}$, $a-c\notin\{1,-1\}$. Therefore $p_2\in\{c-b,b-c\}$ where $\{b,c\}=\{0,1\}$. Since $-p_2(\alpha_i=b)=p_3(\alpha_i=b)$ and since p_1, p_2, and p_3 are relatively prime, it follows that $a-c$ is the only normal factor of p_3. Similarly substituting c for a into $p_1-p_2=p_3$, $a-b$ is the only normal factor of p_1. It is now easily seen that that the multiplicity of both $a-b$ in p_1 and $a-c$ in p_3 is 1. Furthermore if $p_1=a-b$, then $p_2=c-b$. Also if $p_1=b-a$, then $p_2=b-c$. Hence if p_2 has no normal factors, then the result holds.

Assume that p_2 has exactly one normal factor. Then either $c-b$ is a factor of p_2, in which case $a-c$ is a normal factor of p_3, or $a-c$ is the only normal factor of p_2, in which case $c-b$ is a factor of p_3. Assume that the former case holds. There are two possibilities to consider. Assume first that $c-b$ is not normal. Since $-p_2(\alpha_i=b)=p_3(\alpha_i=b)$ and since p_1, p_2, and p_3 are relatively prime polynomials, it follows that there is an element d in $\{\alpha_1,\alpha_2,\ldots,\alpha_k\}-\{a\}$ such that either $b-d$ or $a-d$ is the only normal factor of p_2. If $b-d$ is a normal factor of p_2, then $a-d$ is a normal factor of p_3. If $a-d$ is a normal factor of p_2, then $b-d$ is a normal factor of p_3. We now show that $b-d$ is not a normal factor of p_2. If it was normal factor, then, by substituting c for a into $p_1-p_2=p_3$, we see that $b-d$ is a factor of p_1. But then the fact that p_1 and p_2 are relatively prime is contradicted. Hence $a-d$ is the only normal factor in p_2. Therefore $b-d$ is a normal factor in p_3. Using the fact that $-p_2(\alpha_i=b)=p_3(\alpha_i=b)$ again, it follows that $a-c$ and $b-d$ are the only normal factors of p_3. Substituting c for a into $p_1-p_2=p_3$, it follows that $c-d$ must be a factor of p_1. Moreover it also follows that $a-b$ and $c-d$ are the only normal factors of p_1. Again it is easily seen that all the normal factors of p_1, p_2, and p_3 have multiplicity 1. If $p_1=(a-b)(c-d)$, then $p_2=(c-b)(a-d)$. If $p_1=(b-a)(c-d)$, then $p_2=(b-c)(a-d)$. Therefore for this possibility the result holds. Now assume that $c-b$ is normal. Then, arguing as before, $a-c$ is the only normal factor of p_3 and $a-b$ is the only normal factor of p_1. Again it is easily seen that all normal factors of p_1, p_2, and p_3 have multiplicity 1. If $p_1=a-b$, then $p_2=c-b$. If $p_1=b-a$, then $p_2=b-c$. Therefore for this possibility the result holds. The case that $a-c$ is the only normal factor of p_2 is treated similarly, completing the proof.

Assume that p_2 has at least two normal factors. Assume that $c-b$ is a factor of p_2. Then, using the argument in the proof of (1), there is an element d in $\{0,1,\alpha_1,\alpha_2,\ldots,\alpha_k\}-\{a,b,c\}$ such that $a-d$ is a normal factor of p_2 and $b-d$

is a factor of p_3. Since $p_1 - p_3 = p_2$, it follows that if $a - c$ is a normal factor of p_2, then $b - d$ is a factor of p_2 and $a - d$ is a normal factor of p_3.

Sublemma 2. *If p_2 has exactly two normal factors, then either $p_1 = (a-b)(c-d)$ and $p_2 \in \{(c-b)(a-d), (a-c)(b-d)\}$ or $p_1 = (b-a)(c-d)$ and $p_2 \in \{(b-c)(a-d), (c-a)(b-d)\}$.*

Proof. Assume first that $c - b$ is a factor of p_2. We first show that $c - b$ must be a normal factor of p_2. If not, then $a - d$ is a normal factor of p_2 and both $a - c$ and $b - d$ are normal factors of p_3. Since $-p_2(\alpha_i = b) = p_3(\alpha_i = b)$ and since p_1, p_2, and p_3 are relatively prime, it follows that there is an element e of $\{\alpha_1, \alpha_2, \ldots, \alpha_k\} - \{a, d\}$ such that either $e - b$ or $a - e$ is a normal factor in p_2. Using an argument similar to that in the proof of Sublemma 1, it follows that $e - b$ cannot be a normal factor in p_2. Therefore $a - e$ is a normal factor of p_2. Substituting b for d into $p_1 - p_2 = p_3$, we see that $a - e$ is also a normal factor in p_1. This contradicts the fact that p_1 and p_2 are relatively prime. Therefore $c - b$ must be a normal factor in p_2. From the proof of Sublemma 1, it follows that either $p_1 = (a-b)(c-d)$ and $p_2 = (c-b)(a-d)$ or $p_1 = (b-a)(c-d)$ and $p_2 = (b-c)(a-d)$. Therefore if $c - b$ is a factor of p_2, then the result holds. Since $p_1 - p_3 = p_2$, it follows that if $a - c$ is a normal factor in p_2, then the result also holds.

It now readily follows from the proof of Sublemma 2 that p_2 has at most two normal factors. A similar argument also shows that p_1 has at most two normal factors. Therefore all pairs of polynomials p_1 and p_2 have been found. The theorem follows on combining Sublemmas 1 and 2, and appropriately interchanging the roles of the elements a, b, c, and d if necessary.

4 Main result

The next result is needed as a lemma for Theorem 7. We note that if $z_1, z_2 \in \mathbf{R}_\mathbf{k}^*$, then $z_1 - z_2 \in \mathbf{R}_\mathbf{k}$ if and only if $z_1/z_2 - 1 \in \mathbf{R}_\mathbf{k}$. The proof is a routine case analysis using this observation in combination with Theorem 5.

Lemma 6. *Let z_1 and z_2 be distinct fundamental elements in $\mathbf{R}_\mathbf{k}$ such that $z_1, z_2 \notin \{0, 1\}$. Then $z_1 - z_2$ is defined if and only if $\{z_1, z_2\}$ is equal to one of the following sets:*

(i)

$$\left\{\frac{a_1 - b}{c - b}, \frac{a_2 - b}{c - b}\right\}$$

where a_1, a_2, b, and c are distinct elements of $\{0, 1, \alpha_1, \alpha_2, \ldots, \alpha_k\}$.

(ii)

$$\left\{\frac{a - b_1}{c - b_1}, \frac{a - b_2}{c - b_2}\right\}$$

where a, b_1, b_2, and c are distinct elements of $\{0, 1, \alpha_1, \alpha_2, \ldots, \alpha_k\}$.

(iii)

$$\left\{\frac{a-b}{c_1-b}, \frac{a-b}{c_2-b}\right\}$$

where a, b, c_1, and c_2 are distinct elements of $\{0,1,\alpha_1,\alpha_2,\ldots,\alpha_k\}$.

(iv)

$$\left\{\frac{a-b}{c-b}, \frac{(a-b)(c-d)}{(c-b)(a-d)}\right\}$$

where a, b, c, and d are distinct elements of $\{0,1,\alpha_1,\alpha_2,\ldots,\alpha_k\}$.

(v)

$$\left\{\frac{(a-b)(c-d_1)}{(c-b)(a-d_1)}, \frac{(a-b)(c-d_2)}{(c-b)(a-d_2)}\right\}$$

where a, b, c, d_1, and d_2 are distinct elements of $\{0,1,\alpha_1,\alpha_2,\ldots,\alpha_k\}$.

Before stating and proving the main result of this paper we make the following observation. Let $\varphi : \{\alpha_1,\alpha_2,\ldots,\alpha_k\} \to \mathbf{R_k}$ be a map. Suppose we can extend φ to an automorphism τ of $\mathbf{R_k}$. Then it follows that

$$\tau(\pm\prod_{i=1}^{k}\alpha_i^{l_i}\prod_{i=1}^{k}(\alpha_i-1)^{m_i}\prod_{i,j\subseteq\{1,2,\ldots,k\},i\neq j}(\alpha_i-\alpha_j)^{n_{i,j}})$$

$$=\pm\prod_{i=1}^{k}(\varphi(\alpha_i))^{l_i}\prod_{i=1}^{k}(\varphi(\alpha_i)-1)^{m_i}\prod_{i,j\subseteq\{1,2,\ldots,k\},i\neq j}(\varphi(\alpha_i)-\varphi(\alpha_j))^{n_{i,j}}.$$

Hence every automorphism of $\mathbf{R_k}$ is determined by its action on $\{\alpha_1,\alpha_2,\ldots,\alpha_k\}$.

Theorem 7. *Let $\varphi : \{\alpha_1,\alpha_2,\ldots,\alpha_k\} \to \mathbf{R_k}$ be a map. Then φ extends to an automorphism of $\mathbf{R_k}$ if and only if $\{\varphi(\alpha_1),\varphi(\alpha_2),\ldots,\varphi(\alpha_k)\}$ is equal to one of the following sets:*

(i)

$$\left\{\frac{a_1-b}{c-b}, \frac{a_2-b}{c-b}, \ldots, \frac{a_k-b}{c-b}\right\}$$

where $\{a_1,a_2,\ldots,a_k,b,c\} = \{0,1,\alpha_1,\alpha_2,\ldots,\alpha_k\}$;

(ii)

$$\left\{\frac{a-b_1}{c-b_1}, \frac{a-b_2}{c-b_2}, \ldots, \frac{a-b_k}{c-b_k}\right\}$$

where $\{a,b_1,b_2,\ldots,b_k,c\} = \{0,1,\alpha_1,\alpha_2,\ldots,\alpha_k\}$;

(iii)

$$\left\{\frac{a-b}{c_1-b}, \frac{a-b}{c_2-b}, \ldots, \frac{a-b}{c_k-b}\right\}$$

where $\{a,b,c_1,c_2,\ldots,c_k\} = \{0,1,\alpha_1,\alpha_2,\ldots,\alpha_k\}$;

(iv)

$$\left\{\frac{a-b}{c-b}, \frac{(a-b)(c-d_1)}{(c-b)(a-d_1)}, \frac{(a-b)(c-d_2)}{(c-b)(a-d_2)}, \ldots, \frac{(a-b)(c-d_{k-1})}{(c-b)(a-d_{k-1})}\right\}$$

where $\{a, b, c, d_1, d_2, \ldots, d_{k-1}\} = \{0, 1, \alpha_1, \alpha_2, \ldots, \alpha_k\}$.

Proof. If φ extends to an automorphism, then, using Lemma 6, it is clear that $\{\varphi(\alpha_1), \varphi(\alpha_2), \ldots, \varphi(\alpha_k)\}$ is equal to one of the sets (i)–(iv) in the statement of the theorem. Conversely, suppose that $\{\varphi(\alpha_1), \varphi(\alpha_2), \ldots, \varphi(\alpha_k)\}$ is equal to one of these sets. We need to show that φ extends to an automorphism of $\mathbf{R_k}$. Consider the function $\tau : \mathbf{R_k} \to \mathbf{R_k}$ defined by $\tau(0) = 0$ and

$$\tau(\pm \prod_{i=1}^{k} \alpha_i^{l_i} \prod_{i=1}^{k} (\alpha_i - 1)^{m_i} \prod_{i,j \subseteq \{1,2,\ldots,k\}, i \neq j} (\alpha_i - \alpha_j)^{n_{i,j}})$$

$$= \pm \prod_{i=1}^{k} (\varphi(\alpha_i))^{l_i} \prod_{i=1}^{k} (\varphi(\alpha_i) - 1)^{m_i} \prod_{i,j \subseteq \{1,2,\ldots,k\}, i \neq j} (\varphi(\alpha_i) - \varphi(\alpha_j))^{n_{i,j}}.$$

Observe that φ extends to an automorphism if and only if τ is an automorphism. Therefore it suffices to show that τ satisfies the properties of Theorem 3. Evidently τ satisfies (3)(ii). We next show that τ is a bijection. Assume first that $\{\varphi(\alpha_1), \varphi(\alpha_2), \ldots, \varphi(\alpha_k)\}$ is equal to set (i) in the statement of the theorem. Then, for all distinct $i, j \in \{1, 2, \ldots, k\}$, $\tau(\alpha_i - 1) = \varphi(\alpha_i) - 1 = (a_i - c)/(c - b)$ and $\tau(\alpha_i - \alpha_j) = \varphi(\alpha_i) - \varphi(\alpha_j) = (a_i - a_j)/(c - b)$. Furthermore, as $\{a_1, a_2, \ldots, a_k, b, c\} = \{0, 1, \alpha_1, \alpha_2, \ldots, \alpha_k\}$, $a_i - b$, $a_i - c$, $a_i - a_j$, and $c - b$ are all distinct. Therefore exactly one of $a_i - b$, $a_i - c$, $a_i - a_j$, and $c - b$ is an element of $\{1, -1\}$ and the other elements are exactly the generators α_i, $\alpha_i - 1$, and $\alpha_i - \alpha_j$ of $\mathbf{R}^*_\mathbf{k}$. From these observations one can now readily check that in this case τ is a bijection. The cases that $\{\varphi(\alpha_1), \varphi(\alpha_2), \ldots, \varphi(\alpha_k)\}$ is equal to one of the sets (ii)–(iv) is treated similarly. Hence τ satisfies (3)(i).

Lastly we show that τ satisfies (3)(iii). Suppose $z \in \mathbf{R_k}$ such that $z - 1$ is defined. Using the fact that $\tau(0) = 0$ and $\tau(1) = 1$, it is easily checked that if $z \in \{0, 1\}$, then (3)(iii) holds. So assume that $z \notin \{0, 1\}$. Assume first that z is equal to $(a - b)/(c - b)$ where a, b, and c are distinct elements of $\{0, 1, \alpha_1, \alpha_2, \ldots, \alpha_k\}$. Then

$$\begin{aligned}
\tau(z) - 1 &= \tau\left(\frac{a-b}{c-b}\right) - 1 \\
&= \frac{\tau(a) - \tau(b)}{\tau(c) - \tau(b)} - 1 \\
&= \frac{\tau(a) - \tau(c)}{\tau(c) - \tau(b)} \\
&= \tau\left(\frac{a-c}{c-b}\right).
\end{aligned}$$

Since the expression in the last line is defined, $\tau(z)-1$ is defined. The argument in the case that z is equal to $[(a-b)(c-d)]/[(c-b)(a-d)]$ where a, b, c, and d are distinct elements of $\{0,1,\alpha_1,\alpha_2,\ldots,\alpha_k\}$ is similar and is omitted.

Now suppose that $\tau(z)-1$ is defined. Assume first that $\tau(z)$ is equal to $(a-b)/(c-b)$ where a, b, and c are distinct elements of $\{0,1,\alpha_1,\alpha_2,\ldots,\alpha_k\}$. Since τ is a bijection, it has a unique inverse τ^{-1} such that $\tau(\tau^{-1}(p))=p$. Using this fact one readily checks that, for all $p,q\in \mathbf{R_k}$, $\tau^{-1}(pq)=\tau^{-1}(p)\tau^{-1}(q)$ and, whenever $\tau(p-q)=\tau(p)-\tau(q)$, $\tau^{-1}(p-q)=\tau^{-1}(p)-\tau^{-1}(q)$. From this we get

$$\begin{aligned} z-1 &= \tau^{-1}(\tau(z))-1 \\ &= \tau^{-1}\left(\frac{a-b}{c-b}\right)-1 \\ &= \frac{\tau^{-1}(a)-\tau^{-1}(b)}{\tau^{-1}(c)-\tau^{-1}(b)}-1 \\ &= \frac{\tau^{-1}(a)-\tau^{-1}(c)}{\tau^{-1}(c)-\tau^{-1}(b)} \\ &= \tau^{-1}\left(\frac{a-c}{c-b}\right). \end{aligned}$$

Since the expression in the last line is defined, $z-1$ is defined. The argument in the case that $\tau(z)$ is equal to $[(a-b)(c-d)]/[(c-b)(a-d)]$ where a, b, c, and d are distinct elements of $\{0,1,\alpha_1,\alpha_2,\ldots,\alpha_k\}$ is similar and is omitted. Moreover in all cases $\tau(z-1)=\tau(z)-1$. Hence (3)(iii) holds, and the theorem is proved.

It was noted in Section 3 that the matroid $U_{3,6}$ is not 2–regular. We now show that this is indeed the case. This shows that the class of 2–regular matroids is properly contained in the class of matroids representable over all fields of size at least 4.

Corollary 8. *The matroid $U_{3,6}$ is not 2–regular.*

Proof. Assume that $[I_r|A]$ is an $\mathbf{R_2}$–representation of $U_{3,6}$. Using the results of [4, Section 3], we can assume that A is

$$\begin{bmatrix} 1 & 1 & 1 \\ 1 & a & c \\ 1 & b & d \end{bmatrix}$$

where a, b, c, and d are non-zero elements of $\mathbf{R_2}$. It follows from Theorem 7 that $U_{2,5}$ is uniquely representable over $\mathbf{R_2}$ and so we may also assume that $a=\alpha_1$ and $c=\alpha_2$. Since $U_{3,6}$ has no 3–circuits, it follows that $b-1$, $d-1$, $b-a$, $d-b$, and $d-c$ are all non-zero and defined. Using Lemma 6 we get that

$$\begin{aligned}(b,d)\in\{&(\alpha_2,\alpha_1),(\alpha_2,(\alpha_1-\alpha_2)/(\alpha_1-1)),(\alpha_2,\alpha_2/\alpha_1),(\alpha_2,\alpha_2(\alpha_1-1)/(\alpha_1-\alpha_2)),\\ &(-(\alpha_1-\alpha_2)/(\alpha_2-1),\alpha_1),(\alpha_1/\alpha_2,\alpha_1),(-\alpha_1(\alpha_2-1)/(\alpha_1-\alpha_2),\alpha_1)\}.\end{aligned}$$

Furthermore, as $[I_r|A]$ is an $\mathbf{R_2}$–representation, the 3×3 determinants $ad - cb$ and $ad - cb - d + b + c - a$ are non-zero and defined. But routine checking shows that no choice of (b, d) gives both these determinants being non-zero and defined. Hence $[I_r|A]$ is not a 2–regular representation for $U_{3,6}$. We conclude that $U_{3,6}$ is not 2–regular.

Acknowledgements

I thank Geoff Whittle for the helpful discussions and whose constructive comments led to a significant improvement in the exposition.

References

1. Cohn, P. M. (1991). *Algebra, Vol. 3.* Second edition. John Wiley and Sons, Chichester.
2. Geelen, J., Gerards, A. M. H., Kapoor, A. In preparation.
3. Oxley, J. G. (1992). *Matroid Theory.* Oxford University Press, New York.
4. Semple, C. A., Whittle, G. P. (1996). Partial Fields and Matroid Representation. *Adv. in Appl. Math.* **17**, 184–208.
5. Semple, C. A. (1995). *Matroid Representation Over Partial Fields.* MSc Thesis, Victoria University of Wellington.
6. Seymour, P. D. (1979). Matroid representation over $GF(3)$. *J. Combin. Theory Ser. B.* **26**, 305–359.
7. Tutte, W. T. (1958). A homotopy theorem for matroids, I, II. *Trans. Amer. Math. Soc.* **88**, 144–174.
8. Tutte, W. T. (1965). Lectures on matroids. *J. Res. Nat. Bur. Standards Sect. B.* **69B**, 1–47.
9. Whittle, G. (1995). A characterisation of the matroids representable over $GF(3)$ and the rationals. *J. Combin. Theory Ser. B.* **65**, 222–261.
10. Whittle, G. On matroids representable over $GF(3)$ and other fields. To appear in *Trans. Amer. Math. Soc.*

Representations versus Numberings: On Two Computability Notions*

Dieter Spreen

Fachbereich Mathematik, Theoretische Informatik
Universität–GH Siegen, D–57068 Siegen, Germany
Email: spreen@informatik.uni-siegen.de

Abstract. This paper gives an answer to Weihrauch's question [26] whether and, if not always, when an effective map between the computable elements of two represented sets can be extended to a (partial) computable map between the represented sets.
Examples are known showing that this is not possible in general. A condition is introduced which in the case of countably based topological T_0-spaces forces the effective map to be extendable. The requirement is satisfied in the standard cases of effectively given separable metric spaces and continuous directed-complete partial orders, in which the extendability is already known.

Introduction

In a series of papers K. Weihrauch has developed a general approach to study constructivity in analytical mathematics (cf. e.g. [26]). The essential insight was that all sets studied there are equipped with some notion of approximation and that in important cases their elements can be *represented* as limits of certain sequences of points or (open) subsets approximating them. Moreover, in these cases the elements of the sequences can be coded by natural numbers in an easy way. Thus, every element of the space can be *named* by an arithmetic function, namely the one which generates the approximating sequence. The appropriate morphisms of such spaces are those maps which are defined by a computable transformation of the approximating sequences, i.e., by a recursive operator on the name space. They are called computable.

Elements that can be approximated by a computable sequence, i.e., which can be named by a total recursive function, are considered as computable. By coding the programs that compute the sequences (functions) one obtains a canonical indexing of the computable points. Indexed sets have been studied in great depth by Mal'cev [15] and Eršov [8, 9, 10]. The canonical morphisms of such sets are those maps which are defined by a recursive function on the indices. In our case this means that they are defined by operations which effectively transform the programs generating the approximating sequences. These morphisms are called effective.

* This research has partially been supported by ESPRIT BRA 7232

As it is easy to see, the restriction of a computable map to the subspace of all computable points is effective. For certain special cases such as effectively given separable metric spaces and continuous directed-complete partial orders it is known that the converse of this implication is also true, i.e., every effective map on the computable elements can be extended to a (partial) computable map. In [26] Weihrauch asks whether and, if not always, when such an extension result holds in general.

From the literature, examples are known which show that in general effective maps are not extendable to a computable map (cf. [11, 13, 14, 18, 28, 29]). Thus the computability notion for mappings between represented spaces is stronger than the computability notion for maps between indexed spaces. The reason is that in the first case computability involves continuity, which is not true, in general, in the second case.

In this paper we show under a fairly general and natural assumption that each effective map between the subspaces of all computable elements of a countably based separable T_0-space and another countably based T_0-space has a partial computable extension, if it has a *witness for noninclusion*, which means: if some basic open set in the domain is *not* mapped into some basic open set in the codomain, then we must be able to effectively find a witness for this, that is, an element of the basic open set in the domain which is mapped outside the basic open set in the codomain. This result contains the before mentioned extension theorems as special cases: As follows from [22, 24] effective maps always have a witness for noninclusion, if their domain of definition is domain-like or their codomain is an effectively given separable metric space.

In [26] the theory of representations has been developed in great generality. Nevertheless considering only countably based T_0-spaces is not a real restriction. Such spaces appear quite naturally in computer science. Scott [20] and Smyth [21] pointed out that data types can be thought of as T_0-spaces the basic open sets of which are the finitely describable properties of the data objects. Most structures considered in programming language semantics are equipped with a canonical topology. Prominent examples are metric spaces, Scott domains, and A- and f-spaces [5, 6, 7, 9]. As is shown by Stoltenberg-Hansen and Tucker [25], many algebraic structures, e.g., all term algebras over a finite signature, can be canonically embedded in complete ultrametric spaces as well as in Scott domains.

This paper is organized as follows: Section 1 contains basic definitions and facts from computability, numeration, and representation theory.

In Section 2 countably based T_0-spaces and various representations with their derived numberings are studied. Moreover, some important, standard examples such as effectively given separable metric spaces, continuous directed-complete partial orders, A-, and f-spaces are considered.

In Section 3, finally, it is shown that under some very general effectivity assumption, which is satisfied by all examples studied in Section 2, each effectively continuous map between the subspaces of all computable elements of a countably based separable T_0-space and another countably based T_0-space can be extended to a partial computable map between these spaces. As is shown in [24] (and in a slightly more general framework in [22]), each effective map between such subspaces which has a witness for noninclusion is effectively continuous. From these two results the before mentioned answer to Weihrauch's question follows.

1 Basic definitions and properties

In what follows, let $\langle\ ,\ \rangle : \omega^2 \to \omega$ be a recursive pairing function. Moreover, let $P^{(n)}$ ($R^{(n)}$) denote the set of all n-ary partial (total) recursive functions, and let W_i be the domain of the ith partial recursive function φ_i with respect to some Gödel numbering φ. We let $\varphi_i(a){\downarrow} \in A$ mean that the computation of $\varphi_i(a)$ stops with value in A. If $A \subseteq \omega$ is not empty and (relatively) recursively enumerable (r.e.), A_s is the finite subset of A which can be enumerated in s steps with respect to some fixed (relative) enumeration of A, i.e., $A_s = \{f(0), \ldots, f(s-1)\}$, where f is the fixed (relative) enumeration function. The cardinality of a set A is denoted by $|A|$. If F is a partial map from a set X into a set Y this is written as $F: X \rightharpoonup Y$.

Let S be a nonempty set. A *(partial) numbering* ν of S is a partial map $\nu: \omega \rightharpoonup S$(onto) with domain $\text{dom}(\nu)$. The value of ν at $n \in \text{dom}(\nu)$ is denoted, interchangeably, by ν_n and $\nu(n)$.

Definition 1. Let ν and κ be numberings of the set S.

1. $\nu \leq_m \kappa$, read ν is *many-one reducible* to κ, if there is a function $g \in P^{(1)}$ such that $\text{dom}(\nu) \subseteq \text{dom}(g)$, $g(\text{dom}(\nu)) \subseteq \text{dom}(\kappa)$, and $\nu_m = \kappa_{g(n)}$, for all $n \in \text{dom}(\nu)$.
2. ν is *many-one equivalent* to κ, if $\nu \leq_m \kappa$ and $\kappa \leq_m \nu$.

A subset X of S is *completely enumerable*, if there is an r.e. set W_n such that $\nu_i \in X$ if and only if $i \in W_n$, for all $i \in \text{dom}(\nu)$. Set $C_n = X$, for any such n and X, and let C_n be undefined, otherwise. Then C is a numbering of the class of all completely enumerable subsets of S. If W_n is recursive, X is said to be *completely recursive*. A relation $R \subseteq S \times S$ is *completely enumerable*, if there is an r.e. set A such that $(\nu_i, \nu_j) \in R$ if and only if $\langle i, j\rangle \in A$, for all $i, j \in \text{dom}(\nu)$.

Numbered sets form a category [8]. Morphisms are the effective maps, where for two nonempty sets S and S' with numberings ν and ν', respectively, a map $F: S \rightharpoonup S'$ is called *effective*, if there is a function $f \in P^{(1)}$ such that for all $n \in \nu^{-1}(\text{dom}(F))$, $f(n){\downarrow} \in \text{dom}(\nu')$ and $F(\nu_n) = \nu'_{f(n)}$.

Obviously, only countable sets have numberings. In order to study the computability of mappings between sets which may have the cardinality of the continuum, Weihrauch considered sets the elements of which are no longer represented by numbers, but by arithmetic functions.

Let $\mathbb{B}$ be the set of all functions $p: \omega \to \omega$, and let T be a nonempty set. A *representation* δ of T is a partial map $\delta: \mathbb{B} \rightharpoonup T$(onto) with domain $\text{dom}(\delta)$. As in the case of numberings the value of δ at $p \in \text{dom}(\delta)$ is denoted by δ_p and/or $\delta(p)$.

Definition 2. Let δ and η be representations of the set T.

1. $\delta \leq_c \eta$, read δ is *(computably) reducible* to η, if there is a recursive operator [19] Γ such that for all $p \in \mathrm{dom}(\delta)$, $\Gamma(p) \in \mathrm{dom}(\eta)$, and $\delta_p = \eta_{\Gamma(p)}$.
2. $\delta \equiv_c \eta$, read δ is *(computably) equivalent* to η, if $\delta \leq_c \eta$ and $\eta \leq_c \delta$.

For $p \in \mathbb{B}$ define $\mathbb{M}_p$ to be the set of all numbers i with $i + 1 \in \mathrm{range}(p)$. Then $\mathbb{M}$ is a representation of the powerset of ω.

Let T' be a further nonempty set and δ' be representation of T'. A map $F: T \rightharpoonup T'$ is called *computable*, if there is a recursive operator Γ with $\Gamma(p) \in \mathrm{dom}(\delta')$ and $F(\delta_p) = \delta'_{\Gamma(p)}$, for all $p \in \delta^{-1}(\mathrm{dom}(F))$.

There is a natural way to introduce computability on a represented set. An element z of T is said to be *computable*, if there is some $p \in R^{(1)}$ with $z = \delta_p$. Let T_c be the set of all computable elements of T. Obviously, the computability of an element is invariant under the equivalence of representations. Associated with each representation δ of T there is a canonical numbering ν^δ of the corresponding set T_c of computable elements, called the *derived numbering*: $\nu_i^\delta = \delta(\varphi_i)$, if $\varphi_i \in \mathrm{dom}(\delta)$; otherwise ν_i^δ is undefined. As it is easy to see, the derived numberings of two equivalent representations of T are many-one equivalent.

Lemma 3. *Let δ and δ', respectively, be representations of the sets T and T'. Moreover, let $F: T \rightharpoonup T'$ be computable. Then $F(T_c) \subseteq T'_c$ and the restriction of F to T_c is effective with respect to the derived numberings.*

The proof of this lemma is given in [26], where Weihrauch also asks when the converse is true, i.e.:

> *Given an effective map $F: T_c \to T'_c$, is there a computable map $\tilde{F}: T \rightharpoonup T'$ which extends F?*

As follows from the examples in the literature, there is no such extension in general. In Section 3 we present a condition which in the case of countably based separable topological T_0-spaces and certain natural representations ensures the existence of a computable extension.

2 Topological spaces and standard representations

Let $\mathcal{T} = (T, \tau)$ be a topological T_0-space with a countable basis $\mathcal{B}$, and let $B: \omega \to \mathcal{B}$(onto) be a total numbering of $\mathcal{B}$. In the applications we have in mind the basic open sets can be described in some finite way. The indexing B is then obtained by an encoding of the finite descriptions. Moreover, in these cases there is a canonical relation between the (code numbers of the) descriptions which is stronger than the usual set inclusion between the described sets. This relation is r.e., which, in general, is not true for set inclusion. One has to use this stronger relation in order to derive the result we talked about in the introduction.

Definition 4. Let $\prec_B$ be a transitive binary relation on ω. We say that:

1. $\prec_B$ is a *strong inclusion*, if for all $m, n \in \omega$ from $m \prec_B n$ it follows that $B_m \subseteq B_n$.
2. $\mathcal{B}$ is a *strong basis*, if $\prec_B$ is a strong inclusion and for all $z \in T$ and $m, n \in \omega$ with $z \in B_m \cap B_n$ there is a number $a \in \omega$ such that $z \in B_a$, $a \prec_B m$ and $a \prec_B n$.

If one follows Smyth [21] and thinks of open sets as properties of data objects, then strong inclusion relations can be considered as 'definite refinement' relations. For what follows we assume that $\prec_B$ is an r.e. strong inclusion with respect to which $\mathcal{B}$ is a strong basis.

Since in a T_0-space every point y is uniquely determined by (a base of) its neighbourhood filter $\mathcal{N}(y)$, there is a canonical way to define representations of such spaces.

Definition 5. Let $\mathcal{H}$ be a filter. A nonempty subset $\mathcal{F}$ of $\mathcal{H}$ is said to be a *strong base* of $\mathcal{H}$ if the following two conditions hold:

1. $(\forall m, n \in \omega)[B_m, B_n \in \mathcal{F} \Rightarrow (\exists a \in \omega) B_a \in \mathcal{F} \wedge a \prec_B m \wedge a \prec_B n]$
2. $(\forall m \in \omega)[B_m \in \mathcal{H} \Rightarrow (\exists a \in \omega) B_a \in \mathcal{F} \wedge a \prec_B m]$.

Definition 6. For $p \in \mathbb{B}$ let

1. ξ_p be the unique point $y \in T$ such that $B(\mathbb{M}_p)$ is a strong base of $\mathcal{N}(y)$,
2. ρ_p be the unique point $z \in T$ such that $\mathbb{M}_p$ is the set of all numbers n with $z \in B_n$,

if there are such points, and let ξ_p and ρ_p be undefined, otherwise. The map ξ is called *standard representation* of T.

Lemma 7. $\rho \equiv_c \xi$.

Proof. As $\mathcal{B}$ is a strong basis, we have for every point $z \in T$ that the set of all B_n with $z \in B_n$ is a strong base of $\mathcal{N}(z)$. Thus, the identity on $\mathbb{B}$ witnesses that $\rho \leq_c \xi$.

In order to show that $\xi \leq_c \rho$, let $p \in \mathbb{B}$ and consider the set

$$A[p] = \{ n \in \omega \mid (\exists a \in \mathbb{M}_p) a \prec_B n \} .$$

Obviously, $A[p]$ is r.e. in p. Hence, there is some recursive operator Θ with $\mathbb{M}_{\Theta(p)} = A[p]$. Assume that $p \in \text{dom}(\xi)$ and let $n \in A[p]$. Then $\xi_p \in B_n$. Now, to the contrary, let the number m be so that $\xi_p \in B_m$. Since $B(\mathbb{M}_p)$ is a strong base of $\mathcal{N}(\xi_p)$, there is a number $a \in \mathbb{M}_p$ with $a \prec_B m$. It follows that $m \in A[p]$. This shows that $\xi_p \in B_e$ if and only if $e \in A[p]$. Thus $\xi_p = \rho_{\Theta(p)}$. □

A representation ζ of T with $\zeta \equiv_c \xi$ is called *admissible*. For the remainder of this paper we assume that the spaces under consideration are admissibly represented. Moreover, the computability of a point is always meant with respect to such a representation.

Observe that the notions of a standard representation and of admissibility used in this paper differ slightly from Weihrauch's: In [26] filter bases are not required to be strong.

Let us now introduce some well known examples of T_0-spaces with a strong basis. We first give the definition of the space and then discuss properties of their admissible representations.

Example 1 (Effectively Given Metric Spaces). Let $\mathcal{M} = (M, d)$ be a separable metric space with dense subset M_0. $\mathcal{M}$ is said to be *effectively given*, if there exists a total indexing β of M_0 such that the restriction of the distance function d to M_0 has only computable values and is effective. (Here, computability is understood with respect to the standard representation of the real number set $\mathbb{R}$, and $\mathbb{R}_c$ is enumerated by the corresponding derived numbering. As basis of the usual topology on $\mathbb{R}$ the set of all open intervals with rational end points is taken, enumerated in a canonical way.)

As is well known, the collection of all sets $B_{\langle i,m\rangle} = \{ y \in M \mid d(\beta_i, y) < 2^{-m} \}$ $(i, m \in \omega)$ is a basis of the canonical Hausdorff topology on M. Define

$$\langle i,m\rangle \prec_B \langle j,n\rangle \Leftrightarrow d(\beta_i, \beta_j) + 2^{-m} < 2^{-n} \ .$$

Using the triangular inequation it is easily verified that $\prec_B$ is a strong inclusion and the collection of all $B_{\langle i,m\rangle}$ is a strong basis. Since the less-than relation on the computable real numbers is completely enumerable [16], we moreover have that $\prec_B$ is also r.e.

Another widely known fact is that each element of M is the limit of a normed Cauchy sequence of elements of the dense subset M_0, where a sequence $(y_a)_{a\in\omega}$ of elements of M_0 is said to be *normed* if $d(y_m, y_n) < 2^{-m}$ for all m; $n \in \omega$ with $m \le n$. For $p \in \mathbb{B}$ define γ_p to be the limit of the sequence $(\beta_{p(a)})_{a\in\omega}$, if this sequence is normed and the limit exists. In any other case let γ_p be undefined. The map γ is called *Cauchy representation* of M.

Lemma 8. *The Cauchy representation is admissible.*

Proof. For $p \in \mathbb{B}$ let $A[p] = \{ \langle a,m,n\rangle \mid a \in \mathbb{M}_p \wedge a \prec_B m \wedge a \prec_B p(n) - 1 \}$. Obviously, $A[p]$ is r.e. in p. Hence, there is some recursive operator Ω with $\mathbb{M}_{\Omega(p)} = A[p]$. Define the operator $\Phi\colon \mathbb{B} \rightharpoonup \mathbb{B}$ by

$$\begin{aligned}\Phi(p)(0) &= p(0)\\ \Phi(p)(n+1) &= 1 + \pi_1^3(\Omega(p)(\mu e : \pi_2^3(\Omega(p)(e) - 1) = \Phi(p)(n) - 1\\ &\qquad\qquad \wedge \pi_3^3(\Omega(p)(e) - 1) = n + 1) - 1) \ .\end{aligned}$$

Clearly, Φ is recursive.

Now, assume that $p \in \mathrm{dom}(\xi)$. Then $B(\mathbb{M}_p)$ is a strong base of $\mathcal{N}(\xi_p)$. Therefore it follows by induction that $\Phi(p)(n)$ is defined, for all $n \in \omega$. By the definition of Φ we moreover have that $\mathbb{M}_{\Phi(p)} \subseteq \mathbb{M}_p$ and $\Phi(p)(n+1) - 1 \prec_B \Phi(p)(n) - 1$. Set $\hat{\Phi}(p)(n) = \pi_1(\Phi(p)(n) - 1)$. Then we obtain from the definition of $\prec_B$ that $(\beta_{\hat{\Phi}(p)(n)})_{n\in\omega}$ is a normed Cauchy sequence of elements of the dense subset M_0 which converges to ξ_p. Thus $\xi_p = \gamma_{\hat{\Phi}(p)}$. This shows that $\xi \le_c \gamma$.

For the proof that also $\gamma \leq_c \xi$ let

$$\Psi(p)(n) = 1 + \langle p(n+1), n\rangle \ .$$

Then Ψ is recursive. Moreover, we have for $p \in \mathrm{dom}(\gamma)$ and $n \in \omega$ that

$$\begin{aligned} d(\beta_{\pi_1(\Psi(p)(n+1)-1)}, \beta_{\pi_1(\Psi(p)(n)-1)}) + 2^{-\pi_2(\Psi(p)(n+1)-1)} \\ &= d(\beta_{p(n+2)}, \beta_{p(n+1)}) + 2^{-(n+1)} \\ &< 2^{-(n+1)} + 2^{-(n+1)} \\ &= 2^{-\pi_2(\Psi(p)(n)-1)} \ . \end{aligned}$$

Thus $\Psi(p)(n+1) - 1 \prec_B \Psi(p)(n) - 1$, for all $n \in \omega$. In order to obtain that $B(\mathbb{M}_{\Psi(p)})$ is a strong base of $\mathcal{N}(\gamma_p)$ it remains to show that for each $B_m \in \mathcal{N}(\gamma_p)$ there is some $a \in \omega$ with $\Psi(p)(a) - 1 \prec_B m$.

Let $\gamma_p \in B_m$ and let a be the smallest number n with $2^{1-n} \leq 2^{-\pi_2(m)} - d(\gamma_p, \beta_{\pi_1(m)})$. Since $(\beta_{p(n)})_{n\in\omega}$ is normed and converges to γ_p, we have that

$$\begin{aligned} d(\beta_{\pi_1(\Psi(p)(a)-1)}, \beta_{\pi_1(m)}) + 2^{-\pi_2(\Psi(p)(a)-1)} \\ &= d(\beta_{p(a+1)}, \beta_{\pi_1(m)}) + 2^{-a} \\ &\leq d(\beta_{p(a+1)}, \gamma_p) + d(\gamma_p, \beta_{\pi_1(m)}) + 2^{-a} \\ &< 2^{-(a+1)} + 2^{-a} + d(\gamma_p, \beta_{\pi_1(m)}) \\ &< 2^{-\pi_2(m)} \ . \end{aligned}$$

It follows that $\gamma_p = \xi_{\Psi(p)}$, i.e., $\gamma \leq_c \xi$. □

Note that an analogous result has been shown by Weihrauch [26] for separable metric spaces without isolated points. But as has already been mentioned his notion of a standard representation is slightly different from ours.

A standard example of an effectively given metric space is *Baire space*, that is, the set $\mathbb{B}$ of all total arithmetic functions with the metric

$$d^B(p,q) = \begin{cases} 0 & \text{if } p = q, \\ 2^{-\mu a: p(a)\neq q(a)} & \text{otherwise.} \end{cases}$$

The functions that are eventually zero form a dense subset. Let α be a canonical encoding of these functions (cf. eg. [26, p. 399]). The indexing of the basic open balls considered above is denoted by H in this case. With the help of Lemma 8 it is readily verified that the identity mapping on $\mathbb{B}$ is an admissible representation.

Example 2 *(Effectively Given Continuous Directed-Complete Partial Orders).* Let $\mathcal{Q} = (Q, \sqsubseteq)$ be a partial order with smallest element $\perp$. A nonempty subset S of Q is *directed* if for all y_1, $y_2 \in S$ there is some $u \in S$ with y_1, $y_2 \sqsubseteq u$. $\mathcal{Q}$ is a *directed-complete* partial order (dcpo) if every directed subset S of Q has a least upper bound $\sup S$ in Q. Let $\ll$ denote the *way-below relation* on Q, i.e., let $y_1 \ll y_2$ if for directed subsets S of Q the relation $y_2 \sqsubseteq \sup S$ always implies the existence of a $u \in S$ with $y_1 \sqsubseteq u$.

A subset Z of Q is a *basis* of $\mathcal{Q}$, if for any $y \in Q$ the set $Z_y = \{ z \in Z \mid z \ll y \}$ is directed and $y = \sup Z_y$. A dcpo that has a basis is called *continuous.* As is well known, on each dcpo there is a canonical T_0 topology: the *Scott topology.* A subset X of Q is open if X is upwards closed with respect to $\sqsubseteq$ and with each $u \in X$ there is some $y \in X$ with $y \ll u$. If $\mathcal{Q}$ is continuous, this topology is generated by the sets $O_z = \{ y \in Q \mid z \ll y \}$ with $z \in Z$. Moreover, Z is dense in $\mathcal{Q}$.

Let $\mathcal{Q}$ be continuous, then it is called *effectively given*, if there exists a total indexing β of Z such that the restriction of the way-below relation to Z is completely enumerable. Set $B_n = O_{\beta(n)}$ and define

$$m \prec_B n \Leftrightarrow \beta_n \ll \beta_m \ .$$

Then $\prec_B$ is an r.e. strong inclusion and the collection of all B_n is a strong basis.

As is well known, on T_0-spaces there is a canonical partial order, the specialization order, which we denote by $\leq_\tau$.

Definition 9. Let $\mathcal{T} = (T, \tau)$ be a T_0-space, and $y,\ z \in T$. $y \leq_\tau z$ if $\mathcal{N}(y) \subseteq \mathcal{N}(z)$.

Note that in the case of a dcpo with the Scott topology the specialization order coincides with the partial order.

Example 3 (Effectively Given A-Spaces). A- and f-spaces have been introduced by Eršov [5, 6, 7, 9] as a more topologically oriented approach to domain theory. They are not required to be complete. An example of an f-space which is not directed-complete can be found in [12].

Let $\mathcal{Y} = (Y, \sigma)$ be a topological T_0-space. For a subset X of Y, $\mathrm{int}(X)$ is its interior. Moreover, for $y,\ z \in Y$ define $y \ll z$ if $z \in \mathrm{int}(\{ u \in Y \mid y \leq_\sigma u \})$. Then $\mathcal{Y}$ is an *A-space* if there is a subset Y_0 of Y satisfying the following three properties:

1. Any two elements of Y_0 which are bounded in Y with respect to the specialization order have a least upper bound in Y_0.
2. The collection of sets $\mathrm{int}(\{ u \in Y \mid y \leq_\sigma u \})$, for $y \in Y_0$, is a basis of topology σ.
3. For any $y \in Y_0$ and $u \in Y$ with $y \ll u$ there is some $z \in Y_0$ such that $y \ll z$ and $z \ll u$.

Any subset Y_0 of Y with these properties is called *basic subspace.*

The A-space $\mathcal{Y}$ with basic subspace Y_0 is *effectively given*, if there is a total indexing β of Y_0 such that the restriction of the relation $\ll$ to Y_0 is completely enumerable. For $m,\ n \in \omega$ set $B_n = \mathrm{int}(\{u \in Y \mid \beta_n \leq_\rho u \})$ and define

$$m \prec_B n \Leftrightarrow \beta_n \ll \beta_m \ .$$

Then $\prec_B$ is an r.e. strong inclusion. Moreover, the collection of all B_n is a strong basis.

Example 4 (Effectively Given f-Spaces). Let $\mathcal{Y} = (Y, \sigma)$ be again an arbitrary topological T_0-space. An open set V is an *f-set*, if V is nonempty and there is a some element $z_V \in V$ such that $V = \{ y \in Y \mid z_V \leq_\sigma y \}$. The uniquely determined element z_V is called *f-element.* $\mathcal{Y}$ is an *f-space*, if the following two conditions hold:

1. If U and V are f-sets with nonempty intersection, then $U \cap V$ is also an f-set.
2. The collection of all f-sets is a basis of topology σ.

An f-space is *effectively given*, if the set of all f-elements has a total numbering β such that the restriction of the specialization order to this set and the boundedness of two f-elements are completely recursive and there is a function $su \in R^{(2)}$ such that in the case that β_n and β_m are bounded, $\beta_{su(n,m)}$ is their least upper bound.

Obviously, every f-space is an A-space with basic subspace the set of all f-elements. Moreover, for y, $z \in Y$ such that y or z is an f-element, $y \ll z$ if and only if $y \leq_\sigma z$.

A common property of the above examples is that the spaces contain a dense indexed subset such that for each basic open set B_n one can effectively and uniformly generate all dense points included in it.

Definition 10. A T_0-space $\mathcal{T} = (T, \tau)$ with a countable basis $\mathcal{B}$ and a total indexing B of $\mathcal{B}$ is *effectively separable*, if there exists a dense subset T_0 of T and a total numbering β of T_0 such that $\{ \langle i, n \rangle \mid \beta_i \in B_n \}$ is r.e.

A further property of the spaces in Examples 2, 3 and 4 is that their canonical topology has a basis with every basic open set being an upper set which has a lower bound in a certain neighbourhood of it.

Definition 11. We say that $\mathcal{T}$ is *pointed*, if there is a subset Pt of T and a numbering χ of Pt such that for all $n \in \omega$ with $B_n \neq \emptyset$, $n \in \mathrm{dom}(\chi)$, $\chi_n \leq_\tau z$, for all $z \in B_n$, and $\chi_n \in B_m$ if and only if $n \prec_B m$, for all $m \in \omega$.

Let $\mathrm{hl}_n(B_n) = \bigcap \{ B_m \mid n \prec_B m \}$. Then $\chi_n \in \mathrm{hl}_n(B_n)$. Note that

$$B_n \subseteq \{ z \in T \mid \chi_n \leq_\tau z \} \subseteq \mathrm{hl}_n(B_n) \ .$$

As it is readily verified, Pt is dense in T. Since strong inclusion is assumed to be r.e., we obtain that if $\mathcal{T}$ is pointed then it is also effectively separable.

Obviously, every effectively given continuous dcpo with the Scott topology, every effectively given A-space and every effectively given f-space is pointed.

For pointed spaces it can be shown that every element y is the least upper bound of all χ_n with $y \in B_n$, where the least upper bound is taken with respect to the specialization order [23]. Thus, we have for the representation ρ that $\rho_p = \sup \chi(\mathbb{M}_p)$ $(p \in \mathrm{dom}(\rho))$.

As is well known, (the restriction of) each recursive operator (to Baire space) is effectively continuous. The same can be shown for computable maps. To this end we show that the representation ρ is both effectively continuous and effectively open.

Definition 12. Let $\mathcal{T} = (T, \tau)$ and $\mathcal{T}' = (T', \tau')$, respectively, be T_0-spaces with countable bases $\mathcal{B}$ and $\mathcal{B}'$ and total numberings B and B' of $\mathcal{B}$ and $\mathcal{B}'$. A map $F: T \rightharpoonup T'$ is called

1. *effectively continuous*, if there is a function $g \in R^{(1)}$ such that for all $n \in \omega$, $F^{-1}(B'_n) = \bigcup \{ B_a \cap \mathrm{dom}(F) \mid a \in W_{g(n)} \}$.
2. *effectively open*, if there is a function $h \in R^{(1)}$ such that for all $n \in \omega$, $F(B_n) = \bigcup \{ B'_a \cap \mathrm{range}(F) \mid a \in W_{h(n)} \}$.

Lemma 13. *The representation ρ is effectively continuous.*

Proof. For $p \in \mathrm{dom}(\rho)$ and $n \in \omega$ we have

$$\begin{aligned} \rho_p \in B_n &\Leftrightarrow n+1 \in \mathrm{range}(p) \\ &\Leftrightarrow (\exists i, j, a \in \omega) p \in H_{\langle i,a \rangle} \wedge j \leq a \wedge \alpha_i(j) = n+1 \ . \end{aligned}$$

Let the function $g \in R^{(1)}$ be such that $W_{g(n)} = \{ m \mid (\exists j \leq \pi_2(m)) \alpha_{\pi_1(m)}(j) = n+1 \}$. Then it witnesses the effective continuity of ρ. □

It follows that all admissible representations of T are effectively continuous.

Lemma 14. *The representation ρ is effectively open.*

Proof. Let $z \in T$ and i, $n \in \omega$. Then we have

$$\begin{aligned} z \in \rho(H_{\langle i,n \rangle}) &\Leftrightarrow (\forall a \leq n)[\alpha_i(a) > 0 \Rightarrow z \in B_{\alpha_i(a)-1}] \\ &\Leftrightarrow (\exists m \in \omega) z \in B_m \wedge (\forall a \leq n)[\alpha_i(a) > 0 \Rightarrow m \prec_B \alpha_i(a) - 1] \ . \end{aligned}$$

The last equation holds as $\mathcal{B}$ is a strong basis. It follows that ρ is effectively open. □

Now, let $\mathcal{T}' = (T', \tau')$ be a further T_0-space with countable basis $\mathcal{B}'$ and a total numbering B' of $\mathcal{B}'$, then we obtain as a consequence.

Proposition 15. *Every computable map $F: T \rightharpoonup T'$ is effectively continuous.*

As is widely known, the set $R^{(1)}$ of all unary total recursive functions is dense in Baire space. It follows that if it is also dense in the domain of an admissible representation (with respect to the induced topology), then the subset of all computable points is dense in T.

Let us now assume that $\mathcal{T}$ is effectively separable with dense subset T_0 and total numbering β of T_0. Moreover, let x be the numbering of T_c derived from an admissible representation. Without restriction suppose that it is derived from the representation ρ. Then the function $g \in R^{(1)}$ with $\mathbb{M}(\varphi_{g(i)}) = \{ n \mid \beta_i \in B_n \}$ witnesses that $\beta \leq_m x$.

In [22] a countably based countable T_0-space $\hat{\mathcal{T}} = (\hat{T}, \hat{\tau})$ with numberings $\hat{x}$ and $\hat{B}$, respectively, of $\hat{T}$ and the topological basis $\hat{\mathcal{B}}$ such that for some function $g \in R^{(1)}$ with $\mathrm{range}(g) \subseteq \mathrm{dom}(\hat{x})$ the set $\hat{x}(\mathrm{range}(g))$ is dense in $\hat{\mathcal{T}}$ is called *recursively separable*. Moreover, $\hat{\mathcal{T}}$ is said to be *effectively pointed*, if there is a function $pd \in P^{(1)}$ such that for all $n \in \mathrm{dom}(\hat{B})$ with $\hat{B}_n \neq \emptyset$, $pd(n)\downarrow \in \mathrm{dom}(\hat{x})$, $\hat{x}_{pd(n)} \in \mathrm{hl}_n(\hat{B}_n)$ and $\hat{x}_{pd(n)} \leq_{\hat{\tau}} z$, for all $z \in \hat{B}_n$.

Let τ_c be the induced topology on T_c. Then it follows that $\mathcal{T}_c = (T_c, \tau_c)$ is recursively separable, if $\mathcal{T}$ is effectively separable. Moreover, $\mathcal{T}_c$ is effectively pointed, if $\mathcal{T}$ is pointed.

3 A sufficient condition

As we have seen above computable mappings are effectively continuous. In [11] Friedberg presents an example of an effective map $F: R^{(1)} \to P^{(1)}$ which is not effectively continuous (see also [17]). By Proposition 15 it follows that it cannot be extended to a computable map. This shows that the converse of Lemma 3 is not true, in general, and hence that working with computable elements only and maps which are computed by transforming the programs that compute the approximations of the points results in a more general computability notion than working with maps which are computed by transforming the approximations itself. From the literature also other examples of effective maps that are not effectively continuous are known [13, 14, 18, 28, 29]. The importance of Friedberg's example lies in the fact that it provides an example of a total, effective, and discontinuous map on a natural space.

In this section we present a condition which forces an effective map between the subspaces of computable elements of a countably based represented effectively separable T_0-space and a countably based represented T_0-space to be extendable to a computable map. We proceed in two steps: First we show that each effectively continuous mapping between such subspaces is extendable to a computable mapping, and then we introduce the additional assumption under which it is shown in [22, 24] that effective maps are effectively continuous.

Let $\mathcal{T} = (T, \tau)$ and $\mathcal{T}' = (T', \tau')$, respectively, be topological T_0-spaces with countable bases $\mathcal{B}$ and $\mathcal{B}'$, let B and B', respectively, be total numberings of $\mathcal{B}$ and $\mathcal{B}'$, and let x be the derived numbering of an admissible representation of T.

Proposition 16. *Let $F: T \rightharpoonup T'$ be effectively continuous. Then a computable map $\tilde{F}: T \rightharpoonup T'$ can be constructed which extends F.*

Proof. Without restriction assume that T and T', respectively, are represented by ρ and ρ'. Moreover, let the function $g \in R^{(1)}$ witness the effective continuity of F. For $p \in \mathbb{B}$ set $A[p] = \{ a \in \omega \mid W_{g(a)} \cap \mathbb{M}_p \neq \emptyset \}$. Then $A[p]$ is r.e. in p. Hence, there is some recursive operator Γ with $\mathbb{M}_{\Gamma(p)} = A[p]$. Obviously, for p, $q \in \mathbb{B}$ with $\mathbb{M}_p = \mathbb{M}_q$ we have that $\mathbb{M}_{\Gamma(p)} = \mathbb{M}_{\Gamma(q)}$. It follows for p, $q \in \mathrm{dom}(\rho) \cap \Gamma^{-1}(\mathrm{dom}(\rho'))$ with $\rho_p = \rho_q$ that $\rho'_{\Gamma(p)} = \rho'_{\Gamma(q)}$.

For $y \in T$ set $\tilde{F}(y) = \rho'_{\Gamma(p)}$, for some $p \in \rho^{-1}(y) \cap \Gamma^{-1}(\mathrm{dom}(\rho'))$, if this set is not empty, and let $\tilde{F}(y)$ be undefined, otherwise. Then $\tilde{F}$ is computable.

It remains to show that on $\mathrm{dom}(F)$, $\tilde{F}$ coincides with F. Let to this end $p \in \rho^{-1}(\mathrm{dom}(F))$. Then $a \in A[p]$ if and only if $F(\rho_p) \in B'_a$. Hence $\Gamma(p) \in \mathrm{dom}(\rho')$ and $\tilde{F}(\rho_p) = \rho'_{\Gamma(p)} = F(\rho_p)$. □

The next step is to show that effective maps are effectively continuous. As follows from Friedberg's example, this is not true in general. But it holds, if the effective map has a witness for noninclusion.

Definition 17. A map $F: \mathcal{T}_c \to \mathcal{T}'_c$ has a *witness for noninclusion*, if there exist functions $s \in P^{(2)}$ and $r \in P^{(3)}$ such that for all $i \in \text{dom}(x)$ and $m, n \in \omega$ the following hold:

1. If $F(x_i) \in B'_m$ then $s(i,m)\downarrow \in \text{dom}(C)$ and $x_i \in C_{s(i,m)} \subseteq F^{-1}(B'_m)$.
2. If, in addition, $F(B_n) \not\subseteq B'_m$ then also $r(i,n,m)\downarrow \in \text{dom}(x)$ and $x_{r(i,n,m)} \in \text{hl}_n(B_n) \setminus C_{s(i,m)}$.

Note that such a condition appears quite naturally. As follows from [1, 2, 3, 4], the effective continuity of effective maps can only be proved indirectly. But assuming that a map is not continuous means assuming that some basic open set in the domain is not mapped into any basic open set in the codomain. And since in most cases only the elements of a basic open set, but not of its complement are effectively listable, a witness for this not being mapped into cannot be found effectively, in general, if the map is only effective.

The next result is proved in [22, 24]. Observe that the requirements on the numberings of the spaces $\mathcal{T}_c$ and $\mathcal{T}'_c$ used in [22, 24] are satisfied in case that the numberings are derived from admissible representations.

Theorem 18. *Let $\mathcal{T}_c$ be recursively separable. Then every effective map $F: \mathcal{T}_c \to \mathcal{T}'_c$ that has a witness for noninclusion must be effectively continuous.*

As follows from the proof, the recursive function witnessing that F is effectively continuous is constructable from the partial recursive functions showing that F is effective and has a witness for noninclusion.

With Proposition 16 we now obtain the main result of this paper.

Theorem 19. *Let $\mathcal{T}$ be effectively separable. Moreover, let $F: \mathcal{T}_c \to \mathcal{T}'_c$ be effective and let it have a witness for noninclusion. Then a computable map $\tilde{F}: \mathcal{T} \rightharpoonup \mathcal{T}'$ can be constructed which extends F.*

Note that, on the contrary, by Proposition 15 the effective continuity of the map F is also a consequence of this theorem.

Remark. In the above theorems the map F is required to be total. If it is only partial, but the witness for noninclusion exists in $\text{dom}(F)$ and $\text{dom}(F)$ is effectively separable with respect to the induced topology τ^F, one can apply the results to $\mathcal{T}^F = (\text{dom}(F), \tau^F)$ instead of $\mathcal{T}$. Set $B^F_n = B_n \cap \text{dom}(F)$ and let $m \prec_{B^F} n$, if $m \prec_B n$. Then $\prec_{B^F}$ is an r.e. strong inclusion with respect to which the set of all B^F_n is a strong basis.

There are important special cases in which total effective maps *have* a witness for noninclusion. As follows from [22, 24], this holds if the domain of F is the subset of all computable elements of a pointed space or the codomain is the subset of all computable elements of an effectively given metric space. Note that in the second case at least the elements of a substantial part of the complement of a basic open set B_n, namely of its exterior, are effectively listable, since

$$x_a \in \text{ext}(B_{\langle i,n\rangle}) \Leftrightarrow (\exists\langle j,m\rangle) x_a \in B_{\langle j,m\rangle} \wedge d(\beta_i,\beta_j) > 2^{-n} + 2^{-m} .$$

As a special case we thus obtain the following result (cf. [26]).

Theorem 20. *Let $\mathcal{M}$ and $\mathcal{M}'$ be effectively given metric spaces. Then every effective map $F: M_c \to M'_c$ has a computable extension $\tilde{F}: M \rightharpoonup M'$.*

In [19, p. 370, Ex. 15-30(b)] an effective map $F: R^{(1)} \to \omega$ is given which is *not* the restriction of a *total* computable map between the effectively given metric spaces $\mathbb{B}$ and ω. This shows that the above theorem as well as Theorem 19 cannot be improved. Moreover, an example of Myhill [14, p. 293, Remark 2] shows that the above theorem is false if F is only a partial map (see also [17]).

The next result displays an important case in which the extension $\tilde{F}$ constructed in Proposition 16 is total. Moreover, it is uniquely determined in this case.

Definition 21. A T_0-space $\mathcal{T} = (T, \tau)$ with countable strong basis $\mathcal{B} = \{ B_a \mid a \in \omega \}$ is *strongly complete*, if each normed enumeration of nonempty basic open sets converges.

In the case of an effectively given metric space, it follows from the proof of Lemma 8 that the space is strongly complete if and only if every normed sequence of elements of the dense subset converges.

Proposition 22. *Let $\mathcal{T}$ be pointed and $\mathcal{T}'$ be strongly complete. Moreover, let $F: T \rightharpoonup T'$ be effectively continuous with $Pt \subseteq \text{dom}(F)$. Then the computable extension $\tilde{F}$ of F constructed in Proposition 16 is total and uniquely determined.*

Proof. Let $y \in T$ and $p \in \rho^{-1}(y)$. We first show that $A[p]$ is filtered with respect to $\prec_{\mathcal{B}'}$. (Note that a nonempty set is filtered with respect to some relation, if it is directed with respect to the converse relation.)

Let to this end $a, c \in A[p]$. Then there are $m \in W_{g(a)}$ and $n \in W_{g(c)}$ such that $y \in B_m \cap B_n$. Since $\mathcal{B}$ is a strong basis, there is some $i \in \omega$ with $y \in B_i$, $i \prec_{\mathcal{B}} m$ and $i \prec_{\mathcal{B}} n$. It follows that $\chi_i \leq_\tau y$ and $\chi_i \in B_m \cap B_n$. Hence $F(\chi_i) \in B'_a \cap B'_c$. As $\mathcal{B}'$ is also a strong basis, this implies that there exists an index e such that $F(\chi_i) \in B'_e$, $e \prec_{\mathcal{B}'} a$ and $e \prec_{\mathcal{B}'} c$. Thus $\chi \in B_b$, for some $b \in W_{g(e)}$. Since open sets are upwards closed under the specialization order, we obtain that $y \in B_b$ too, i.e., $e \in A[p]$.

It is now easy to construct a sequence $(a_i)_{i\in\omega}$ of elements of $A[p]$ which is decreasing with respect to $\prec_{\mathcal{B}}$ such that for every $c \in A[p]$ there is some index i with $a_i \prec_{\mathcal{B}} c$. As we have just seen, all basic open sets B_c with $c \in A[p]$ are nonempty. Because $\mathcal{T}'$ is strongly complete, it follows that there is some point $z \in T'$ such that $\{ B'_a \mid a \in A[p] \}$ is a strong base of $\mathcal{N}(z)$. This implies that $a \in A[p]$ if and only if $z \in B'_a$. Thus $\Gamma(p) \in \text{dom}(\rho')$, which shows that $\tilde{F}(y)$ is defined, for every $y \in T$.

It remains to show that $\tilde{F}$ is uniquely determined. As has already been mentioned, it is shown in [23] that for every $y \in T$, $y = \sup\{\chi_n \mid y \in B_n\}$. In a similar way it can be proved for continuous maps $G: T \to T'$ that $G(y) = \sup\{G(\chi_n) \mid y \in B_n\}$. With this it follows that F has at most one computable extension. □

As we have previously seen, pointed spaces are effectively separable with dense base Pt. Moreover $Pt \subseteq T_c$. With Theorem 19 we therefore have the following result.

Theorem 23. *Let T be pointed and T' be strongly complete. Moreover, let $F: T_c \to T'_c$ be effective. Then F has a uniquely determined total computable extension $\tilde{F}: T \to T'$.*

Obviously, every effectively given continuous dcpo is strongly complete. As a special case we thus obtain the well known Myhill/Shepherdson Theorem (cf. [26, 27]). But since the domain of the map is not required to be strongly complete, we obtain that already every total effective map from the computable elements of an effectively given A-space into the computable elements of an effectively given dcpo or complete metric space has a uniquely determined total computable extension.

Acknowledgement

The author is grateful to Prof. Klaus Weihrauch for fruitful discussions. Thanks for useful comments are also due to the referees.

References

1. Beeson, M.: The nonderivability in intuitionistic formal systems of theorems on the continuity of effective operations. *J. Symbolic Logic* **40** (1975) 321–346
2. Beeson, M.: The unprovability in intuitionistic formal systems of the continuity of effective operations on the reals. *J. Symbolic Logic* **41** (1976) 18–24
3. Beeson, M.: Continuity and comprehension in intuitionistic formal systems. *Pacific J. Math.* **68** (1977) 29–40
4. Beeson, M., Ščedrov, A.: Church's thesis, continuity, and set theory. *J. Symbolic Logic* **49** (1984) 630–643
5. Eršov, Ju. L.: Computable functionals of finite type. *Algebra i Logika* **11** (1972) 367–437 [In Russian]; English transl., *Algebra and Logic* **11** (1972) 203–242
6. Eršov, Ju. L.: Continuous lattices and A-spaces. *Dokl. Akad. Nauk. SSSR* **207** (1972) 523–526 [In Russian]; English transl. *Soviet Math. Doklady* **13** (1972) 1551–1555
7. Eršov, Ju. L.: The theory of A-spaces. *Algebra i Logika* **12** (1973) 369–416 [In Russian]; English transl., *Algebra and Logic* **12** (1973) 209–232
8. Eršov, Ju. L.: Theorie der Numerierungen I. *Zeitschr. f. math. Logik Grundl. d. Math.* **19** (1973) 289–388

9. Eršov, Ju. L. Theorie der Numerierungen II. *Zeitschr. f. math. Logik Grundl. d. Math.* **21** (1975) 473–584
10. Eršov, Ju. L.: Theorie der Numerierungen III. *Zeitschr. f. math. Logik Grundl. d. Math.* **23** (1977) 289–371
11. Friedberg, R.: Un contre-exemple relatif aux fonctionelles récursives. *Compt. rend. Acad. Sci. (Paris)* **62** (1958) 852–854
12. Giannini, P., Longo, G.: Effectively given domains and lambda-calculus models. *Inform. and Control* **62** (1984) 36–63
13. Helm, J.: On effectively computable operators. *Zeitschr. f. math. Logik Grundl. d. Math.* **17** (1971) 231–244
14. Kreisel, G., Lacombe, D., Shoenfield, J.: Partial recursive functionals and effective operations. *Constructivity in Mathematics* (Heyting, A., ed.), 290–297. Amsterdam: North-Holland (1959)
15. Mal'cev, A. I.: *The Metamathematics of Algebraic Systems.* Collected Papers: 1936–1967 (Wells III, B. F., ed.). Amsterdam: North-Holland (1971)
16. Moschovakis, Y. N.: Recursive analysis. Ph.D. Thesis. Univ. of Wisconsin, Madison, Wis. (1963)
17. Odifreddi, P.: *Classical Recursion Theory.* Amsterdam: North-Holland (1989)
18. Pour-El, M. B.: A comparison of five 'computable' operators. *Zeitschr. f. math. Logik Grundl. d. Math.* **6** (1960) 325–340
19. Rogers, H., Jr.: *Theory of Recursive Functions and Effective Computability.* New York: McGraw-Hill (1967)
20. Scott, D.: Lectures on a mathematical theory of computation. Techn. Monograph PRG-19. Oxford Univ. Comp. Lab. (1981)
21. Smyth, M. B.: Power domains and predicate transformers. *Automata, Languages and Programming* (Diaz, J., ed.), 662–675. Lecture Notes in Computer Science **154**. Berlin: Springer (1983)
22. Spreen, D.: On effective topological spaces. To appear in the *J. of Symbolic Logic*
23. Spreen, D.: On some decision problems in programming. *Inform. and Computation* **122** (1995) 120–139
24. Spreen, D., Young, P.: Effective operators in a topological setting. *Computation and Proof Theory, Proc. Logic Colloquium Aachen 1983, Part II* (Richter, M. M. et al., eds.), 437–451. Lecture Notes in Mathematics **1104**. Berlin: Springer (1984)
25. Stoltenberg-Hansen, V., Tucker, J. V.: Algebraic and fixed point equations over inverse limits of algebras. *Theoret. Comp. Sci.* **87** (1991) 1–24
26. Weihrauch, K.: *Computability.* Berlin: Springer (1987)
27. Weihrauch, K., Deil, T.: Berechenbarkeit auf cpo's. Schriften zur Angew. Math. u. Informatik Nr. 63. Rheinisch–Westfälische Technische Hochschule Aachen (1980)
28. Young, P.: An effective operator, continuous but not partial recursive. *Proc. Amer. Math. Soc.* **19** (1968) 103–108
29. Young, P., Collins, W.: Discontinuities of provably correct operators on the provably recursive real numbers. *J. Symbolic Logic* **48** (1983) 913–920

The Role of Kripke Semantics in Reasoning of Knowledge and Time

Mouw-ching Tjiok

Galgenbergstr. 61
71034 Böblingen, Germany

my parents, in memoriam

Abstract. Much of the current research on distributed processing adopts the *situated-automata* approach based on ascribing knowledge on the basis of the information carried by the state of the machine. The standard approach to define knowledge is Hintikka's possible worlds model, and applying Kripke semantics to possible worlds model. This allows one to start with the local state and instant, thus the application of modal and temporal operators, both operates with indistinguishability structure and reachability relation. Using the local state(s), we then describe the applications or process as global state. However, the description of this approach is not simple, as it requires various tuples denoting the local and global states, and besides, maintaining their interrelationships. In this paper, we apply the algebraic semantics approach—using partially ordered set and Heyting algebra—to describe Kripke semantics. In mathematics, there is a formalism for handling local and global properties and its transition from global-to-local and vice versa within a well-defined framework of topological and analytical problems, which is generally known as sheaf theory. In this paper, we introduce sheaf theory to maintain the transition of local information to global information.

1 Introduction

Current research in distributed systems and security logic has recognized that, reasoning about knowledge is a useful and powerful approach in designing, analyzing, and understanding complex systems composed of interacting parts ([17], [8], [9], [13]). In the area of artificial intelligence (AI), two main approaches to ascribing knowledge to machines or components of systems have been established ([28]). The classical AI approach—the *interpreted-symbolic-structures* approach—which ascribes knowledge on the basis of the information stored in certain data structures (such as semantic nets, frames, or data structures to encode formulas of predicate logic). The second—the *situated-automata* approach—can be viewed as ascribing knowledge on the basis of the information carried by the state of the machine. Much of the research on distributed processing adopts the situated-automata approach ([2], [9]), as this is more appropriate to analyze the logical structure as well as dynamical behavior of the system, in which we are interested. In this paper we only consider the *external* knowledge, which does not involve any cognitive activity on the part of the process or component. We use Hintikka's approach to define knowledge using *possible worlds* model, which essentially says, a subject is said to *know* a formula if the formula is true in all worlds considered possible by the subject and formulate the possible worlds semantics using Kripke structures according to Sato ([21], [22], [29]). Semantical analysis becomes crucial to understand the systems and its properties in order to use this notion of knowledge to describe and analyze systems. In most cases, the knowledge are of local nature, and it satisfies all the axioms of the classical modal logic *S5*; and the abstract models used to capture this notion have been variants of the classical Kripke-style *possible worlds* model for *S5*. In this approach one defines a set of "worlds"—the so-called *possible worlds*—and a family of indistinguishability relations, one for each subject, that define which worlds a subject cannot tell apart. This is based on semantical analysis developed by Kripke, which comprises of two essential parts: the semantical analysis for modal logic and the semantical analysis for intuitionistic logic. A logical approach based on modal logic formalism is defined in [13] for handling security problems. The advantage is that there already exists a well-understood theory of how modalities interact with ordinary logical connectives ([4]). Essentially, a modal logical formalism that defines knowledge in terms of indistinguishability of states is close to many realistic scenarios. Thc basic idea is that a subject has a partial view of a global state. A subject "knows" something in a particular global state if that thing is true in all the states that are indistinguishable (by the subject) from the global state. In [17], Halpern and Moses define the *classical* model for knowledge and belief, the so-called *possible-worlds* model for the analysis of distributed protocol. This is done in terms of the states of knowledge of the processors involved. In describing security protocol, transaction processing in a distributed database and in temporal database systems, we encounter the intrinsic relationship between knowledge and time, more precisely, between knowledge and the execution of a system. This can be handled by temporal logic. Temporal logic is a well-understood formalism for reasoning about systems. In defining a temporal theory, there are two contrasting interpretations regarding the nature of time: *linear* and *branching*. In the linear interpretation, one talks about truth of formulas with respect to a *given* computation sequence. The branching interpretations allows one to talk about a world and *all* of its future worlds [6, 7]. In both interpretations, the following two properties are crucial: the *persistence* property—the persistence in time of truth (see **6** and **7**)—and the *liveness* property—the reachability relation (see **2.2** and **9**).

The purpose of this paper is to provide a mathematical framework to describe the Kripke structure and modal logic; based on the idea of Kripke, we then investigate the semantics, or more precisely, the locally truth, and its algebraic semantics using Heyting algebra. At the same time, using reachability relation we investigate the semantics along a certain time-path, this allows us to introduce the temporal notion to the system. The main ingredient of the mathematical tools is the sheaf theory. We first formulate the algebraic semantics of Kripke structure (or sometimes, also Kripke frame) using Heyting algebra, using *monad* we describe the semantical transition from non-modal sentences to modal sentences. The semantics arises in Kripke structure is mainly taken through the valuation function (with respect to a certain hereditary set P). The model is then developed around this specific partially-ordered set $(P, \subseteq)$, we denote this by $\mathbf{M} = (\mathbf{P}, \mathbf{V})$ (in **6**.) $\mathbf{M}$ is defined as a certain sentence φ is true in $\mathbf{M}$ at a particular stage p (which is verified by the valuation function.) This *stage* (or, also *state*) has a very local nature; we then define a set around the *stage p* at which φ is true (in the local model $\mathbf{M}$ (or, more precisely, $\mathbf{M}_p$).) We thus define an equivalence class, which will be used in Kripke structure, on which the modal operator K_i operates. This approach is intrinsically related to the concepts of "locally true" and specific equivalence class (which we call "germ"). Together with the definition of *stalk* we introduce the notion of sheaf theory, which gives us geometric insight of the "locally true", and the globalization of local context, and vice versa. Sheaf can be considered as an *abstraction* of the real-world space (which we called the "base space" in topology), in this abstraction we can extract the data or information from the base space, by restricting the data/information to a certain specific "view" (as we shall show in the example "New-Yorker" in **4**); conversely, we can also construct a global view, which consists of the selected local views using a certain "patching" procedure with respect to a certain compatibility condition. This geometric construct provides a convenient global view to consider the various local view or processes, which are indistinguishable to the global view. The transition from one global state to another one can then be considered as sheaf morphism. Sheaf arises in this way is called the sheaf of germs, which has among others, one strong feature, being the *subobject classifier* of some specific topos. Subobject classifier is a powerful formalism to construct and verify a specific equivalence class according to some given predicate (see Section **4**). This construct and the additional naming convention gives us a semantically rich variety for defining "semantic objects". Besides, the sheaf of germs carries a rich algebraic structure, for example, as abelian group, or as a ring, the algebraic structure can be inherited from the global view (or, the sheaf itself, in this case). We note that, using sheaf-theoretic approach, we can also link the Fourman-Scott ([11]) approach, and put the entire discussion of Kripke semantics back on the logical footing (see also [32]). As applications in database systems, we show the following: (i) database applications involving temporal logic, (ii) temporal completeness information commitment ([33]), (iii) transaction processing in a distributed environment, using semantic knowledge ([12], [17], [9], also [3])—due to space constraint, we will treat this in the future paper.

We first present the logic of knowledge following the line of [13] and [8]. Subsequently, we consider the process as time-function, run as global state, and the state of a dynamical system is related to this context. We then carry the idea of equivalence class and subset over to define *subobject*, this is used in the specification of describing "locally true" around some local predicate, say p. This gives rise to the concept of *germ*. We briefly introduce the semantics in relation to the "naming" of some equivalence class. These will be given in more detail in section **8**, after we formally described Kripke semantics in sections **5**, **6**, and **7**. In **9** we treat database applications involving temporal logic, which we relate our result in [33] on temporal completeness information commitment for providing information completeness within a certain time-interval or time-slice. In **11**, we apply the machinery developed and semantic knowledge to describe transaction processing in a distributed environment and contribute to optimization of the query processing. The detail will be presented in the future. While the approach of the theory of knowledge using *possible-worlds model* following Hintikka and Kripke is well accepted in general, however, the application of sheaf theory to treat Kripke semantics is barely known outside the mathematicians' circle ([15], [16]). In this paper, I insist the introduction of sheaf theory for the following reasons:

- a construct which provides an easy transition between local and global informations, and vice versa;
- a geometric construct which enables sufficient abstraction level to consider the specific object under focused, with relative simple notations;
- a formalism which enables the naming systems to the objects, hence more appropriate to Kripke's approach [23];
- last not least, the affinity of sheaf theory to intuitionistic logic as demonstrated by Fourman-Scott [11].

I hope this will contribute to the simplification both in conceptual representation and notations of the Kripke-Sato approach as sheaf-theoretic approach contributes to pure mathematics.

2 A logic of knowledge

We apply the standard theory for reasoning about knowledge using a *possible-worlds model*, which asserts a subject is said to *know* a formula if the formula is true in all worlds considered possible by the subject; and the subsequent formulation of the possible worlds semantics using Kripke structures. This theory is an *S5* modal logic. The language for reasoning about knowledge consists of

- a set of subjects U labelled $1, \ldots, n$,
- a set of primitive propositions Σ; and
- for all subjects i, a modal operator K_i.

A proposition of the form $K_i\varphi$ states that subject i knows that proposition φ is true. The notion of possible worlds can be formalized using Kripke structures. In a system with n subjects, the Kripke structure M is a tuple $M = (S, \pi, \kappa_1, \ldots, \kappa_n)$ where

1. S is the set of possible worlds.
2. The relation κ_i for a subject i is an equivalence relation (reflexive, symmetric, and transitive) on S. We say that two worlds are *indistinguishable* to subject i if they belong to the same equivalence class in κ_i.

3. A formula is defined to be *true* or *false* in a possible world. We write $\models_s \varphi$ to denote that formula φ is true in world s. For each world $s \in S$ and each primitive proposition $\varphi \in \Sigma, \pi$ assigns a truth value to φ in s (that is, $\pi(s, \varphi) \in \{true, false\}$).

We define the notion of a formula being true in a world via the $\models$ (reads "models") relation as follows: For all $s \in S$, and $\varphi, \psi \in \Sigma$:

1. $M \models_s \varphi \iff \pi(s, \varphi) = true$;
2. $M \models_s \varphi \wedge \psi \iff M \models_s \varphi$ and $M \models_s \psi$;
3. $M \models_s \neg\varphi \iff \neg M \models_s \varphi$; and
4. $M \models_s K_i\varphi \iff$ for all s' such that $(s, s') \in \kappa_i, M \models_{s'} \varphi$.

A formula φ is said to be *valid* if it is true in all worlds $s \in S$. We denote this as $\models \varphi$. It is well known that the following set of axioms and inference rules, which goes back to Hintikka, provides a sound and complete axiomatization for the logic of knowledge just defined with respect to the class of Kripke structures for knowledge.

- (K1) $K_i\varphi \rightarrow \varphi$.
- (K2) $K_i(\varphi \rightarrow \psi) \rightarrow (K_i\varphi \rightarrow K_i\psi)$.
- (K3) $K_i\varphi \rightarrow K_iK_i\varphi$.
- (K4) $\neg K_i\varphi \rightarrow K_i\neg K_i\varphi$.

Axiom (K1) (the knowledge axiom) states that a subject cannot know anything that is false. This distinguishes knowledge from belief. Axiom (K2) (the consequence closure axiom) states that a subject knows all things that can be deduced from its knowledge. The final two axioms (the positive and negative introspection axioms) state that a subject has introspective ability. The theory includes the proof rules of propositional logic as well as a rule that states that if a formula is valid (true in all possible worlds), then a subject knows that formula:

Rule 1 If φ is valid then so is $K_i\varphi$.

Rule 2 From φ and $\varphi \Rightarrow \psi$ infer ψ (modus ponens).

2.1 Knowledge and time

The temporal logic for knowledge presented in this paper is based on a branching time structure. In particular, we use the logic *UB*: the unified system of branching time to develop our theory. In the *UB* system, the underlying model consists of a branching tree of all possible paths of reachable possible worlds. Temporal operators are defined that correspond to quantification over these paths. As well as the modal operators K_i for knowledge, we include three of the six temporal operators of *UB*. These are the operators: $\forall\Box$ always; $\forall\Diamond$ eventually, and $\exists\Diamond$ sometimes. The first symbol in these operators denotes the quantification over paths, while $\Box$ and $\Diamond$ denote temporal quantification over the selected path. Given a world s, a path defined as a sequence of worlds, and a formula φ, we interpret the temporal operators applied to φ, informally, as follows:

1. $M \models_s \forall\Box\varphi \iff \varphi$ is true in all worlds along all paths initiated at s;
2. $M \models_s \forall\Diamond\varphi \iff \varphi$ is true for some world along all paths initiated at s;
3. $M \models_s \exists\Diamond\varphi \iff \varphi$ is true for some world along some path initiated at s.

Given a set S of possible "worlds"; henceforth we call them states. These are to be thought of as possible system states. We have a set $\mathcal{R}$ of finite or infinite sequences of members of S called *runs*; these are the dynamical aspect described in discrete sequence. If r is a typical member of $\mathcal{R}$, we write $r[i]$ for the i-th member of the sequence r. There is a binary relation R on the possible states S derived from the set of runs $\mathcal{R}$. We say that for any two states s and s', $(s, s') \in R$ if and only if there is a run r such that for some positive integers i and j, $r[i] = s$ and $r[i+j] = s'$, run r is parametrized by time variables, if we consider the positive integers i and j as t_1 and t_2 correspondingly, with $t_1 \leq t_2$ (see the definition of *run* as a function in **3**). The states in S are equipped with the indistinguishability relations. Thus one has the *indistinguishability structure* that allows one to model knowledge, and the *reachability relation* needed to model temporal notions. This class of models has been used to model distributed systems; in this system, one has a set of autonomous computing agents that communicate in some way, the system evolves in time and the processes exchange messages. The global state of the system encodes the information of the local states of all the processes. Each process is only aware of its local state, and there is a natural notion of indistinguishability for each process—namely a process i cannot distinguish two global states that have the same local state for i. The runs, of course, correspond to the executions of the system.

2.2 The combined logic for time and knowledge

The alphabet of the language consists of

1. a denumerable set Σ of primitive proposition letters $\varphi, \psi, \ldots,$
2. the logical symbols $\top$ (truth) and $\bot$ (falsehood), and the logical connectives $\neg$ (negation), $\wedge$ (conjunction), $\vee$ (disjunction) and $\rightarrow$ (implication);
3. the modal operators K_i (knowledge) for all subjects i; and
4. the temporal operators $\forall\Box$ (always), $\forall\Diamond$ (eventually) and $\exists\Diamond$ (sometimes).

The set of well-formed formulas for the language is the smallest set W such that

1. every proposition letter in Σ, $\top$, and $\bot$ are in W;
2. if φ and ψ are in W, then so are $\neg\varphi, \varphi \wedge \psi, \varphi \vee \psi, \varphi \rightarrow \psi$ amd (φ);
3. if φ is in W, then so are $K_i\varphi$ (for all subjects i), $\forall\Box\varphi, \forall\Diamond\varphi$, and $\exists\Diamond\varphi$.

We now describe how the modal operators are modeled. In other words, we give the semantics for the logic. In the following we assume that the quantification of r is over runs in the model and the quantification of s, s' and s'' is over possible states.

1. $M \models_s \forall\Box\varphi \iff$ for all $(s, s') \in R, M \models_{s'} \varphi$;

2. $M \models_s \forall\Diamond\varphi \iff$ for all r, if $r[i] = s$, then there exists s' and nonnegative integer j such that $s' = r[i+j]$ and $M \models_{s'} \varphi$;
3. $M \models_s \exists\Diamond\varphi \iff$ for some $(s, s') \in R$, $M \models_{s'} \varphi$;

The knowledge operators are modeled exactly as described before. In most cases, time enters the context *implicitly*, that is, not as a clock function, but more or less in an ordered time-sequence of possible event occurrences, or also called time-path (see also [3]).

3 Process as time-function

We consider the system to be evolving over time—as a *run*—for each system, we assume that there is a finite set of processes, say $\{1, \ldots, n\}$ (if n is the total number of processes). We assume, a set L of local states that the processes can take on, a set E of environment states. The set G of global states consists of tuples $(e, l_1, \ldots, l_n)$, with $e \in E$ and $l_i \in L$ for $i = 1, \ldots, n$. We then have $G = \{(e, l_1, \ldots, l_n) \mid e \in E, l_i \in L \text{ for } i = 1, \ldots, n\}$. Formally, a *run* is a function $r : \mathbf{N} \to G$ that associates with each time m a global state $r(m)$, where $\mathbf{N}$ denotes the set of natural numbers. We may refer to $r(0)$ as the initial global state of the run r, or simply the initial state. Following [17], we may refer to a pair (r, m) consisting of a run r and a time m as a point. Points, rather than global states, are our "possible worlds." We shall then define what it means for a formula to be true at a point, rather than what it means for a formula to be true at a global state. Finally, a system is a set $\mathcal{R}$ of runs. We consider a distributed system consists of a collection of processes, say $1, \ldots, n$, connected by a communication network. We think of these processes as running some protocol. At any time in the execution of such protocol, the system is in some global state, which is a tuple of the form $\langle s_e, s_1, \ldots, s_n \rangle$, where s_i is the local state of process i, and s_e is the state of environment. We identify a system with a set of runs (these can be thought of as the possible runs of the system when running a particular protocol). If $\mathcal{R}$ is a system (a set of runs), then we say that $s = (e, l_1, \ldots, l_n)$ is a global state of $\mathcal{R}$ if $s = r(m)$ for some run $r \in \mathcal{R}$ and some time m; we call e the environment component of s and $(l_1, \ldots, l_n)$ the process component. Let $s = (e, l_1, \ldots, l_n)$ and $s' = (e', l'_1, \ldots, l'_n)$ be two global states in $\mathcal{R}$. We say s and s' are indistinguishable to process i, written $s \sim_i s'$, if $l_i = l'_i$. If (r, m) and (r', m') are points, then we say that (r, m) and (r', m') are are indistinguishable to process i, written $(r, m) \sim_i (r', m')$ if $r(m) \sim_i r'(m')$. If s and s' are two global states (or two points), and if $s \sim_i s'$ for every process i, then we say s and s' are process equivalent, written $s \sim s'$. Thus, two global states s and s' are process equivalent precisely if they have the same process component. Process i's history in run r up to time m is the sequence $\langle s_1, \ldots, s_k \rangle$ of states that process i takes on in run r up to time m, with consecutive repetitions omitted. Suppose we fix a set Φ of primitive propositions. We define an interpreted system $\mathbf{I}$ to be a pair $(\mathcal{R}, \pi)$, where $\mathcal{R}$ is a system (set of runs), and π is a truth assignment to the primitive propositions of Φ at every point in $\mathcal{R}$. With this definition, it is easy to view an interpreted system as a Kripke structure, where the points play the role of states and the K_i relation is given by $\sim_i$. Truth is now defined with respect to a point (r, m) in an interpreted system **I**. In particular, we have $\mathbf{I} \models_{(r,m)} K_i\varphi \iff \mathbf{I} \models_{(r',m')} \varphi$ for all (r', m') such that $(r', m') \sim_i (r, m)$. Since $\sim_i$ is an equivalence relation, it is to check that all the axioms of *S5* hold for this interpretation of knowledge. We note that this representation allows us to distinguish semantically equivalent processes and those with identical and/or different environment components; this additional information is crucial if the sharing of resources has been taken into consideration; likewise for the semantically equivalent processes having independent environment components, that is, resources, which enables performance improvement and parallel processing ([12]).

3.1 Dynamical system

A *dynamical system* is a way of describing the passage in time of all points of a given space $\mathcal{S}$. The space $\mathcal{S}$ could be thought of, for example, as the space of states of some process [31]. A dynamical system on $\mathcal{S}$ tells us for x in $\mathcal{S}$, where x is 1 unit of time later, 2 units of time later, and so on. We denote these new positions of x by x_1, x_2, respectively. At time zero, x is at x or x_0 (the initial point). If one extrapolates to fill up the real numbers (or, in our previous case, with natural numbers—for the sake of discrete events), one obtains the trajectory x_t for all time t. The map $\mathbf{R} \rightarrow \mathcal{S}$, which sends t into x_t, is a curve in $\mathcal{S}$ that represents the life history of x as time runs through a certain time-interval $[a, b]$. The image of x_t $x_t(\mathbf{R})$ is called a *path* or *orbit* of x; this is what we called run in the previous section, if we consider the image as $x_t(\mathbf{N})$, which also causes our system to be discrete. It is assumed that the map from $\mathbf{R} \times \mathcal{S} \rightarrow \mathcal{S}$ defined by $(t, x) \rightarrow x_t$ is well defined. The map $\phi_t : \mathcal{S} \rightarrow \mathcal{S}$ that takes x into x_t is defined for each t and from our interpretation as states moving in time, it is reasonable to expect ϕ_t to have as an inverse ϕ_{-1}. Also, ϕ_0 should be the *identity* and $\phi_t(\phi_u(x)) = \phi_{t+u}(x)$ is a natural condition (we recall $\phi_t(x) = x_t$). In our consideration here, we are merely interested at the transition from a state (which is described by the position coordinate and time parameter t), and the snap-shot of the run through a certain time-interval, that forms an ordered-sequence. This allows us to put the reachability relation (along the path), and apply the temporal operators in **9**. Thus, the map $\phi_t : \mathcal{S} \rightarrow \mathcal{S}$ is considered in our specific case as an accessibility function that for each state gives a unique next state; and the i-th state in the sequence (or path) $s, \phi_{t1}(s), \phi_{t1}(\phi_{t2}(s)), \phi_{t1}(\phi_{t2}(\phi_{t3}(s))), \ldots$ of successors of a state s is written $\phi_{t1+t2+t3+\ldots+ti}(s)$ or, simply $\phi_t^i(s)$. Using the interpreted system **I** defined in the previous section, we can capture and maintain the semantic consistency along the path (see also **9**.) In order to do this, it is convenient to consider the space $\mathcal{S}$ as a lattice; where $\mathcal{S}$ itself is the underlying partially ordered set is $(\mathcal{S}, \subseteq)$ with set-theoretic operations $\cap, \cup$, and complement $(')$; we call this lattice the *state-space lattice (SSL)*: $\langle |\mathcal{S}|, \subseteq, \cap, \cup, ' \rangle$. In our approach, $\mathcal{S}$ can then be defined as our possible worlds, applied with the related Kripke structure (see **2**).

4 Equivalence class, germ and "locally true"

In his remarkable book, "*Die Idee der Riemannschen Fläche*" [35], Hermann Weyl describes a general principle and its psychological origin about the ability of human intelligence to comprehend abstraction. It is based on dealing with equivalence classes, and taking a *representative* of an equivalence class which essentially characterizes the object of this specific class. Given a name to the representative, then this name also characterizes the meaning of equivalence relation, for example, if we denote the citizen of United States of America living, working, and having residence in the metropolitan area of New York as *New-Yorker*, then this name also gives the same meaning to the individual belonging to the class. Equivalence class treated this way is called *germ* in mathematics, which goes back to Karl Weierstrass in his investigation of complex-analytic functions. Incorporated with the sheaf theory developed in the 1950's by Jean Leray, the approach becomes a powerful tool in dealing with complicated mathematical problems, such as, algebraic geometry, complex geometry, and algebraic topology; in particular, in solving problems with intrinsic properties between local and global aspects. We note also, the technical approach is similar to the "*small model property*" and the quotient construction as proposed by Fischer-Ladner [10, 7]. We present here briefly the formal aspect—which will be treated in more detailed in **7**—of *subobject* which is the categorical version of the set-theoretical subset and the simple construction of *subobject classifier* as sheaf of germs, that gives semantics through true valuation and the subset as equivalence class as well. Later, we shall incorporate the idea of "Naming" to the specific germ of possible world—that is, equivalence class—to introduce semantics, according to Kripke ([23]). The main idea of subobject is to regard a subset A of a given object B as a monomorphism $f : C \to B$ (intuitively, a monomorphism f such that $f(C) = A$.) Two sets S and T are equivalent at p if there is some neighborhood U of p with $S \cap U = T \cap U$. Again this means that S and T are locally equal at p, that is, that "$S = T$" is true when relativized to points close to p. And we have the equivalent statements:

- (a) α is locally true at $p \iff \alpha$ is true at all points close to p; and
- (b) α is locally true at $p \iff \alpha$ is true throughout some neighborhood of p.

By conditions (a) and (b), and the equivalence of both conditions, the *locally true* property stated can be represented by means of the equivalence class of some neighborhood i, this is the notion which we called *germ* previously. Using this notion we define the *stalk* as $F_x \equiv \lim_{\to x \in U} F(U) = \bigcup_{x \in U} F(U)/\sim_x$. Here, for $s \in F(U)$ and $t \in F(V)$, and $s \sim_x t$ means $s\mid_W = t\mid_W$ for some open $W \subseteq U \cap V$ such that $x \in W$. We define $|Y_F| \equiv \bigcup_{x \in X} F_x$ is the disjoint union of stalks F_x (over X). If $[s]_x$ denotes the equivalence class of s modulo $\sim_x$, we have $|Y_F| \equiv \{(x, [s]_x) \mid \exists_{U \in O(X)} (x \in U \land s \in F(U))\}$. The topology on Y_F is generated by basic open sets of the form $V(s, U) \equiv \{(x, [s]_x) \mid x \in U\}$, where $U \in O(X)$ and $s \in F(U)$. This is one way to define the sheaf Y_F over X. Recalling the global state over local processes in **3**, we consider the relation between the set of global states, G, and the set of local states that the processes can take on L, and the set of environment states, E, as sheaf (G, π, L) (or, $(G, \pi, L \times E)$ if we want to take E also into consideration), then we have the following If we consider the equivalence class as subset of a certain set—as *subobject*—we can define the equivalence class as follows. Given a subset $A \subseteq D$, we define the function $\chi_A : D \to \mathbf{2}$, called the *characteristic function* of A, by the rule $\chi_A(x) = 1$ if $x \in A$ and $\chi_A(x) = 0$ if $x \notin A$. The assignment of χ_A to A is injective from $\mathbf{P}(D)$—the powerset—to $\mathbf{2}^D$, that is, if $\chi_A = \chi_B$ then $A = B$. It is also surjective, for if $f \in \mathbf{2}^D$, then $f = \chi_{A_f}$, where $A_f = \{x : x \in D$ and $f(x) = 1\}$. This correspondence between subset and characteristic function can be "captured" by a pullback diagram. The set A_f just defined, is the inverse image under f of the subset $\{1\}$ of $\{0, 1\}$, that is, $A_f = f^{-1}(\{1\})$, and

$$\begin{array}{ccc} \mathbf{A_f} & \xrightarrow{\subseteq} & D \\ \downarrow{\scriptstyle !} & & \downarrow{\scriptstyle f} \\ 1 & \xrightarrow{true} & \mathbf{2} \end{array}$$

is a pullback square, that is, A_f arises by pulling back $\{1\} \xrightarrow{\subseteq} \mathbf{2}$ along f., The technique presented here states that, the sheaf of germs is the *subobject classifier* of the corresponding topos ([14]). If we consider the function f as database query-function, then $A_f = f^{-1}(\{1\})$ can be interpreted as the database access with specific WHERE clause Q_f = database-query $f^{-1}(\{1\})$, for example, in Customer-database, query Customer with LastName = "Mueller", in case the result is non-void, we have an equivalence class with LastName = "Mueller", this is the common property shared by different individuals. We consider the following example: the "Intercontinental Hotel Inc." database. We suppose the database is organized as distributed database and connected through network. The database consists of a set of local databases { NewYork, Frankfurt, London, Berlin, Paris, Tokyo, ... , Beijing }. These local databases contain information on room occupation, availability, room-rate, statistics, etc. This database forms a *base space*, with certain topological structure [30]. Whenever an application, say the reservation system, requires data from NewYork, then this will be retrieved from NewYork database. From the view of application, the database contains room information about the location "NewYork"; that is, all the rooms available are in NewYork. From the global view, we project into the base space, and denote the projection by π, using π we could "reach" every local views or objects in the base space. Now, when we are interested in data from the local view "NewYork", we merely need to consider the projection π to base space, B restricted to "NewYork", written π |'NewYork', in the same way, we could pull the data from "NewYork" up to the global view, using the inverse mapping $\pi^{-1}(B$ |'NewYork'), that is, the inverse mapping from the base space restricted to "NewYork". Roughly, the global view we put over the base space B with the projection π is a sheaf over B, and $\pi^{-1}(B$ |'NewYork'), is called the *stalk* of the sheaf over B. Note that, there is also a similar description to specify for unique query retrieval in database using unique key approach.[1] If we recall the definition of Kripke structure in **3**, the equivalence classes defined in κ_i, are nothing but the germs in this context. The global states which are indistinguishable to process i, defined in **3**, consists basically out of the stalk of i-neighborhood. We can consider the germs as the core semantics of this process,—considering as the steady-state process (without considering the action of time). Now using the *reachability relation* of **2.1**, and the *run* of **3**, we can treat the temporal approach—in which the time-dependent processes can be treated. In **6**, we use the basic fact, that *PL*-sentences are interpreted as subsets of a poset. On the poset we impose the hereditary condition, and perform the *valuation* on the hereditary set. The model derived inherits a natural ordering of the poset.

5 The essentials of Kripke semantics

During 1960's, Saul Kripke develops a semantical analysis for modal logic, and subsequently, a formal semantics for intuitionistic logic in which *PL*-sentences are interpreted as subsets of a poset [21], [22]. This was further refined later in the works of E.J. Lemmon and R. Goldblatt [24], [15], [16]. Briefly, modal logic is concerned with the concept of necessity, and on the propositional level uses the language *PL* enriched by a *connective* whose interpretation is "*it is necessarily the case that*". The appropriate algebraic *models* here are *BA*'s (Boolean Algebra) with an additional operation for this new connective. One attractive feature of this theory is that its structures, apart from being generally more tractable than the algebraic ones, have an informal interpretation that accords well with the intuitionistic account of the nature of validity. In the latter, truth is temporally conditioned. A sentence is not true or false per se, as in classical logic, but is only so at certain times, namely, those times at which it has been constructively determined. Each moment of time is associated with a particular *stage*, or *state* of knowledge. This comprises all the facts that have been constructively established at that time. Sentences then true are so in view of the existing state of knowledge. We thus speak of sentences as being "*at a certain stage*" or "*true at a certain state of knowledge*". The collection of all states of knowledge is ordered by its temporal properties. We speak of one state as coming after, or being later than, another state in time. A sentence true at a certain stage will be held to be true at all later (future) stages. This embodies the ideas that constructive knowledge, once established, exists forever more. Having proven α, we cannot later show α to be false (see [18])[2]. Now the temporal ordering of states is a partial ordering, not necessarily linear. The states we consider do not always follow one another in a linear sequence because they are *possible* states of

[1] This sounds trivial for single location database, however, in distributed database, either we apply the retrieve from the specific located database—as we know the where the information is available; or we call an "agent"—who knows where to get the appropriate information from the specific database, and in this case, we must give him/her a "name" or "token" to match his/her "knows" [1].

[2] This is also called *stable predicates* in distributed systems ([3]).

knowledge, not just those that do actually occur. The collection of possible states of knowledge is a poset under the ordering of time. A sentence corresponds to a particular subset of this poset, consisting of the states at which the sentence is true. In view of the persistence of truth in time, this set has a special property: given a particular state in the set, all states in the future of that state belong to the set as well. We refer to [14], [11] for the interrelationship between *elementary topoi* and higher-order intuitionistic logic; and the subsequent results of Scott, which describes a system of higher-order logic which may be interpreted in any topos, and Fourman-Scott ([11]) describe the models of this logic given by sheaves over a complete Heyting algebra (**cHa**). These sheaf models subsume the more familiar Beth, Kripke and topological interpretations of intuitionistic logic, which correspond to interpretations in sheaves over the appropriate **cHa** of *truth values*. In [32], [33] the sheaf-theoretic approach has been used to construct the *local property* of database and introduce Kripke semantics using topological approach, in which, local information can be passed to global information and vice versa, from global to local using the *restriction* operation. The algebraic semantics is captured through Heyting algebra ([14], [16]), moreover, the temporal ordering of states as a partial ordering and the persistence of truth in time, imposed by the "hereditary" condition, add the temporal aspect to our approach. Together with the *locally true* property and the monadic operator, we then show the Kripke structure obtained in this way, can be used in conjunction with agents of [8], [17] to solve our problems in dealing with global query process and temporal database issues.

6 Kripke semantics—a formal approach

First we give the formal approach of Kripke, following [15], [16], and deal with an object language in which sentences are constructed from a denumerable set Ψ_0 of sentence letters, together with the constants $\perp$ (False), by means of the connectives $\wedge$ (and), $\vee$ (or), $\rightarrow$ (implies) and ∇ ("*it is locally the case that*"). Negation, the biconditional, and the constant $\top$ are defined by $\sim \varphi = \varphi \rightarrow \perp$, $\varphi \equiv \psi = (\varphi \rightarrow \psi) \wedge (\psi \rightarrow \varphi)$, $\top = \perp \rightarrow \perp$. The set of all sentences will be denoted by Ψ. Φ denotes the subset of all non-modal sentences, that is, those with no occurrence of ∇. The Kripke semantics for Φ-sentences employs as its basic structure a partially-ordered set (poset) $\mathbf{P} = (P, \subseteq)$—also called a *frame* in this context. A set $A \subseteq P$ is *hereditary* in $\mathbf{P}$ if it is closed "upwards" under $\subseteq$, that is, if we have that whenever $p \in A$ and $p \subseteq q$, then $q \in A$. The collection of hereditary subsets of $\mathbf{P}$ will be denoted $\mathbf{P}^+$. A $\mathbf{P}$-*valuation* is a function $\mathbf{V} : \Psi_0 \rightarrow \mathbf{P}^+$ assigning to each π_i an hereditary subset $\mathbf{V}(\pi_i) \subseteq P$. A *model based on* $\mathbf{P}$ is a pair $\mathbf{M} = (\mathbf{P}, \mathbf{V})$, where P is a collection of stages of knowledge temporally ordered by $\subseteq$. $\mathbf{V}(\pi_i)$ is the set of stages at which π_i is true. The requirement that $\mathbf{V}(\pi_i)$ be hereditary formalizes the "*persistence in time of truth*". We now extend the notion of truth at a particular stage to all sentences. The expression "$\mathbf{M} \models_p \varphi$" is to be read "$\varphi$ *is true in* $\mathbf{M}$ *at* p," and is defined inductively as follows:

- (1) $\mathbf{M} \models_p \pi_i$ iff $p \in \mathbf{V}(\pi_i)$,
- (2) $\mathbf{M} \models_p \varphi \wedge \psi$ iff $\mathbf{M} \models_p \varphi$ and $\mathbf{M} \models_p \psi$,
- (3) $\mathbf{M} \models_p \varphi \vee \psi$ iff $\mathbf{M} \models_p \varphi$ or $\mathbf{M} \models_p \psi$,
- (4) $\mathbf{M} \models_p \sim \varphi$ iff for all $p \subseteq q$, not $\mathbf{M} \models_p \varphi$,
- (5) $\mathbf{M} \models_p \varphi \supset \psi$ iff for all $p \subseteq q$, if $\mathbf{M} \models_q \varphi$ only if $\mathbf{M} \models_q \psi$.

Thus at stage p, $\sim \varphi$ is true if φ is never established at any later stage, and $\varphi \supset \psi$ is true if ψ holds at all later stages that φ is true at. φ is *true* (holds) in the model $\mathbf{M}$, denoted $\mathbf{M} \models \varphi$, if $\mathbf{M} \models_p \varphi$ for every $p \in P$. φ is *valid* on the frame $\mathbf{P}$, $\mathbf{P} \models \varphi$, if φ is true in every model $\mathbf{M} = (\mathbf{P},\mathbf{V})$ (based on $\mathbf{P}$). If we denote by $\mathbf{M}(\varphi)$ the set of points at which φ is true in $\mathbf{M}$, that is, $\mathbf{M}(\varphi) = \{p : \mathbf{M} \models_p \varphi\}$, then the semantic clauses (1), (2) and (3) can be expressed as become

- (1′) $\mathbf{M}(\pi_i) = \mathbf{V}(\pi_i)$,
- (2′) $\mathbf{M}(\varphi \wedge \psi) = \mathbf{M}(\varphi) \cap \mathbf{M}(\psi)$,
- (3′) $\mathbf{M}(\varphi \vee \psi) = \mathbf{M}(\varphi) \cup \mathbf{M}(\psi)$.

To re-express (4) and (5) we define, for hereditary S, T, $\neg S = \{p :$ for all q such that $p \subseteq q, q \notin S\}$, $S \Rightarrow T = \{p :$ for all q with $p \subseteq q$, if $q \in S$ then $q \in T\}$. We then have

- (4′) $\mathbf{M}(\sim \varphi) = \neg \mathbf{M}(\varphi)$,
- (5′) $\mathbf{M}(\varphi \supset \psi) = \mathbf{M}(\varphi) \Rightarrow \mathbf{M}(\psi)$.

The intersection and union of two hereditary sets are both hereditary, so the poset $\mathbf{P}^+ = (\mathbf{P}^+, \subseteq)$ of hereditary sets under the inclusion ordering is a (bounded distributive) lattice with meets and joins given by $\cap$ and $\cup$. $\mathbf{P}^+$ is indeed a Heyting algebra, with $S \Rightarrow T$ being the pseudo-complement of S relative to T. We have

- $U \subseteq S \Rightarrow T$ iff $S \cap U \subseteq T$, all hereditary U, and
- $\neg S = S \Rightarrow \emptyset$, the pseudo-complement of S.

Algebraic semantics for Φ uses the notion of a *Heyting algebra* (**Ha**) [11, 20, 25]. Any **Ha** has a greatest element 1, with $1 = (x \Rightarrow x)$ for any $x \in H$, where H is a lattice with least element 0. An *$\mathcal{H}$-valuation* is a function $\mathbf{V} : \Psi_0 \to H$. Such a function is lifted canonically to $\mathbf{V} : \Phi \to H$ by stipulating that

- (6) $\mathbf{V}(\bot) = \emptyset$,
- (7) $\mathbf{V}(\varphi \wedge \psi) = \mathbf{V}(\varphi) \cap \mathbf{V}(\psi)$,
- (8) $\mathbf{V}(\varphi \vee \psi) = \mathbf{V}(\varphi) \cup \mathbf{V}(\psi)$,
- (9) $\mathbf{V}(\varphi \to \psi) = \mathbf{V}(\varphi) \Rightarrow \mathbf{V}(\psi)$,
- and hence $\mathbf{V}(\sim \varphi) = \neg \mathbf{V}(\varphi)$, where $\neg : H \to H$ is defined by $\neg x = x \Rightarrow 0$ (x is called the *pseudo-complement of x*.)

Sentence φ is *$\mathcal{H}$-valid*, written $\mathcal{H} \models \varphi$, if $\mathbf{V}(\varphi) = 1$ for every $\mathcal{H}$-valuation $\mathbf{V}$. Now given a poset $\mathbf{P}$, we obtain the corresponding algebra $(\mathbf{P}^+, \cap, \cup, \Rightarrow, \emptyset)$ of hereditary set. $\mathbf{P}^+$ is in fact an **Ha** with greatest element P. The $\mathbf{P}^+$-valuations $\mathbf{V} : \Psi_0 \to \mathbf{P}^+$ correspond exactly to the models $\mathbf{M} = (\mathbf{P}, \mathbf{V})$ based on $\mathbf{P}$.

When such a valuation is extended to $\mathbf{V} : \Phi \to \mathbf{P}^+$ by clauses (6)–(9), we find that for all $\varphi \in \Phi$, $\mathbf{M}(\varphi) = \mathbf{V}(\varphi)$ (using induction on (1′)–(5′)). Hence $\mathbf{M} \models \varphi \iff \mathbf{V}(\varphi) = P = 1$. Since this is true for all $\mathbf{V}$, we have for all $\varphi \in \Phi$ that $\mathbf{P} \models \varphi$ iff $\mathbf{P}^+ \models \varphi$, that is Kripke-validity on the frame $\mathbf{P}$ in the same sense as **Ha**-validity on the algebra $\mathbf{P}^+$. From this it is immediate that a sentence valid on all **Ha**'s will be valid on all posets and hence be an intuitionistic theorem. All such theorems are **Ha**-*valid*. This contributes to the verification of the basic characterization theorem for frame validity, which is that for any $\varphi, \vdash_{IL} \varphi$ iff φ is valid on every frame,. where $\vdash_{IL}$ is a provability relation that takes the intuitionistic theorems into account. For the soundness part, we note that if $\vdash_{IL} \varphi$ then φ is **Ha**-valid, so for any $\mathbf{P}$, $\mathbf{P}^+ \models \varphi$, hence $\mathbf{P} \models \varphi$. One way of proving the completeness part would be to use the representation theory of Stone to turn **Ha**'s into frames. The construction of $\mathbf{P}^+$ assigns to each poset a semantically equivalent **Ha**. In general there is only a partial converse to this. From an **Ha**, $\mathcal{H}$ we obtain via the representation theory of Stone the poset $\mathcal{H}_+ = (P_\mathcal{H}, \subseteq)$, where $P_\mathcal{H}$ is the set of prime filters in $\mathcal{H}$, and $\subseteq$ is set inclusion. The function $x \mapsto \{p \in P_{calH} : x \in p\}$ is an isomorphism between $\mathcal{H}$ and a subalgebra of the **Ha** $(calH_+)^+$ of of $(calH_+)$-hereditary sets: $calH \approx (calH_+)^+$. From this it can be shown that $(calH_+)^+ \models \varphi$ implies $calH \models \varphi$ but the converse need not obtain. The construction does however give a proof that a sentence valid on all posets will be valid on all **Ha**'s ([15], [16]). From topological point of view, the *local* property and the *connective* whose interpretation is "*it is necessarily the case that*" is related to the concept of germs in sheaf theory; while the temporal ordering of states as a partial ordering, which forms a frame, has mainly created a strong link to Heyting algebra.

7 Germs and "locally true"

We start with the definition of the equivalence relation $\sim_i$ that defines the germ $[U]_i$ of an open set $U \in \Theta$ at i in the sheaf Ω [20, 25]. We have $U \sim_i V$ iff U and V have the same intersection with some i-neighborhood of i. We interpret this to mean that the statement "$U = V$" or "$x \in U$ iff $x \in V$" is locally true at i, that is, true throughout some neighborhood of i. This in turn represents the intuitive notion that the statement holds for all points *close* to i. By conditions (a) and (b) of **4**, and the equivalence of both conditions, the *locally true* property stated can be represented by means of the equivalence class of some neighborhood i. This local property is essential for the sheaf-theoretic approach in the Kripke semantics, and obtain a natural formalism to treat *locally true* property, hence the modal operators. Given now a poset $\mathbf{P}$ we introduce a binary relation $p \prec q$ on P, with the reading "*q is close to p*". Then given a model $\mathbf{M} = (\mathbf{P},\mathbf{V})$ based on $\mathbf{P}$ the connective ∇ can be semantically interpreted as $\mathbf{M} \models_p \nabla\varphi$ iff $p \prec q$ implies $\mathbf{M} \models_q \varphi$ –(1)–, thereby formalizing condition (a). Writing $\mu(p) = \{q : p \prec q\}$ for the "monad" of p, this clause becomes $\mathbf{M} \models_p \nabla\varphi$ iff $\mu(p) \subseteq \mathbf{M}(\varphi)$ –(2)–, where $\mathbf{M}(\varphi) = \{q : \mathbf{M}_q\varphi\}$ –(3)–. Define $j_\mu : \mathbf{P}^+ \to \mathbf{P}^+$ by $j_\mu(S) = \{p : \mu(p) \subseteq S\}$ –(4)– which does indeed make $j_\mu(S)$ $\mathbf{P}$-hereditary, we find that $\mathbf{M}(\nabla\varphi) = j_\mu(\mathbf{M}(\varphi))$ –(5)–, and from this it can be shown that $\mathbf{M}(\varphi)/in\mathbf{P}^+$, for all $\varphi \in \Psi$. The notions $\mathbf{M} \models \varphi$ and $\mathbf{P} \models \varphi$ of truth in a model and validity on a frame are again as for posets above. Amongst the sentences that are valid on all frames we cite

- (6) $\nabla(\varphi \to \psi) \to (\nabla\varphi \to \nabla\psi)$,
- (7) $\nabla(\varphi \wedge \psi) \to (\nabla\varphi \wedge \nabla\psi)$,
- (8) $\nabla(\varphi \vee \psi) \to (\nabla\varphi \vee \nabla\psi)$,

while amongst the validity-preserving rules there are "if $\mathbf{P} \models \varphi$ then $\mathbf{P} \models \nabla\varphi$," and "if $\mathbf{P} \models \varphi \to \psi$ then $\mathbf{P} \models \nabla\varphi \to \nabla\psi$." We say that μ, or more loosely $\mathbf{P}$, is increasing if it satisfies $\mu(p) \subseteq [p)$, for all $p \in P$ –(9)–, that is, $p \prec q$ implies $p \subseteq q$. Then we have $\mathbf{P}$ is *increasing* iff $\mathbf{P} \models \varphi \to \nabla\varphi$. Further, using the topology $\mathbf{P}^+$ (the upper order topology), we can show every frame is semantically equivalent to one with hereditary monads, that is, that $\mu(p) \in \mathbf{P}^+$ for all $p \in P$.

8 Equivalence class and the meaning of naming

We recall the definition of A_f in **4**, which arises by pulling back $\{1\} \xrightarrow{\subseteq} \mathbf{2}$ along f., in diagram (1), in this way the equivalence class is considered as subset of a given set. This definition is closely related to the formulation in **6**, where we say that φ is *true* (holds) in the model $\mathbf{M}$, denoted $\mathbf{M} \models \varphi$, if $\mathbf{M} \models_p \varphi$ for every $p \in P$. This is tantamount to the following $\mathbf{M} \models \varphi \iff \mathbf{V}(\varphi) = P = 1$. Recall the "*New-Yorker*" example in **4**, for an observer, say x, based on this property, he is unable to distinguish the individuals (that is, the element of the equivalence class). The naming "New-Yorker" to this equivalence class then assigned a meaning in the sense of φ. This is based on the definition of the *cluster concept theory of names* (which is discussed in [23], not without controversial), and we have $x \to$ 'New-Yorker' : $\varphi, \varphi =$ "the citizen of United States of America living ...." This naming approach is used in **10** for defining *semantic object* (sometimes, we also use the word *tokenizing*.)

8.1 Some definitions of sheaf theory and examples

Let B be a base space with a certain topological structure, F a topological space, and $\pi : F \to B$ a locally topological mapping. Then $\mathbf{F} = (F, \pi)$ is called a sheaf of sets over B. If $x \in B$, then we call $F_x := \pi^{-1}(x)$ the stalk of $\mathbf{F}$ over x (we shall denote the sheaf by F instead of $\mathbf{F}$).

- Definition: Let (F, π) be a sheaf over B, $\tilde{F} \subset F$ open and $\tilde{\pi} := \pi \mid \tilde{F}$. Then $(\tilde{F}, \tilde{\pi})$ is called a *subsheaf* of F. If $W \subset B$ is open, $F \mid W := \pi^{-1}$, then then $(F \mid W, \pi \mid (F(W))$ is also a sheaf, the restriction of F to W.
- Definition: Let (F, π) be a sheaf over $B, W \subset B$ open and $s : W \to F$ a continuous mapping with $\pi \circ s = id_W$. Then s is called a *section* of F over W, the set of all sections of F over W is denoted by $\Gamma(W, F)$.

- Theorem 1: Let (F, π) be a sheaf over $B, W \subset B$ open and $s \in \Gamma(W, F)$. Then $\pi : s(W) \to W$ is topological and $s = (\pi \mid s(W))^{-1}$.
- Theorem 2: Let (F, π) be a sheaf over $B, W \subset B$ open. If for two sections $s_1, s_2 \in \Gamma(W, F)$ there is a point $x \in W$ with $s_1(x) = s_2(x)$, then there is an open neighborhood $V(x) \subset W$ with $s_1 \mid V = s_2 \mid V$.

Examples:

1. *Database as topological object* Database as the base space of the (topological) sheaf, with the application view as œtale space (display space) as sheaf [32]. The application view in **4** can be shown easily as a subsheaf of the sheaf of (global) database application, using the *restriction* of π over local-database "New-York".
2. *DatabaseQuery as Sheaf* also called QuerySheaf over the base space (the collection of databases). (see also [36]).

9 Database applications involving temporal logic

One of the immediate applications of Kripke semantics in database is certainly those related with temporal database models [5]. Since temporal logic deals with time, we specify the model of time that we will be working with, starting with the most general model represents time as an arbitrary set with a partial order imposed on it. In **6**, we associate to each moment of time with a particular *stage* or *state* of knowledge. This is the model of time generally considered by *historical* and *temporal* data models. The syntax of a predicate temporal logic is obtained from the first-order logic by adding various temporal operators to it. We consider the *US* logic, that is, the temporal logic with *until* and *since* temporal operators. The semantics of a temporal logic formula is defined with a temporal structure, which comprises the values of all of its predicates at *all* of the time instants. Formally, let $\pi_1, \ldots, \pi_k$ be a finite set of predicates considered in the predicate (temporal) language. A **P**-*valuation* **V** assigning to each π_i an hereditary subset $\mathbf{V}(\pi_i) \subseteq P$. $\mathbf{V}(\pi_i)$ is the set of stages at which π_i is true. In temporal data model, this is called a temporal structure, which is a mapping $K : T \to \pi_1 \times \ldots \times \pi_k$, where T is a set of time instants and π_i is the set of all the possible interpretations of predicate (the possible worlds). Thus the valuation **V** is a selection function for the truth value (at a particular stage p). The expression "$\mathbf{M} \models_p \varphi$" ("*$\varphi$ is true in* **M** *at* p")—p may be set to a particular time instant of T in this case. Now the mapping K assigns to each time instant an interpretation of each of the predicates $\pi_1, \ldots, \pi_k$ at that time. We use K_t to denote the value of temporal structure K at time t. We make an assumption, that the domains of predicates do not change over time, which is the requirement that $\mathbf{V}(\pi_i)$ be hereditary, and formalizes the *persistence in time of truth*. From a database perspective, a temporal structure K is most naturally looked at as a mapping of each moment of time t into a state of the database, that is, into instants of each of the database relations at time t. Therefore, each predicate in a temporal structure determines a historical relation, that is, a relation that changes over time. An instant of a historical relation R at time t, $K_t(R)$, is denoted by R_t. Using Kripke-Beth semantics [14, 34][3], and the formulas in **6**, **7**, we can describe the temporal operators that correspond to quantification over all possible paths of reachable possible worlds (see **2.1**, **2.2**). A *path* is a sequence $(s_0, s_1, \ldots)$ of states such that $(s_i, s_{i+1}) \in R$ that is *maximal* (that is, either infinite or whose last state has no R-successor). We recall, given a state s in M, and a state formula φ, we write $\mathbf{M} \models_s \varphi$ which means φ is true of *state* s in M. Given a path x in M, and a path formula φ', we write $\mathbf{M} \models_x \varphi'$ which means φ' is true of *path* x in M. We extend $\models$ defined in **6** (1)–(5) by the following modal operators F, X, E, A, (all unary operators) and U (binary operator) [6, 7].

- (P1) If $x = (s_0, s_1, \ldots)$ is a path, then $M \models_x F\varphi$ iff for some s_i on $x, M \models_{s_i} \varphi$; $M \models_x X\varphi$ iff $M \models_{s_1} \varphi$;
- (P2) $M \models_s E\varphi$ if for some path x starting at $s, M \models_x \varphi$;
- (P3) $M \models_s A\varphi$ if for all path x starting at $s, M \models_x \varphi$;
- (P4) $M \models_x (\varphi U \psi)$ if for some initial prefix $(s_0, s_1, \ldots, s_k)$ of $x, M \models_{s_k} \psi$ and $M \models_{s_i} \varphi$ for all $i < k$.
- (P5) If φ, ψ are path formulas $M \models_x \varphi \wedge \psi$ iff $M \models_x \varphi$ and $M \models_x \psi$ $M \models_x \neg\varphi$ iff $\neg M \models_x \varphi$

In (P1), if φ is a state formula then $F\varphi$ is path formula, which intuitively say that at *some* state on the path φ holds. In (P2), if φ is a path formula then $E\varphi$ is a state formula, which says *some* path satisfies φ. In (P3), if φ is a path formula then $A\varphi$ is a state formula, which says *all* paths satisfy φ. In (P4), if φ, ψ are state formulas then $(\varphi U \psi)$ is a path formula, which says there is *some* state on the path which satisfies ψ, and *all* states before it satisfy φ, that is, φ holds *until* ψ. In (P5), if φ, ψ are path formulas, then so are $\varphi \wedge \psi$ and $\neg\varphi$. These modal operators given by state formulas and path formulas are essential for the determination and representation of temporal sequencing and time-interval.

10 Temporal completeness information commitment

In [26], Motro introduces among others the *completeness constraints*, he also notices that, in general the enforcement of these constraints requires human supervision. With this motivation we define the notion of *Temporal Completeness Information Commitment (TCIC)* in [33] by setting the *completeness checkpoint* to confirm the *temporal completeness information* of a certain elementary facts data stored in the database.[4] We consider the model $\mathbf{M}$ and if φ is *true* $\mathbf{M}$, and the model $\mathbf{M} = (\mathbf{P},\mathbf{V})$ (based on $\mathbf{P}$) defined in **6**. Denoting the set of points at which φ is true in $\mathbf{M}$, by $\mathbf{M}(\varphi)$, that is, $\mathbf{M}(\varphi) = \{p : \mathbf{M} \models_p \varphi\}$. Now taking

[3] Kripke-Beth semantics basically extends the Kripke structure along a specific path.
[4] this is in analogy to the `COMMIT` for database-integrity maintenance in the transaction processing; however, in this *completeness checkpoint* we are aiming for the (semantic) completeness within a certain database "semantic objects", like location "`New York`", or Car "`Ford Taurus`", that is, they are semantic meaningful objects grouped under this common label and stored in the database.

the valuation of non-modal sentences (or predicates) by clauses (6)–(9) (in **7**), we find that for all $\varphi \in \Phi$, $\mathbf{M}(\varphi) = \mathbf{V}(\varphi)$. We have $\mathbf{M} \models \varphi \Leftrightarrow \mathbf{V}(\varphi) = P = 1$. This is the criteria in order $\mathbf{M}(\varphi)$ to be *complete* at a particular time instant, $p(p \in P)$, we denote this by $\mathbf{M_p}(\varphi)$. In case of *TCIC*, we have $\overline{\mathbf{M_p}}(\varphi) = \{p : \mathbf{M} \models_p \varphi\} \cup \{TCIC\}$. Using the idea of subobject classifier in **4**, and the pullback square diagram (1) the condition for completeness: $\overline{\mathbf{M_f}} = f^{-1}(1)$. Basically, the procedure can be described as follows: (i) p can be determined to a particular time instant, or a particular time interval; (ii) formulate φ; (iii) evaluate φ to *true* in $\mathbf{V}(\varphi) = P$. The commitment of *TCIC* is set whenever a certain event (milestone) occurred or a user-decision is made. Using the criteria above and *TCIC*, we indeed define an equivalence class locally or for a particular time instant (or both), which specified by certain properties bearing semantics. We call this process a *local completion* of (local) information. *TCIC* assures a local *CWA*.[5]

11 Knowledge in a distributed environment

In [17], Halpern and Moses define the *classical* model for knowledge and belief for the analysis of distributed protocol (see also [2]). In a Kripke structure for knowledge, the "*possible worlds*" can be viewed as a collection of stages of knowledge temporally ordered by $\subseteq$, one corresponding to each *known* or *agent*. Two possible worlds are joined for agent i exactly if they are indistinguishable—equivalence class—as far as agent i is concerned. By the definition of K_i given in **2**, applying to the process defined in **3**, this asserts for any predicate φ, some process i knows φ if i can be determined solely from its local state that φ holds. Suppose we have agents $1, \ldots, n$. The corresponding Kripke structure M is a tuple $(P, \pi, K_1, \ldots, K_n)$, where P is a set of *states*, $\pi(s)$ is a truth assignment to the primitive propositions for each state $s \in P$, and K_i is an equivalence relation on P. Intuitively, $(s,t) \in K_i \iff s$ and t are indistinguishable as far as agent i's knowledge is concerned. The definition for a formula φ to be *satisfied* at a state s of M, is the one defined in **6** ((1)–(5)), and the modalities of $K_i\varphi$ can be defined along the line of **7**, considering "*t is close to s*" and $K_i\varphi$ as $\nabla\varphi$. Now the map j_μ and (5) in **7**, we have $\mathbf{M}(\nabla\varphi) = j_\mu(\mathbf{M}(\varphi))$. The modality K_i has all the properties discussed in the previous section. The reflexivity of K_i gives us $K_i\varphi \Rightarrow \varphi$, transitivity gives us $K_i\varphi \Rightarrow K_iK_i\varphi$, and symmetry and transitivity together give us $\neg K_i\varphi \Rightarrow K_i\neg K_i\varphi$. Now if we consider K of a process set, say Q, K_Q as a predicate transformer[6] By direct verification of (6)–(9) of **7** for K_Q, we obtain the basic properties of K_Q as formulated in [2] and [3]. They show that locality and knowledge are equivalent notions, which is also being implied by the results in **6** and **7**. The knowledge predicate (for process i) K_i operates on the a possible state—that is, an equivalence class of

[5] We recall the *closed world assumption* (*CWA*), which asserts the negation of any atomic formula can be inferred from the inability to infer the atomic formula [27].

[6] we recall a *predicate transformer* is a function from Φ to Φ, which also implies in our case the mapping $\mathbf{P}^+ \to \mathbf{P}^+$

local process components—in other words, this equivalence class can be considered as germ of local process components, and denote this by L_1. Locally we have $K_i(L_1)$, if we want to put this setting in global, we take the inverse of the mapping π, $\pi^{-1}(K_i(L_1))$. Likewise, for process j, having the local process components L_1, we write $\pi^{-1}(K_j(L_1))$. The global states so specified are equivalent, and we have $\pi^{-1}(K_i(L_1)) \sim_i \pi^{-1}(K_j(L_1))$; and this could be identified with $s \sim_i s'$ (we can take i, as $i \sim j$.) Since we are interesting in semantics of the process, we may apply the "Naming Systems" as discussed in **8** to identify the germ of specific class [1]. For instance, we set the following locations { Broadway, Times-Square, Empire-State, Wall-Street, Tian-Ann-Men-Square, Schlossplatz, Brandenburger-Tor, ...}, we assert that "New-Yorker" should know the locations of the landmarks: $K_{\mathbf{NewYorker}}$('Broadway'), $K_{\mathbf{NewYorker}}$('Times-Square'), $K_{\mathbf{NewYorker}}$('Empire-State'), $K_{\mathbf{NewYorker}}$('Wall-Street'). Considering K as predicate transformer we can obtain the position (or coordinates) of the object. In dealing with distributed environment, we are strongly interested to know the location of the process takes place, the state at that location, and probably the event, or time it occurs (related to other events, before, after, while, ...), in this case, if one knows the landmark(s), then one can also allocate the position or coordinates. In many cases, we are interested in the discrete local properties, which are vital to the global state.

12 Acknowledgements

I would like to express my thanks to my brother Ronnie for his constant support and encouragement, and to my son Jeffrey for his cheerful presence. Last not least, to the excellent support of the IBM Central Library at the Schönaich Laboratory.

References

1. M. Bowman, S.K. Debray and L.L. Peterson, "Reasoning About Naming Systems", ACM Trans. on Progr. Lang. and Sys., Vol. 15, No. 5, November, 1993, pp. 795–825.
2. K.M. Chandy and J. Misra, "How Processes Learn", Distri. Comput., Vol. 1, 1986, pp. 40–52.
3. B. Charron-Bost, C. Delporte-Gallet, and H. Fauconnier "Local and Temporal Predicates in Distributed Systems", ACM Trans. on Progr. Lang. and Sys., Vol. 17, No. 1, January, 1995, pp. 157–179.
4. B.F. Chellas, *Modal Logic*, [Cambridge University Press, Cambridge, U.K., 1980].
5. J. Clifford, A. Croker, and A. Tuzhilin, "On Completeness of Historical Relational Query Languages", ACM Trans. on Database Sys., Vol. 19, No. 1, March 1994, pp. 64–116.
6. E.A. Emerson, "Temporal and Modal Logic", in "Handbook of Theoretical Computer Science", Chapter 16, Edited by J. van Leeuwen; Elsevier Science Publishers B.V., Amsterdam 1990. pp. 996–1072.
7. E.A. Emerson and J.Y. Halpern, "Decision Procedures and Expressiveness in the Temporal Logic of Branching Time", J. Comput. Sys. Sci., 30, (1985), pp. 1–24.

8. R. Fagin, J.Y. Halpern, and M.Y. Vardi, "A Model-theoretic Analysis of Knowledge", Journal of the ACM, Vol. 38, No. 2, April 1991, pp. 382–428.
9. R. Fagin, J.Y. Halpern, and M.Y. Vardi "What Can Machines Know? On the Properties of Knowledge in Distributed Systems", Journal of the ACM, Vol. 39, No. 2, April 1992, pp. 328–376.
10. M.J. Fischer and R.E. Ladner, "Propositional Dynamic Logic of Regular Programs", J. Comput. Sys. Sci., 18, No. 2 (1979), pp. 194–211.
11. M.P. Fourman and D.S. Scott, "Sheaves and Logic", In *Applications of Sheaves*, ed. by M.P. Fourman, C.J. Mulvey, and D.S. Scott. [LN in Mathematics 753, Springer Verlag, Berlin-New York, 1979].
12. H. Garcia-Molina, "Using Semantic Knowledge for Transaction Processing in a Distributed Database", ACM Trans. on Database Sys., Vol. 8, No. 2, June 1983, pp. 186–213.
13. J. Glasgow, G. MacEwen, and P. Panangaden, "A Logic for Reasoning about Security", ACM Trans. on Comp. Sys., Vol. 10, No. 3, August 1992, pp. 226–264.
14. R. Goldblatt, *Topoi*, (second edition), Vol. 98 of Studies in Logic. [North-Holland, Amsterdam, 1984].
15. R. Goldblatt, *Mathematics of Modality*, [CSLI Publications, Stanford, California, 1993].
16. R. Goldblatt, "Metamathematics of Modal Logic", In [Gold2 pp. 9–79 1993].
17. J.Y. Halpern, and Y. Moses, "Knowledge and Common Knowledge in a Distributed Environment", Journal of the ACM, Vol. 37, No. 3, July 1990, pp. 549–587.
18. H. Hermes, *Einführung in die Verbandstheorie*, [Springer Verlag, Berlin-Heidelberg-New York, 1967].
19. J. Hintikka, *Knowledge and Belief*, [Cornell University Press, Ithaca, NY. 1962].
20. P.J. Johnstone, *Stone Spaces*, [Cambridge University Press, Cambridge U.K., 1982].
21. S.A. Kripke, "Semantic Analysis of Modal Logic I", Zeitschr. für math. Logik und Grundlagen d. Mathematik 9 (1963), pp. 67–96.
22. S.A. Kripke, "Semantic Analysis of Intuitionistic Logic I", In J.N. Crossley and M.A.E. Dummett (eds.) *Formal systems and recursive functions*, [North-Holland, Amsterdam (1965) pp. 92–130].
23. S.A. Kripke, "Naming and Necessity", *Lectures Given to the Princeton University Philosophy Colloquium. 1970*. In Davidson and Harman (eds.) *Semantics of Natural Language*, (1972) pp. 253–355.
24. E.J. Lemmon, "Algebraic Semantics for Modal Logic, I, II", Journal of Symbolic Logic, 31 (1966), pp. 46–65; pp. 191–218.
25. S. MacLane and I. Moerdijk, *Sheaves in Geometry and Logic. A First Introduction to Topos Theory*, [Springer Verlag, New York Berlin Heidelberg, 1992].
26. A. Motro, "Completeness Information and its Application to Query Processing", In *Proceedings of the Twelfth International Conference on Very Large Data Bases*, Kyoto, Japan. 1986, pp. 170–178.
27. R. Reiter "On Closed World Data Bases", In *Logic and Databases*, H. Gallaire and J. Minker (Eds.), [Plenum Press, New York, 1978, pp. 55–76].
28. S.J. Rosenschein, "Formal Theories of Knowledge in AI and Robotics", New Generation Comput., 3 1985, pp. 345–357.
29. M. Sato, "A Study of Kripke-style Methods of Some Modal Logics by Gentzen's Sequential Method", *Publications of the Research Institute for Mathematical Sciences, Kyoto University*, 13:2, 1977.

30. M.B. Smyth, "Topology", In *Handbook of Logic in Computer Science*, Volume 1 *Background: Mathematical Structures*, Edited by S. Abramsky, D.M. Gabbay and T.S.E. Maibaum, [Clarendon Press, Oxford 1992. pp. 641–761].
31. R. Thom, "Structural Stability and Morphogenesis: An Outline of General Theory of Models." [Addison-Wesley Publishing Company, Inc., Reading, Mass., 1972/1989].
32. M. Tjiok, "An Application of Sheaf Theory in Computer Science", In "European Colloquium of Category Theory (ECCT 1994)" Tours, France.
33. M. Tjiok, "Completeness Information revisited—a distributed database version", Preprint (to be published).
34. A.S. Troelstra and D. van Dalen, *Constructivism in Mathematics: an Introduction*, volume II. Vol. 123 of *Studies in Logic and the Foundations of Mathematics*, [North-Holland, Amsterdam, 1988].
35. H. Weyl, "Die Idee der Riemannschen Fläche" (The idea of Riemann Surface), [Teubner, Berlin, 1913; 3rd edition, Teubner, Stuttgart, 1955]
36. C. Yu and W. Meng, "Progress in Database Search Strategies", IEEE Software, May 1994, pp. 11–19.

Springer Series in

Discrete Mathematics and Theoretical Computer Science Series

Editors

Advisory Editorial Board

Springer-Verlag Singapore announces a new series, *Discrete Mathematics and Theoretical Computer Science*, produced in cooperation with the Centre for Discrete Mathematics and Theoretical Computer Science of the Universities of Auckland and Waikato, New Zealand. This series will bring to the research community information about the latest developments on the interface between mathematics and computing, especially in the areas of artificial intelligence, combinatorial optimization, computability and complexity, and theoretical computer vision. It will focus on research monographs and proceedings of workshops and conferences, aimed at graduate students and professional researchers, and on textbooks, primarily for the advanced undergraduate or lower graduate level.

For details of forthcoming titles, please contact the publisher at:

Springer-Verlag Singapore Pte. Ltd.
#04-01 Cencon I
1 Tannery Road
Singapore 347719
Tel: (65) 842 0112
Fax: (65) 842 0107
e-mail: ianjs@cyberway.com.sg